The Potato

Botany, Production and Uses

The Potato

Botany, Production and Uses

Edited by

Roy Navarre

USDA-ARS, Washington, USA

and

Mark J. Pavek

Washington State University, USA

www.cabi.org

CABI is a trading name of CAB International

CABI
Nosworthy Way
Wallingford
Oxfordshire OX10 8DE
UK

Tel: +44 (0)1491 832111
Fax: +44 (0)1491 833508
E-mail: info@cabi.org
Website: www.cabi.org

CABI
38 Chauncy Street
Suite 1002
Boston, MA 02111
USA

Tel: +1 800 552 3083 (toll free)
E-mail: cabi-nao@cabi.org

A catalogue record for this book is available from the British Library, London, UK.

Library of Congress Cataloging-in-Publication Data

The potato : botany, production and uses / edited by Roy Navarre, USDA-ARS, Washington, USA, and Mark J. Pavek, Washington State University, USA.
pages cm
Includes bibliographical references and index.
ISBN 978-1-78064-280-2 (alk. paper)
1. Potatoes. I. Navarre, Roy. II. Pavek, Mark J.

SB211.P8P774 2014
635'.21--dc 3

2014006677

ISBN-13: 978 1 78064 280 2

Commissioning editor: Charlotte Hammond
Editorial assistant: Alexandra Lainsbury
Production editor: James Bishop

Typeset by SPi, Pondicherry, India.
Printed and bound by CPI Group (UK) Ltd, Croydon, CR0 4YY

Contents

Contributors		vii
Preface		ix
Acknowledgments		xi
1	**A History of the Potato** *Charles Raymond Brown and Jan-Willem Henfling*	1
2	**Taxonomy** *Ronald G. van den Berg and N. Groendijk-Wilders*	12
3	**Potato Utilization and Markets** *Jennifer K. Bond*	29
4	**Tuber Development** *Bjorn Kloosterman and Christian Bachem*	45
5	**Plant Growth and Development** *Carrie H. Wohleb, N. Richard Knowles and Mark J. Pavek*	64
6	**Commercial Potato Production and Cultural Management** *Mark J. Pavek*	83
7	**Plant–Water Relations and Irrigation Management of Potato** *David Levy and Warren K. Coleman*	103
8	**Seed Potato Production** *Robert D. Davidson and Kaiyun Xie*	115
9	**Insect Pests of Potato** *Timothy D. Waters and Andrew S. Jensen*	133
10	**Plant Parasitic Nematodes of Potato** *Juan E. Palomares-Rius, Clandio M.G. Oliveira and Vivian C. Blok*	148
11	**Fungal and Bacterial Disease Aspects in Potato Production** *William W. Kirk and Phillip S. Wharton*	167

12 **Virus Disease Problems Facing Potato Industries Worldwide: Viruses Found, Climate Change Implications, Rationalizing Virus Strain Nomenclature, and Addressing the *Potato Virus Y* Issue** 202
Roger A.C. Jones

13 **Weed Control Strategies for Potato (*Solanum tuberosum*)** 225
Ian C. Burke and Wesley J. Everman

14 **Tuber Physiological Disorders** 237
Loretta Mikitzel

15 **Postharvest Storage and Physiology** 255
Paul C. Bethke

16 **Traditional Breeding and Cultivar Development** 272
Richard Novy

17 **Molecular Breeding of Potato in the Postgenomic Era** 290
Richard E. Veilleux and T. Boluarte-Medina

18 **Nutritional Characteristics of Potatoes** 310
Duroy Navarre, Aymeric Goyer, Raja Payyavula and Hanjo Hellmann

19 **Potato Flavor** 345
Mark A. Taylor

Index 361

Contributors

Christian Bachem, Laboratory of Plant Breeding, Wageningen University and Research Centre, PO Box 386, 6700 AJ Wageningen, the Netherlands. E-mail: christian.bachem@wur.nl

Paul Bethke, USDA Agricultural Research Service and University of Wisconsin Department of Horticulture, 1575 Linden Drive, WI 53706, USA. E-mail: paul.bethke@ars.usda.gov

Vivian Blok, Cell and Molecular Sciences, James Hutton Institute, Invergowrie, Dundee, DD2 5DA, Scotland, UK. E-mail: vivian.blok@hutton.ac.uk

T. Boluarte-Medina, 544 Latham Hall, Virginia Tech, Blacksburg, VA 24061, USA. E-mail: tboluart@vt.edu

Jennifer K. Bond, USDA-Economic Research Service, 1400 Independence Avenue SW, Mail Stop 1800, Washington, DC 20250-0002, USA. E-mail: jkbond@ers.usda.gov

Charles Raymond Brown, USDA-ARS, 24106 N. Bunn Road, Prosser, WA 99350, USA. E-mail: chuck.brown@ars.usda.gov

Ian C. Burke, Weed Science, Washington State University, Department of Crop and Soil Science, 171 Johnson Hall, Pullman, WA 99164-6420, USA. E-mail: icburke@wsu.edu

Warren K. Coleman, Potato Research Centre (retired), Agriculture and Agri-Food Canada, 850 Lincoln Road, Fredericton, New Brunswick, Canada E3B 4Z7. E-mail: shadow1@nbnet.nb.ca

Robert D. Davidson, Department of Horticulture and Landscape Architecture, 205 Shepardson Building, Colorado State University, Fort Collins, CO 80523-1173, USA. E-mail: robert.davidson@colostate.edu

Wesley J. Everman, Department of Crop Science, North Carolina State University, 4402E Williams Hall, Campus Box 7620, Raleigh, NC 27695-7620, USA. E-mail: wesley_everman@ncsu.edu

Aymeric Goyer, Oregon State University, Hermiston Agricultural Research and Extension Center, Hermiston, OR 97838, USA. E-mail: aymeric.goyer@oregonstate.edu

N. Groendijk-Wilders, Biosystematics Group, Wageningen University, 6708 PB Wageningen, the Netherlands. E-mail: nynke.groendijk-wilders@wur.nl

Hanjo Hellmann, Plant Stress Physiology, Washington State University, Abelson 435/437, Pullman, WA 99164, USA. E-mail: hellmann@wsu.edu

Jan-Willem Henfling, Erfgooiersstraat 512, 1222DK Hilversum, the Netherlands. E-mail: jan-willem@henfling.nl

Andy Jensen, Northwest Potato Research Consortium, Idaho, Oregon, and Washington Potato Commissions, 4032 N Ballantyne Lane, Eagle, ID 83616, USA. E-mail: ajensen@potatoes.com

Roger A.C. Jones, School of Plant Biology and Institute for Agriculture, Faculty of Science, University of Western Australia, 35 Stirling Highway, Crawley, Western Australia 6009. E-mail: roger.jones@uwa.edu.au

William Kirk, Department of Plant Soil and Microbial Sciences, 35 Plant Biology Building, Michigan State University, East Lansing, MI 48824-1325, USA. E-mail: kirkw@msu.edu

Bjorn Kloosterman, (current address) Keygene NV, PO Box 216, 6700 AE Wageningen, the Netherlands. E-mail: bjorn.kloosterman@keygene.com

N. Richard Knowles, Department of Horticulture, Washington State University, PO Box 646414, Pullman, WA 99164-6414, USA. E-mail: rknowles@wsu.edu

David Levy, Institute of Plant Sciences and Genetics in Agriculture, R.H. Smith Faculty of Agriculture, Food and Environment, The Hebrew University of Jerusalem, Rehovot, Israel 76100. E-mail: levy.david@mail.huji.ac.il

Loretta Mikitzel, New Brunswick Department of Agriculture, Aquaculture and Fisheries, Potato Development Centre, 39 Barker Lane, Wicklow, New Brunswick, Canada E7L 3S4. E-mail: loretta.mikitzel@gnb.ca

Duroy Navarre, USDA-ARS, 24106 North Bunn Road, Prosser, WA 99350, USA. E-mail: roy.navarre@ars.usda.gov

Richard Novy, USDA-ARS, 1693 S 2700 W, Aberdeen, ID 83210, USA. E-mail: rich.novy@ars.usda.gov

C.M.G. Oliveira, Instituto Biológico, Caixa Postal 70, CEP 13001-970, Campinas, SP, Brazil. E-mail: marcelo@biologico.sp.gov.br

J.E. Palomares-Rius, Institute for Sustainable Agriculture (IAS), Spanish National Research Council (CSIC), Alameda del Obispo s/n, Apdo. 4084, 14080 Córdoba, Spain. E-mail: palomaresje@ias.csic.es

Mark J. Pavek, Department of Horticulture, Washington State University, PO Box 646414, Pullman, WA 99164-6414, USA. E-mail: mjpavek@wsu.edu

Raja Payyavula, IAREC, Washington State University, Prosser, WA 99350, USA. E-mail: rajasekhar78@yahoo.com

Mark A. Taylor, Cell and Molecular Sciences, James Hutton Institute, Invergowrie, Dundee, DD2 5DA, Scotland, UK. E-mail: mark.taylor@hutton.ac.uk

Ronald G. van den Berg, Biosystematics Group, Wageningen University, Droevendaalsesteeg 1, 6708 PB, Wageningen, the Netherlands. E-mail: ronald.vandenberg@wur.nl

Richard E. Veilleux, 544 Latham Hall, Virginia Tech, Blacksburg, VA 24061, USA. E-mail: potato@vt.edu

Timothy D. Waters, Franklin and Benton Co, Washington State University, 404 W. Clark, Pasco, WA 99301, USA. E-mail: twaters@wsu.edu

Phillip S. Wharton, University of Idaho, Aberdeen Research and Extension Center, Aberdeen, ID 83210, USA. E-mail: pwharton@uidaho.edu

Carrie H. Wohleb, Regional Vegetable Specialist, Grant and Adams Co, Washington State University, 35 C Street NW, Ephrata, WA, USA. E-mail: cwohleb@wsu.edu

Kaiyun Xie, CIP Beijing Liaison Office, Beijing 100081, China. E-mail: k.xie@cgiar.org

Preface

Potatoes are the most important non-cereal in the world and are an important part of the diet of over 1.5 billion people across the globe. Despite the continued expansion of the world population, resources needed to grow food are limited. Also pressuring food production will be increased meat and dairy consumption in countries with rising affluence, and competition with biofuels. Some estimate that crop production must double by 2050 to meet needs (Ray *et al.*, 2013). Potatoes yield substantially more calories per acre than maize, wheat, rice, or soybean (Chapter 18), and produce >50% more protein per unit of land area than wheat and approximately 80% more than rice. No other food, not even soybeans, can match the potato for production of food, energy, and food value per unit of land.

Aside from being highly land-use efficient, potatoes offer consumers a wide array of complex carbohydrates, protein, vitamins, and phytonutrients. For these reasons, and others, many scientists believe potatoes will have a key role in providing "food security", which can be defined as "when all people, at all times, have access to sufficient, safe, and nutritious food to meet their dietary needs for an active and healthy life". The definition specifies "nutritious food", which potatoes are. A medium-sized (~170 g) potato can supply an adult with close to half of their daily requirement of vitamin C (Pennington and Wilkening, 1997). Potatoes are the third largest contributor of antioxidants in the American diet (Chun *et al.*, 2005), and high-antioxidant potatoes have been shown to reduce inflammation and lower blood pressure in separate human feeding studies (Kaspar *et al.*, 2011; Vinson *et al.*, 2012). A study of nutritional value concluded that potatoes and beans provided the most nutrients per dollar out of 98 vegetables evaluated (Drewnowski and Rehm, 2013). Affordability is a primary consideration for much of the planet's population and is a key component of food security.

A gradual increase in global potato production will help address future food supply and nutrition issues, especially when combined with modern breeding techniques. The major focus of many potato breeding programs is to improve yield and quality while using fewer production resources, like water, fertilizer, and land, than current varieties. These programs are also focused on identifying varieties with resistance to major diseases and insect pests in order to reduce the need for pesticides, thereby maintaining or improving the health of the environment. Advances in marker-assisted selection, gene mapping, and genomics have allowed breeders to identify desirable and undesirable traits in new cultivars more quickly; improving the chances of finding superior varieties while reducing the time it takes to get them to market. In conjunction with modern breeding efforts, the industry is continuously improving production, storage, processing, and marketing operations to offer consumers high-quality, nutritious potatoes year-round.

This book discusses many of the items above and explains the origin of the domesticated potato, its uses, nutritional contributions, production, storage and markets, plant anatomy, physiology and development, pest identification and management, and traditional and biotechnological cultivar development techniques. By incorporating intensive literature review with the applied expertise of the authors, this book was designed to educate and enlighten consumers, students, growers, industry personnel, and the scientific community.

Duroy "Roy" A. Navarre and Mark J. Pavek

References

Chun, O.K., Kim, D.O., Smith, N., Schroeder, D., Han, J.T. and Lee, C.Y. (2005) Daily consumption of phenolics and total antioxidant capacity from fruit and vegetables in the American diet. *Journal of the Science of Food and Agriculture* 85, 1715–1724.

Drewnowski, A. and Rehm, C.D. (2013) Vegetable cost metrics show that potatoes and beans provide most nutrients per penny. *PLoS One* 8, e63277.

Kaspar, K.L., Park, J.S., Brown, C.R., Mathison, B.D., Navarre, D.A. and Chew, B.P. (2011) Pigmented potato consumption alters oxidative stress and inflammatory damage in men. *Journal of Nutrition* 141, 108–111.

Pennington, J.A.T. and Wilkening, V.L. (1997) Final regulations for the nutrition labeling of raw fruits, vegetables, and fish. *Journal of American Dietetic Association* 97, 1299–1305.

Ray, D.K., Mueller, N.D., West, P.C. and Foley, J.A. (2013) Yield trends are insufficient to double global crop production by 2050. *PLoS One* 8, e66428.

Vinson, J.A., Demkosky, C.A., Navarre, D.A. and Smyda, M.A. (2012) High-antioxidant potatoes: acute *in vivo* antioxidant source and hypotensive agent in humans after supplementation to hypertensive subjects. *Journal of Agricultural and Food Chemistry* 60, 6749–6754.

Acknowledgments

We thank the authors for their generous contribution of the chapters of this book. Their work represents state-of-the-art expertise on diverse aspects of potato biology. We respectfully acknowledge those previous generations of potato specialists who came before us and upon whose efforts the current generation benefits and builds. It is humbling to think back to the nameless discoverers of the first primitive tubers in the Andes thousands of years ago, and how their discovery went on to change world history. Today, thousands of years later, civilizations have risen and vanished, but potatoes are still here, better than ever, and offer the hope of food security to future generations. Indeed, this ancient discovery even shaped the lives of those writing and reading this book! We are privileged to be a small part of this story of the potato and are fortunate that the extended potato community (researchers, growers, consumers, and industry workers) are such wonderful people with whom it is a pleasure to work.

We thank the many reviewers for their generous efforts to improve the book further. We also thank and appreciate Alyson R. Kral for her editorial assistance, and Jeanne Debons and Dylan Vermeul for their artwork. Finally, we are deeply beholden to our parents, families, and friends, living and passed on, whose love and friendship is priceless.

1 A History of the Potato

Charles Raymond Brown[1]* and Jan-Willem Henfling[2]
[1]*USDA-ARS, Prosser, Washington, USA;* [2]*Hilversum, the Netherlands*

1.1 Domesticating the Potato Crop

The popularity of the potato has fluctuated over the years and it is therefore appropriate to consider the history of the potato leading up to modern times. About 7000 years ago, inhabitants of the Andes in South America were predominantly hunter-gatherers and tended semi-wild herds of native camelids (llamas, vicuñas, and alpacas), yet they began to take an interest in a curious plant (Fig. 1.1). It flowered and produced inedible seed balls, but also produced starchy underground tubers. The tubers were produced at the end of underground stems, oftentimes located a fair distance from the mother plant. The tubers were large enough for a mouthful after cooking and were energy rich. Furthermore, they acted as big seeds, and once planted, they produced potato plants, which in turn produced more tubers. Because the seeds were large, they had enough stored carbohydrates to restart plant growth initially inhibited by a killing frost (International Potato Center, 2008).

The tubers were storable and transportable, and provided nourishment to a society that was in constant motion. Some of the tubers were bitter tasting, but a palatable solution was found. Specific clay soils were used to render the potatoes non-bitter by adsorbing glycoalkaloids into the clay (Johns, 1996). Cultivated potato, *Solanum tuberosum* Group Stenotomum, is now known to have been selected from the *Solanum brevicaule* complex, which gave rise to today's potato (Spooner *et al.*, 2005).

By the time the Spanish arrived in South America, Andean societies were highly organized and had developed agricultural systems, including a collective possession of thousands of types of potato and maize. One of the first descriptions of potato was made by a Spanish soldier in the highlands of Colombia in 1535. A drawing of the potato harvest (Fig. 1.2) comes from the handwritten book scribed by Felipe Guamán Poma de Ayala, who sought to document the societal interactions of the Spanish and native Quechua-speaking peoples. The drawing depicts harvesting potatoes using the Andean foot plow, or *chaquitaccla*, an implement still used in the Andes of Peru.

Guamán's manuscript was taken to Spain in 1616, ostensibly as a report to King Phillip II, but was given as a gift to the Ambassador from Denmark. It disappeared from view and resurfaced in the Royal Danish Library in 1908. Years of scholarship were required to interpret its archaic Spanish mixed with Quechua (Adorno, 1986). The first facsimile appeared in 1936, produced by the Institute of Ethnography in Paris (Guamán Poma de Ayala, 1615, 1936, 1944). The book is available online, with

* E-mail: chuck.brown@ars.usda.gov

Fig. 1.1. Pastoral hunter-gatherers took interest in a plant that yielded edible starchy underground stems about 7000 years ago. (Reprinted with permission from CIP International Potato Center, Lima, Peru.)

Fig. 1.2. A drawing by Felipe Guamán Poma de Ayala depicting the potato harvest, from his historical documentation of Andean society under Spanish rule, *Nueva Crónica y Buen Gobierno*, 1616.

translations: http://www.kb.dk/permalink/2006/poma/info/en/frontpage.htm.

Although this remarkable tract documented the major role that potato played in the Andean environment, application of this knowledge to the newly introduced potato in Europe did not occur.

Introducing the potato to Europe

The earliest introductions of the potato in Europe were to the Canary Islands in 1567 and Seville, Spain, in 1570 (Hawkes and Francisco-Ortega, 1992). Sister Teresa de Avila, founder of the Order of the Barefoot Carmelites, eschewed material wealth and took a vow of poverty. Such behavior was considered challenging and embarrassing to the church hierarchy, and she was punished by home imprisonment. While under imprisonment in 1578 in Seville, and having been taken ill, she remarked in a letter that she was fed potatoes and recovered her strength (Oliemans, 1988).

It was also recorded that potato was fed to patients in the Carmelite hospital in Seville, with remarkably curative results. From these references, it can be surmised that potato was grown in some fashion and recognized as a health-restoring food within a few years after its introduction (Salaman, 1949). Yet a vision of potato as a famine preventer did not emerge.

In the Piedmont region of northern Italy, under French control in the late 1170s, came reports of a Protestant group called the Waldensians. The Waldensians were strict adherents to their faith and openly critical of Roman Catholics. Their existence was threatened at various times with extremely violent suppression. Depicted as satanic heretics, they were burned at the stake, forced to flee their homeland, and eventually spread across Europe.

Serendipitously, they had become potato farmers. The potato was more productive than alternative crops and able to support more people on less land than other grains or roots of the time. With each action against them, the Waldensians spread north, seeking the protection of Protestant enclaves as Europe became one large battlefield (Oliemans, 1988).

In a short time, the potato became a major source of sustenance for the Waldensians as they fled north along the Swiss border into France with their potatoes. Potatoes had an additional benefit of escaping taxation, since taxation based on agricultural production was levied at the grain mill. Potatoes did not require milling and thus were not taxed. Waldensians moved into the welcoming Protestant areas of France and Germany, and also the Netherlands (Reader, 2008).

A description of potato was included in Gerard's *Herbal* in 1597, which correctly named its origin as Perú (Gerard, 1631). At roughly the same time, the Swiss botanist, Gaspard Bauhin, described the potato in the following passage:

> The root is of an irregular round shape, it is either brown or reddish-black, and one digs them up in the winter lest they should rot, so full are they of juice. One put them in the earth once more in spring: should it happen that one leaves them in the sun, in the springtime they will sprout of themselves. Further at the base of the stem close to the roots there spring long fibrous radicles on which are borne the very small round roots. The root itself generally rots when the plant is fully developed. We have judged it our duty to call this plant Solanum by reason of the resemblance of its leaves with that of tomato, and its flowers with those of the Aubergine, its seed with that of Solanums and because of its strong odor which is common to these latter. It is called by some the Pappas of the Spanish and by others Pappas of the Indies. We have further learned that this plant is known under the name of Tartouffoli, doubtless because of its tuberous root, seeing that this the name by which one speaks of Truffles in Italy where one eats these fruits in a similar fashion to truffles.
>
> (Bauhin, 1596)

It is apparent this observer was unaware that the plant was a major food crop underpinning the civilization of the Native Peoples of the Andes. This was as a result of the secrecy maintained around all information coming back from the New World.

Frederick the Great, the young ruler of Prussia in the 1780s, spent considerable time in the Netherlands, studying naval architecture. He came upon the newly arrived potato and sent batches back to Prussia (Salaman, 1949). A French nobleman, A.A. Parmentier, having eaten potatoes while a prisoner of war, promoted its adoption by famously leaving a royal potato planting unguarded at night so that the locals could be introduced to it by stealing it (Parmentier, 1781). Parmentier recognized the stabilizing effect the potato could have on food supply, even when failure of grain crops could lead to famine. The stage was set for an explosion of potato farming and the ability to feed the masses in an agricultural changeover, accompanied by religious revolution and persecution. Far from remaining a botanical curiosity, the potato was becoming the food of the poor.

Even more intriguing was that where the soil and climate were appropriate for potato culture, a startling increase in population occurred in the rural areas and adjacent cities. After the introduction of the potato to the diet of the French Army, records show that the average height of soldiers increased by one-half of an inch (Nunn and Qian, 2011).

In Swiss, the potato was called "erdäpfel", while in Italian it was called truffle, or "tartouffli". In France, it became known as "pomme de terre", and in the Netherlands "aardappel", earth fruit and earth apple, respectively. In German and Russian, it was called "kartoffel", a possible sound-alike of "tartouffli". In Spain, it was called "patata", again, a sound-alike of the already adopted "batata" or sweet potato. In Great Britain, the word "potato" was used for both potato and sweet potato during a confusing introduction period. Sir Francis Drake wrote indistinctly of potatoes on the island of Chiloé and of potatoes being grown by escaped slaves in the jungles of Panama, referring to *Solanum* potato in the former and *Ipomoea* potato in the latter. The next period in potato evolution had a major impact on human nutrition and the global economy.

Cultivation of the potato crop in North America

Barely half a century had passed between the first European appearance of the potato and the potato arriving in the newly founded colony of Virginia, USA. It is remarkable that little mention of the introduction of South American potato germplasm to Europe or North America occurred in the 17th, 18th, and 19th centuries. A partial explanation is that the Spanish were not interested in local foods on arriving in the western hemisphere. Rather, they were intent on recreating Old Spain in New Spain, which meant searching for environments where wheat could be grown, and beef cattle and sheep would prosper. The suitability of cultivating Old World food was, in fact, one of the criteria for establishing a town (Reader, 2008).

Another reason behind the secrecy of potato cultivation was competition and fear of war between England and Spain (Cook, 1973). Nearly all information gleaned from observers and chroniclers of the New World was regarded as a military secret. An important scientific and economic question, also the subject of constant conjecture and debate, was the location of the Northwest Passage (Cook, 1973; Wagner, 2002). Both the British and Spanish governments lived in fear the other would find the Northwest Passage and lay claim to it. As a result, free dissemination of the attributes of the diets, clothing, and technology of the people of South America did not occur for many more centuries.

In the course of establishing sovereignty over crucial latitudes on the west coast of North America, the Spanish founded forts in modern-day Canada and the USA. The locations were predominantly Nootka and Makah native villages and harbors in Nootka Sound, now known as Vancouver Island (Canada), and Neah Bay and Cape Flattery on the north-west tip of Washington State's Olympic Peninsula. In 1792, a naturalist, Mariano Moziño, born in modern-day Mexico (New Spain), traveled as the accompanying naturalist on a voyage from San Blas, New Spain (Mexico). He made an inventory of the plants and animals around Nootka Sound and included *S. tuberosum* on his list (Swan, 1868; Cook, 1973; Moziño, 1991).

The Makah have long known they received their famous Ozette potato from the Spanish (Swan, 1868; McDonald, 1972; Cook, 1973; Moziño, 1991; Zhang *et al.*, 2010). Documentation of potatoes in Puget Sound, Washington State, among native peoples was recorded by several sources (Swan, 1868; Suttles, 1951; McDonald, 1972; Gill, 1983). Suttles (1951) concluded that the Coastal Salish Indians obtained potatoes from the British-owned Hudson Bay Company, while he noted the Makah Ozette Potato was peculiar and probably had been obtained earlier from a different source (i.e. Spanish mariners). In south-east Alaska, Gibson (1999) noted the potatoes grown by the Haida and Tlingit were cultivated under contract to supply the Russian sea otter fleets in the first half of the 19th century.

To the south in California, the potato 'Bodega Red' was instrumental in feeding the hordes of gold miners during the California gold rush of the mid-1800s. This was a potato variety brought from Chile and cultivated around Bodega Bay, California (Burbank, 1914; Burke, 2007).

Although the aforementioned varieties are of eclectic interest as representatives of an alternative path of potato germplasm diffusion, they pose a question regarding the importance of Andean and Chilean cultivars as sources. Hosaka and Hanneman (1988) found, almost without exception, potatoes now growing in long-day summer adaptation had a specific deletion in their chloroplast genome, denoted as T-type cytoplasm. All of the potatoes collected from the west coast of North America and south-east Alaska have T-type cytoplasm, which indicates some recent ancestry of Chilean potatoes or direct rendition from Chile itself.

The potato in Ireland

After its introduction to Europe through Spain, the potato was incorporated into the Irish diet, and by 1790, the Irish were described as having embraced the tuber completely (Salaman, 1949). The widespread acceptance of the potato in Ireland, spurred on by economic repression exercised by English landowners, provided the impetus to grow the potato as a sustenance crop. For the Irish, the crop was not counted as taxable production of the English estates.

In 1843, potato late blight surfaced in the USA and later in Belgium, leading to crop failure on a scale rarely seen before. By 1845, crop failure and famine occurred in Ireland and continued for an additional year. The germ theory of diseases had yet to be developed, and the cause of late blight became a target of endless conjecture of plant scientists of the day (Bourke, 1993). The environment was usually attributed the largest share of the causation of disease.

The Irish Potato Famine was one of the worst catastrophes of the 19th century. During 2 years of severe food shortage, it inspired one of the first national debates in England over the role of the state in coming to the aid of the suffering masses. The Irish were seen as a population out of harmony with its resources and needing adjustment by natural (albeit genocidal) processes. Arguments against helping the Irish included the notion they were lazy, and the potato, which required little care and investment, was a lazy race's food. It was argued in Parliament that famine food aid would simply make the recipients indolent and unable to sustain themselves. Additionally, the 1840s was a decade of public discourse on religious principles. Emerging from this discussion was a declaration that the famine was a visitation from God. Arguments to the contrary were raised, but largely ignored.

Malthus's discourse on controlling the growth of populations (overpopulation being the expected mechanism for natural selection and evolution toward a fitter population) provided a basis for popular support for the idea that competition for space and sustaining resources occurred not only among species of the natural world but also among humans (Malthus, 1798). In this mode of thought, the fittest would survive, and the Irish were simply a casualty of natural laws of survival and extinction.

1.2 Potato Late Blight and Selecting for Disease Resistance

The potato late blight epidemic started in North America and culminated in the complete destruction of the potato crop in Ireland in 1845. This event has been presumed to mark a clear separation of potato bloodlines. One of the most successful potato varieties, persisting for over 100 years, is Russet Burbank. Russet Burbank remains the first variety in production in the USA.

The origin of Russet Burbank begins with the introduction of Rough Purple Chili from the Panamanian Consulate by Reverend Chauncey Goodrich in 1850 (Goodrich, 1863a,b). The Reverend took open-pollinated berries from his plants and produced Garnet Chili. Later, other breeders produced Early Rose, which was the genetic mother of Burbank.

In the late 19th century, Early Rose was crossed extensively in nearly all potato breeding programs around the world. It is unclear which traits Early Rose contributed to its progeny to make it a universal ancestor; however, breeders listed disease resistance among the most important traits they were seeking to incorporate into new varieties. Lou Sweet selected a russeted variant of the Burbank potato, leading to the modern-day variety, Russet Burbank (Dreyer, 1993).

From breeding efforts, the Early Rose became the founding ancestor of nearly all potato germplasm in Europe. One would presume Rough Purple Chili brought with it resistance to late blight, yet little evidence of this exists. Using the marker for T-type cytoplasm, Ames and Spooner (2008) analyzed the DNA of herbarium specimens in Europe. They determined Chilean germplasm was extant in Europe before the Irish Potato Famine, and after the famine, the Andean cytoplasm persisted for decades, but eventually disappeared (Fig. 1.3).

Evidence from the cytoplasm of heirloom potatoes in the Canary Islands and India indicates a mixture of Chilean and Andean cytoplasm in modern potatoes (Spooner *et al.*, 2005; Ríos *et al.*, 2007). The disappearance of Andigena cytoplasm could have occurred due to selection for long-day adaptation, or due to the nuclear genome entering an irrevocable state of male sterility when crossed with the T-type cytoplasm of the Chilean germplasm (Vilaró *et al.*, 1989).

As a relic of post-famine breeding efforts, Russet Burbank shows a distinct and repeatable level of resistance, which is described more properly as less susceptibility (Inglis *et al.*, 1996). For the mid-1800s, and in comparison to the susceptibility of germplasm under wide cultivation at that period in history, Russet Burbank and its

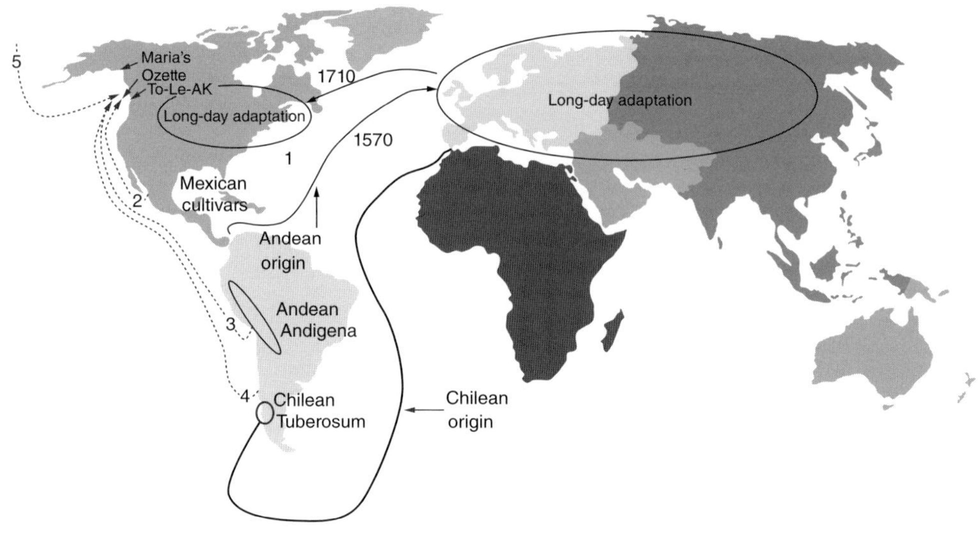

Fig. 1.3. The potato's journey from the New World (South America) to the Old World (Europe) and back to North America.

progenitor, Early Rose, showed late blight resistance. Furthermore, it is proven that crosses utilizing pollen from Group Andigena led to sterile male progeny (Vilaró *et al.*, 1989). This feature would limit the breeding pool severely, since only Andigena pollen would be available to pollinate male sterile progeny, leading to a conversion of the cytoplasm to a T-type, regardless of better day-length adaptation afforded by genes in the nuclear genome (Fig. 1.4).

Breeding and cultivating new varieties

By the 1800s, a number of potato breeders existed, contributing to the development of numerous varieties of potato with narrow genetic parentage. One of the first people to take on directed breeding of the potato crop was Knight, in 1807 (Knight, 1807; Glendinning, 1983). The science of Gregor Mendel's genetics would not be dusted off until the early 20th century. Emerging from varietal development was a competitive potato sales environment, where purveyors of seed and commercial growers vied for recognition and market share, and a bushel of seed of a new variety could cost US$900 on first release at present-day values (Best, 1870; Burbank, 1914).

The production of new varieties was the sole province of the private breeder in the late 1800s to early 1900s. The science of disease resistance was inadequate, and the explanation of decline in performance and vigor in seedstock was unavailable to the horticulturalist. The introduction of new varieties occurred at a rapid pace. Where controlled crosses were undertaken, a study of the parentage was often illustrative, but some breeders simply kept open-pollinated berries from the field. The occasional introduction of potatoes from South America was unproductive, due to their adaptation to short days during the growing season, which made intercrossing among existing cultivars superior.

A collection of tuber-bearing *Solanum* species was kept and grown at the Edinburgh Botanic Garden in Scotland. Late blight had not subsided after the Irish Potato Famine and continued be a serious disease wherever potato crops were grown. In 1911, a late blight epidemic swept through the botanic garden, killing all the potatoes, wild and domestic, except for a few accessions from Mexico. It was discovered the clones that survived were derivatives of wild species named *Solanum demissum* (*Solanum* × *edinense*) (Salaman, 1949).

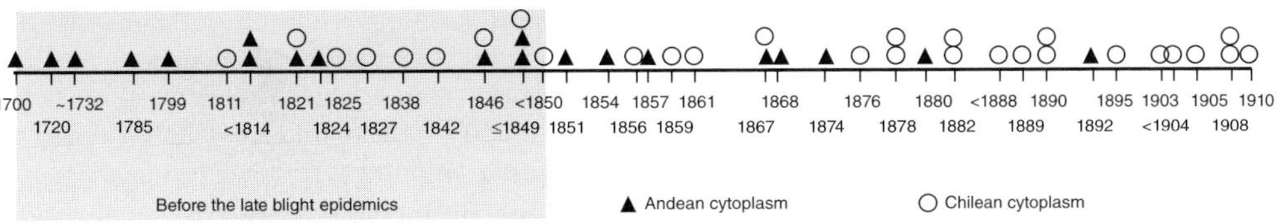

Fig. 1.4. Composition of potato cytoplasm based on DNA extraction of dried foliage of herbarium specimens. (From Ames and Spooner, 2008; reproduced with permission from the *American Journal of Botany*.)

For many decades, private breeders had sought to produce commercial blight-resistant varieties. The Mexican specimens at the botanic garden were a source of resistance; it was also the beginning of publicly funded research programs for the breeding of new varieties of crop species. Over the next 40 years, numerous blight-resistant varieties were developed, by introducing genes from *S. demissum*.

Crop geneticists also began to examine the idea of returning to the center of origin of crops. Returning to these locations and collecting crop specimens was a positive and successful way to expand the gene pool and introduce new varieties with unique traits.

Resistances to *Potato virus Y*, PVX, and *Globodera* (Cyst nematode spp.) were found first in collections of wild and primitive cultivar potato taken to home countries and maintained as collections (Ellenby, 1948; Ross, 1954, 1958, 1986). Meanwhile, comprehensive taxonomic treatments were emerging, due to the efforts of Americans, Europeans, and South Americans.

1.3 Botanical Collections and Breeding Research

The plant collector, George Bitter, was among the first to travel to South America, although he was not deeply influential on potato taxonomy (Bitter, 1912). Potato and wild potato biosystematics began to take shape, with expeditions to South America by Russian and British plant collectors (Jucepczuk and Bukasov, 1929; Bukasov, 1933; Hawkes, 1941, 1990). Additional expeditions undertook surveys of wild potato species that were resistant to disease and pests, which yielded immediate results and contributed to the publication of taxonomic compendia (Hawkes and Hjerting, 1969; Ochoa, 1990, 1999).

The incorporation of traits from wild species became a large component of most national breeding and genetic programs (Ellenby, 1948; Ross, 1954, 1958, 1986). Two reviews of resistance breeding in potato by Jansky (2000) and Pavek and Corsini (2001) provide a thorough overview of active research fields, while numerous instances of exotic ancestry in modern varieties are noted by Plaisted and Hoopes (1989).

Breeding programs initially bred potatoes towards modern needs, resulting in taming the extravaganza of shapes in native varieties. Similarly, the number and depth of each potato eye was selected through breeding. The native varieties in South America were selected in a cultural context that valued diversity and eccentricity over convenience. The modern-bred varieties are devised for an industrialized society where ease of peeling reigns, and a controlled size and shape is packed easily in boxes, stored, and transported to consumers.

Douches *et al.* (1996) compared groups of old, intermediate and modern breeds in agronomic trials. First, improvement in total yield, comparing the old to the intermediate and modern varieties, was not seen. This could be attributed to genetic impoverishment of the breeding pool. Alternatively, the number of traits that required optimal performance as the potato went from a household-prepared item to factory processed were more important than yield itself. Yet improvement in overall appearance, depth of eye, and common scab resistance was found. Extreme pressure for higher solids achieved a high specific gravity.

The necessity for a specific variety to store well for long periods without a rise in reducing sugars means superior performance in yield, or a percentage of large tuber sizes might not be selected for. Tuber appearance has improved significantly, because a uniform shallow-eyed tuber is best for mechanical peeling, used in manufacturing potato chips and French fries. The need to select for an attractive russeted skin in oblong-shaped potatoes suitable for French fries is highly limiting in the full range of expression of other traits, including total yield and specified size categories.

In modern breeding, the desire to maintain genetic diversity for heterosis and vigor in sexual progeny has prevailed. A number of strategies have been developed to inject exotic alleles into the breeding process. The Neotuberosum breeding strategy was extracted from the Central Andean short-day adapted gene pool and selected recurrently for long-day adapted maturity (Glendinning, 1975). Sexual polyploidization by use of 2n pollen-producing diploids gave convenient access to cultivated and wild diploid species (Mendiburu and Peloquin, 1977). However, by the late 20th century, the breeding pool from which new varieties were emerging was still remarkably depauperate (Mendoza and Haynes, 1974). While efforts have been made in the last 70 years to expand the germplasm of the potato, the active breeding pool remains restricted and closely related, leading to the inability to maximize heterozygosity and increase yield.

The last few decades in India and China have seen potato production emerge as a significant engine of development. India and China are now the first and third largest producers of potato in the world, respectively. The density of potato production of the world has shifted down the mountain and into hotter production zones. In China, the potato is considered a major player in future increases in food supply. Up until 2005, the potato in China was subsidized, yet today farmers are exempt from paying taxes on their earnings. In Fig. 1.5, the trends of production are shown over the past 40 years. By 2010, China was producing 23% of the world's potatoes, while contributing 18% of the world's population. Production in China and India has increased steeply year after year, while in the USA it has stagnated (Walker *et al.*, 1999; Scott and Suarez, 2012).

1.4 Summary

Seven thousand years ago, pastoral hunter-gatherer families discovered a subterranean root with immediate appeal for a population that, until then, struggled to consume sufficient calories to sustain an active lifestyle. On the introduction of the potato to each new culture,

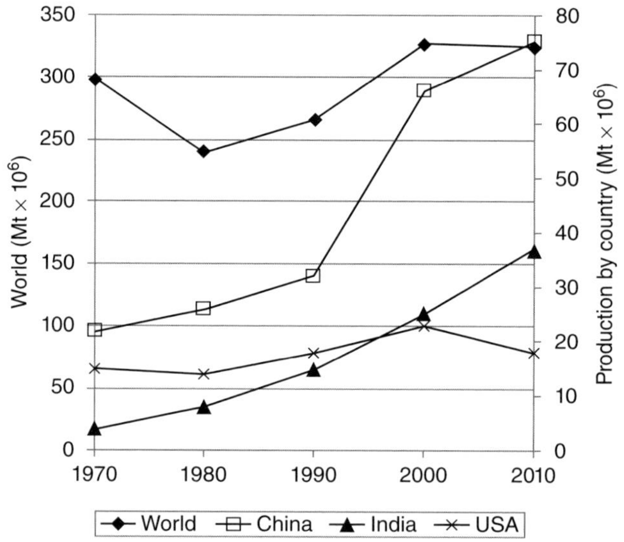

Fig. 1.5. Production statistics from China, India, and the USA illustrate the changing paradigms of potato utilization. While China and India reflect steep increases in potato production, the USA presents a relatively static pattern. By 2010, China was producing 23% of the world's potatoes, while contributing 18% of the world's population. (Reproduced with permission from FAOSTAT, Rome, Italy, 2013.)

the recipients found tremendous nutrient benefits, which satisfied a need in lives dominated by high amounts of physical labor.

When the potato was adopted by the Makah, Quillayute, Haida, and Tlingit native peoples of North America and south-east Alaska, it was inserted into a diet that had no rich carbohydrate source (Swan, 1868; Suttles, 1951; McDonald, 1972; Cook, 1973). These peoples spent much of each day gathering food from the sea and actively traveling out in canoes to capture fish and mammals. The potato was the perfect food to fuel a full day of rowing in a canoe, kept for months in storage, and could be used as a trade item or commodity, as the Haida and Tlingit did by growing potatoes to sell to the Russian sea otter fleets.

In Europe, potato agriculture engendered population growth and economic development wherever it was feasible to grow productively (Nunn and Qian, 2011). During the 20th century, a scientific understanding of the origin of the potato was developed slowly through the work of many plant explorers and taxonomists. Further, the use of primitive cultivars and wild species in breeding has become commonplace, and has resulted in disease and pest resistances not found previously in long-day adapted potato. The development of a draft of the potato genome certainly will help in the extraction of greater genetic treasure from wild potato in the future.

Throughout the years, consumer preferences have changed with trends and food technology innovations; however, the antioxidant-rich potato remains a global powerhouse of valuable and important vitamins and minerals (Brown, 2005).

References

Adorno, R. (1986) *Guamán Poma: Writing and Resistance in Colonial Peru*. University of Texas Press, Austin, Texas.

Ames, M. and Spooner, D.M. (2008) DNA from herbarium specimens settles a controversy about origins of the European potato. *American Journal of Botany* 95, 252–257.

Bauhin, G. (1596) *Phytopinax*. Geneva, Switzerland.

Best, G.W. (1870) *Best's Potato Book*. Reprint 2013, Repressed Publishing, Provo, Utah.

Bitter, G. (1912) Solanaceae andinae. *Beiblatt zu den Botanisk jahrbuch* No 119, 54, 5–8.

Bourke, A. (1993) *The Visitation of God? The Potato and the Great Irish Famine*. Lilliput Press Ltd, Dublin.

Brown, C.R. (2005) Antioxidants in potato. *American Journal of Potato Research* 82, 163–172.

Bukasov, S.M. (1933) The potatoes of South America and their breeding possibilities. Prilozhenie 58-e K. Trudam po prikladnoi botanike, genetike I selektsii, Leningrad, Russia.

Burbank, L. (1914) The potato itself – Who will improve it further? In: Whitson, J., John, R. and Williams, H.S. (eds) *Luther Burbank: His Methods and Discoveries and Their Practical Application*, Vol 7. Luther Burbank Press, New York, pp. 267–302.

Burke, R.M. (2007) *An Intimate History of Bodega Country and the McCaughey Family*. Tomales Regional History Center, Tomales, California.

Cook, W.L. (1973) *Flood Tide of Empire: Spain and the Pacific Northwest, 1543–1819*. Yale University Press, London.

Douches, D.S., Maas, D., Jastrzebski, K. and Chase, R.W. (1996) Assessment of potato breeding progress in the USA over the last century. *Crop Science* 36, 1544–1552.

Dreyer, S. (1993) *A Gardener Touched with Genius: The Life of Luther Burbank*. Luther Burbank Home and Gardens, Santa Rosa, California.

Ellenby, C. (1948) Resistance to potato root eelworm. *Nature* 162, 704.

FAO (2013) *FAO Statistical Yearbook*. (http://www.fao.org/docrep/015/i2490e/i2490e00.htm, accessed 2 June 2014).

Gerard, J. (1631) *The Herbal or General Historie of Plants*. 2008 Reprint. Velluminous Press.

Gibson, J.R. (1999) *Otter Skins, Boston Ships, and China Trade: The Maritime Fur Trade of the Northwest Coast, 1785–1841*. University of Washington Press, Seattle, Washington.

Gill, S.J. (1983) Ethnobotany of the Makah and Ozette people, Olympic Peninsula, Washington (USA). Doctoral Dissertation. Department of Botany, Washington State University, Pullman, Washington.

Glendinning, D.R. (1975) Neotuberosum: new potato breeding material. 1. The origin, composition, and development of the Tuberosum and Neo-Tuberosum gene pools. *Potato Research* 18, 256–261.
Glendinning, D.R. (1983) Potato introductions and breeding up to the early 20th century. *New Phytology* 94, 479–505.
Goodrich, C.E. (1863a) The origination and test culture of seedling potatoes. *Transcripts of the New York State Agricultural Society* 23, 89–102.
Goodrich, C.E. (1863b) The potato—its diseases—with incidental remarks on its soils and culture. *Transcripts of the New York State Agricultural Society* 23, 103–134.
Guamán, F. Poma de Ayala (1944) *Nueva crónica y buen gobierno* (written between 1584 and 1614). A. Poznansky (ed.), Instituto Tiahuanacu de Antropología, Etnografía y Prehistoria, La Paz, Bolivia.
Hawkes, J.G. (1941) *Potato Collecting Expeditions in Mexico and South America*. Bulletin of the Imperial Bureau of Plant Breeding and Genetics. School of Agriculture, Cambridge, UK.
Hawkes, J.G. (1990) *The Potato: Evolution, Biodiversity, and Genetic Resources*. Belhaven Press, London.
Hawkes, J.G. and Francisco-Ortega, J. (1992) The potato in Spain during the late 16th century. *Economic Botany* 46, 86–97.
Hawkes, J.G. and Hjerting, J.P. (1969) *The Potatoes of Argentina, Brazil, Paraguay and Uruguay: A Biosystematic Study.* Clarendon Press, Oxford, UK.
Hosaka, K. and Hanneman, R.E. Jr (1988) The origin of the cultivated tetraploid potato based on chloroplast DNA. *Theoretical and Applied Genetics* 76, 172–176.
Inglis, D.A., Johnson, D.A., Legard, D.E., Fry, W.E. and Hamm, P.B. (1996) Relative resistances of potato clones in response to new and old populations of *Phytophthora infestans*. *Journal of Plant Disease* 80, 575–578.
International Potato Center (2008) *The Odyssey of Potato*. International Potato Center, Lima.
Jansky, S. (2000) Breeding for disease resistance in potatoes. *Plant Breeding Reviews* 87, 69–151.
Johns, T. (1996) *The Origins of the Human Diet and Medicine.* University of Arizona Press, Tucson, Arizona.
Jucepczuk, S.W. and Bukasov, S.M. (1929) A contribution to the question of the origin of the potato. *TrVses Azeda Genet Selek* (with English summary) 3, 592–611.
Knight, T.A. (1807) On raising of new and early varieties of the potato (*Solanum tuberosum*). *Transactions of the Horticultural Society of London* 1, 57–59.
McDonald, L. (1972) *Swan Among the Indians*. Binford and Mort, Portland, Oregon.
Malthus, T. (1798) *An Essay on the Principle of Population (Published Between 1798 and 1826)*. Empire Books, London.
Mendiburu, A.O. and Peloquin, S.J. (1977) The significance of 2n gametes in potato breeding. *Theoretical and Applied Genetics* 49, 53–61.
Mendoza, H.A. and Haynes, F.L. (1974) Genetic relationship among potato cultivars grown in the United States. *HortScience* 9, 328–330.
Moziño, J.M. (1991) *Noticias de Nutka: An Account of Nootka Sound in 1792*. Translated by Iris H. Wilson Engstrand. University of Washington Press, Seattle, Washington.
Nunn, N. and Qian, N. (2011) The potato's contribution to population and urbanization: Evidence from a historical experiment. *The Quarterly Journal of Economics* 126, 593–550.
Ochoa, C.M. (1990) *The Potato of South America: Bolivia*. Cambridge University Press, Cambridge, UK.
Ochoa, C.M. (1999) *Las Papas de Sudamérica: Perú*. International Potato Center, Lima.
Oliemans, H.O. (1988) *Het brood van de armen* (The bread of the poor). SDU uitgiveri, s'Gravenhage, the Netherlands.
Parmentier, A.A. (1781) *Recherches sur les vegetaux nourissans, qui, dans les temps de disette peuvent remplacer les aliments ordinaires: Avec de nouvelles observations sur la culture des pommes de terre*. Imprimerie Royale, Paris, In-8, pp. (2), XVI, 599.
Pavek, J.J. and Corsini, D.L. (2001) Utilization of potato genetic resources in variety development. *American Journal of Potato Research* 78, 433–441.
Plaisted, R.L. and Hoopes, R.W. (1989) The past record and future prospects for the use of exotic germplasm. *American Potato Journal* 66, 603–627.
Reader, J. (2008) *Potato: A History of the Propitious Esculent*. Yale University Press, New Haven, Connecticut.
Ríos, D., Ghislain, M., Rodriguez, F. and Spooner, D. (2007) What is the origin of the European potato? Evidence from Canary Island landraces. *Crop Science* 47, 1271–1280.
Ross, H. (1954) Die Vererbunger der "Immunität" gegen das X-Virus in tetraploiden *Solanum acaule*. In: *Proceedings of the 9th International Congress of Genetics,* Bellagio, 1953. *Caryologia* 6 (Suppl), 1128–1132.

Ross, H. (1958) Inheritance of extreme resistance to virus Y in *Solanum stoloniferum* and its hybrids with *Solanum tuberosum*. In: Quak, F., Dijkstra, J., Beemster, A.B.R. and van der Want, J.P.H. (eds) *Proceedings of the Third Conference on Potato Virus Diseases*, Lisse-Wageningen, 1957. Veenman and Zonen, Wageningen, the Netherlands, pp. 204–211.

Ross, H. (1986) Potato breeding—problems and perspectives. In: Brandes, J., Bartels, R., Volk, J. and Wetter, C. (eds) *Advances in Plant Breeding.* Verlag Paul Parey, Berlin, pp.1–20.

Salaman, R.N. (1949) *The History and Social Influence of the Potato*. Cambridge University Press, Cambridge, UK.

Scott, G. and Suarez, V. (2012) From Mao to McDonald's: Emerging markets for potato products in China 1961–2007. *American Journal of Potato Research* 89, 216–231.

Spooner, D.M., McLean, K., Ramsay, G., Waugh, R. and Bryan, G.J. (2005) A single domestication for potato based on multilocus AFLP genotyping. *Proceedings of the National Academy of Sciences* 120, 14694–14699.

Suttles, W. (1951) The early diffusion of the potato among the Coast Salish. *Southwestern Journal of Anthropology* 7, 272–288.

Swan, J.G. (1868) *The Indians of Cape Flattery at the Entrance to the Strait of Juan de Fuca, Washington Territory*. Smithsonian Institution "Contributions to Knowledge", Facsimile Reproduction 1964, Washington, DC, 108 pp.

Vilaró, F.L., Plaisted, R.L. and Hoopes, R.W. (1989) Comparison of cytoplasmic male sterilities in progenies of Tuberosum × Andigena and Tuberosum × Neo-Tuberosum crosses. *American Potato Journal* 66, 13–24.

Wagner, H.R. (2002) *Spanish Explorations in the Strait of Juan de Fuca*, 2nd edn. Martino Publishing, Eastford, Connecticut.

Walker, T.S., Schmiediche, P.E. and Hijmans, R.J. (1999) World trends and patterns in the potato crop: An economic and geographic survey. *Potato Research* 42, 241–264.

Zhang, L., Brown, C.R., Culley, D., Baker, B., Kunibe, E., Denney, H., *et al.* (2010) Inferred origin of several Native American potatoes from the Pacific Northwest and Southeast Alaska using SSR markers. *Euphytica* 174, 15–29.

2 Taxonomy

Ronald G. van den Berg* and N. Groendijk-Wilders
Biosystematics Group, Wageningen University, the Netherlands

The taxonomy of "the potato", *Solanum* L. section *Petota* Dumort, encompassing both the wild relatives of the cultivated potato and the cultigen itself, has a long, convoluted history. In the present chapter, we will summarize this history briefly, describe the current state of affairs that is influenced heavily by the use of molecular data, and look at future developments, especially the impact of genomic data. We will first deal with what is known about the wild potatoes, and then look at the various ways the potato crop has been classified. It should become apparent to what extent these two taxonomic situations have become clear through recent research, and to what extent a standardization has been reached that is generally accepted. Finally, as more and more genomic data become available, we need to discuss possible ways the taxonomy of the potato can benefit from these.

2.1 Classification of the Wild Relatives of the Potato

The potato is a crop with a large number of wild relatives, a group of more than 100 tuber-bearing *Solanum* species, mostly in the Andes mountain range. In order to use these related species as sources of traits useful for the improvement of the crop, the classification and phylogeny of these species must be clear. Different taxonomists have, however, arrived at different results, and such controversies can hamper the utilization of the crop's wild relatives and pose problems for gene bank management. Here, we look briefly at the history of the taxonomic classification of the wild relatives and discuss remaining problems and the current situation.

A brief outline of the history of potato taxonomy

The genus *Solanum* L. is a large genus, with approximately 900 species (Hawkes, 1990). The tuber-bearing species attracted attention early on and were recognized as a separate group: section *Tuberarium*, with representatives from the southern part of the USA to Chile. The number of species assigned to this section grew rapidly since Walpers (1844–1845) described ten species, especially with the work of Bitter (1912), who added more than 50 species. Russian taxonomists, Bukasov (1933), for example, added even more, recognizing minor variants as species. In an early publication, Hawkes (1944), applying the same "splitter" approach, recognized 150 wild and 20 cultivated species. This approach

* E-mail: ronald.vandenberg@wur.nl

was criticized by Correll (1962) in his monograph of the group, where he himself recognized a total of 191 species within subsection *Hyperbasarthrum* (a name pointing at a common characteristic of the tuber-bearing species, the presence of an articulation above the base of the pedicel). Hawkes followed his lead by publishing a series of treatments (Hawkes, 1956, 1963, 1990), finally recognizing 232 species, including 7 cultivated species, within the group that was now called section *Petota* Dumort.

Hawkes and his Peruvian colleague, Ochoa, proceeded to dominate the field for decades, producing a number of landmark studies (Hawkes and Hjerting, 1969, 1989; Hawkes, 1990; Ochoa, 1990, 2004). Interestingly, Hawkes and Hjerting (1989) and Ochoa (1990) focused on the same material, the wild species of Bolivia, enabling a comparison between their approaches. Spooner and Van den Berg (1992a) highlighted the apparent difficulty these experts experienced in consistently identifying herbarium specimens, especially when dealing with taxa that were closely related to the cultivated potato. Spooner and collaborators contributed a number of detailed studies of parts of section *Petota* (e.g. Spooner and Van den Berg, 1992b; Giannattasio and Spooner, 1994a,b; Spooner *et al.*, 1997; Van den Berg *et al.*, 1998, 2002; Miller and Spooner, 1999), criticizing the use of a "splitter" approach and a typological species concept.

Still, Hawkes's 1990 treatment became the standard one, used extensively by gene banks and potato breeders. The book gives keys to and a short description of all recognized species, and classifies them in 21 series that are arranged in two "superseries" based on the outline of the flowers (stellate or rotate). It recognizes many hybrids.

Since 1990, a wide range of molecular markers has been used for testing the validity and interrelationships of potato species. This has resulted in a gradual reduction of the number of species recognized. Spooner and Hijmans (2001) reduced the number of species to 206, and Spooner and Salas (2006) recognized 188 wild and one cultivated species. More recently, Spooner and co-workers have contributed a final revision to the Solanaceae Source website (http://solanaceaesource.org/), with 166 accepted taxa (156 species).

The group of wild relatives of the cultivated potato displays a polyploid series from diploid to hexaploid. Since these polyploids are interpreted as allopolyploids, progenitor–derivative relationships can be hypothesized, and these can be expressed using genome formulae (Matsubayashi, 1991). The diploid species occurring in South America share a common genome that is also present in *Solanum verrucosum*, a species occurring in Mexico. This common genome is indicated with AA by Matsubayashi (1991). The polyploid species present in Central America and Mexico resulted from crosses, and the contributions of the parental species can be expressed in the genome formula of the polyploids, e.g. AABB for tetraploid species of the Mexican series *Longipedicellata* and $AADDD^dD^d$ for the hexaploid *Solanum demissum*. In this way, evidence could be provided for the evolutionary relationships between the diploid and polyploid species. In addition to this, based on their crossability and the ratio of maternal and paternal genomes, an endosperm balance number (1, 2, 4) has been assigned to the species, indicating their membership to crossability groups (Johnston and Hanneman, 1978).

Summarizing the results of the quite extensive phylogenetic studies that have been performed, a rather robust phylogeny reconstruction is apparent: almost all of the produced trees have a similar topology with the following order of appearance of groups within section *Petota*:

- 1 endosperm balance number (EBN) Mexican diploids
- 1 EBN South American diploids
- 2 EBN South American diploids
- 2 EBN South American polyploids
- 4 EBN Mexican polyploids

This conforms with Hawkes's (1990) scenario of an origin in Central America, migration to South America, followed by diversification in the microhabitats provided in the Andes mountain range and a re-migration towards the north, accompanied by polyploidization. However, the data of Volkov *et al.* (2003), for example, did not support this scenario. These authors, working with rDNA external transcribed spacers concluded that all "primitive" representatives of section *Petota* originated in South America, and that later primitive species with stellate flowers migrated to Central America and Mexico. Also, the high similarity of resistance (R) genes found in *Solanum bulbocastanum* and *Solanum stoloniferum* strongly suggested that the diploid species *S. bulbocastanum* was one of the progenitors of the tetraploid *S. stoloniferum* (Wang *et al.*, 2008). This would indicate a different, more direct relationship between the Mexican diploids and polyploids.

Remaining problems in potato taxonomy

A large number of scientists have devoted a considerable amount of time and effort to arrive at a classification of the group of wild relatives of the potato. Still, controversies about species delimitations and competing classifications have arisen. What are the causes of these problems in potato taxonomy? On the one hand, as mentioned above, some of the potato taxonomists displayed an unfortunate splitter approach, creating a multitude of names and extensive synonymy. But also, the plasticity of potato plants, causing their morphological characteristics to be influenced easily by environmental factors, has contributed to the problem. Notwithstanding this plasticity, there is not that much variation in the morphology of leaves and flowers, leaving the taxonomist with only a few characters to work with. As most of these characters display an almost continuous variation in character states, it is very hard to draw boundaries between species. At least part of section *Petota* consists of a large group of closely related entities, many of which are able to hybridize, again confounding species boundaries. Finally, since these species are related to the cultivated potato, human influence is likely to increase the taxonomic problems. All this led Spooner *et al.* (2000) to conclude "a practical taxonomic treatment of the group using morphology likely will remain an elusive goal".

A number of classification problems are apparent: how to define species; how many species should be recognized; how to subdivide species; how to cluster species in natural groups; and how to deal with hybridization.

Species concepts

Paradoxically, the species—the basic building block used by taxonomists to produce classifications—has proven to be difficult to define in the potato. Potato taxonomists relying on morphological characters have found that there is too much overlap to draw boundaries easily between species. Spooner and Hetterscheid (2005) pointed at the lack of species-specific characters and advocated a polythetic morphological species concept, which defined species by the statistical covariance of characters that were deemed important, not all of which were necessarily displayed by all individuals of that species (Mayden, 1997). Species are thus characterized by (unique) combinations of characters. Cladistic analyses of molecular data sets made it possible to circumscribe cladistics species concepts based on monophyly, and Olmstead (1995) proposed the terms "apospecies" for the monophyletic daughter species and "plesiospecies" for the remnant paraphyletic progenitor species. This terminology was also used in potato taxonomy; for example, by Castillo and Spooner (1997), in a paper on series *Conicibaccata*, and by Van den Berg *et al.* (2002), in a paper about the reduction of the number of species in series *Longipedicellata*.

Camadro *et al.* (2012), in a recent paper on the species problem in wild potatoes, argue that—since barriers against hybridization are incomplete, introgression within and between populations takes place, and populations are at different stages of genetic divergence—the taxonomic species concept cannot be applied. And since potatoes can propagate both sexually and asexually, the overlapping of uni- and biparental generations hinders the application of the biological species concept. They make the point that potato taxonomists often work with gene bank accessions and that it is not certain at all that these accessions are representative samples of the genetic diversity of the originally sampled populations. In taxonomic studies in potato, genetic, reproductive, and environmental aspects must be taken into account. The authors indicate the need for a set of genetic markers at various exploratory levels and for sampling strategies to capture the dynamics of wild populations.

Overclassification

There has been a tendency to recognize too many species, distinguishing minor variants at the species level on the basis of morphology. Many morphological characters, however, proved to be influenced easily by environmental factors, such as the effect of shade on leaf dissection

(Spooner *et al.*, 2004), and in a number of morphometric studies, extensive overlap of character states was observed in many of these characters. The extremes can be distinguished easily from each other, but they are connected by a continuous series of intermediate character states, making the delimitation of species highly arbitrary. Two examples of the need for a reduction of species numbers are the studies of the so-called *brevicaule* complex (Van den Berg *et al.*, 1998; Miller and Spooner, 1999; Alvarez *et al.*, 2008) and of the Mexican series *Longipedicellata* (Spooner *et al.*, 2000; Van den Berg *et al.*, 2002). Both studies relied on extensive morphometric characterization of gene bank accessions, planted in a common garden plot, supplemented with molecular data.

The *brevicaule* complex consists of a group of more than 30 species placed in series *Tuberosa* by Hawkes (1990). These species are the closest relatives of the cultivated potato, are extremely similar to each other, and are distributed from southern Peru, via Bolivia, to northern Argentina, with different species names being applied in different countries. Hawkes (1990) suggested that the extreme speciation of this complex in the Peruvian Andes could be accounted for by the large number of closed valley systems in that mountain range, where many populations became relatively isolated after the last glaciation. Phenetic results (Van den Berg *et al.*, 1998) did not support many distinct taxa in this complex. Only two or three broadly defined taxa could be recognized. Miller and Spooner (1999), in a study of the molecular data of this complex, concluded that their molecular data concurred with the data of Van den Berg *et al.* (1998). Two geographically based groups were indicated, one from Peru and the other from Bolivia and Argentina. The repeat experiment in Peru (Alvarez *et al.*, 2008) came to the same conclusion: the collapse of many of the species of the *Solanum brevicaule* complex was unavoidable.

Similarly, a morphological study (Spooner *et al.*, 2000), followed by a study of the data from three molecular markers (amplified fragment length polymorphism (AFLPs); random amplified polymorphic DNA (RAPDs); and chloroplast simple sequence repeats (cpSSRs)) (Van den Berg *et al.*, 2002), supported only two species in the series *Longipedicellata*, instead of the six recognized by Hawkes (1990). The morphological characters that had been used to distinguish species displayed substantial overlap in their character states, making equivocal identifications impossible, especially between *Solanum fendleri* and *S. stoloniferum*. The molecular data did not separate the species recognized by Hawkes, but allowed the distinction of *S. stoloniferum* s.l. and *Solanum hjertingii*.

The study also included South American species that were morphologically similar. The reproductive isolation of the Mexican *Longipedicellata* species (based on their 4× (2EBN) crossability and AABB genome) from these morphologically very similar South American species (especially from series *Tuberosa*) demonstrates a conflict between a morphological and a biological species concept. Apparently, morphology does not always allow us to recognize biologically meaningful entities.

There are many similar examples of overclassification. Castillo and Spooner (1997) suggested a reduction in series *Conicibaccata* to only 16 (or even 8) taxa, instead of the 40 species of Hawkes (1990). Van den Berg *et al.* (2001) reduced the four species in the small series *Circaeifolia* to only one species. The difficulty in separating extremely similar species and in producing and using identification keys shows the need for a drastic reduction of the total number of species recognized in section *Petota*. Hawkes (1990) admitted that identification keys were difficult to construct and use, and added: "I included them [in his various revisions]. Having had no feedback from users I can only conclude that these keys were either satisfactory or that no one had the courage to use them!" The multitude of very similar wild potato species has bewildered users, and a reduction of the number of species, as, for example, advocated by the Solanaceae Source website, will indeed be welcomed.

Infraspecific taxa

In a situation where the delimitation of species is extremely difficult to outright impossible (e.g. within the *brevicaule* complex), it is not a good idea—to put it mildly—to recognize infraspecific taxa. Jacobs *et al.* (2011) concluded that most of the subspecies and varieties in this section did not receive support, with the exception of *Solanum commersonii* subsp. *commersonii* and *S. commersonii* subsp. *malmeanum* (which have now been raised to the species level on the Solanaceae Source website). Refraining from using infraspecific classification certainly lessens the nomenclatural burden that is already excessive due to the extensive synonymy created.

Series classification

Not all of the many series recognized within section *Petota* are natural units and the system is far from stable, with Correll (1962) recognizing 26 series versus Hawkes (1990) with 21 series. Spooner and collaborators produced cpDNA phylogenies, using material of 17 of the 21 series recognized by Hawkes in section *Petota*, which showed four main clades (Spooner and Castillo, 1997). Recent AFLP studies (Jacobs *et al.*, 2008) could not find support for most of the series, nor for the four clades of Spooner and Castillo (1997). Some of the series seem to be good, however, especially the diploid and tetraploid Mexican groups that are apparent in many phylogenetic studies. Spooner *et al.* (2004), studying the Central American representatives of section *Petota*, decided to recognize 11 informal "species groups" rather than the formal taxon "series". Most of the South American series are probably artificial. This was also shown by Jacobs *et al.* (2008), who found a lack of phylogenetic structure, especially in the part of the trees that contained the South American species.

Hybrids

Crosses between species that share the same EBN are, in principle, possible (even between different ploidy levels), and hybridization is a common phenomenon. The majority of the diploid representatives of section *Petota* are outcrossing species that are able to cross with each other, occasionally resulting in the establishment of a hybrid population that maintains itself. A number of these have been recognized as "nothospecies" and have been given Linnean names. In other cases, populations have been indicated with a hybrid formula indicating their parental species. Backcrosses and crosses between different generations may produce complicated hybrid swarms, while backcrosses to one of the parental species may cause introgression.

Hawkes and Hjerting (1969), in their treatment of the potatoes of Argentina, Brazil, Paraguay, and Uruguay, devoted much attention to the occurrence of natural hybrids and also performed many crossing experiments to check the parentage of presumed hybrids. They made the point that the presence of these natural hybrids led to difficulties in classification and identification, and stressed the necessity of studying living material. They made the following observation:

> Perhaps at first sight it will be thought that we have seen hybrids and introgressed forms everywhere! We can assure the reader that we have tried to resist this temptation as far as possible, and have attempted not to fall into the common trap of labelling everything as a hybrid that does not fit into a preconceived narrow concept of what each species should conform to.
> (Hawkes and Hjerting, 1969, pp. 119–120)

However, Spooner and Van den Berg (1992a) noted that the main problem in the taxonomy of section *Petota* was the lack of documentation of morphological variability within taxa and that most authors (e.g. Correll, 1962; Hawkes, 1990; Ochoa, 1990) indicated the necessity to use a broad species concept, but that, in practice, many taxa were narrowly defined by minor and often overlapping character states and variability frequently either was not documented or was ignored.

Spooner and Van den Berg (1992b) examined the status of the species *Solanum berthaultii* and *Solanum tarijense*. These were treated as distinct, and even as belonging to different superseries, by Hawkes (1990), and the many intermediate forms were ascribed to extensive hybridization. The morphological data of Spooner and Van den Berg (1992b) suggested the possibility that the two species were the extremes of a continuous range of variability and that they should be treated as one species, and Spooner *et al.* (2007a) confirmed this with AFLP and chloroplast restriction site data.

Present situation

We here compare two recent initiatives to arrive at an updated taxonomy of the potato and its wild relatives, viz. the CBSG project (Centre of Biosystem Genomics) and the Solanaceae Source website. The CBSG project, conducted from 2003 onwards, aimed to provide a phylogenetic backbone of *Solanum* section *Petota* to facilitate the search for novel sources of resistance to *Phytophthora infestans*. The Solanaceae Source website publishes the results of the National Science Foundation (NSF) project to produce a worldwide taxonomic monograph of all species in the genus *Solanum*, divided into about 13 major clades, with molecular data from chloroplast NADH dehydrogenase F (ndhF) sequences (Bohs, 2005). Spooner and collaborators contributed data for section *Petota*.

Within the CBSG project, AFLP patterns of 4929 genotypes of more than 950 gene bank accessions were generated, the largest data set ever constructed for section *Petota*. The results were reported in Jacobs *et al.* (2008 and 2011).

Jacobs *et al.* (2008) constructed a neighbor-joining (NJ) tree of these 4929 genotypes (Fig. 2.1), but due to the size of the data set, they were only able to perform NJ and maximum parsimony (MP) jackknife analyses on a reduced set consisting of one genotype from each available accession (951 accessions, representing 196 taxa). Large parts of both jackknife trees consisted of a polytomy, but branches with jackknife support were used as the basis for informal species groups. These partly coincided with a number of series from Mexico and Central America, but most of the South American material could not be subdivided. Jacobs *et al.* (2008) consider it likely that the polytomy is indicative of the real biological situation in section *Petota*. As reasons why the phylogenetic links between species are so difficult to establish, they put forward the relatively fast spread of tuber-bearing *Solanum* species over South America, due to the geographic conditions in the Andes, combined with high levels of hybridization.

Jacobs *et al.* (2011) used population genetic tools to establish the genetic structure in the group of South American representatives of section *Petota* (566 accessions representing 89 taxa). They established the optimal partitioning of the group, combining a Structure analysis with maximizing the F-statistics (Fst) values among subgroups, and found that the 89 species arrangement had a lower Fst value than an arrangement of the material in 16 clusters. Within these clusters, they obtained support for additional groups, arriving at a total of 44 groups. They concluded that the South American part of section *Petota* was overclassified and provided a table with an estimate of the level of support of all the species names in current use.

Recently, Vleeshouwers *et al.* (2011) released an online database, Sol*R*gene, which contained the CBSG data on resistance to *P. infestans* and the presence of R genes and R gene homologs in *Solanum* section *Petota*. Phylogenetic data of the entire Sol*R*gene collection are included, and an interactive AFLP-based NJ tree is presented that allows generating phylogenetic trees of selected groups of germplasm. The collection is classified according to the nomenclature of Hawkes (1990), using his designations of subsections and (super)series.

The NSF established the Planetary Biodiversity Inventory (PBI) program in 2002 (http://www.nsf.gov/news/news_summ.jsp?cntn_id=103065) with the aim to accelerate taxonomic research focused on the species-level taxonomy of a monophyletic group and present the results online. The Solanaceae Source website presents the results of the PBI project "*Solanum*: a worldwide treatment" (http://solanaceaesource.org/content/pbi-solanum-worldwide-treatment). Taxonomic information on the whole genus *Solanum* will be uploaded, including names lists, species descriptions, specimen collections and publications. The following statement on the site explains the status of the information: "The names list on Solanaceae Source is a work in progress. Synonymy and nomenclature information are considered authoritative only for names that are accompanied by a full species description. For accepted names that do not yet have a species description, synonymy should be considered provisional." In the potato clade, Spooner and co-workers recognize 156 species (http://solanaceaesource.org/content/pbi-solanum-worldwide-treatment). There is no classification of these species in series, clades, or species groups.

To highlight agreement and disagreement between the approaches of the Solanaceae Source website and the CBSG AFLP study, we discuss two examples: first, the latest and perhaps "final" comment on the classification of the *brevicaule* complex; and second, the treatment of a South American group of species, series *Megistacroloba*.

The brevicaule complex

This complex is a prime example of overclassification, as discussed above. More than 30 names have been associated with material considered to belong to this complex. Spooner and collaborators

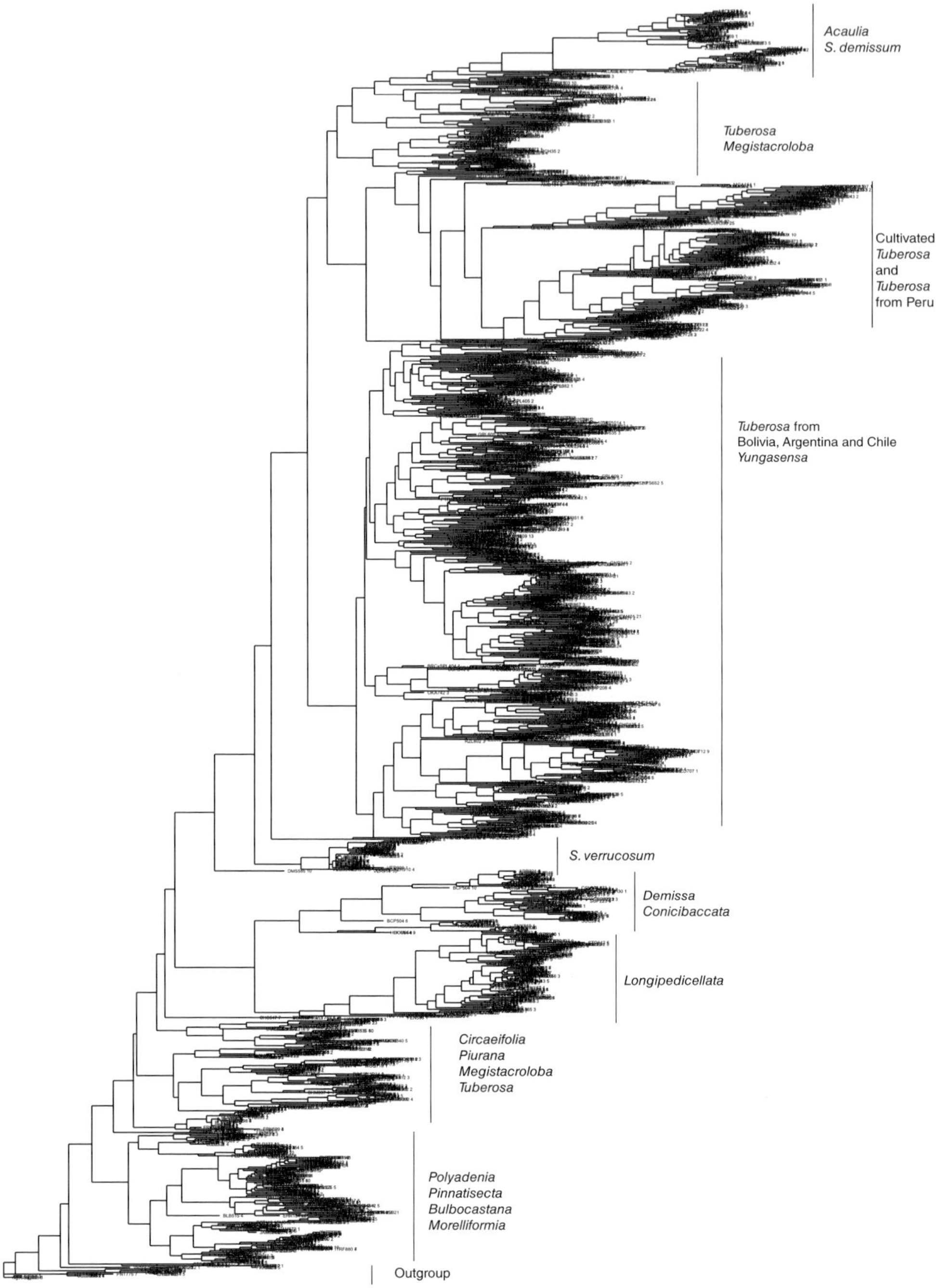

Fig. 2.1. Neighbor-joining tree from Jacobs *et al.* (2008) showing the clustering pattern of 4929 potato genotypes using AFLPs. A number of species groups can be distinguished: the North and Central American diploids, the North and Central American polyploids, *Solanum verrucosum* (the only representative of the *Tuberosa* group occurring in Central America), the South American representatives of the *Tuberosa* group, divided in a southern and northern group with the cultivated potatoes joining the northern group, and the polyploid *Acaulia* group with representatives in both Central (*Solanum demissum*) and South America (*Solanum acaule*, *Solanum albicans*).

recognize only two species in this complex, one from the "northern" part of the distribution area (*Solanum candolleanum* from central and southern Peru and north-western Bolivia), the other from the "southern" part (*S. brevicaule* from northern Bolivia to central Argentina). However, even these two species are extremely similar, differing primarily in their geography, and are difficult to key out. A key is announced, but although based on the insights from the replicated field trials (Van den Berg *et al.*, 1998; Alvarez *et al.*, 2008) and a thorough study of herbarium specimens, the authors themselves warn that determinations will often fail. The extensive synonymy proposed on the Solanaceae Source website is based on the impracticality of morphological data to distinguish these species and the low support they receive in molecular phylogenetic studies.

This solution to the problem of the *brevicaule* complex, already envisaged by Ugent (1970), who reduced the whole complex to one species, *S. brevicaule*, was anticipated by several studies (Van den Berg *et al.*, 1996, 1998; Miller and Spooner, 1999). Although a certain infrastructure of the complex can be recognized, leading to a series of geographically restricted "species" from north to south, the morphometric studies mentioned above clearly showed that many of the taxa involved could not be separated from each other reliably, and extensive synonymy was unavoidable. The Solanaceae Source website provides a workable taxonomic solution and it would be wise to heed their warning:

> We suspect that our synonymy will be questioned by those using molecular marker data, but we suggest that such workers wishing to resurrect our synonymized names do so only with morphological data needed to support the identification of these species in realistic and practical ways. Without such data, we envision the perpetuation of alternative taxonomies, variant identifications by different workers, and the maintenance of unstable names and taxonomic confusion in these important but taxonomically confusing species related to the cultivated potato.
>
> (http://solanaceaesource.org/)

Series Megistacroloba

Hawkes (1990) recognized 11 species in series *Megistacroloba*, mostly relying on leaf dissection characters (e.g. simple leaves in *Solanum boliviense* and *Solanum astleyi* and enlarged terminal leaflet in most other species). Giannattasio and Spooner (1994a,b) reduced *Solanum toralapanum* to a subspecies of *Solanum megistracrolobum*, and Spooner *et al.* (1997) reduced *S. astleyi* to a subspecies of *S. boliviense*. The treatment on the Solanaceae Source website reduces most of the former species names to synonyms of *S. boliviense*, accepts *Solanum sogarandinum* as a separate species, but notes that it is extremely difficult to distinguish it from *S. boliviense* and indicates that the status of other species still has to be designated.

Our own AFLP data (Van den Berg and Groendijk-Wilders, 2007) and the CBSG data (Jacobs *et al.*, 2011) more or less clearly distinguish a complex consisting of *S. boliviense* and *S. astleyi* (which are indistinguishable from each other) from the rest of the species of series *Megistacroloba*. The synonymizing of *S. astleyi* as subspecies of *S. boliviense* is not in line with the results of Jacobs *et al.* (2011), who in their Structure analysis found no support for a subspecies level within *S. boliviense*. Reducing *S. astleyi* to a synonym of *S. boliviense* seems more logical. The AFLP data also show an unexpected closer relationship between *S. megistacrolobum* and *Solanum sanctae-rosae*, with *S. toralapanum* sister to both, which would be in line with the Solanaceae Source treatment. Jacobs *et al.* (2011) found only weak evidence for species status for the combination of *S. megistacrolobum* and *S. toralapanum*, and for *S. sanctae-rosae*, but the material with these species names clustered separately from that of an *astleyi*/*boliviense* cluster. All this leaves some doubt whether the synonymy arrived at on the Solanaceae Source website is too extreme.

The gene bank predicament

As the cultivated potato is an important crop and its wild relatives are valuable potential sources of useful traits to improve the crop, much energy and time have been invested to collect, maintain, and evaluate germplasm of the wild species. These collections are maintained in gene banks, with the most important collections at CIP (International Potato Center, Lima, Peru), NRSP-6 (Potato Introduction Station, US Potato Genebank, Sturgeon Bay, USA), and CGN (Centre for Genetic Resources, Wageningen, the Netherlands).

For gene bank managers, a workable taxonomy of the material they maintain is crucial, and the presence of competing classifications is problematic. Gene bank managers, who usually observe at least part of their material during yearly rejuvenation cycles, get to know the species quite well, but do not have enough time and resources to be able to decide on taxonomic controversies. So, how should gene banks deal with competing taxonomies? It is virtually impossible to choose the "correct" classification, and practice has shown that taxonomists are not likely to reconcile their views. The only practical solution is to record the name under which the material was originally collected and add later taxonomic opinions in additional fields in the database. In this way, it is possible to cross-reference alternative names in gene bank databases. However, it would be advantageous if some kind of standardization could be arrived at.

A new situation has arisen recently, now the Solanaceae Source website has provided a recent, comprehensive taxonomic treatment of the species to be recognized in section *Petota*. Especially, the fact that this new taxonomy has been adopted by the Germplasm Resources Information Network (GRIN, 2014) may have the desired effect of standardization, although gene banks, by their own admission, are conservative in accepting name changes; see, for example, the website of the CGN, the Netherlands, where certain changes proposed by the Solanaceae Source website have not yet been adopted.

2.2 Classification of the Cultivated Potato

The cultivated potato originated somewhere from central Peru to central Bolivia (Hawkes, 1990), possibly near Lake Titicaca, on the border of Peru and Bolivia. In its center of origin, thousands of landraces were grown before the arrival of the Spanish conquistadores. The introduction of South American material into Europe led to the adaptation of this material to long-day circumstances, and was accompanied with a number of morphological changes in their leaves, stolons, and tubers. After the late blight epidemic in Europe in the 1840s, new introductions took place, but the nature and place of origin of this material was debated. Plant breeding activities produced a wide variety of cultivars, and the crop spread over the rest of the world. Here, we look briefly at the history of the taxonomic classification of the cultivated potato and discuss remaining problems and the current situation.

A brief outline of the history of the taxonomy of the cultivated potato

Because of its importance as a crop, the cultivated potato has attracted the attention of scientists since its arrival in Europe at the end of the 16th century, and there is a huge amount of literature describing the history of its taxonomy. Hawkes (1990) gives a chronological table listing the early records and pre-Linnean descriptions. Ovchinnikova *et al.* (2011) summarize the historical developments in the classification of the cultivated potato, starting with Linnaeus (1753), who included the potato as *Solanum tuberosum* L. in his *Species Plantarum*, probably based on material that was still closely similar to the cultivated potatoes grown in South America, later designated as *Solanum andigenum* Juz. & Bukasov and *S. tuberosum* subsp. *andigenum* (Juz. & Bukasov) Hawkes. Russian taxonomists divided the crop into up to 18 cultivated species and described hundreds of infraspecific taxa (subspecies, convarieties, varieties, and forma). Also, the Peruvian taxonomist, Carlos Ochoa, published numerous infraspecific taxa for cultivated material, recognizing varieties and forma to accommodate the landraces of Bolivia (Ochoa, 1990). Dodds (1962), observing the lack of morphological support of the cultivated species, maintained only the hybrids *Solanum* × *curtilobum* and *Solanum* × *juzepczukii* at the species level and proposed five groups within *S. tuberosum*: Group Stenotomum (2×), Group Phureja (2×), Group Chaucha (3×), Group Andigena (4×), and Group Tuberosum (4×). The standard treatment by Hawkes (1990) lists seven species: *Solanum ajanhuiri* Juz. & Bukasov, *S. chaucha* Juz. & Bukasov, *S. curtilobum* Juz. & Bukasov, *S. juzepczukii* Buk., *S. phureja* Juz. & Bukasov, *S. stenotomum* Juz. & Bukasov, and *S. tuberosum* L., with subspecies within *S. phureja*, *S. stenotomum*, and *S. tuberosum*.

More recently, Huaman and Spooner (2002) extended Dodd's classification and classified all cultivated potatoes within the single species

S. tuberosum, with eight cultivar groups: Ajanhuiri, Andigenum, Chaucha, Chilotanum, Curtilobum, Juzepczukii, Phureja, and Stenotomum. Arguments to use International Code of Nomenclature for Cultivated Plants (ICNCP) categories rather than International Code of Botanical Nomenclature (ICBN) categories were discussed by Spooner and Hetterscheid (2005), but Spooner *et al.* (2007b) classified the cultivated potatoes again at the species level, with four species (*S. ajanhuiri*, *S. juzepczukii*, *S. curtilobum*, and *S. tuberosum*). Within *S. tuberosum*, two cultivar groups are recognized: Group Andigenum and Group Chilotanum.

A very useful paper is that of Ovchinnikova *et al.* (2011), who compiled all the epithets associated with the cultivated potato (many of which have appeared only in the Russian literature) and placed them in synonymy of the four species names they accepted. They selected type specimens (lectotypes) for as many of the validly published names as possible, using—if available—material cited by the original authors. They also provided descriptions of the cultivated potato species and cultivar groups, and a key, where the "relatively modern varieties" within *S. tuberosum* were also included.

Problems/controversial issues in the taxonomy of the potato

The taxonomic treatment of the South American landraces and their relationship to the wild species has been the subject of many studies, sometimes with contrasting results. We review briefly controversies about the origin and further development of the cultivated potato, whether there is a single or multiple origin(s), the nature of the material introduced into Europe before and after the late blight epidemics of the 1840s, and the modern views on how to subdivide the crop.

The origin and further development of the cultivated potato

The situation about the origin of the cultivated potato seems clear enough. Many studies have pointed to the *brevicaule* complex as harboring the closest relatives of the crop. Unfortunately, the taxonomy of this complex was somewhat confused, but the recent synonymy proposed by Spooner and collaborators on the Solanaceae Source website made it unnecessary to establish which species could be the actual progenitor of the cultivated material. The complex has often been found to be separated into a northern and a southern group, and in phylogenetic studies the cultivated material is usually associated with the northern representatives of the complex, indicating a Peruvian origin (Kardolus *et al.*, 1998; Spooner *et al.*, 2005; Jacobs *et al.*, 2008). The AFLP data of Spooner *et al.* (2005) showed that the cultivated potato material they investigated grouped in one clade, nested within the clade of the northern part of the *brevicaule* complex. Since, in their view, that part of the *brevicaule* complex should be reduced to one species, they consider the origin of the potato to be a single domestication event. However, other studies have pointed out that the presence of several chloroplast haplotypes in the cultivated material, many of which also occur in wild species, suggest a number of successive domestication events (Sukhotu *et al.*, 2004; Sukhotu and Hosaka, 2006). Hosaka (2003), also based on chloroplast haplotypes, suggested that perhaps other species, like *S. tarijense*, could have played a role in the pedigree of the cultivated potato. In any case, defining "single origin" as "derived from one species" is not very informative about the way the cultivated potato arose from the *brevicaule* complex, which would certainly have involved multiple events and different progenitor genotypes.

If the problem of the progenitor species seems to be solved, there remain the questions as to what constituted the first cultivated material and how did our modern cultivated potato arise? Hawkes (1990) presented the following account, explaining the origin of the seven cultivated species he recognized. The diploid *S. stenotomum* is considered to be the most primitive cultigen, and in its turn, the progenitor of all other cultivated material, including the tetraploid *S. tuberosum* subsp. *andigenum*, the group of South American landraces from which the modern cultivated potato, *S. tuberosum* subsp. *tuberosum*, was derived. Crosses between *S. stenotomum* and *S. megistacrolobum*, *Solanum acaule* and *S. tuberosum* subsp. *andigenum* led to the diploid *S. ajanhuiri* and the triploids *S. juzepczukii* and *S. chaucha*, respectively. Crosses between *S. juzepczukii* and *S. acaule* led to the pentaploid *S. curtilobum*. *S. phureja* was derived from *S. stenotomum* and is characterized by its lack of tuber dormancy. After *S. stenotomum* originated through selection from its wild progenitors in the *brevicaule* complex, the tetraploid *S. tuberosum* subsp. *andigenum* arose through hybridization between *S. stenotomum* and another member of the *brevicaule* complex, accompanied by polyploidization and selection. Sukhotu and Hosaka (2006) suggested that the initial *S. tuberosum* subsp. *andigenum* populations arose by multiple origin from *S. stenotomum*. Finally, transporting South American tetraploid material to Europe caused it to adapt to long-day circumstances, producing *S. tuberosum* subsp. *tuberosum*. Hawkes (1990) suggested that the origin of *S. tuberosum* subsp. *tuberosum* from *S. tuberosum* subsp. *andigenum* actually occurred twice, the first time by transporting South American landraces from the Bolivian Andes southward towards Chile, where subsp. *tuberosum* still occurs at low altitudes.

The late blight epidemic in Europe in the 1840s virtually wiped out the originally introduced material and this was replaced by new introductions. Already, the origin of the first introductions was a matter for debate, with Juzepczuk and Bukasov (1929) arguing that these were subsp. *tuberosum* from Chile, while most other authors assumed these to be subsp. *andigenum*, probably from Colombia (Grun, 1990; Hawkes, 1990). Also, the material introduced after the late blight epidemic has been identified as either subsp. *tuberosum* or subsp. *andigenum*. Spooner *et al.* (2005) looked at nuclear and chloroplast markers of potato varieties in India that presumably resembled the original subsp. *andigenum* introductions, because late blight did not arrive in India until 1870. They found that the early introductions to India already contained material of both subspecies. Also, the landraces of the Canary Islands—where the potato was first introduced—contained both types, indicating that the introductions before the late blight epidemic already contained material from both the Andes and Chile (Rios *et al.*, 2007). Ames and Spooner (2008) studied historical herbarium specimens and concluded that although the Andean potato appeared in Europe first, the Chilean potato was already present long before the late blight epidemics. Also, the Solanaceae Source website summarizes the situation as follows:

> The Andean potato first appeared in Europe around 1700 and persisted until 1892, long after the late blight epidemics, while the Chilean potato first appeared in Europe in 1811, fully 34 years before the late blight epidemics and persisted until the present day, when over 99% of extant modern potato cultivars possess Chilean cytoplasm. These results support original introductions from the Andes, but refute the idea that the late blight epidemics beginning in Europe in 1845 stimulated introductions of Chilean germplasm as breeding stock to combat this disease, or eliminated the Andean potato.
>
> (http://solanaceaesource.org/)

Utility of polyploidy to subdivide the crop and the use of ICBN/ICNCP categories

Since the description of the cultivated potato as a species, *S. tuberosum*, (Linnaeus, 1753), taxonomists dealing with the crop have added a number of species to accommodate the various types of cultivated material found in South America. Some of this material arose through hybridization between cultivated material and wild species of different ploidy levels, leading to diploid, triploid, tetraploid, and pentaploid cultivated potatoes. All of these received Linnean taxon names, while the species *S. tuberosum* was subdivided in two subspecies (subsp. *tuberosum* and subsp. *andigenum*), to emphasize the distinction between the "modern", "western", "common" potato and the South American landraces.

As mentioned above in the paragraph on the history of the taxonomic classification of the cultivated potato, Dodds, in a chapter of Correll's 1962 treatment, proposed the extremely sensible solution to consider all the material of the potato crop as one cultigenic species, *S. tuberosum* (besides the two hybrid species, *S.* × *juzepczukii* and *S.* × *curtilobum*) and to recognize five groups within this species for the different types and ploidy levels, applying the cultivar-group classification advocated in the ICNCP (Brickell *et al.*, 2009), rather than the formal taxa in use under the ICBN (McNeill *et al.*, 2006). His approach took into account the difference between taxa like species and subspecies (suitable to classify populations of wild plants) and classificatory units designed to deal with the man-made variation of a crop (like cultivar and cultivar group), which could be termed "culta" instead of taxa (Hetterscheid and Brandenburg, 1995; Spooner and Hetterscheid, 2005).

This sensible approach was also applied by Huaman and Spooner (2002), who proposed eight cultivar groups for the cultivated material of South America. These authors refrained from completing the classification with a cultivar group for the modern potatoes, since that assortment should probably be subdivided into several groups; for example, according to the various uses to which the material is put. In a "cultonomic" classification, there is no need to classify all of the material, only groups that are deemed useful should be circumscribed.

An emphasis on ploidy as the criterion to subdivide the crop has always determined the classification of cultivated potatoes. Unfortunately, ploidy determinations have not been carried out for most of the accessions present in potato gene banks, and names of cultivated species have often been applied to this material without information on chromosome numbers. Ghislain *et al.* (2006) reported that 32 out of 102 accessions of the presumably diploid cultivated species *S. phureja* turned out to be either triploid or tetraploid, and Spooner *et al.* (2007b) found many exceptions of clustering by ploidy in the set of potato landraces that they studied with nuclear SSRs. They concluded that the gene pool structure of potato landraces should be reevaluated, and felt the necessity to reintroduce the species category for some of the variants of cultivated potatoes. Within *S. tuberosum*, two cultivar groups are recognized (Andigenum and Chilotanum), but the rest of the crop is classified in the species *S. ajanhuiri* (2×), *S. juzepczukii* (3×), and *S. curtilobum* (5×). On the other hand, the species *S. phureja* was obliterated, because the material with this name was not consistently diploid.

While Ghislain *et al.* (2006) still advocated the use of a cultivar group (rather than a species) classification, Spooner *et al.* (2007b) thus opted for the reclassification of the cultivated potatoes at the species level. Their NJ tree did show groups clustering by ploidy, but with many exceptions and without statistical support. They propose to recognize the species *S. ajanhuiri*, *S. juzepczukii*, and *S. curtilobum*, because only that part of the tree shows bootstrap support values above 50%. However, the tree presented in Spooner *et al.* (2007b) does not really show support for these species. There is 100% support for a branch uniting *S. curtilobum*, and some of the accessions of *S. ajanhuiri* and *S. juzepczukii*, and 89% support for a group of all other accessions of *S. ajanhuiri*. Thus, it is unclear why these species should be maintained and the other groups (admittedly showing less than perfect clustering by ploidy) should not. Especially, the decision to remove *S. phureja* is highly debatable. There is a clear group of material with this name and most of this material is characterized by the lack of tuber dormancy. This is exactly a situation where a cultivar group could be applied successfully, according to the definition cited by the authors: "Cultivar Groups are taxonomic categories used by the International Code of Nomenclature of Cultivated Plants to associate cultivated plants with traits that are of use to agriculturists and are not meant to represent natural groups or species in any classification philosophy." If a certain trait is deemed important to group material, other characteristics—like chromosome number—may vary within that group.

The solution of Spooner *et al.* (2007b) leaves us with an unfortunate mix of ICBN (McNeill *et al.*, 2006) and ICNCP (Brickell *et al.*, 2009) categories to describe the structure of the potato crop, while it is highly debatable that the species level is suitable for cultigens. The drawback of this approach is the inherent instability of a system based on Linnean taxon names that must conform to the rules of the ICBN, and must be changed when the rules of the code prescribe this. Since name changes are undesirable (especially for cultivated material), a set of rules governing the conservation of names has been included in the ICBN. These allow the continued use of a name that turns out not to be the oldest valid name. The inclusion of these rules shows an awareness that there is a potential problem in applying ICBN categories to cultivated material. On the other hand, the ICBN still contains the Hybrid Appendix that governs the naming of nothotaxa (taxa resulting from hybridization), while these hybrids consist of cultivated plants in most of the cases presented. In all matters concerning the nomenclature of cultivated plants, it would be better to use the appropriate code, the ICNCP.

2.3 The Impact of Genomic Data

As outlined above (in the section on the remaining problems in potato taxonomy), morphological characters have proven to be problematic, due to their plasticity and often insufficient variation, and are unable to provide evidence for a clear-cut classification of section *Petota*. As soon as molecular data became available, potato taxonomists applied them to reconstruct the phylogeny of the group. The markers used included restriction fragment length polymorphism (RFLPs), RAPDs, AFLPs, SSRs, various sequences and single nucleotide polymorphism (SNPs). The analyses of these molecular data sets, although sometimes giving contradicting results, have produced a rather robust phylogeny reconstruction of section *Petota* as a whole, and many detailed studies of subgroups (Van den Berg and Jacobs, 2007), but the markers used could not always provide sufficient resolution and details of the phylogeny-based classification remain debatable. Will this situation improve when we turn to the recent developments in genome sequencing?

With respect to chloroplast genomes, there are a few papers reporting the application of genome sequencing in species-level comparisons within section *Petota*. Daniell *et al.* (2006) generated complete chloroplast genome sequences of *S. bulbocastanum* and *S. lycopersicum*, and compared these with the genome organization of two other representatives of the *Solanaceae* with completely sequenced chloroplast genomes, *Nicotiana* and *Atropa*. They compared these taxa for gene content and gene order, and distribution and location of repeated sequences. Not surprisingly, they found that sequence divergence corresponded to the phylogenetic relationships, with the two most closely related species, potato and tomato, having the lowest divergence values for all classes of genes. Gargano *et al.* (2012) compared the plastome sequence of the cultivated potato (*S. tubersosum* subsp. *tuberosum*) with that of *S. bulbocastanum*, to characterize the variability in different genomic regions and to develop markers. They selected four intergenic regions and compared these among 16 tuber-bearing *Solanum* species. All species showed a distinct haplotype, suggesting good resolution at the species level for these four markers. Their results largely confirmed the relationships found in earlier studies (e.g. Spooner and Castillo, 1997), with the 1 EBN diploid species from Central America displaying higher divergence in comparison to the other species and showing high similarity between *S. verrucosum* and the polyploid Mexican species, confirming the accepted hypothesis on the origin of these polyploids.

Visser *et al.* (2009) described the sequencing effort of the Potato Genome Sequencing Consortium (PGSC), a collaboration of 13 research groups, which at that time had completed about 30% of the 840 Mb genome, announcing a draft of the complete genome by the end of 2009. A recent paper by the PGSC (2011) presented that draft and discussed the insights in genome evolution, tuber biology, and occurrence of disease resistance that could be derived from the genome sequence. The availability of an ultra high-density (UHD) genetic map was instrumental for assigning sequences to each of the 12 chromosomes. The authors expect that the knowledge of the complete genome sequence will be invaluable for the identification of genes and novel alleles of genes for traits of interest for potato breeders.

The recent special issue of the *American Journal of Botany* (Volume 99(2), 2012) gives an excellent overview of the applications of next-generation sequencing (NGS) in plant biology. Unfortunately, none of the contributions deal with the potato or its wild relatives, even though potato taxonomist David Spooner is one of the organizers of this special issue. In their introduction, Egan *et al.* (2012) point at the future possibilities of NGS technologies. Summarizing the research presented in the issue, they describe the possibilities for the development of molecular markers, the detection of introgression, transcriptome investigations, phylogenetic and ecological studies, polyploidy genetics, and applications for gene bank collections. They note that the application of NGS to large-scale population and phylogenetic studies is hindered by large genome sizes, variation in the proportion of organellar DNA in total DNA, polyploidy, and gene number/redundancy. An interesting application is discussed by Kane *et al.* (2012), who suggest an ultra bar-coding approach to examine variation below the species level systematically, assembling whole organellar genomes and large (greater than 5 kb) contiguous portions of the nuclear genome, and expanding the traditional bar-coding regions to their full, many-kilobase length, making it far more sensitive than traditional DNA bar coding. So far, this has not yet been applied to potato, and Spooner (2009), looking at ordinary bar coding, even claims that bar coding does not work in section *Petota* and ascribed this to the complexity of the group and not to the suspect nature of many wild potato "species". Bar coding aims at distinguishing species, and in a group where the species status of the entities that have been recognized is debatable, the failure of bar coding does not reflect on the approach but on the nature of the studied group. Whether "ultra bar coding" might improve the situation remains to be seen.

If, through the power of NGS, we could identify and characterize the genes actually coding for the characters used in taxonomic classification (e.g. leaf dissection, corolla shape), a high-resolution phylogeny reconstruction could become possible, with genes underlying key traits either agreeing or disagreeing with the neutral markers usually used for phylogeny reconstruction. Complete sequences will also be helpful to study synteny at different levels; for example, giving information on the sequences of the breakpoints involved in chromosome segment rearrangements. Since phenomena like inversions are rare events, they can be phylogeny informative (Szinay *et al.*, 2012). Undoubtedly, (next-generation) genome sequence work can be instrumental for the development of markers that might be useful to elucidate further the phylogenetic relationships within section *Petota*. Thus far, such new markers have not led to new insights in the structure of this group of *Solanum* species, and the availability of complete genome sequences has not been exploited in other ways for this purpose. Genome sequences are a potential source of relevant comparisons, but so far their use has been aimed mainly at plant breeding strategies, like the isolation of disease-resistance genes and the elucidation of tuberization (PGSC, 2011), rather than taxonomic applications.

References

Alvarez, N.M.B., Peralta, I.E., Salas, A. and Spooner, D.M. (2008) A morphological study of species boundaries of the wild potato *Solanum brevicaule* complex: replicated field trials in Peru. *Plant Systematics and Evolution* 274, 37–45.

Ames, M. and Spooner, D.M. (2008) DNA from herbarium specimens settles a controversy about origins of the European potato. *American Journal of Botany* 95(2), 252–257.

Bitter, G. (1912) Solana nova vel minus cognita. *Feddes Repertorium* 10, 531–532.

Bohs, L. (2005) Major clades in *Solanum* based in ndhF sequences. In: Keating, R.C., Hollowell, V.C. and Croat, T.B. (eds) *A Festschrift for William G. D'Arcy: The Legacy of a Taxonomist*. Monographs in Systematic Botany from the Missouri Botanical Garden, Missouri Botanical Garden Press, St Louis, Missouri 104, pp. 27–49.

Brickell, C.D., Alexander, C., David, J.C., Hetterscheid, W.L.A., Leslie, A.C., Malecot, V., *et al.* (eds) (2009) *International Code of Nomenclature for Cultivated Plants*. International Society for Horticultural Science, Leuven, Belgium (Regnum Vegetabile 151).

Bukasov, S.M. (1933) The potatoes of South America and their breeding possibilities. *Trudy po Prikladnoj Botanike, Genetike I Selekcii* Supplement 58, 1–192.

Camadro, E.L., Erazzu, L.E., Maune, J.F. and Bedogni, M.C. (2012) A genetic approach to the species problem in wild potato. *Plant Biology* 14, 543–554.

Castillo, R. and Spooner, D.M. (1997) Phylogenetic relationships of wild potatoes, *Solanum* series *Conicibaccata* (sect.*Petota*). *Systematic Botany* 22, 45–83.

Correll, D.S. (1962) *The Potato and Its Wild Relatives, Section Tuberarium of the Genus Solanum*. Texas Research Foundation, Renner, Texas.

Daniell, H., Lee, S.-B., Grevich, J., Saski, C., Quesada-Vargas, T., Guda, C., *et al.* (2006) Complete chloroplast genome sequences of *Solanum bulbocastanum*, *Solanum lycopersicum* and comparative analyses with other *Solanaceae* genomes. *Theoretical and Applied Genetics* 112, 1503–1518.

Dodds, K.S. (1962) Classification of cultivated potatoes. In: Correll, D.S. (ed.) *The Potato and its Wild Relatives, Section Tuberarium of the Genus Solanum*. Texas Research Foundation, Renner, Texas, pp. 517–539.

Egan, A.N., Schlueter, J. and Spooner, D.M. (2012) Applications of next-generation sequencing in plant biology. *American Journal of Botany* 99, 175–185.

Gargano, D., Scotti, N., Vezzi, A., Bilardi, A., Valle, G., Grillo, S., *et al*. (2012) Genome wide analysis of plastome sequence variation and development of plastidial CAPS markers in common potato and related *Solanum* species. *Genetic Resources and Crop Evolution* 59, 419–430.

Germplasm Resources Information Network (GRIN) (2014) USDA, ARS, National Genetic Resources Program. National Germplasm Resources Laboratory, Beltsville, Maryland (http://www.ars-grin.gov/cgi-bin/npgs/html/paper.pl?language=en, accessed 16 April 2014).

Ghislain, M., Andrade, D., Rodriguez, F., Hijmans, R. and Spooner, D.M. (2006) Genetic analysis of the cultivated potato *Solanum tuberosum* L. Phureja Group using RAPDs and nuclear SSRs. *Theoretical and Applied Genetics* 113, 1515–1527.

Giannattasio, R.B. and Spooner, D.M. (1994a) A reexamination of species boundaries between *Solanum megistacrolobum* and *S. toralapanum* (*Solanum* sect. *Petota*, series *Megistacroloba*): morphological data. *Systematic Botany* 19, 89–105.

Giannattasio, R.B. and Spooner, D.M. (1994b) A reexamination of species boundaries and hypotheses of hybridization concerning *Solanum megistacrolobum* and *S. toralapanum* (*Solanum* sect. Petota, series *Megistacroloba*): molecular data. *Systematic Botany* 19, 106–115.

Grun, P. (1990) The evolution of cultivated potatoes. In: Bretting, P.K. (ed.) New Perspectives on the Origin and Evolution of New World Domesticated Plants. *Economic Botany* (Supplement 3) 44, 39–55.

Hawkes, J.G. (1944) *Potato Collecting Expeditions in Mexico and South America. II. Systematic Classification of the Collections*. Imperial Bureau of Plant Breeding and Genetics, Cambridge, UK.

Hawkes, J.G. (1956) A revision of the tuber-bearing Solanums (first edition). In: *Annual Report*. Scottish Plant Breeding Station, pp. 37–109.

Hawkes, J.G. (1963) A revision of the tuber-bearing Solanums (second edition). In: *Annual Report*. Scottish Plant Breeding Station, pp. 76–181.

Hawkes, J.G. (1990) *The Potato, Evolution, Biodiversity and Genetic Resources*. Belhaven Press, London.

Hawkes, J.G. and Hjerting, J.P.H. (1969) *The Potatoes of Argentina, Brazil, Paraguay and Uruguay*. Clarendon Press, Oxford, UK.

Hawkes, J.G. and Hjerting, J.P.H. (1989) *The Potatoes of Bolivia*. Clarendon Press, Oxford, UK.

Hetterscheid, W.L.A. and Brandenburg, W.A. (1995) Culton versus taxon: conceptual issues in cultivated plant systematics. *Taxon* 44, 161–175.

Hosaka, K. (2003) T-type chloroplast DNA in *Solanum tuberosum* L. ssp. *tuberosum* was conferred from some populations of *S. tarijense* Hawkes. *American Journal of Potato Research* 80, 21–32.

Huaman, Z. and Spooner, D.M. (2002) Reclassification of landrace populations of cultivated potatoes (*Solanum* sect. *Petota*). *American Journal of Botany* 89(6), 947–965.

Jacobs, M.J.M., Van den Berg, R.G., Vleeshouwers, V.G.A.A., Visser, M., Mank, R., Sengers, M., *et al*. (2008) AFLP analysis reveals a lack of phylogenetic structure within *Solanum* section *Petota*. *BMC Evolutionary Biology* 8,145 (http://www.biomedcentral.com/1471-2148/8/145, accessed 16 April 2014).

Jacobs, M.J.M., Smulders, M.J.M., Van den Berg, R.G. and Vosman, B. (2011) What's in an name; genetic structure in *Solanum* section *Petota* studied using population-genetic tools. *BMC Evolutionary Biology* 11, 42 (http://www.biomedcentral.com/1471-2148/11/42/abstract, accessed 29 April 2014).

Johnston, S.A. and Hanneman, R.E. (1978) Endosperm balance factors in some tuber-bearing *Solanum* species. *American Potato Journal* 55, 380.

Juzepczuk, S.W. and Bukasov, S.M. (1929) A contribution to the question of the origin of the potato. Proceedings of the USSR Congress of Genetics. *Plant and Animal Breeding* 3, 592–611.

Kane, N., Sveinsson, S., Dempewolf, H., Yang, J.Y., Zhang, D., Engels, J.M.M., *et al*. (2012) Ultra-barcoding in cacao (*Theobroma* spp.; *Malvaceae*) using whole chloroplast genomes and nuclear ribosomal DNA. *American Journal of Botany* 99, 320–329.

Kardolus, J.P., Van Eck, H.J. and Van den Berg, R.G. (1998) The potential of AFLPs in biosystematics: a first application in *Solanum* taxonomy. *Plant Systematics and Evolution* 210, 87–103.

Linnaeus, C. (1753) *Species Plantarum: a facsimile of the first edition 1753*. Ray Society 1957, 1959, London.

McNeill, J., Barrie, F.R., Burdet, H.M., Demoulin, V., Hawksworth, D.L., Marhold, K., *et al*. (2006) *International Code of Botanical Nomenclature (Vienna Code)*, adopted by the Seventeenth International Botanical Congress Vienna, Austria, July 2005. A.R.G. Gantner Verlag KG (Regnum Vegetabile 146).

Matsubayashi, M. (1991) Phylogenetic relationships in the potato and its related species. In: Tsuchiya, T. and P.K. Gupta (eds) *Chromosome Engineering in Plants: Genetics, Breeding, Evolution, Part B*. Elsevier, Amsterdam, pp. 93–118.

Mayden, R.L. (1997) A hierarchy of species concepts: the denoument in the saga of the species problem. In: Claridge, M.F., Dawah, H.A. and Wilson, M.R. (eds) *Species: The Units of Diversity*. Chapman and Hall, London, pp. 381–423.

Miller, J.T. and Spooner, D.M. (1999) Collapse of species boundaries in the wild potato *Solanum brevicaule* complex (*Solanaceae*, *S.* sect. *Petota*): molecular data. *Plant Systematics and Evolution* 214, 103–130.

Ochoa, C.M. (1990) *The Potatoes of South America: Bolivia*. Cambridge University Press, New York.

Ochoa, C.M. (2004) *The Potatoes of South America: Peru (Part 1: The Wild Species)*. International Potato Center (CIP), Lima.

Olmstead, R.G. (1995) Species concepts and plesiomorphic species. *Systematic Botany* 20, 623–630.

Ovchinnikova, A., Krylova, E., Gavrilenko, T., Smekalova, T., Zhuk, M., Knapp, S., *et al*. (2011) Taxonomy of cultivated potatoes (*Solanum* section *Petota*: *Solanaceae*). *Botanical Journal of the Linnean Society* 165, 107–155.

PGSC (Potato Genome Sequencing Consortium) (2011) Genome sequence and analysis of the tuber crop potato. *Nature* 475, 189–197.

Rios, D., Ghislain, M., Rodriguez, F. and Spooner, D.M. (2007) What is the origin of the European potato? Evidence from Canary Island landraces. *Crop Science* 47, 1271–1280.

Spooner, D.M. (2009) DNA barcoding will frequently fail in complicated groups: an example in wild potatoes. *American Journal of Botany* 96, 1177–1189.

Spooner, D.M. and Castillo, R.T. (1997) Reexamination of series relationships of South American wild potatoes (*Solanaceae*: *Solanum* sect. *Petota*): evidence from chloroplast DNA restriction site variation. *American Journal of Botany* 84, 671–685.

Spooner, D.M. and Hetterscheid, W.L.A. (2005) Origins, evolution, and group classification of cultivated potatoes. In: Motley, T.J., Zerega, N. and Cross, H. (eds) *Darwin's Harvest, New Approaches to the Origins, Evolution, and Conservation of Crops*. Columbia University Press, New York, pp. 285–307.

Spooner, D.M. and Hijmans, R. (2001) Potato systematics and germplasm collecting, 1989–2000. *American Journal of Potato Research* 78, 237–268.

Spooner, D.M. and Salas, A. (2006) Structure, biosystematics, and genetic resources. In: Gopal, J. and Khurana, S.M.P. (ed.) *Handbook of Potato. Production, Improvement, and Postharvest Management*. Haworth Press, Binghamton, New York, pp. 1–39.

Spooner, D.M. and Van den Berg, R.G. (1992a) An analysis of recent taxonomic concepts in wild potatoes (*Solanum* sect. *Petota*). *Genetic Resources and Crop Evolution* 39, 23–37.

Spooner, D.M. and Van den Berg, R.G. (1992b) Species limits and hypotheses of hybridization of *Solanum berthaultii* Hawkes and *Solanum tarijense* Hawkes: morphological data. *Taxon* 41, 685–700.

Spooner, D.M., Ugarte, M.L. and Skroch, P.W. (1997) Species boundaries and interrelationships of two closely related sympatric diploid wild potato species, *Solanum astleyi* and *S. boliviense*, based on RAPDs. *Theoretical and Applied Genetics* 95, 764–771.

Spooner, D.M., Van den Berg, R.G. and Miller, J.T. (2000) Species and series boundaries of *Solanum* series *Longipedicellata (Solanaceae)* and phenetically similar species in ser. *Demissa* and ser. *Tuberosa*: implications for a practical taxonomy of sect. *Petota*. *American Journal of Botany* 87, 113–130.

Spooner, D.M., Van den Berg, R.G., Rodríguez, A., Bamberg, J., Hijmans, R.J. and Lara-Cabrera, S.I. (2004) Wild potatoes (*Solanum* section *Petota*) of North and Central America. *Systematic Botany Monographs* 68, 1–209.

Spooner, D.M., McLean, K., Ramsay, G., Waugh, R. and Bryan, G.J. (2005) A single domestication for potato based on multilocus amplified fragment length polymorphism genotyping. *Proceedings of the National Academy of Sciences of the United States of America* 102, 14696–14699.

Spooner, D.M., Fajardo, D. and Bryan, G.J. (2007a) Species limits of *Solanum berthaultii* Hawkes and *S. tarijense* Hawkes and the implications for species boundaries in *Solanum* sect. *Petota*. *Taxon* 56(4), 987–999.

Spooner, D.M., Nunez, J., Trujillo, G., Rosario Herrera, M. del, Guzman, F. and Ghislain, M. (2007b) Extensive simple sequence repeat genotyping of potato landraces supports a major reevaluation of their gene pool structure and classification. *Proceedings of the National Academy of Sciences of the United States of America* 104, 19397–19403.

Sukhotu, T. and Hosaka, K. (2006). Origin and evolution of Andigena potatoes revealed by chloroplast and nuclear DNA markers. *Genome* 49, 636–647.

Sukhotu, T., Kamijima, O. and Hosaka, K. (2004) Nuclear and chloroplast DNA differentiation in Andean potatoes. *Genome* 47, 46–56.

Szinay, D., Wijnker, E., van den Berg, R.G., Visser, R.G.F., de Jong, H. and Bai, Y. (2012) Chromosome evolution in *Solanum* traced by cross-species BAC FISH. *New Phytologist* 195, 688–698.

Ugent, D. (1970) The potato: what is the botanical origin of this important crop plant, and how did it first become domesticated? *Science* 170, 1161–1166.

Van den Berg, R.G. and Groendijk-Wilders, N. (2007) AFLP data support the recognition of a new tuber-bearing *Solanum* species. *Plant Systematics and Evolution* 269, 133–143.

Van den Berg, R.G. and Jacobs, M.J.M. (2007) Molecular taxonomy. In: Vreugdenhil, D. (ed.) *Potato Biology and Biotechnology Advances and Perspectives*. Elsevier, Amsterdam, pp. 55–76.

Van den Berg, R.G., Groendijk-Wilders, N. and Kardolus, J.P. (1996) The wild ancestors of cultivated potato: the *brevicaule* complex. *Acta Botanica Neerlandica* 45, 157–172.

Van den Berg, R.G., Miller, J.T., Ugarte, M.L., Kardolus, J.P., Villand, J., Nienhuis, J., *et al.* (1998) Collapse of morphological species in the wild potato *Solanum brevicaule* complex (*Solanaceae*: sect. *Petota*). *American Journal of Botany* 85, 92–109.

Van den Berg, R.G., Groendijk-Wilders, N., Zevenbergen, M.J. and Spooner, D.M. (2001) Molecular systematics of *Solanum* series *Circaeifolia (Solanum* section *Petota)* based on AFLP and RAPD markers. In: Van den Berg, R.G., Barendse, G., Van der Weerden, G.W. and Mariani, C. (eds) *Solanaceae V: Progress in Taxonomy and Utilization*. Nijmegen University Press, Nijmegen, the Netherlands, pp. 72–84.

Van den Berg, R.G., Bryan, G.J., Del Rio, A. and Spooner, D.M. (2002) Reduction of species in the wild potato *Solanum* section *Petota* series *Longipedicellata*: AFLP, RAPD and chloroplast SSR data. *Theoretical and Applied Genetics* 105, 1109–1114.

Visser, R.G.F., Bachem, C.W.B., De Boer, J.M., Bryan, G.J., Chakrabati, S.K., Feingold, S., *et al.* (2009) Sequencing the potato genome: outline and first results to come from the elucidation of the sequence of the world's third most important food crop. *American Journal of Potato Research* 86, 417–429.

Vleeshouwers, V.G.A.A., Finkers, R., Budding, D., Visser, M., Jacobs, M.J.M., Van Berloo, R., *et al.* (2011) Sol*R*gene: an online database to explore disease resistance genes in tuber-bearing *Solanum* species. *BMC Plant Biology 2011* 11, 116 (http://www.biomedcentral.com/1471-2229/11/116, accessed 16 April 2014).

Volkov, R.A., Komarova, N.Y., Panchuk, I.I. and Hemleben, V. (2003) Molecular evolution of rDNA external transcribed spacer and phylogeny of sect. *Petota* (genus *Solanum*). *Molecular Phylogenetics and Evolution* 29, 187–202.

Walpers, V.G. (1844–1845) Ordo CLII. *Solanaceae* Bartl. *Repertorium Botanices Systematicae* VIII, 163–168.

Wang, M., Allefs, S., Van den Berg, R.G., Vleeshouwers, V.G.A.A., Van der Vossen, E. and Vosman, B. (2008) Allele mining in *Solanum*: conserved homologues of Rpi-blb1 are identified in *Solanum stoloniferum*. *Theoretical and Applied Genetics* 116, 933–943.

3 Potato Utilization and Markets

Jennifer K. Bond*
USDA-Economic Research Service, Washington, USA

The potato is ranked by the Food and Agriculture Organization of the United Nations (FAO) as the world's fourth most important food plant behind rice, wheat, and maize (FAO, 2006). As one of the most diverse and nutritious crops on Earth, potatoes can be grown on most continents and come in an array of shapes, sizes, and colors (CIP, 2012a).

More than 4000 individual varieties are thought to exist worldwide, yet modern cultivars share a single origin limited to the southern regions of Peru and the north-east sections of Bolivia (CIP, 2012a). Potatoes found growing in this area were cultivated between 7000 and 10,000 years ago, although it was not until the 16th century that Spanish explorers brought potatoes back to Europe (Kiple and Ornelas, 2000).

Once the tuber arrived in Europe, it slowly gained popularity. Initially, it was a novelty shared primarily among royal courts, and later as an inexpensive and scurvy-preventing food source for sailors. Subsequently, the general population overcame misgivings about this relative of the poisonous nightshade plant and accepted the potato as a nutritious and hardy staple crop (Kiple and Ornelas, 2000).

In the past 20 years, considerable changes have taken place in world potato markets (Fig. 3.1). Production has shifted away from some traditional potato producing regions, including Europe and North America, to the developing world, where the need for nutritionally rich staple crops is arguably greater. New and significant suppliers, most notably China, have emerged as production leaders, while worldwide demand for processed potato products has grown steadily (FAO, 2008a).

3.1 Production Growth in Emerging and Developing Countries

According to the FAO, potato production attributed to developed nations has declined by approximately 1% each year for the past 20 years, and an increasing share of production is used for the processing market (FAO, 2008a). In contrast, cultivation and consumption in developing countries focuses on fresh (table-stock) production and has grown by about 1% per year, although, collectively, these countries remain net importers of the tuber (FAO, 2008b; FAOSTAT, 2013).

In 1990, emerging and developing countries, which included both Russia and China, accounted for a slight majority of production (55.5%). By 2010, emerging and developing country production comprised nearly two-thirds of total

* E-mail: jkbond@ers.usda.gov

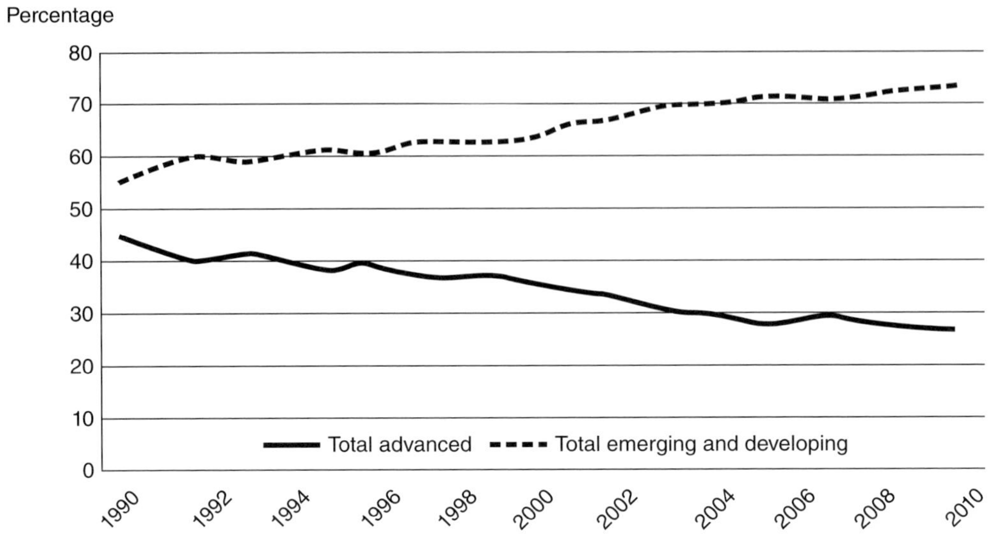

Fig. 3.1. Percentage of total world potato harvest by advanced versus emerging and developing counties, 1990–2010. (From USDA Economic Research Service based on data of the United Nations Food and Agriculture Organization, FAOSTAT; accessed 20 May 2013.)

world production (73.5%), relative to the 26.5% production share of advanced countries (FAOSTAT, 2013).

A near 20% increase in the world production share over a 20-year period is a testament to the growing popularity of the crop in low-income and newly industrialized markets. Increased cultivation of potatoes in economically challenged and food-insecure countries stems from the nutritious qualities of the tuber, providing a sizeable amount of micro- and macronutrients for each medium-sized potato serving (CIP, 2012a). Potatoes are also easy to grow on small land parcels, and can be eaten with minimal preparation and cooked without expensive equipment (CIP, 2012a).

Potatoes also rank highly on FAO's nutrition productivity index, a measure relating water use and resultant calorie production. Every cubic meter of water applied to the potato crop in cultivation produces an average of 5600 calories of dietary energy. By comparison, maize, wheat, and rice respectively produce 3860, 2300, and 2000 calories with the same water application (FAO, 2008c). Finally, some of the growing popularity in developing countries may be explained by rising potato yields, which stands in contrast to stagnant and declining yield gains for major cereals (World Bank, 2008).

3.2 Shifting Regional Cultivation

Shifts in regional supply patterns reflect the fact that emerging and developing countries are cultivating a growing proportion of total world potato production. Historically, most potatoes were grown in Europe and North America. However, in 2011, Asia grew more potatoes than any other single region and produced close to 50% of total world output (FAOSTAT, 2013).

Europe continues to contribute a sizeable (35% in 2011), though generally declining, share of total production. In 2000, European production accounted for 45.5% of total production, down from 60.1% in 1990. North American potato production has generally declined for the past 15 years, and the area now supplies just slightly more than 6% of world total production, a 2% decline since 2000.

Despite gradual reductions in total production and harvested area over the past 10 years, North America generated the highest yields of any region at 15.87 t acre^{-1}; 202% more than the world average of 7.87 t acre^{-1} in 2011 (FAOSTAT, 2013). In the same year, yields from Asia averaged 7.40 t acre^{-1}, and while this figure is near the world average, North American production is significantly more efficient. For example, for every 1 acre of potatoes cultivated in

North America, the most recent production figures imply that 2.14 acres would need to be planted in Asia to realize the same output.

Improvements in cultivation practices and seed stock are likely to make Asian production more efficient in the future. For other developing regions such as Africa, where relative production inefficiencies are greater, improvements in cultivation practices have the potential to increase harvests, without proportional increases in cultivated area, while enhancing food security (CIP, 2012a).

3.3 Production Rankings and Totals

A comparison of production rankings from 2000 to 2011 provides a synopsis of major industry developments. Consistent with its status as an emerged production leader, in 2011 top-ranked China produced 88.4 million tonnes (Mt) and approximately 24% of all potatoes grown in the world (Table 3.1) (FAOSTAT, 2013).

In 2000, China grew 64.6 Mt of potatoes, representing 20% of total world production (Table 3.1) (FAOSTAT, 2013). The trajectory of Chinese potato production is clear and provides evidence of the important role potatoes play in feeding China's growing population. The International Potato Center (CIP) reports that Chinese officials "(e)xpect fully 50% of the increased food production it will need to meet demand in the next 20 years will come from potatoes" (CIP, 2012a).

In China, potatoes are grown as a staple, as well as a cash crop. In the mountainous northern and south-west regions, production provides cash income. Potatoes are grown primarily for household subsistence consumption in the lower plains (FAO, 2008d; CIP, 2012b).

India, the second largest producer in 2011, grew less than half as many potatoes as top-ranked China, while still accounting for 11% of total production and 42.3 Mt (Table 3.1). In a little over a decade, Indian production has experienced dramatic growth, with production expanding by more than 90% and 20 Mt between 2000 and 2011. However, for more than 20 years, Indian production has been trending towards expansion, as harvested hectares and yields have both improved. Similar to many developing countries, expanded potato consumption (and hence production) in India is sustained by a growing number of citizens living in urban areas, where incomes are rising (CIP, 2012b). Urban populations are demanding convenient and ethnically diverse foods, as well as Western-style dining experiences, many of which prominently feature processed potato products (FAO, 2008b,d; CIP, 2012b). Consumption in rural areas of India is lower, as potatoes are not considered to be a staple and are typically grown as a cash crop (FAO, 2008d; CIP, 2012b).

While India and China have increased their share of world production, Poland has experienced notable production declines. Polish potato output between 2000 and 2011declined nearly 58%, due to periods of low prices and damaging floods (FAOSTAT, 2013). Collectively, the share

Table 3.1. Top potato producing countries, 2000 and 2011.

Rank in 2000	Country	2000	Country	2011
		Tonnes		Tonnes
1	China	64,596,119	China	88,350,220
2	Russian Federation	34,965,200	India	42,339,400
3	India	22,242,700	Russian Federation	32,681,500
4	USA	19,852,500	Ukraine	18,705,000
5	Poland	19,378,900	USA	19,361,500
6	Ukraine	17,343,500	Germany	11,800,000
7	Germany	11,916,800	Bangladesh	8,326,390
8	Belarus	7,767,600	Poland	8,196,700
9	Netherlands	7,115,000	France	8,016,230
10	UK	6,649,000	Netherlands	7,333,470

(From USDA Economic Research Service based on data of the United Nations Food and Agriculture Organization, FAOSTAT; accessed 20 May 2013.)

of production by the historically dominant regions of the USA, Russia, and Western Europe declined from 27% in 2000 to 22% in 2011 (FAOSTAT, 2013)

Russia was unseated as the second largest potato growing nation in 2011, but remained the largest European producer. Russian output has declined in recent years and occasionally dropped significantly, as in 2010 when pest pressure and weather challenges combined to produce a 20-year low level of production (PotatoPro, 2010; FAOSTAT, 2013). US production has slipped slightly from 2000, and its global ranking fell to the fifth position by 2011. The decline is consistent, with a trend towards gradually decreasing North American production and consumption.

3.4 World Potato Consumption

Average world potato consumption was estimated around 32.6 kg per person in 2009 (Table 3.2) (FAOSTAT, 2013). Despite being a leading production region, the Asian population annually consumes a relatively modest 26.2 kg per person (Table 3.2) (FAOSTAT, 2013). In contrast, Europe produces fewer potatoes, but with a relatively smaller population and supplementary imports, the region enjoys the largest average annual per capita consumption at 85.5 kg (Table 3.2) (FAOSTAT, 2013). Per capita consumption in Africa is relatively low, although developing nations on the continent are cultivating increasing amounts of potatoes and consumption may increase over time as cultivation practices improve.

Table 3.2. World potato consumption by region, 2009. (From USDA Economic Research Service, based on data of the United Nations Food and Agriculture Organization, FAOSTAT; accessed 20 May 2013.)

Region	Total potato consumption (tonnes)	Kg per capita
Africa	14,823,633	15.9
Asia	105,934,808	26.2
Europe	63,003,549	85.5
South America	10,771,097	27.7
North America	18,749,149	54.9
WORLD	217,253,003	32.6

In addition to fresh potato consumption, regional per capita consumption numbers are influenced by the demand for processed potato products, including French fries (also known as chips). Processed potato demand has grown in Europe and the USA; however, the greatest expansion has been in the population centers of the Middle East, South America, and East Asia, including China (FAO, 2008d). The tripling of the US French fry export volume between 1990 and 2010, and the doubling of European exports in the past 10 years, are evidence of substantial growth in world demand for frozen potatoes (USDA-FAS, 2013).

Consumption levels in advanced economies have been supported by the increased availability of novel varieties, packaging improvements, and convenient preparations (USPB, 2011b). Specialty cultivars have become common features in many US grocery stores and may reintroduce potential consumers to the spud. Easy-mash and pre-made refrigerated potatoes make it simple for convenience-seeking consumers to enjoy potato products (USPB, 2011b). In the years to come, domestic consumption may be strengthened by a growing appreciation for products, like potatoes, that contain high levels of resistant starch. On the other hand, potential consumer concerns about the presence of acrylamide in cooked starchy foods may nominally impact potato demand, creating further consumer education opportunities for the sector.

3.5 Global Potato Trade

Approximately 6% of total world potato production is estimated to be traded (FAO, 2008b). The inherent bulkiness of tubers makes it relatively expensive to transport, and perishability concerns complicate international movement (FAO, 2008c). Despite these impediments, international potato trade has been increasing in recent years, doubling in volume and rising by fourfold, in terms of dollar value, since the mid-1980s until the present (USDA-FAS, 2011; FAOSTAT, 2013).

Much of the trade growth is attributed to the distribution of frozen potato products (Fig. 3.2). The growth in imports is located primarily in

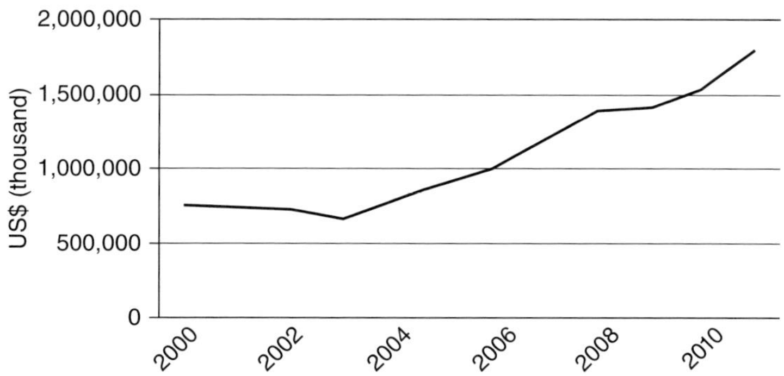

Fig. 3.2. World total exports of frozen potato products, 2000–2011. (From US Census Bureau, Foreign Trade Statistics.)

emerging and developing markets, which are distanced from concentrations of processing potato capacity in sections of the European Union (EU) and North America (USDA-FAS, 2011). To satisfy demand for processed products, the frozen potato export market is likely to become increasingly important to the USA, Canada, and select EU countries, including Belgium and the Netherlands.

Belgium is one of the world's largest manufacturers and exporters of frozen potato products (FAO, 2008d). These factors, in combination with high per capita consumption, put Belgium on top as the largest importer of fresh potatoes in the world (Table 3.3). Russian potato imports fluctuate in negative correlation with domestic production. In 2010, when persistent drought conditions reduced the local harvest an estimated 25–40%, Russia imported the second highest dollar value of potatoes (PotatoPro, 2010; FAOSTAT, 2013). As potato production in Europe declines and demand for processed potato products remains strong, both domestically and in export markets, European countries may increase imports of processing and table-stock potatoes.

For several decades, the Netherlands has remained a major potato producing country and the world's top potato exporting country (in terms of value) (Table 3.4). Assisted by favorable growing conditions, access to major shipping channels, and diversity in potato varieties, US$790 million worth of seed, fresh, and processed potatoes were exported in 2010 (USDA-FAS, 2013).

Other European countries including France, Germany, and Belgium are ranked among the top ten potato exporters (by dollar value). These EU members harvest substantial production volumes, which serve to increase the supply available for fresh sales, processing, and exportation. In France, exports include a sizeable amount of seed potatoes, while Germany and Belgium export relatively more processed potatoes (FAO, 2008d). Canada and the USA frequently rank among the top five exporting countries, due to the combination of sales abroad of processed, fresh, and seed potatoes.

For more than a decade, China has ranked as the world's largest producer of potatoes. Historically, however, the country has not enjoyed similar status as a top exporter. For example, in 2000, China did not rank among the top 20 potato exporters, and only occasionally appeared on top ten lists during the 1990s. In the last decade, Chinese production expansions have supported growth in potato export activities. In 2011, China was ranked as the tenth largest potato exporting nation (in value terms), a remarkable rise in prominence in a relatively short period (Table 3.4) (FAOSTAT, 2013).

3.6 Fresh and Processed Products

Potatoes are marketed in one of seven broad product categories. Seed variety selection, management practices, weather, relative prices, and other micro and macro market and environmental conditions determine the suitability of a given producer's crop to a specific market. Prior to planting, producers will determine whether they are growing potatoes for one or more of

Table 3.3. Top potato importing countries ranked by value of imports, 2010. (From USDA Economic Research Service, based on data of the United Nations Food and Agriculture Organization, FAOSTAT; accessed 20 May 2013.)

		Quantity	Value	Unit value
Rank	Country	Tonnes	US$ (thousand)	US$/tonne
1	Belgium	1,382,592	2,908,089	216
2	Russia	664,348	272,861	411
3	Spain	777,942	248,088	319
4	Germany	573,769	237,133	413
5	Netherlands	1,220,236	233,054	191
6	Italy	706,995	232,713	329
7	USA	415,411	148,046	356
8	UK	248,841	119,340	480
9	United Arab Emirates	38,387	111,196	2,897
10	Egypt	146,787	108,283	738
11	France	414,729	107,880	260
12	Portugal	308,856	99,496	322
13	Canada	234,658	88,858	379
14	Algeria	113,263	74,428	657
15	Greece	132,643	67,565	509
16	Malaysia	155,951	65,046	417
17	Poland	235,438	63,792	271
18	Iraq	180,535	61,963	343
19	Lebanon	113,203	48,313	427
20	Czech Republic	151,626	43,934	290

Table 3.4. Top potato exporting countries ranked by value of exports, 2010. (From USDA Economic Research Service, based on data of the United Nations Food and Agriculture Organization, FAOSTAT; accessed 20 May 2013.)

		Quantity	Value	Unit value
Rank	Country	Tonnes	US$ (thousand)	US$/tonne
1	Netherlands	1,883,601	789,735	419
2	France	2,318,680	613,074	264
3	Germany	1,592,520	306,905	193
4	Canada	493,657	172,606	350
5	USA	386,176	165,005	427
6	UK	336,699	158,290	470
7	Belgium	784,858	152,667	195
8	Egypt	298,557	129,562	434
9	Iran (Islamic Republic of)	262,973	104,621	398
10	China	258,683	104,277	403
11	Italy	168,099	98,756	587
12	Spain	240,527	98,460	409
13	Israel	206,640	82,656	400
14	Denmark	171,480	69,829	407
15	Syrian Arab Republic	112,123	51,604	460
16	Pakistan	245,329	50,663	207
17	Cyprus	58,188	44,111	758
18	India	173,864	28,624	165
19	Saudi Arabia	135,488	26,838	198
20	Austria	84,069	26,508	315

seven general potato markets: fresh, frozen, chip, dehydrated, starch, seed, or other industrial categories.

Fresh

The history of potatoes is primarily an account of fresh, also known as table-stock, varieties. With the exception of early methods of potato preservation, such as Incan freeze-drying, most potatoes were transported, stored, and purchased fresh (CIP, 2012a). Fresh potatoes are not cooked, and are sold whole. Minimal handling occurs to remove field debris and to sort and package the product for market. Table-stock potatoes are graded according to size and type. Once in the grocery stores, fresh potatoes are frequently organized by variety and package.

Bulk potatoes can be sold loose and purchased on a price per unit of measurement basis. Polybags, containing varying amounts of potatoes, are an economical way to buy larger quantities. Specialty packaging, such as clear plastic clamshells, re-sellable bags, and polybags, add value and are typically used for premium and specialty varieties including creamers, fingerlings, and petites.

On average, the American consumer buys 16.47 kg of fresh potatoes annually, with Russet varieties comprising the largest portion of table-stock sales (USDA-ERS, 2011). However, sales can vary significantly by region, and it is believed non-Russet varieties comprise a sizeable share of display space in many supermarkets, particularly in the South and Mid-Atlantic and New England regions of the country (Bond and Richards, 2008).

In most regions of the USA, Russet potato sales remain higher than any other potato variety; however, collectively, sales of non-Russet varieties are similar to Russet sales in several regions, and represent a significant segment of the potato market across the USA (Bond and Richards, 2008). In fact, consumers in the New England and South Atlantic regions exhibit strong demand for red, white, and yellow potatoes, which drives the percentage of total non-Russet volume and dollar sales above those of the commodity potato (Bond and Richards, 2008). Consumer acceptance of specialty potatoes has been increasing for several years, with heightened demand for yellow flesh, red or purple skin varieties, and those with unusual shapes such as the banana fingerling (Naegely, 2002; Olsen *et al.*, 2003; Bond and Richards, 2008).

Frozen

Today, most of the frozen potatoes sold around the globe are French fry-style products that begin with fresh potatoes, which are peeled, sliced, blanched, air-dried, and partially fried, before being blast frozen and packaged for shipment (McCain Foods, 2013). French fry-style frozen potatoes are distributed primarily to restaurants, where they are fried again before being served to customers (McCain Foods, 2013).

The origins of sliced and deep-fried potatoes are open to debate, with Belgium and France both claiming to have popularized the dish. In the USA, the invention of the mechanical potato peeler in the 1920s helped make French fries (and potato chips/crisps) more widely available and among the most popular snacks in the nation (Panati, 1989). Quick-serve restaurants (QSR), such as McDonald's, catalyzed the popularity of French fries by making them a staple on the menu of their rapidly expanding franchise restaurants (Makki and Plummer, 2005).

The global popularity of Western-style diets, which prominently feature fast foods like French fries, has increased both global consumption and production (Makki and Plummer, 2005). Therefore, growth in demand for French fries is correlated strongly with the expansion of fast food and QSRs around the globe (Makki and Plummer, 2005). For years, expansion of the QSR sector in the USA, the EU, and other Western markets supported sales growth; however, growth in domestic frozen potato consumption has recently declined as the QSR market matures in developed countries (Makki and Plummer, 2005).

While per capita consumption slackened in some Western markets, frozen potato production has risen in Canada and the USA. Available data from the US Department of Agriculture-Foreign Agricultural Service (USDA-FAS, 2013) indicates production in Canada increased by 69% between 1996/97 and 2006/07. During the same period, the USA and the Netherlands experienced modest production growth of 13.6% and 6.6%, respectively, while Japanese production fell by nearly 15% (USDA-FAS, 2013). In 2003/04, USDA-FAS recorded the beginning of frozen potato production in China. For the next 4 years, China averaged a 12.5% annual increase in production. Between 1996/97 and 2006/07, total global frozen potato production expanded by about 21% (USDA-FAS, 2013).

When production figures are compared with consumption data from the same period, opportunities for trade become clear. In 2006/07, Canada produced roughly 925,000 t more frozen potato product than it consumed, providing for a potentially significant exportable surplus. Conversely, Japanese consumption estimates for 2006/ 07 relative to production imply a shortfall of about 300,000 t; a gap filled through trade with the USA, Canada, and other partners.

In December 2012, the US Potato Board (USPB) reported that targeted markets, primarily in Asia, increased their imports of US frozen potatoes by 18%, while sales to other markets were down 7% on the year (USPB, 2012a). A USPB-sponsored potato marketing campaign, focused on promoting American potatoes abroad, was reportedly effective in increasing sales to the Philippines by 76%, to Indonesia by 37%, to South Korea by 50%, and to Mexico by 43% (USPB, 2012a).

Continued growth in the frozen potato category is likely to be dependent on QSR growth in developing countries and sustained demand in Western and developed countries (Makki and Plummer, 2005). Because frozen potatoes are most often sold to restaurants, the economic conditions that support consumption of meals eaten outside the home will have an ongoing influence in the patronization of QSRs, and therefore consumption of French fries. Concerns about the healthfulness of fast food may inhibit growth in demand in the USA and the EU.

Chips

The origin of potato chips (also known as crisps) has been attributed to an American chef working at a resort in Saratoga Springs, New York, in 1850 (SFA, 2012). The first chips were thin sliced and crunchy, in order to appease customers who were unhappy with soggy French-fried potatoes. The chips, known as Saratoga Chips, were popular with guests and demand expanded quickly (SFA, 2012).

A potato chip is a thin-sliced piece of potato, fried until crunchy, and then coated with seasonings ranging from salt to powdered cheese, vinegar, or any number of regionally popular flavors and spices. Transporting the fragile chips is challenging, and limits the economic appeal of trade in finished chip products. Consequently, much chip production occurs regionally or locally. There is thought to be notable trade in fresh chipping variety potatoes, which, once imported, are processed into chips locally; however, exact figures are difficult to find, as chipping potato shipments are aggregated with fresh potato shipment numbers.

In 2011, the US Department of Agriculture-Economic Research Service (USDA-ERS) reported that the average domestic consumer purchased 6.7 kg of potato chips annually. Eighty-one facilities processed 2.8 t of fresh potatoes (54.7 million cwt) as chips and snacks in 2010, an increase of 29% from the previous year's production figures (USDA-ERS, 2011). Processing facilities are distributed across the country; however, a concentration of chip plants is located near the population centers of the eastern seaboard (USDA-ERS, 2011).

According to the USPB, the USA manufactures more potato chips than any country, which reflects the domestic popularity of the snack food (USPB, 2012b). Year-round demand necessitates year-round chip production, and thus a supply of chipping varieties, often from geographically diverse locations, to maintain chip stock. Chipping potatoes are grown throughout the country, and major domestic varieties include: Alturas, Andover, Atlantic, Chipeta, Dakota Pearl, Ivory Crisp, Kennebec, LaChipper, Marcy, Megachip, NorValley, Norwis, Pike, Reba, and Snowden (USPB, 2012b).

Dehydrated

The process of removing moisture from a potato, known as dehydration, increases the tuber's storability and, arguably, its versatility. Dehydrated potatoes are lightweight and less bulky than fresh potatoes. When rehydrated, 1 kg of dehydrated product will yield approximately 5 kg of potato product (USPB, 2012c). Unlike frozen potatoes, no refrigeration is necessary to preserve the integrity of dehydrated products (USPB, 2012c).

Flakes, flours, granules of various sizes, shreds, noodles, slices, and more, are possible to manufacture through the dehydration process. Once processed, dehydrated potatoes typically enjoy a shelf life of 18–24 months, if stored in airtight packaging (USPB, 2012c).

Potato flour has gained favor in the marketplace for its ability to extend foods and to enhance the nutritional properties of breads, soups, and other food products (USPB, 2012c). Items made with potato flour are gluten-free, though potato flour is frequently combined with other flours when used in baked goods. At Passover, when grains are not eaten, potato flour is an important substitute for wheat flours. Potato flour can also be used as a thickener when added to sauces and soups.

USDA-ERS reported in 2010, 1.7 Mt (34.1 million cwt) of fresh potatoes were dehydrated in US processing facilities. These facilities provide a diversity of products that include potato pancake mix, specific types of potato chips, mashed potatoes, and some canned soup products (USDA-ERS, 2011).

Starch

Potatoes provide food for human consumption, but many varieties also produce high levels of starch that can be removed through an extraction process and used as a component in industrial and pharmaceutical products (Bayer Crop Science, 2008). Fresh potatoes can be used in starch production, though starch may also be extracted from the peelings and other potato waste generated from the processing of other potato products. Once the raw material is gathered from fresh potatoes or waste, the starch is extracted and manufactured into one of seven common starch products: paper and cardboard manufacturing; biotechnological raw materials or substrate; textiles; pastes, glues, and gums; pharmaceuticals and cosmetics; detergents; and biologically degradable materials (Bayer Crop Science, 2008).

To extract starch from potatoes, the potatoes must first be crushed. Crushing the potato damages individual cell walls, creating a slurry of both liquid and starch cell contents. Starch grains, also called leucoplasts, are then rinsed and dried, ultimately creating a powdered starch (Bayer Crop Science, 2008). On average, potatoes contain about 20% starch by volume (Bayer Crop Science, 2008). The minimal presence of fat and protein, 2% and 1%, respectively, results in starch that is white in color, neutral in taste, has good clarity, demonstrates a high binding strength, and is not prone to foaming during the cooking process (Bayer Crop Science, 2008). High moisture retention, a lack of gluten, and high phosphorous content, which enhances the viscosity of liquids augmented with potato starch, are additional appealing aspects.

In March 2010, the Amflora potato, a genetically modified (GM) potato, was cultivated in Europe (BASF, 2010). The GM variety has a modified starch composition that improves its suitability for industrial applications. A lack of political and public support is a likely reason why the variety's developer, BASF Plant Science, stopped marketing the Amflora potato in European markets in January 2012.

Seed

The seed potato market is vitally important. Replanting a portion of the crop each year with certified seed and seed from improved varieties helps to minimize the risk of disease contamination, as well as helping to preserve yields (FAO, 2008e; CIP, 2012c). Potatoes are typically reproduced through vegetative propagation, a form of asexual reproduction in plants. CIP estimates that 5–15% of each year's harvested potato crop (depending on quality) is saved for the purpose of planting the next season. Small sections of the saved tubers are planted once they have sprouted, and the cycle continues (CIP, 2012c). Farmers in developing countries are less likely to invest in sourcing certified seed potatoes than producers in developed countries (CIP, 2012c). As such, production in developing countries may be more susceptible to disease and yield loss.

Other and industrial

The remaining category of potato products includes minor products such as livestock feed, packaged mashed potatoes, potato pancakes/pirogues, potato vodka and schnapps, peeled potatoes, pasteurized potato products, potato protein, potato fiber, potato salad, and canned potatoes. Industrial potato products are similarly diverse, and may include bioplastic, glazes, imitation cheese starch or rennet casein replacement, starch fiber, and gum (CIP, 2012c).

3.7 Potato Characteristics and Varieties

Years of potato cultivation and breeding have resulted in potatoes that vary significantly in shape, size, color, flavor, and nutritional content (CIP, 2012a). In fact, there are more than 4000 distinct varieties worldwide, but only about 100 are suitable for commercial use and are consistently available in consumer markets (USPB, 2011a).

When selecting a potato in the grocery store, it is helpful to understand the differences in the suitability of particular varieties to a specific cooking application. Potato starch and sugar content generally determine which preparations are most appropriate; for processing applications including baking, frying, and mashing, a starch level of at least 13% or higher is preferred (Stark and Love, 2003). A higher level of starch in baked or mashed potatoes leads to an appealing dry or floury texture; however, the same potato may fall apart in soups or salads (USPB, 2011a). When a firmer potato with a smooth texture is needed, waxy or boiling varieties, which exhibit lower starch content, are more suitable (USPB, 2011a).

A potato variety's characteristic sugar content will also determine its suitability for various preparations. For example, sugar content affects the color of fried potatoes. During the frying process, potatoes are exposed to high temperatures; Stark and Love (2003) note that at high temperatures, sugars are known to "combine with amino acids and other compounds to form the dark color and flavor [that] was associated with 'burned' foods". When potatoes with high sugar levels are fried, the result is likely to be less appealing to consumers, and these varieties are ill suited to the process.

3.8 Popular Domestic and International Varieties

According to the International Potato Center (CIP), potatoes are grown in more than 100 countries and under a variety of climate conditions ranging from "sea level up to 4700 meters above sea level and from southern Chile to Greenland" (CIP, 2012a). The geographic and climactic diversity of potatoes lends itself to the cultivation of a wide range of potato varieties. Most commercial Western varieties can be described as belonging to one of seven subcategories: Russet, yellow, white, red, fingerling, purple and blue, and petites/other (USPB, 2011a).

In 2012, the majority of all potatoes grown in the USA were Russet varieties (Table 3.5) (USDA-NASS, 2013). Approximately 44.5% of fall acres were planted to the famous Russet Burbank. In Idaho, the top potato producing state, fully 52.5% of all potato acres were planted to Russet Burbank and 90% of total acres were planted to Russet varieties (Table 3.5) (USDA-NASS, 2013).

In the top seven potato growing states, the majority of potato acres are cultivated to Russets, including the Russet Burbank, Russet Norkotah, Umatilla Russet, Ranger Russet, and others; however, significant acres of yellow and white varieties are being grown alongside Russets (Table 3.5) (USDA-NASS, 2013). Acres planted to specialty and premium varieties have increased in the last 25 years, as changing consumer preferences are influencing diversity in potato production and supporting more colorful and abundant potato product options in the grocery store (Bond and Richards, 2008).

Internationally, many countries grow native and non-standard potato varieties, making the categorization of cultivated potatoes more challenging (CIP, 2012a). Several South American and African countries do, however, import seed stock from Canada, the USA, the Netherlands, and other parts of Europe (CIP, 2012a; USPB, 2012d). Popular varieties of imported seed stock potatoes include familiar cultivars such as Kennebec—a Russet-like variety that is noted for its frying properties—, Red Pontiac/Dakota Chief—a red-skinned variety with white flesh—, Cardinal—a high-yielding variety with appealing processing characteristics—, Spunta—a white/yellow flesh variety that is ideal for

Table 3.5. Percent of fall potatoes planted to major varieties in the USA during 2012[a]. (From USDA-NASS, *Crop Production*, December 2012.)

Variety	Planted acres (%)	Variety	Planted acres (%)	Variety	Planted acres (%)
Russet Burbank	44.5	Silverton	0.6	Red Pontiac	0.1
Russet Norkotah	13.1	Superior	0.6	Rio Grande Russet	0.1
Ranger Russet	9.4	Dakota Pearl	0.6	Classic	0.1
Umatilla Russet	5.1	Western Russet	0.5	Norwis	0.1
Frito-Lay	3.2	Alpine	0.5	Mega Chip	0.1
Norland	3.1	Atlantic	0.5	Defender	0.1
Alturas	2.6	Modoc	0.5	Kennebec	0.1
Shepody	1.8	Cal White	0.3	Satina	0.1
Goldrush	1.2	Cascade	0.3	Keuka Gold	0.1
Snowden	1.1	Red LaSoda	0.3	Katahdin	0.1
Chieftain	1.0	Pike	0.3	NorDonna	0.1
Yukon Gold	0.8	Blazer Russet	0.3	Reba	0.1
Premier	0.8	Dakota Crisp	0.2	Agata	0.1
Bannock	0.7	Dakota Rose	0.2	All Blue	0.1
Prospect	0.6	La Ratte	0.2	Canela	0.1
Ivory Crisp	0.6	Bintje	0.1	Other	2
Innovator	0.6	Ontario	0.1		

Note: [a]Seven state total includes: Idaho, Maine, Minnesota, North Dakota, Oregon, Washington, and Wisconsin.

baking and boiling—, Radosa—a white-/yellow-fleshed variety with floury properties—, Alpha—a yellow skin/flesh fresh market variety, among others (CIP, 2012a; ECPD, 2013).

The FAO provides examples of many additional popular cultivars from various regions of the world. These include the Peru-bred Atahualpa, the Dutch-developed Nicola, the Finnish Lapin puikula, the Canadian-bred Yukon Gold, the CIP-cultivated Tubira—intended for cultivation in West Africa—, the Mondial—popular in Australia—, and the red-skinned, Netherlands-bred Désirée, among others (FAO, 2008f).

A number of developing countries work with the CIP to develop locally suitable potato stock. One such example is the new cultivar, Cooperation 88, a variety developed jointly by the CIP and the National Agricultural Research System of China (CIP, 2012b). In addition to working with partner countries to develop new cultivars, the CIP works with many developing countries to improve seed saving and cultivation practices.

Russets

Russets and similar types are among the most commonly grown and consumed potato varieties in North America and much of the world. These bulky, oval potatoes are characterized by netted, brown skin and white flesh. In general, Russet-type varieties exhibit high starch content, containing between 60 and 70% starch, making them especially appealing to frying and chipping segments of the processing industry. Potatoes with high levels of solids and correspondingly high starch content reduce the amount of raw product needed, and where applicable, lessen the amount of oil for frying. The relatively high starch level of Russets produces the fluffy and dry-textured baked and mashed potatoes preferred by most consumers (USPB, 2011a).

Russets also contain low moisture levels. The lack of moisture enhances the potato's ability to absorb liquids and flavors during the cooking process, but also makes the potatoes prone to falling apart during long periods of boiling. Varieties with higher moisture contents will hold their shape better when boiled (USPB, 2011a).

Common Russet varieties are: Russet Burbank, Kennebec, Russet Norkotah, Canela Russet, Ranger Russet, Umatilla Russet, Classic Russet, and Premier Russet (USDA-NASS, 2013).

Yellows

The yellow potato category is gaining popularity for its diversity, versatility, and typically appealing flavor. Yellow potatoes vary in size from 1 oz to larger sizes comparable to a Russet. The category gets its name from the yellow interior flesh, though the exterior skin is typically light brown to a yellow-golden color. The interior of a common yellow variety will be smooth and creamy, and sometimes described as waxy when cooked (USPB, 2011a).

Moderate levels of starch and sugar make many yellows a popular choice for a variety of uses. Most yellows hold their shape well during the boiling process and are a popular addition to soups and potato salads. Many yellows are also well suited for mashing, grilling, and roasting cooking methods, which enhance their natural sweetness and often-perceived buttery flavor (USPB, 2011a).

Common yellow varieties are: Yukon Gold, Satina, and Dutch Baby (USDA-NASS, 2013).

Whites

Whites are typically small–medium size and many varieties are oval or oblong in shape. The interior flesh is white and the delicate exterior skin can range from near colorless to shades of tan. Common white varieties frequently have a slightly creamy, dense texture and average starch levels (USPB, 2011a).

A mild to slightly sweet flavor is associated with these low-sugar varieties, making them well suited to frying preparations including chips and French fries. White varieties typically hold their shape well and are additionally compatible with boiling and steaming cooking methods. Atlantic and Snowden varieties are popular chipping potatoes (USPB, 2012b).

Common white varieties are: Atlantic, Snowden, Dakota Pearl, Superior, Kennebec, Shepody, Cal White, Reba, and Pike (USDA-NASS, 2013).

Reds

In the USA, a concentration of red potato production can be found in the Red River Valley of North Dakota. However, red potatoes can be grown commercially nearly anywhere that potato production takes place. Internationally, varieties of red potatoes are popular in sections of the UK and Canada (USPB, 2011a). Red potatoes are sometimes referred to as "new potatoes", since a relatively small proportion of the crop goes into long-term storage, whereas long-term storage is a common practice for most yellow and Russet varieties (USPB, 2011a).

The skin of red potatoes tends to be thin, and the tubers are often smaller and rounder than other varieties. The interior flesh is commonly cream or white in color and is firm and waxy. Red potatoes are typically cooked, boiled, or roasted with their skin on. Leaving on the skin adds to the nutritional and fiber content of the potato. The color of the potato and the ability to retain its shape when boiled encourages the use of red potatoes in potato salads and soups. Boiled and roasted red potatoes are regularly characterized as having a slightly sweet flavor. Higher levels of sugar and lower starch levels in many red varieties inhibit significant chip and fry production in favor of home-cooking options (USPB, 2011a).

Common red varieties are: Sangre, Norland, Dakota Rose, Chieftain, Red La Soda, Red Pontiac, Cherry Red, and Modoc (USDA-NASS, 2013).

Fingerlings

Fingerling varieties are named for their comparatively slender and long appearance. Average fingerlings are approximately 10 cm in length, although sizes vary greatly by variety and desired end use. The term "fingerling" characterizes potatoes of a particular shape and size, rather than a specific subset of varieties. Thus, there is more diversity in color, texture, flavor, nutrition, and starch and sugar content within this category of potatoes compared to others (e.g. Russet, white, etc.).

External fingerling colors range from brown/tan, red, orange, yellow, white, and even purple. Interior colors are often cream or white. The cooked texture can range from waxy and firm to dry and floury. Fingerlings encompass a spectrum of flavors, including tobacco or earthy notes, buttery elements, or nutty flavors. One common fingerling variety, the Russian Banana, has low to medium sugar content and is suited to halving and then pan frying or roasting (USPB, 2011a).

Common fingerling varieties are: Russian Banana, French, Rose Finn Apple, and Purple Peruvian (WSU, 2012).

Purples and blues

These colorful potatoes are an eye-catching addition to the dinner table, and come in a range of hues of blue, purple, and even red. The skin and the flesh of these varieties tend to be highly pigmented and contain relatively high levels of antioxidants. Category members tend to be small to medium size and may range in shape from round to oblong, and even fingerling size.

Cooked texture depends on the variety. Most varieties can be prepared using all the typical cooking methods, but to maintain the vibrancy of the deeply hued purple and blues, microwave cooking is recommended (USPB, 2011a).

Common purple and blue varieties are: Purple Majesty, Purple Peruvian, All Blue, Mountain Rose, Adirondack Blue, Congo Blue, and True Blue (WSU, 2012; USDA-NASS, 2013).

Petites and others

The category of "Petites" refers to smaller versions of full-sized potatoes. The unifying characteristic is the C-size class distinction, which is no larger than 3.8 cm in diameter, roughly marble or pearl-sized. Category members range significantly in terms of starch and sugar content, flavor, color, and texture. Attractive medleys of yellow, red, and purple or blue flesh are increasingly available in US grocery stores (USPB, 2011a).

Similar to fingerlings, the small size of petites discourages peeling or significant chopping. Therefore, cooking and preparation time are reduced and the resulting product is similar to larger versions of the variety, but frequently has a more concentrated flavor. Some potato varieties, such as the Dutch Baby or Rosara/Klondike Rose, have been cultivated to be smaller in size and nearly all production will grade out at a size C or creamer size (USPB, 2011a).

Common petite varieties are: smaller sizes of most potato varieties, Dutch Babies, and Rosara/Klondike Rose.

3.9 Trends in Potato Consumption and Marketing

Trends shaping the international and US domestic marketplace have been evident throughout this chapter. The most prominent trends include: the significant growth in Chinese production and consumption; worldwide growth in the consumption of processed potato products (primarily fry products); and the increasing use of potatoes as a staple and cash crop in developing countries. In this section, some of the more nuanced trends are shared, as well as projections for future market developments.

Demand for fry products has grown at a slower pace in the EU and the USA, while the greatest growth has been in the emerging population centers of the Middle East, South America, and East Asia including China (USDA-FAS, 2013). The tripling of US fry exports (tonnes) between 1990 and 2010, and the doubling of European (EU-4) exports in the last 10 years, both signify growth in world demand for frozen potatoes (USDA-FAS, 2011).

Population growth and growth in demand for potatoes is expected to be greater in Asia, Latin America, Africa, and India than in the potato production centers of North America and Europe (FAO, 2008b). To accommodate growth in demand for potatoes and potato products, the export market is likely to becoming increasingly important to the USA, Canada, and select EU countries.

In developed countries, potato yields are rising more rapidly than demand for potatoes and potato products (Huffaker, 2005; World Bank, 2008). This presents production challenges in areas of high productivity such as the USA and Canada, and may create the need to find alternative domestic and/or international markets for surplus production.

In response to increasing pressure to use water, fuel, and fertilizer efficiently, potato breeders will be tasked with developing and improving drought hardy and possibly GM potato varieties (Huffaker, 2005). In developing countries in particular, improvements to potato varieties and production practices have great potential to enhance yields (CIP, 2012a). Research from the CIP in Lima, Peru, has contributed towards cultivation and germplasm improvements in partner countries. However, more research and education has the potential to affect food security positively, especially in countries where population growth is expected to be significant in the years to come (FAO, 2008b; CIP, 2012a).

3.10 Consumer Trends

Despite retaining its ranking as the top side dish to serve with protein, potatoes have encountered increasing competition on the dinner plate. According to the 2008 NPD Group National Eating Trends study, annual in-home potato servings per capita between 1985 and 2007 declined 33% from 105.2 servings to 70.8 servings per year. Driving this decline was a significant reduction in fresh potato consumption (NPD, 2008). The instances where fresh potatoes were eaten at home declined by 23.7 occasions between 1993 and 2009 (NPD, 2010). The same study identified three primary reasons for the observed decline of per capita consumption: concerns about the healthfulness of potatoes; changing consumer preferences; and the increased availability and appeal of potato substitutes (NPD, 2010).

Recent preliminary experimental data indicate that consumers were least knowledgeable about the specific nutritional qualities associated with potatoes relative to other qualities such as price and preparation method (Bond *et al.*, 2011). Thus, concerns about the healthfulness of potatoes may be attributable to a lack of knowledge of the macro- and micronutrient properties of the tuber. There is some evidence that consumer awareness of the positive nutritional aspects of potatoes is improving. A 2011 USPB-funded study found that fewer people were actively avoiding potatoes, fewer consumers perceived potatoes to be fattening, and health-related concerns were cited less frequently as a reason to avoid purchase (USBP, 2011b).

New varieties, packaging, and preparations have the potential to entice demand from consumers in developed markets where demand has recently stagnated (USPB, 2011b). Specialty varieties have become common features in many US grocery stores and may reintroduce consumers to the potato. Easy-mash and pre-made, refrigerated potatoes are also making it easy for convenience-seeking consumers to enjoy potato products.

Additional consumer preference trends include growth in acrylamide sensitivity, increased awareness of the glycemic index, and consumer interest in foods high in resistant starch. The global snack market is expected to grow annually by 7% through 2015, which may favor growth in the chip and dehydrated potato product sector. Customization of snack flavors to local preferences will assist potato snack products in dominating local markets.

3.11 Summary

The past 20 years have brought dramatic changes to the global potato marketplace. Enhanced cultivation techniques and improved access to disease-free seed stock have assisted in increasing production in emerging and developing countries. In particular, China has become the worldwide leader in potato production and is making strides towards becoming a front-runner in the export market. As production has surged in lower-income countries, developed countries have experienced some stagnation in fresh market demand. Fried products have maintained their popularity in advanced economies and are finding a receptive audience among consumers in transitional countries. New products, preparations, and varieties continue to demonstrate the diversity and usefulness of the humble potato.

References

BASF, Corp. (2010) The story: Amflora makes paper and yarn glossier and stronger (http://www.basf.com/group/corporate/en/news-and-media-relations/science-around-us/amflora/story, accessed 15 December 2012).

Bayer Crop Science (2008) Potato starch—a versatile commodity (http://www.agrokurier.de/bayer/cropscience/cscms.nsf/id/Potato_Starch_Agro/$file/potato_starch.pdf, accessed 15 December 2012).

Bond, C.A., Iverson, D., Bond, J.K. and Kroll, S. (2011) Importance and Consumer Knowledge of Potato Attributes: A Best–Worst Analysis. Presented to the Western Agricultural Economics Association on 30 June 2011, at Banff, Alberta, Canada.

Bond, J.K. and Richards, T. (2008) Generic promotion and demand for non-commodity potato varieties: an empirical investigation of United States Potato Board promotion on non-Russet potato demand. *NICPRE Quarterly* 14(3), 1–4.

CIP (International Potato Center) (2012a) Potato: Facts and Figures (http://cipotato.org/potato/facts, accessed 15 December 2012).

CIP (2012b) The World Potato Atlas (https://research.cip.cgiar.org/confluence/display/wpa/Home, accessed 15 December 2012).

CIP (2012c) Potato: Processing and Uses (http://cipotato.org/potato/processing-uses, accessed 19 June 2013).

ECPD (European Cultivated Potato Database) (2013) Online database of potato varieties cultivated in 8 European Union and 5 Eastern European countries (http://www.europotato.org/menu.php, accessed 19 June 2013).

FAO (Food and Agriculture Organization) (2006) Buried treasure: the potato (http://www.fao.org/potato-2008/en/potato/utilization.html, accessed 15 December 2012).

FAO (2008a) International Year of the Potato: Uses of Potato (http://www.potato2008.org/en/potato/utilization.html, accessed 15 December 2012).

FAO (2008b) International Year of the Potato: Global Potato Economy (http://www.fao.org/potato-2008/en/potato/economy.html, accessed 15 December 2012).

FAO (2008c) International Year of the Potato: Potato and Water Resources (http://www.fao.org/potato-2008/en/potato/water.html, accessed 15 December 2012).

FAO (2008d) International Year of the Potato: Potato World (http://www.fao.org/potato-2008/en/world/index.html, accessed 19 June 2013).

FAO (2008e) International Year of the Potato: Production of Disease-free Seed Tubers (http://www.fao.org/potato-2008/en/potato/seedtubers.html, accessed 19 June 2013).

FAO (2008f) International Year of the Potato: Varieties (http://www.fao.org/potato-2008/en/potato/varieties.html, accessed 6 June 2013).

FAOSTAT (2013) Food and Agriculture Organization Database (http://faostat.fao.org, accessed 19 June 2013).

Huffaker, B. (2005) Processing Potato Outlook. Presentation for the Potato Industry Outlook Summit, Colorado Springs, Colorado, February 2005.

Kiple, K.F. and Ornelas, K.C. (eds) (2000) *The Cambridge World History of Food*. Cambridge University Press, Cambridge, UK.

McCain Foods (2013) History of McCain Foods (http://www.mccain.com/GoodBusiness/Pages/History.aspx, accessed 19 June 2013).

Makki, S.S. and Plummer, C. (2005) Globalization of the frozen potato industry. *Journal of Agribusiness* 23(2), 133–146.

Naegely, S. (2002) An innovative approach to potato marketing. *American Vegetable Grower* 50(6), 10–11.

NPD Group-National Eating Trends (2008) The Potato Market In-Home vs. Restaurant Consumption. Report for United States Potato Board, Denver, Colorado.

NPD Group-National Eating Trends (2010) In-Home Potato Market: Topline. Produced for the United States Potato Board. Pamphlet pp. 1–34.

Olsen, N., Robbins, J., Brandt, T., Lanting, R., Parr, J., Jayo, C., *et al*. (2003) Specialty Potato Production and Marketing in Southern Idaho. Idaho Experimental Station, University of Idaho Extension CIS 1110, Idaho.

Panati, C. (1989) *Panati's Extraordinary Origins of Everyday Things*. Harper Collins, New York.

PotatoPro (2010) Russia considers zero rate potato imports to deal with shortage (http://www.potatopro.com/Lists/News/DispForm.aspx?ID=4214, accessed 15 December 2012).

Snack Food Association (SFA) (2012) The Potato Chip is 150 this Year! (http://www.sfa.org/sfa/news.php?id=60, accessed 15 December 2012).

Stark, J.C. and Love, S.L. (eds) (2003) *Potato Production Systems*. University of Idaho, Idaho, pp. 329–342.

USDA-ERS (USDA Economic Research Service) (2011) Potatoes (http://www.ers.usda.gov/topics/crops/vegetables-pulses/potatoes.aspx, accessed 15 December 2012).

USDA-FAS (USDA Foreign Agricultural Service) (2011) Frozen Potato Products: World Markets and Trade (http://usda01.library.cornell.edu/usda/current/frozpot/frozpot-11-12-2011.pdf, accessed 19 June 2013).

USDA-FAS (2013) Production, supply, and distribution database (http://www.fas.usda.gov/psdonline/psdQuery.aspx, accessed 19 June 2013).

USDA-NASS (USDA National Agricultural Statistics Service) (2013) Crop Production (http://www.nass.usda.gov/Publications/Todays_Reports/reports/crop1212.pdf, accessed 15 May 2013).
USPB (United States Potato Board) (2011a) Potato Types Guide (http://www.ciaprochef.com/potatoes/pdf/PotatoTypes.pdf, accessed 15 December 2012).
USPB (2011b) Shopper Insights. NPD Group Presentation of Consumer Shopping Attitudes Survey. Denver, Colorado, January 2011.
USPB (2012a) Dramatic Increase in US Frozen Potato Exports to USPB Target Markets (http://www.potatopro.com/news/2012/dramatic-increase-us-frozen-potato-exports-uspb-target-markets, accessed 15 December 2012).
USPB (2012b) Chip-Stock Potatoes (http://www.potatoesusa.com/products.php?sec=Chip-Stock%20Potatoes, accessed 15 December 2012).
USPB (2012c) Dehydrated Potatoes (http://www.potatoesusa.com/products.php?sec=Dehydrated%20Potatoes, accessed 15 December 2012).
USPB (2012d) Seed Potatoes: Export Market (http://www.uspotatoes.com/downloads/12SeedPotatoes-Export.pdf, accessed 5 June 2014).
WSU (Washington State University Extension) (2012) Potato Varieties: A Comprehensive List (http://potatoes.wsu.edu/varieties/vars-all.htm, accessed 15 December 2012).
World Bank (2008) Agriculture for Development: World Development Report (http://siteresources.worldbank.org/INTWDRS/Resources/477365327599046334/WDR_00_book.pdf, accessed 15 December 2012).

4 Tuber Development

Bjorn Kloosterman[1,2]* and Christian Bachem[1]

[1]Laboratory of Plant Breeding, Wageningen University and Research Centre, Wageningen, the Netherlands; [2]Keygene NV, Wageningen, the Netherlands (current address)

During their life cycle, potato plants develop tuberous organs from underground stems called stolons, which allow the rapid development of daughter plants from sprouts at the beginning of the new growing season. The formation and growth of a potato tuber is a complex process that can be divided into separate physiological events. In recent years, several scientific breakthroughs in the signaling pathway of tuberization have revealed novel tuber identity genes and regulatory pathways. Different signaling pathways leading to tuber initiation either act independently or intersect, but all seem to be, at least in part, under the control of the circadian clock associated with phase transitions during the plant life cycle. In this chapter, we integrate the different signaling pathways underlying the process of potato tuber initiation and subsequent tuber growth in relation to timing, tuber physiology, gene expression, and the hormonal changes impacting these processes.

4.1 The Physiological Process of Tuber Formation and Morphology

Tubers are formed underground on stem-like structures called stolons. A stolon may be defined as a lateral shoot arising from the basal stem nodes, showing negligible leaf expansion, possessing a hook-shaped tip and growing diageotropically (Booth, 1963). The length and number of stolons and the degree of branching may vary based on genotype or environment. Under environmentally favorable conditions, potato plants are induced to tuberize through the production of a graft-transmissible signal that is transported basipetally from the leaves to the growing stolon tip, where it promotes tuberization (Gregory, 1956; Jackson *et al.*, 1998).

When plants are induced to tuberize, longitudinal stolon growth ceases, and is followed by subapical swelling of the stolon tip (Cutter, 1978; Peterson *et al.*, 1985). The observed initial swelling is a result of increased radial cell expansion, closely followed by cell division in the pith and cortex (Sanz *et al.*, 1996; Xu *et al.*, 1998b). When tubers reach a diameter of around 0.8 cm, more randomly orientated cell division and cell enlargement occurs primarily in the perimedullary region until tubers reach their final diameter (Fig. 4.1).

The perimedullary region forms the major portion of the mature tuber and consists of storage parenchyma cells and vascular tissue. During initial tuber growth, nodal segments of the stolon are incorporated into the main tuber, with the axils of scale-like leaves forming the

---

* E-mail: bjorn.kloosterman@keygene.com

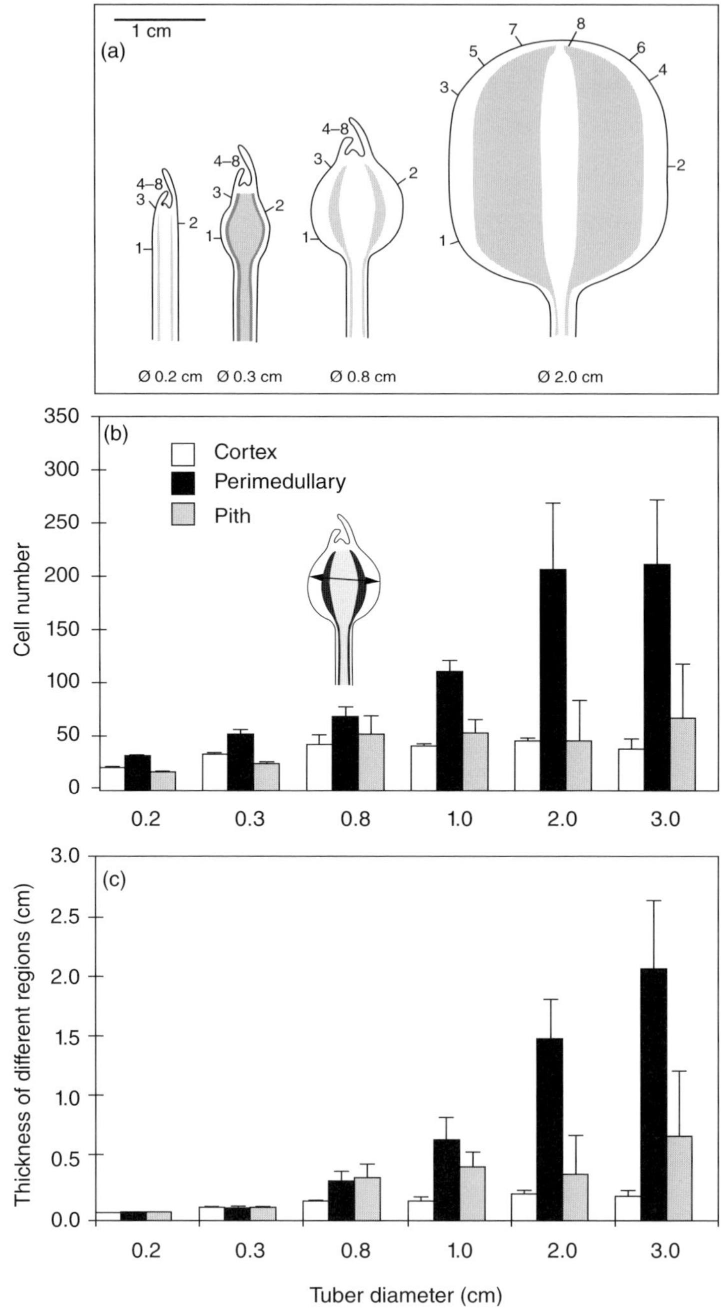

Fig. 4.1. (a) Schematic overview of the different growth stages during tuber formation. From left to right, longitudinal cross sections of a non-tuberizing stolon 0.2 cm, swelling stolon 0.3 cm, small tuber 0.8 cm, and larger tuber of 2 cm. The growth of the perimedullary region is indicated by the gray area, and nodes are indicated numerically. (b) Number of cells along the transversal axis in non-tuberizing stolon (0.2 cm) and large tuber (3 cm). (c) Thickness of the individual regions (cortex, perimedullary, and pith of tuber developmental stages, as shown in (b)). (From Xu *et al.*, 1998b.)

eyebrows and axillary buds forming the dormant eye-buds. The eyes are arranged spirally on the tuber surface and are more concentrated towards the apical end, similar to the main shoot spiral patterning of leaves and axillary buds. Depending on the variety, tuber eye depth may be elevated, superficial, or deep.

When tuber induction is interrupted due to changing conditions (heat, drought stress) or strong competition for assimilates between tubers, early stage tubers can be "reabsorbed" (Walworth and Carling, 2002), leaving visible tuber remnants. Successive switching between an inductive (tuber formation) and noninductive status (continuation of apical stolon growth) can even result in a chain of tubers. The number of stolons and branches is highly relevant with respect to the final number of tubers and their size distribution, as tubers compete for assimilates transported down from the leaves driving tuber growth (Helder *et al.*, 1993). Additional stolons and tubers may still arise during the whole growing season; however, normally, the largest tuber at any stage on the plant will remain the largest at the end of the growing season (Struik *et al.*, 1990). Tubers are generally formed on stolons, but under very strong inductive conditions, may also be sessile, or can even be formed aboveground from axillary buds (Ewing and Struik, 1992).

After tuber initiation and initial growth stages, the tuber enters a bulking phase, in which it incorporates photoassimilates for growth and the synthesis of starch and storage proteins. The control of symplastic and apoplastic transport is thought to be an important regulatory mechanism with respect to nutrient, hormone, and photoassimilate transport and distribution within the growing tuber. In non-swelling stolons, phloem unloading is primarily apoplastic, whereas in swelling stolons, induction of symplastic unloading allows photoassimilates, mainly sucrose, to be unloaded into parenchyma cells (Viola *et al.*, 2001; Geigenberger, 2003). However, the apical meristem of the tuber remains symplastically isolated, maintaining a tightly controlled cellular domain around the bud enforcing tuber dormancy. Breakage of dormancy is associated with symplastic reconnection of the bud meristem to the tuber phloem network (Viola *et al.*, 2007).

During the stages of tuber growth and tuber bulking, and later during storage, the tuber is protected from desiccation and damage through the protective layer of the tuber skin. The original epidermal cell layer of the tuber is replaced early in development with periderm tissue that is made up of three cell types: phellem, phellogen, and phelloderm (Reeve *et al.*, 1969). The phellem (or cork) forms a series of layers at the outermost level of the periderm and is derived from the meristematic phellogen layer (or cork cambium) underneath it. As phellem cells mature, they become suberized, forming a protective layer that is designated as the "skin". Gas exchange in the tuber takes place through lenticels distributed on the surface of the skin. In terms of gene expression, very little information exists on genes involved in controlling skin development and the respiration capacity of the tuber. Tuber dormancy is generally considered to start at the moment of tuber initiation, once stolon apical growth is arrested. The loss of tuber dormancy and the onset of sprout growth is accompanied by numerous biochemical changes and shifts in hormone balances (Suttle, 2004; Hartmann *et al.*, 2011). From its initiation until the end of its dormancy period, the potato tuber remains a metabolically active organ, reacting to internal and environmental cues in which balancing the regulation of gene expression and hormone levels is pivotal.

4.2 Control of Tuber Induction and Signal Transduction

Day length dependent tuberization

Its equatorial origin makes potato essentially short-day (SD) adapted for tuberization. Although most modern-day cultivars have lost their strict need for SD conditions in order to tuberize, decreased day length (or long nights) promotes tuber formation, with the degree of response largely dependent on genotype and physiological age of the plant (Ewing and Struik, 1992). To study the relationships between photoperiod and timing of tuberization, researchers use wild potato genotypes and landraces, such as *Solanum tuberosum* group Andigenum, which still require SD conditions to induce tuberization. Besides the study of photoperiod control on tuberization, use of SD-dependent varieties also provides a synchronous tuber induction system by transferring plants from noninductive long-day (LD) to inductive SD conditions. Provided that plants have obtained sufficient biomass, transfer of plants from LD (16 h) to SD (8 h) conditions generally results in the first visible swellings around 8–10 days after induction. The impact of photoperiod on tuberization becomes evident when interrupting a long night (SD) with a night break of red light, which inhibits tuberization (Jackson, 1999).

Photoperiod perception takes place in the leaves, mediated by photoreceptors including several phytochromes (PHYA and PHYB), the actions of which may be modulated by light quality. Specifically, the action of phytochrome B (PHYB) has been assigned a central role in controlling tuber formation. Antisense *phyB* plants lost their photoperiod-dependent tuberization and tuberized equally well under both LD and SD conditions (Jackson *et al.*, 1996). Reduced PHYA functionality in potato was shown to be involved in photoperiod sensing by resetting the circadian clock, impacting the timing of tuberization (Yanovsky *et al.*, 2000). Grafting experiments showed that PHYB was involved in the transmissible inhibition of tuberization in the aerial parts of the plant, since wild-type plant stocks (SD) were induced to tuberize in LDs when grafted with a scion from an *anti-phyB* plant (Jackson *et al.*, 1998).

4.3 The Tuberization Signal

Kumar and Wareing (1974) proposed the existence of a stimulus signal (named "tuberigen", analogous to the "florigen") based on grafting experiments in which scions grown under SD conditions were able to induce tuberization on potato stocks grown under LD conditions. The precise nature of this mobile signal has long been sought after, and only in recent years have several key determinants been identified, as will be discussed below.

There is a large similarity between photoperiodic control of tuberization with that of flowering. In both flowering and tuberization, inductive conditions (day length) are perceived in the leaves, and under favorable conditions a signal is produced and transported via the phloem to either the shoot apex or the underground stolon apex. Chailakhyan *et al.* (1981) already showed in an early experiment that flowering tobacco shoots were able to induce tuberization, indicating the presence of a mobile signal similar to that of flowering.

In the case of flowering in *Arabidopsis*, a flowering time locus T (FT) was shown to be the mobile florigen protein transported from the leaves to the shoot apex inducing floral transition (Corbesier *et al.*, 2007). Navarro *et al.* (2011) showed that in potato, floral and tuberization transitions were controlled by two different *FT*-like genes (*StSP3D* and *StSP6A*), which responded to independent environmental cues. *StSP6A* gene expression was shown to be induced strongly on transfer to SD conditions, and the protein was demonstrated to be transported to the stolon and capable of inducing tuberization. It is proposed that, during transport, StSP6A is able to promote its own gene expression through an autoregulatory loop, thereby maintaining a plant tuber induction status. Silencing of *StSP6A* severely delays tuber formation in SD conditions, again confirming a crucial role in promoting tuber formation. *StSP6A* expression analyses in commercial cultivars with early (Jaerla), late (Baraka), and intermediate (Kennebec) tuberization periods show that accumulation of this transcript in leaves correlates with the tuberization time of these cultivars (Navarro *et al.*, 2011).

In a diploid potato population segregating for timing of tuber initiation (Celis-Gamboa *et al.*, 2003), *StSP6A* expression levels for early and late genotypes reflect an inductive and noninductive state, and are correlated strongly with tuber formation (Fig. 4.2). In early genotypes of this population (Fig. 4.2), *StSP6A* gene expression is high, and these plants have started setting tubers, while in the late genotype, expression is close to the detection limit and no tuberization is observed (Kloosterman *et al.*, unpublished results). These findings further favor the notion that StSP6A is the main developmental switch in the transition from stolon growth to tuber initiation. StSP6A is, like most identified FT proteins, a member of a family of proteins that contain a phosphatidylethanolamine-binding domain (PEBP) and is not itself a transcription factor (Turck *et al.*, 2008). Instead, StSP6A likely needs one or more partners present in the stolon tip to be able to induce tuberization, analogous to

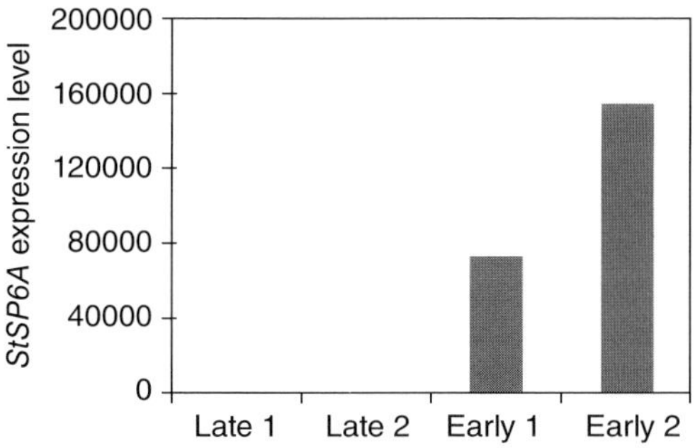

Fig. 4.2. Relative gene expression level comparison for *StSP6A* between two "late" and "early" genotypes of the C × E population segregating for timing of tuber formation. C × E is a diploid potato F1 population descending from *Solanum tuberosum* and *Solanum phureja*. Expression of *StSP6A* in the late genotypes is below reliable detection levels. Only the early genotypes have formed tubers.

flowering locus D (FD) in the apex during flower induction. The identity of the partners remains the subject of further investigation, but several potential downstream targets have been identified and will be discussed below.

4.4 Molecular Regulation of Tuber Induction

If *StSP6A* expression and transport is considered crucial to the developmental switch towards tuber formation, the question that then arises is what sort of mechanism controls *StSP6A* expression in the leaf? In *Arabidopsis*, a major factor controlling flowering is the circadian rhythm-regulated transcription factor, CONSTANS (CO), which acts directly on FT expression in *Arabidopsis* (An *et al.*, 2004). The activity of the CO protein is post-transcriptionally modified under the influence of light through an interaction with the red and blue light receptors, phytochrome A, B, and cryptochrome 2 (Yanovsky and Kay, 2002). Similarly, in potato plants that are grown under LD conditions, PHYB protein is conformationally inhibited from inducing *StCO* gene expression, whereas under SD conditions, PHYB induces *StCO* (Rodriguez-Falcon *et al.*, 2006). Furthermore, Gonzalez-Shain *et al.* (2012) showed that potato plants, which overexpressed *StCO*, tuberized later than wild-type plants under inductive SD photoperiod. *StCO* silencing promoted tuberization under both repressive and weakly inductive photoperiods (Gonzalez-Schain *et al.*, 2012). Constitutive expression of the *Arabidopsis CONSTANS* (*AtCO*) gene in potato also impairs the inductive effect of SDs on tuber initiation (Martinez-Garcia *et al.*, 2002). These results demonstrate that StCO represses tuberization in a photoperiod-dependent manner. The results from González-Schain *et al.* (2012) and also from the laboratory of S. Prat (2012, personal communication), indicate that StCO protein acts as an inducer of an FT homolog, called StSP5G. This protein in turn seems to function as a suppressor of the mobile tuberigen, *StSP6A* (Fig. 4.3). CO expression follows a diurnal rhythm, and its expression in *Arabidopsis* is controlled by members of the cycling DOF factor (CDF) protein family. CDF1 has been shown to be post-transcriptionally controlled by clock gene GIGANTEA (GI) (Park *et al.*, 1999) and a number of blue light receptors such as FLAVIN-BINDING, KELCH REPEAT, F-BOX PROTEIN 1 (FKF1), LOV-KELCH REPEAT PROTEIN 2 (LKP2) and ZEITLUPPE (ZTL) (reviewed in Imaizumi, 2010). In turn, four of the five *Arabidopsis* CDF proteins have been shown to suppress CO. Moreover, it is thought that the FKF1 protein targets CDF proteins for degradation by the proteasome (Fig. 4.3).

The latter finding is of interest as allelic variation was found in a member of the potato CDF gene family (*StCDF1*), resulting in a truncated protein that lost its ability to interact with FKF1, thereby causing an early tuberization phenotype (Kloosterman *et al.*, 2013). In contrast to the wild type, the StCDF1 truncated protein levels remained high throughout the day, resulting in a constitutive repression of two CO genes located as a tandem repeat on chromosome 2 of potato (*StCO1* and *StCO2*). Since it is thought that these *StCO* gene products indirectly release the inhibition on the mobile StSP6A signal, genotypes carrying one or more of the truncated variants of the StCDF1 protein show an early tuberization phenotype. This hypothesis was confirmed by transforming late tuberizing genotypes with the truncated StCDF allele under control of the CaMV 35S promoter, resulting in extreme early tuberization. Interestingly, overexpression of the

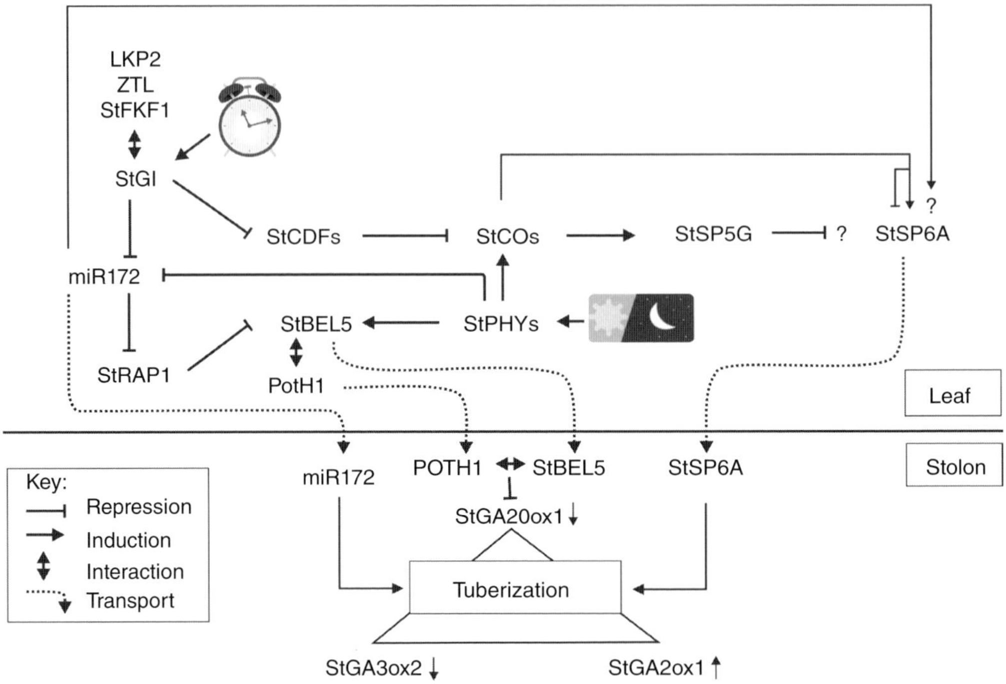

Fig. 4.3. Schematic overview of identified genes associated with timing of tuberization and promoting tuber formation. LKP1, ZTL and FKF1 are blue light receptors that are regulated by the clock genes. FKF1 and GI interact and bind to the transcription regulator CDF. The interaction of GI and FKF1 with CDF targets this protein for degradation by the proteasome in the afternoon. During the early morning of long days, CDF is not yet degraded and inhibits the expression of *CO* genes. CO protein is also a transcription factor thought to be a specific inducer of the *FT* homolog, *StSP5G*. Evidence suggests that StSP5G inhibits the expression of *StSP6A*, which in turn is considered the primary mobile tuberigen transported to the stolon tip. Within the stolon tip, StSP6A activity is expected to trigger expression of tuber identity genes through interaction with a yet unidentified partner. *StBEL5*, *POTH1* have also been shown to be mobile and associated with transport to the stolon tip promoting tuber formation. PHYB plays a central role in the regulation of a number of proteins inducing both *StCO* and *StBEL5* under the correct light conditions (LD), while suppressing *miR172* in stolons under LDs, possibly through StRAP1 (Banerjee *et al.*, 2006; Jung *et al.*, 2007; Martin *et al.*, 2009). At the stolon tip, hormones such as GA play a key role in tuber initiation. It is known that the BEL–KNOX interaction is capable of suppressing GA production by inhibiting *StGA20ox1*. The downregulation of *StGA3ox2* and the upregulation of *StGA2ox1* allows for further rapid reduction of bioactive GA in the stolon associated with tuber formation. Question marks indicate uncertain interactions/mobility or yet-to-be-identified interactors. *Note*: no differentiation has been made between protein and gene in this figure.

same allele also phenocopied a number of other characteristics of early potatoes such as drastically reduced life cycle and more bushy growth habit (Kloosterman *et al.*, 2013). The *StCDF1* gene was identified after fine mapping of the earliness QTL on chromosome 5. The wide variation in plant earliness in modern potato cultivars appears thus to be a result of allelic variation at a single locus. This is likely to be due in part to the selection for *StCDF1* genes that evade the circadian rhythm control by StFKF1, as demonstrated in a genome-wide association study in 83 diverse tetraploid genotypes where a unique and very high association was found between earliness and the CDF1 locus on chromosome 5 (Kloosterman *et al.*, 2013).

The similarity between the flowering and tuberization regulation pathways makes it likely that signaling cascade between the clock output proteins and FT in potato has been re- or neo-functionalized for the regulation of tuberization analogous to beet formation and bolting in sugarbeet (Pin *et al.*, 2010). At the core of the circadian clock are oscillators responsible for generating circadian rhythms. These oscillators can be entrained by cues from the environment, such as daily changes in light and temperature. Potato tuberization thus seems to be controlled through the output of the circadian clock control, together with photoperiod as an input variable affecting oscillation and diurnal patterns of gene expression.

Besides a clear role for StSP6A as a mobile tuberization signal, additional genes and signaling molecules have been shown to be mobile and promote tuber formation to various extents (reviewed in Hannapel, 2013). The micro RNA 172 (*miR172*) is a small RNA molecule recently shown to be able to induce tuberization under noninductive LD conditions (Martin *et al.*, 2009). This transcript has been found to be present in the phloem, and induction of tuberization through *miR172* was shown to be graft transmissible. However, induction of tuber formation was only observed when 35S::miR172 scions were grafted on to wild-type stocks, but not when these plants were used as stocks, suggesting a function of *miR172* in modulating the mobile tuberization signal in the leaves but not as a component of this mobile signal per se (Martin *et al.*, 2009; Abelenda *et al.*, 2011). Furthermore, expression studies showed that *miR172* was expressed in the majority of potato organs and was more abundant in stems, being increased in the stolons at the onset of tuberization. In other plant species, *miR172* abundance is regulated by photoperiod via GI-mediated miRNA processing and controls floral transition by targeting a set of APETALA2-like genes impacting downstream FT expression levels. In *Arabidopsis*, *miR172*-overproducing plants exhibit early flowering under both LDs and SDs, even in the absence of functional CO, indicating that *miR172* promotes photoperiodic flowering through a CO-independent genetic pathway (Jung *et al.*, 2007). In potato, the *RAP1* transcript, encoding an AP-2 related protein, was identified as a likely target of *miR172*, as transcript levels were correlated inversely with *miR172* accumulation (Martin *et al.*, 2009). Expression of *miR172* is regulated by the light receptor PHYB, which, as described above, plays a key role in day length dependent tuberization by inhibiting tuber formation in LDs (Fig. 4.3). In LD-grown *PHYB*-antisense clones, the expression profile of *miR172* is changed similar to wild-type, SD-induced plants. A mechanism of potato *miR172* indirectly controlling *StSP6A* expression in potato plants remains to be resolved.

Interestingly, 35S::miR172 plants show increased levels of the *StBEL5* transcript in all organs. *StBEL5* has been shown to be a mobile transcript able to enhance tuber formation, and its expression is elevated on transfer to inducing SD conditions (Banerjee *et al.*, 2006). RNA detection methods and heterografting experiments demonstrated that *StBEL5* transcripts were present in phloem cells and moved across a graft union to localize in stolon tips, the site of tuber induction. The mobile RNA originates from leaf veins and petioles and its expression is induced by an SD photoperiod and correlated with enhanced tuber production. It has been suggested that PHYB might repress the movement of *StBEL5* mRNA from leaves to stolons under LDs, but this repression is released in SDs. StBEL5 protein function has been linked to regulating gibberellic acid (GA) content in the stolon and cytokinin levels in the shoot, together with its interacting partner, POTH1 (potato homeo box 1). Recently, POTH1 was added to the list of potential long-distance mobile signals transported from the leaf toward the stolon on tuber induction (Mahajan *et al.*, 2012). StBEL5 can form a dimer with POTH1, and gel shift assays have demonstrated that StBEL5 and POTH1 bind to the regulatory region of GA20 oxidase1 (*StGA20ox1*) from potato, a gene encoding a key enzyme in the GA biosynthetic pathway (Chen *et al.*, 2004) (Fig. 4.3). In tandem, StBEL5 and POTH1 had a greater binding affinity for the *StGA20ox1* promoter than either protein alone.

The use of SD-dependent genotypes for tuberization, as a model system to understand photoperiodic control, has provided important insights into the mechanisms underlying long-distance signaling to induce tuber formation. Several genes (*StSP6A*, *StBEL5*, *StPOTH1*, and *miR172*) have all been postulated to be an integral part of the mobile signal transported as RNA or protein from the leaves towards the stolon at tuber initiation. However, the total number of genes that are translocated from the leaves to the stolon tip and affect the process of tuber formation may be even higher. The identification of several RNA-binding proteins differentially present in potato petioles on transfer from LD to SD conditions can provide important evidence in understanding transport of *StBEL5*, *POTH1* and other genes in relation to photoperiodic-dependent tuberization (Banerjee *et al.*, 2009; Shah *et al.*, 2011). Within the identified tuber signaling pathways, a challenging task now awaits to establish the individual importance of the associated genes and their relative position within the signaling network. Generation of complete knockouts for all of these genes should provide more clear evidence as to whether they are crucial for inducing tuber formation, have redundant roles, or merely have a supportive function. The clear link with circadian clock control for several of these signaling genes warrants further analysis of circadian rhythm genes and output of signalling, or so-called "gating". Modern potato cultivars have lost the strict photoperiodic control with respect to tuber formation; however, the signaling components (StCDF1, StSP6A, StBEL5, POTH1 *miR172*) remain under strict control of the circadian clock to regulate seasonal timing of phase transitions such as tuber formation.

4.5 Hormonal Regulation of Potato Tuberization

Gibberellic acid (GA)

The plant hormone, gibberellic acid (GA), has long been implicated to play a key role in tuber formation inhibition (Booth, 1963; Kumar and Wareing, 1974; Ewing, 1987). GAs are cyclic diterpenoid hormones that regulate many plant growth and developmental processes including germination, stem growth, flowering, and fruit and tuber development (Lange, 1998). Applications of biologically active GAs or inhibitors of GA biosynthesis have shown either to delay or to promote tuber formation under tuber-inducing conditions (Vreugdenhil and Struik, 1989; Jackson and Prat, 1996). Xu *et al.* (1998a) quantified endogenous GA levels during various stages of stolon elongation and tuber formation. They found that GA_1 levels were high during the longitudinal elongation of stolons and decreased dramatically prior to first visible swelling. Shibaoka showed that GA_3 might be involved in the orientation of the microtubules and microfibrils in plant cells (Shibaoka, 1993). It was hypothesized that a reduction of GA in inducing conditions, as found in stolon tips, might cause the reorientation of the plane of cell expansion and cell division, resulting in subapical swelling of the stolon (Fujino *et al.*, 1995; Xu *et al.*, 1998a). StGA2ox1 catabolizes bioactive GAs and/or GA precursors, and its expression is strongly induced prior to first visible swelling. Tubers of *StGA2ox1* suppression clones displayed occasionally elongated tuber shapes indicative of incomplete reorientation of the plane of cell divisions (Kloosterman *et al.*, 2007) (Fig. 4.4). 35S overexpression plants exhibit a dwarf phenotype, reduced stolon growth, and earlier *in vitro* tuberization.

The expression of *StGA3ox2*, involved in the last biosynthesis step towards bioactive GAs (GA_1, GA_4), is downregulated in the stolon on tuber induction (Bou-Torrent *et al.*, 2011). Both actions, the upregulation of GA breakdown (*StGA2ox1*) and decrease of GA biosynthesis (*StGA3ox2*), allow for a rapid reduction of GA in the stolon tip. Overexpression of *StGA3ox2* under control of constitutive or leaf-specific promoters results in taller plants, which, surprisingly, tuberize earlier under SD conditions than the controls; whereas *StGA3ox2* tuber-specific overexpression results in non-elongated plants with slightly delayed tuber induction (Bou-Torrent *et al.*, 2011). From these studies, it is interesting to speculate that the differential mobility of GA precursors and bioactive GAs may be important in the control of tuberization. GA_{20} can be transported readily throughout the plant, whereas GA_1 preferentially remains in the vicinity of the cells where it is produced, so that it exerts its action mainly in these and directly adjacent cells (Prat, 2010) (Fig. 4.5). Independent evidence for differential mobility of these two GA molecules has been shown through grafting experiments in pea (Proebsting *et al.*, 1992). Increase of potato GA 3-oxidase activity through *StGA3ox2* overexpression would lead to an increase in the rate of GA_{20} to GA_1 conversion in the aerial tissues, resulting in taller plants but leaving less GA_{20} available for transport towards the stolons. Consequently, GA levels in the stolon are reduced in the overexpressing plants, thereby

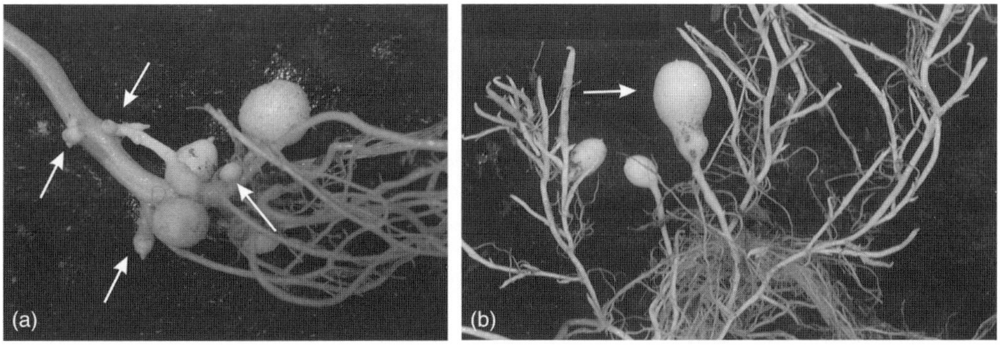

Fig. 4.4. Stolon growth and tuber set in the *StGA2ox1* overexpression (a) and *StGA2ox1* suppression clone (b). The presence of small tuber incipients in the overexpression clone is indicated with white arrows, as well as the slight elongated tuber shape in the suppression clone (Kloosterman *et al.,* 2007).

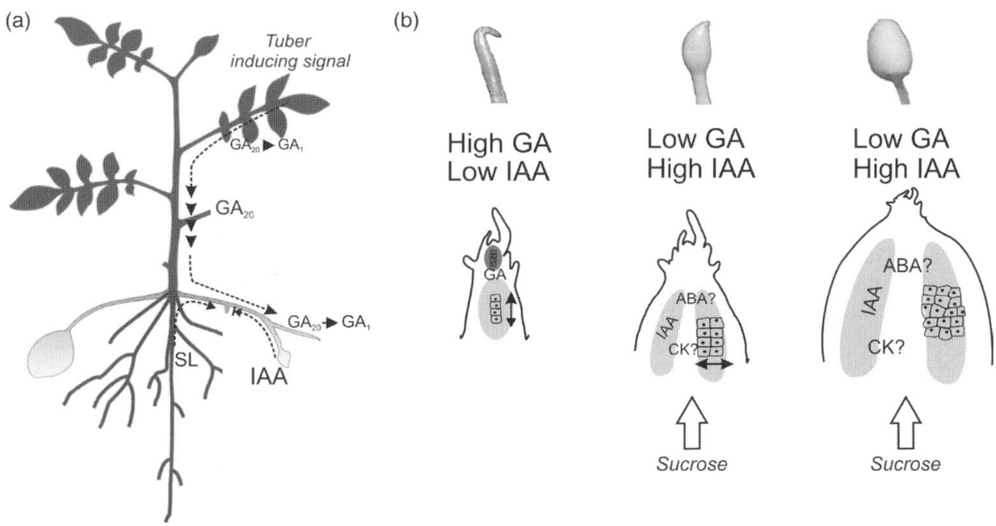

Fig. 4.5. Schematic overview of proposed hormonal influences on potato tuber formation and stolon branching. (a) GA_{20} has a higher mobility than GA_1 and can be transported down to the stolon, where it is converted into bioactive GA_1. During tuber formation, GA_1 levels in the stolon decrease, while auxin is synthesized in the tuber and transported basally to inhibit stolon axillary outgrowth. Strigolactones (discussed below) are produced in the roots and transported to the stolon, which can promote axillary outgrowth. (b) Under noninducing conditions, GA levels in the stolon are high, promoting longitudinal growth. During tuber initiation, GA levels decrease rapidly while indole-3-acetic acid (IAA) content increases. This shift is likely to result in reorientation of the plane of cell division, promoting stolon swelling and subsequent tuber growth. The concentration and localization of cytokinins (CKs) and abscisic acid (ABA) during early tuber development is not clear, but both hormones may be involved in establishing sink strength, impacting sucrose import.

promoting tuber formation. In line with these findings is the observation that when *StGA20ox1* is overexpressed, transgenic plants are taller and show delayed tuber formation under inducing conditions as more GA_{20} becomes available for transport towards the stolon, where it can be converted into bioactive GAs (Carrera *et al.*, 2000).

Several GA metabolism genes have been shown to be under the control of feedback or feedforward regulation. In a previous study, we found that during the early stages of tuber development *StGA20ox1* was upregulated, possibly under the influence of negative feedback regulation as a result of reduced GA_1 levels in the stolon (Kloosterman *et al.*, 2007). However, other studies have shown that StBEL5 is able to reduce *StGA20ox1* expression, together with its partner, POTH1 (Chen *et al.*, 2004), and both genes are highly induced on transfer to SD conditions (Fig. 4.3). These apparently contrasting results may be explained through the cell-type dependent interaction of StBEL5–POTH1 with the *StGA20ox1* promoter in cells surrounding the vascular bundle (Banerjee *et al.*, 2006). In our tuber developmental studies, RNA from whole tuber extracts was collected, which did not allow a distinction between the different cell types present within the tuber.

Auxin

A second plant hormone impacting tuber development is auxin (IAA). This plant hormone has been less studied compared to GA, and its role in tuberization has not always been clear. Recently, it was shown that IAA concentration was elevated in the tuber during early tuber developmental stages in both the stolon apex and the subapical stolon region (Roumeliotis *et al.*, 2012) (Fig. 4.6). Increased transcription levels of a potato IAA biosynthesis gene (*StYUC-like1*) provide further evidence for the increase of auxin during tuber development (Roumeliotis *et al.*, 2012). In earlier studies linking IAA content with tuberization, a small promoting effect for auxin was observed. However, due to lack of accurate auxin measurements within the stolon and tuber tissues, as well as different modes of application, data interpretation was difficult. For example, *in vitro* application of auxin to single nodal potato explants resulted in an earlier tuberization phenotype with sessile and slightly smaller tubers (Xu *et al.*, 1998a), whereas Harmey *et al.* (1966) observed that IAA treatment induced larger tubers at an earlier stage. Interestingly, when auxin is applied apically to *in vitro* stolon explants with their apex removed, tuber formation from the lateral buds is initially reduced or delayed, but the final number of tubers is higher due to lowered inhibition of axillary bud outgrowth. This increase in total number of tubers is probably due to auxin degradation in the medium, because renewal of the IAA-containing medium prolongs the inhibition of tuber formation (Roumeliotis *et al.*, 2012).

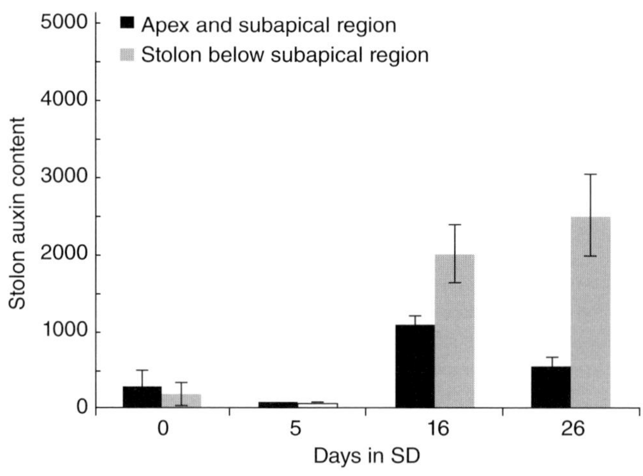

Fig. 4.6. Auxin measurements in the apical and subapical region and the stolon region below the subapical region after the switch to SD conditions. Plants were transferred to SD conditions (day 0) and monitored for tuber formation. On day 5, no visible signs of tuber formation were observed and stolons were harvested. Small tubers were present on day 16 and larger tubers were harvested on day 26, and auxin concentration was determined (Roumeliotis *et al.*, 2012).

Scoring of the transport of ^{14}C-labeled IAA revealed that the efflux of IAA from the stolon tip was 3.5 times higher than stolon influx. This indicated that the main direction of IAA movement in the stolon tissue was from the stolon tip to the basal part of the stolon. Application of 1-N-naphthylphthalamic acid (NPA), a polar transport inhibitor, did not alter the transport of IAA into the stolon tip, but decreased the outflow of IAA to the same levels as the influx, indicating that the polar movement of IAA out of the stolon tip was mediated by active transport (Roumeliotis *et al.*, 2012). These results indicate that stolon tips are a likely site of auxin biosynthesis, and that auxin regulates the process of axillary tuberization in a similar way to axillary shoot growth in the aerial stem (Fig. 4.5). Support for this model comes from the effect of strigolactones (SL) on stolon axillary bud outgrowth. SLs are a new class of plant hormones that are synthesized mainly in the roots and transported upwards via the xylem tissue. It has been proposed that SLs act either directly on axillary bud outgrowth or indirectly via dampening of auxin transport and canalization of auxin from the buds (Brewer *et al.*, 2009; Prusinkiewicz *et al.*, 2009). Application of the synthetic SL-analog from *Gigaspora rosea* GR24 on the basal part of the stolon explants showed a strong inhibitory effect on stolon axillary bud outgrowth (Roumeliotis *et al.*, 2012). Regulation of axillary stolon outgrowth through the interplay of IAA and SL during the growing season may thus control the degree of stolon branching, and indirectly the number of potential tubers formed. It should, however, be noted that the latter experiments were carried out in an *in vitro* tuberization system starting from nodal stem segments, and the model needs to be confirmed *in vivo*.

Localization of auxin maxima within the stolon and tuber through use of an auxin-inducible DR5 promoter driving GUS expression (DR5::-GUS) shows the presence of auxin, primarily around the vascular bundles in the regions where cell divisions normally take place (Fig. 4.7). Polar auxin transport is mediated by members of the PIN efflux carrier family of proteins. Several potato PIN genes have been shown to be upregulated during early tuber developmental stages in line with increased auxin concentration in the swelling stolon (Kloosterman *et al.*, 2008; Navarro *et al.*, 2011; Roumeliotis *et al.*, unpublished results). Promoter analysis of an *St-PIN1*-like gene revealed a strong correlation with sites of auxin accumulation and regions of active cell division (Fig. 4.7). Auxin can have a promoting effect on cell expansion and division. The reorientation of cell division in stolons has historically been attributed to the reduction in GA content. Changing ratios between GA and IAA may be key in this switch as GA levels go down and IAA increases upon tuber induction (Fig. 4.6).

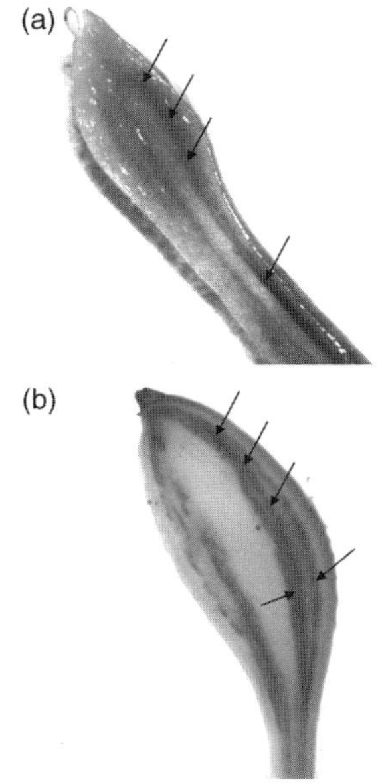

Fig. 4.7. Localization of auxin accumulation in swelling stolons (DR5::GUS) (a) and *Stpin1*-like expression (*Stpin1*-like:GUS). (b) Arrows indicate GUS staining (darker areas) surrounding the vascular bundles (inner and outer phloem) and the enlarging perimedullary region along the axis in the subapical stolon region. Note the absence of staining in the central pith and overlap between auxin accumulation (a) and *pin* expression (b). (From photograph 6A, Roumeliotis *et al.*, 2012.)

Cytokinin and other hormones

Cytokinin (CK) function in plants is often linked to developmental transitions involving meristem establishment and cell proliferation. In this respect, a role for CK in potato tuber development can be expected, as the early stages of tuber formation requires the initiation of cell divisions within the swelling stolon. Palmer and Smith (1969) first reported an increase of tuberization frequency on various CK-containing media. Their observation was supported by further investigations of exogenous CKs (Kumar and Wareing, 1974; Hussey and Stacey, 1984). The addition of CK to the high-sucrose *in vitro* tuberization media has been reported to increase the speed of tuber induction (Xu *et al.*, 1998a). A role of CKs in tuber formation also comes from transgenic potato plants expressing the *Sho* gene from *Petunia hybrida*, a CK biosynthesis gene. The most pronounced effects detectable in these lines were enhanced shoot production, delayed tuber formation, significant reduction in tuber size, and inhibition of tuber dormancy (Zubko *et al.*, 2005). Strong overexpressers of the *Agrobacterium ipt* gene, catalyzing a rate-limiting step in CK biosynthesis, also showed inhibition of tuberization; however, moderate overexpression of the same gene resulted in a promoting effect on tuberization (Gális *et al.*, 1995).

Interestingly, overexpression of the *ipt* gene in flowering tobacco plants induced tuber-like organs from stem axillary buds accumulating starch (Guivarc'h *et al.*, 2002). Overexpression of *POTH1* produced dwarf plants with several pleiotropic effects (Rosin *et al.*, 2003a). *In vitro* tuberization was enhanced in the *POTH1* overexpressing lines and was associated with decreased levels of GAs and a two- to fourfold increase in bioactive CKs measured in the stolon apical meristem.

A secondary role for CKs could be in sink creation, by regulating the expression of genes implicated in assimilate partitioning and source–sink regulation (Roitsch and Ehneß, 2000). Suppression of a potato MADS box gene, *POTM1*, resulted in increased CK content associated with increased cell division and increased starch accumulation in specific regions associated with vegetative meristems (Rosin *et al.*, 2003b). Sergeeva *et al.* (2000) hypothesized that the ratio between auxin and CK was important, with a slight increase in CK promoting tuberization and a larger change resulting in inhibition of tuberization. One could thus imagine that CKs exert their effect on tuber formation by redirecting carbon fluxes towards the stolon cells and promoting cell division in concert with auxin. However, other hormones have been implicated in this process as well. Phytohormone distribution in the pith and cortex was shown to be different in young tubers for IAA, abscisic acid (ABA), and CKs (Fig. 4.5). Enhancement of starch synthesis, accumulation in the pith, and the retardation of these processes in the cortex followed changes in ABA content (Borzenkova and Borovkova, 2003).

The plant hormone, ABA, has also been shown to have a promoting effect on tuber induction. Tuber formation in medium containing ABA and 8% sucrose started earlier than the tuber formation in ABA-free medium with 8% sucrose, in which sessile tubers or tubers on very short stolons were formed (Xu *et al.*, 1998a). However, an ABA-deficient potato mutant forms tubers normally, which indicates that ABA is not essential for tuberization and its function is likely due to antagonistic effects towards GAs (Xu *et al.*, 1998a). The importance of other hormones such as jasmonic acid and ethylene in regulating tuber formation is not clear. Jasmonic acid has been linked with cell expansion in tuber cells (Takahashi *et al.*, 1994), and a clear role in tuberization was demonstrated for an enzyme (POTLX-1) involved in the generation of jasmonic acid precursors (Kolomiets *et al.*, 2001). Plant lipoxygenases (LOXs) catalyze the oxygenation of polyunsaturated fatty acids such as linolenic and linoleic acids. POTLX-1 expression is upregulated during tuberization and is associated with cell enlargement in the region later forming the perimedullary region. Downregulation of POTLX-1 reduced tuber yield and disrupted the process of tuber formation.

Interpretation of the activities and functions of the different endogenous hormones with respect to tuber development remains difficult without precise information on their biosynthesis, sites of accumulation, and shifting balances.

4.6 Gene Expression During Tuber Development

Within the last decade, gene expression studies based on microarrays have yielded large data sets of differentially expressed genes under various biotic and abiotic stresses, developmental stages, or genetic backgrounds (Kloosterman *et al.*, 2005, 2008, 2010; Rensink *et al.*, 2005; Stupar *et al.*, 2007; Ducreux *et al.*, 2008; Vasquez-Robinet *et al.*, 2008). With the completion of the potato genome sequence, additional profiling techniques such as RNAseq and small RNA sequencing have brought new tools to study the process of tuber development and growth. The Potato Genome Sequencing Consortium collected RNAseq expression data from a large number of potato tissues, including tuberous tissue of different developmental stages (PGSC, 2011). When the RNAseq expression data were queried for genes impacting tuber development, 16,812 genes were found with confirmed expression in the tuberous tissues (stolon, young tuber, mature tuber). From this set, 325 genes were upregulated strongly (at least fivefold) in the early phase of tuber growth, and another 1151 genes reached peak expression in the more matured tuber tissues in comparison to the non-tuberizing stolon tissue.

The differential expression of a gene during the transition of a stolon into a tuber may indicate a regulatory role in tuber initiation, growth, or its involvement in one of the many metabolic pathways upregulated as a result of the newly developed storage organ. One could consider the formation of a tuber as a new organ requiring a unique set of genes; however, "true" tuber-specific expression (i.e. only expressed in the tuber) is a rare commodity. This is not surprising, as the tuber can be considered, in its simplest form, a modified underground stem. Although most genes expressed in tubers are also expressed in other parts of the potato plant, one can identify tuber-related adaptations in gene transcriptional regulation to accommodate tuber development, such as very high expression levels of starch biosynthesis genes, and especially storage proteins. The capacity for differential expression in tuber tissue could come from mutations in promoter elements, increased substrate availability for specific enzymatic reactions, shifts in hormonal balance, or as a result of the increased sink strength of the growing tuber.

From the same RNAseq data, a comparison between tissue types of a fully matured tuber (pith, cortex, and peel) can be made. Unique sets of expression profiles can be found for each of the three tissue types, with genes having elevated or reduced expression in one or two of the studied tissues. For example, genes providing the first line of defence against insect attacks or wounding, such as polyphenol oxidases and genes that are part of the glycoalkaloid biosynthesis pathway, were expressed more highly in the peel than in the pith or cortex tissues. As almost half of the predicted genes within the potato genome sequence lack proper functional annotation cell type specific expression may prove a valuable resource for the identification of trait-associated candidate genes.

A more detailed analysis of gene expression changes during the early stages of tuber development using 44 k 60-mer oligo microarray shows a number of common expression patterns that can be linked with tuber developmental stages (Kloosterman *et al.*, 2008) (Fig. 4.8). With respect to tuber initiation and early growth stages, the group exhibiting a temporarily upregulated expression profile is of high interest. This group of genes is highly enriched for genes associated with cell cycle and division, including several members of the cycling kinases and a gene with homology to AINTEGUMENTA (*StANT*-like), associated with the maintenance of meristematic competence. Altered expression of this potato *ANT*-like gene resulted in plants with reduced leaf size and increased tuber initials, but reduced total tuber yield (Kloosterman, unpublished data). A larger set of genes is downregulated on tuber formation. This group of genes is associated with a wide variety of functions likely to be required during stolon growth but no longer necessary during tuber growth. Further characterization of differentially expressed genes during tuber development will enhance our understanding of the regulatory mechanisms impacting commercially interesting traits such as the speed of growth, final tuber size, size distribution, and shape.

4.7 The Potato Tuber as a Storage Organ

Sucrose as a driver for tuber initiation and growth

In the previous paragraphs, the focus has been on physiological changes and gene transcriptional signaling cascades facilitating tuber development.

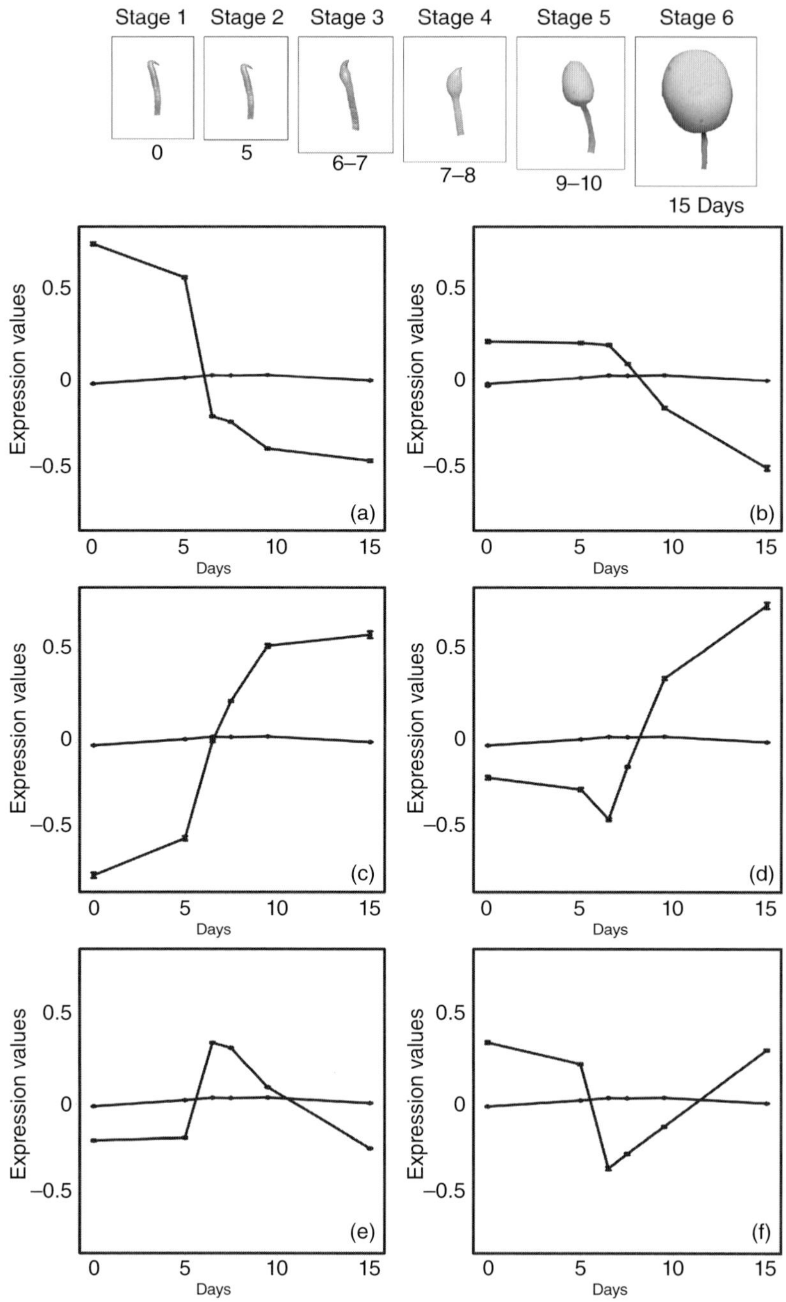

Fig. 4.8. Predominant gene expression profiles found during potato tuber development stages ranging from longitudinal stolon growth to fully grown tuber size (a–f). Early downregulation of gene expression during the switch from longitudinal stolon growth to first visible swelling (a); late downregulation of gene expression after first visible swelling has been observed (b); early upregulation of gene expression during the switch from longitudinal stolon growth to first visible swelling (c); late upregulation of gene expression after first visible swelling has been observed (d); transient upregulation of gene expression during early tuber organogenesis (e); and transient downregulation of gene expression during early tuber organogenesis (f). Average gene expression level of non-differentially expressed genes is plotted as a reference in each profile. (From Kloosterman *et al.,* 2008.)

However, the transport of sucrose from the leaves towards the stolon end is considered an essential force promoting tuber initiation and driving subsequent growth. The removal of a growing tuber from the plant will enhance the tuberization of other stolon tips, indicating tuber initiation is also source driven. The *in vitro* potato tuberization system, implemented in many tuber development studies (Hendriks *et al.*, 1991; Appeldoorn *et al.*, 1997) is characterized by the highly synchronous formation of microtubers from single nodal stem cuttings on medium with high sucrose (8% w/v). The frequency of tuberization is relative to the amount of sucrose in the medium, suggesting an inducing role for sucrose in tuber formation (Xu *et al.*, 1998a). Although the *in vitro* tuberization system has been a very useful research tool in studying early tuber developmental processes, the resulting microtubers remain small and do not exceed a certain diameter, usually ~0.8 cm (Fig. 4.1). This lack of further cell division, and the growth mainly of the perimedullary region, indicate additional source-signaling molecules are absent *in vitro*, where sucrose availability is not a limiting factor.

Lowering sucrose transport in potato plants results in a reduction in tuber formation and tuber yield (Riesmeier *et al.*, 1994; Kuhn *et al.*, 2003). Further evidence for a role of sucrose in tuber initiation comes from the work performed on sucrose transporters (Chincinska *et al.*, 2008). All three studied sucrose transporters (SUT) from potato (*S. tuberosum*), SUT1, SUT2, and SUT4, are co-localized and their RNA levels not only follow a diurnal rhythm but also oscillate in constant light, indicating strong circadian control. The phenotype of StSUT4-RNAi plants includes early flowering and higher tuber production. Sucrose efflux from leaves of transgenic plants was increased at the end of the light period, leading to modified sucrose levels in sink organs. Of high interest was the observation that the inhibition of StSUT4 expression allowed tuber formation in the strictly photoperiodic-dependent cultivars under noninductive LD conditions.

Starch and protein accumulation

The transition of a stolon into a tuber coincides with a strong upregulation of genes associated with starch biosynthesis. Transport of these photoassimilates, mainly in the form of sucrose, from the leaves to the stolon tip is considered one of the driving forces behind tuber growth (Zrenner *et al.*, 1995) (see also previous paragraph). The switch from apoplastic towards symplastic sucrose unloading at tuber organogenesis results in increased cellular sucrose availability (Appeldoorn *et al.*, 1997; Viola *et al.*, 2001). As a result, carbohydrate metabolism and sucrose-linked pathways have to adjust their flux accordingly and are dedicated towards the synthesis of starch. Sucrose synthase (SUSY) gene expression is increased strongly and becomes the primary sugar cleavage pathway increasing the hexose pool entering the starch biosynthesis pathway through the generation and subsequent import of glucose-6-P into the amyloplast (Tauberger *et al.*, 2000).

Within this organelle, plastidic phosphoglucomutase (PGM), subunits of the ADP-glucose pyrophosphorylase (AGPase) protein complex, starch synthases, and several starch-modifying enzymes (branching enzymes, amylases, phosphorylases) all follow an upregulated expression profile during tuber development in support of starch accumulation (Geigenberger, 2003; Kloosterman *et al.*, 2005). The conversion of sucrose to starch is ATP dependent. Tjaden *et al.* (1998) showed that a relatively small decrease in ATP/ADP transporter activity led to a reduced level of total starch content and a lower amylose to amylopectin ratio. The interplay between sucrose mobility, breakdown, and starch biosynthesis, together with increasing cell number and cell size in the tuber, requires the coordinated expression of many genes, involving many cellular processes.

A comparison of a relative modern-day cultivar with a diploid landrace showed a higher expression of starch breakdown genes in the landrace, indicating a strong selection for high starch-containing tubers in the breeding programs for modern European potato cultivars (PGSC, 2011). The expression of starch-modifying enzymes during tuber growth and maturation can alter starch properties, impacting tuber quality and providing different industrial uses. Beside starch, potato storage proteins accumulate to high levels in the tuber. Gene expression profiles for these proteins show strong upregulation throughout all stages of tuber growth and can reach extremely high levels, particularly during the bulking phase. The major potato tuber protein is patatin, a family of 40–42 kDa glycoproteins that can represent around 40% of the soluble protein in tubers (Mignery *et al.*, 1988). Besides starch and other storage proteins, the accumulation of metabolites like amino acids, flavonoids, carotenoids, and other health-related compounds requires additional signaling and gene regulatory networks, and variation between cultivars is often genetically controlled. Similar to the unraveling of the genes driving tuber initiation and growth, the elucidation of genes with functions impacting nutritional content and appearance is needed to understand which genes make the potato tuber.

References

Abelenda, J. A., Navarro, C. and Prat, S. (2011) From the model to the crop: genes controlling tuber formation in potato. *Current Opinion in Biotechnology* 22, 287–292.

An, H., Roussot, C., Suarez-Lopez, P., Corbesier, L., Vincent, C., Pineiro, M., *et al.* (2004) CONSTANS acts in the phloem to regulate a systemic signal that induces photoperiodic flowering of Arabidopsis. *Development* 131, 3615–3626.

Appeldoorn, N.J., de Bruijn, S.M., Koot-Grondsveld, A.M., Visser, R.G.F., Vreugdenhil, D. and van der Plas, L.H.W. (1997) Developmental changes of enzymes involved in the conversion of sucrose to hexose-phosphate during early tuberisation of potato. *Planta* 202, 220–226.

Banerjee, A.K., Chatterjee, M., Yu, Y., Suh, S.G., Miller, W.A. and Hannapel, D.J. (2006) Dynamics of a mobile RNA of potato involved in a long-distance signaling pathway. *Plant Cell* 18, 3443–3457.

Banerjee, A.K., Lin, T. and Hannapel, D.J. (2009) Untranslated regions of a mobile transcript mediate RNA metabolism. *Plant Physiology* 151, 1831–1843.

Booth, A. (1963) The role of growth substances in the development of stolons. In: Ivins, J.D. and Milthorpe, F.L. (eds) *The Growth of the Potato*. Butterworths, London, pp. 99–112.

Borzenkova, R.A. and Borovkova, M.P. (2003) Developmental patterns of phytohormone content in the cortex and pith of potato tubers as related to their growth and starch content. *Russian Journal of Plant Physiology* 50, 119–124.

Bou-Torrent, J., Martinez-Garcia, J.F., Garcia-Martinez, J.L. and Prat, S. (2011) Gibberellin A1 metabolism contributes to the control of photoperiod-mediated tuberization in potato. *PLoS One* 6, e24458.

Brewer, P.B., Dun, E.A., Ferguson, B.J., Rameau, C. and Beveridge, C.A. (2009) Strigolactone acts downstream of auxin to regulate bud outgrowth in pea and Arabidopsis. *Plant Physiology* 150, 482–493.

Carrera, E., Bou, J., Garcia-Martinez, J.L. and Prat, S. (2000) Changes in GA 20-oxidase gene expression strongly affect stem length, tuber induction and tuber yield of potato plants. *Plant Journal* 22, 247–256.

Celis-Gamboa, C., Struik, P.C., Jacobsen, E. and Visser, R.G.F. (2003) Temporal dynamics of tuber formation and related processes in a crossing population of potato (*Solanum tuberosum*). *Annals of Applied Biology* 143, 175–186.

Chailakhyan, M.K., Yanina, L.I., Devedzhyan, A.G. and Lotova, G.N. (1981) Photoperiodism and tuber formation in grafting of tobacco onto potato. *Doklady Botanical Sciences* 257, 1276–1280.

Chen, H., Banerjee, A.K. and Hannapel, D.J. (2004) The tandem complex of BEL and KNOX partners is required for transcriptional repression of ga20ox1. *Plant Journal* 38, 276–284.

Chincinska, I.A., Liesche, J., Krugel, U., Michalska, J., Geigenberger, P., Grimm, B., *et al.* (2008) Sucrose transporter StSUT4 from potato affects flowering, tuberization, and shade avoidance response. *Plant Physiology* 146, 515–528.

Corbesier, L., Vincent, C., Jang, S., Fornara, F., Fan, Q., Searle, I., *et al.* (2007) FT protein movement contributes to long-distance signaling in floral induction of Arabidopsis. *Science* 316, 1030–1033.

Cutter, E.G. (1978) Structure and development of the potato plant. In: Harris, P.M. (ed.) *The Potato Crop*. Chapman and Hall, London, pp. 70–152.

Ducreux, L.J., Morris, W.L., Prosser, I.M., Morris, J.A., Beale, M.H., Wright, F., *et al.* (2008) Expression profiling of potato germplasm differentiated in quality traits leads to the identification of candidate flavour and texture genes. *Journal of Experimental Botany* 59, 4219–4231.

Ewing, E.E. (1987) The role of hormones in potato (*Solanum tuberosum* L.) tuberization. In: Davies, P.J. (ed.) *Plant Hormones and their Role in Plant Growth and Development*. Martinus Nijhoff Publishers, Boston, Massachusetts, pp. 515–538.

Ewing, E.E. and Struik, P.C. (1992) Tuber formation in potato: induction, initiation, and growth. *Horticultural Review* 14, 89–198.
Fujino, K., Koda, Y. and Kikuta, Y. (1995) Reorientation of cortical microtubules in the sub-apical region during tuberization in single note stem segments of potato in culture. *Plant and Cell Physiology* 36, 348–353.
Gális, I., Macas, J., Vlasák, J., Ondřej, M. and Van Onckelen, H. (1995) The effect of an elevated cytokinin level using the *ipt* gene and N^6-benzyladenine on single node and intact potato plant tuberization *in vitro*. *Journal of Plant Growth Regulation* 14, 143–150.
Geigenberger, P. (2003) Regulation of sucrose to starch conversion in growing potato tubers. *Journal of Experimental Botany* 54, 457–465.
Gonzalez-Schain, N.D., Diaz-Mendoza, M., Zurczak, M. and Suarez-Lopez, P. (2012) Potato CONSTANS is involved in photoperiodic tuberization in a graft-transmissible manner. *Plant Journal* 70, 678–690.
Gregory, L.E. (1956) Some factors for tuberization in the potato. *Annals of Botany* 41, 281–288.
Guivarc'h, A., Rembur, J., Goetz, M., Roitsch, T., Noin, M., Schmulling, T., *et al.* (2002) Local expression of the *ipt* gene in transgenic tobacco (*Nicotiana tabacum* L. cv. SR1) axillary buds establishes a role for cytokinins in tuberization and sink formation. *Journal of Experimental Botany* 53, 621–629.
Hannapel, D.J. (2013) A perspective on photoperiodic phloem-mobile signals that control development. *Frontiers in Plant Science* 4, 295, 1–5.
Harmey, M.A., Crowley, M.P. and Clinch, P.E.M. (1966) The effect of growth regulators on tuberization of cultured stem pieces of *Solanum tuberosum*. *European Potato Journal* 9, 146–151.
Hartmann, A., Senning, M., Hedden, P., Sonnewald, U. and Sonnewald, S. (2011) Reactivation of meristem activity and sprout growth in potato tubers require both cytokinin and gibberellin. *Plant Physiology* 155, 776–796.
Helder, H., van der Maarl, A., Vreugdenhil, D. and Struik, P. (1993) Stolon characteristics and tuber initiation in a wild potato species (*Solanum demissum* Lindl.). *Potato Research* 36, 317–326.
Hendriks, T., Vreugdenhil, D. and Stiekema, W.J. (1991) Patatin and four serine proteinase inhibitor genes are differentially expressed during potato tuber development. *Plant Molecular Biology* 17, 385–394.
Hussey, G. and Stacey, N.J. (1984) Factors affecting the formation of *in vitro* tubers of potato (*Solanum tuberosum* L.). *Annals of Botany* 53, 565–578.
Imaizumi, T. (2010) Arabidopsis circadian clock and photoperiodism: time to think about location. *Current Opinion in Plant Biology* 13, 83–89.
Jackson, S.D. (1999) Multiple signaling pathways control tuber induction in potato. *Plant Physiology* 119, 1–8.
Jackson, S.D. and Prat, S. (1996) Control of tuberization in potato by gibberellins and phytochrome B. *Physiologia Plantarum* 98, 407–412.
Jackson, S.D., Heyer, A.G., Dietze, J. and Prat, S. (1996) Phytochrome B mediates the photoperiodic control of tuber formation in potato. *Plant Journal* 9, 159–166.
Jackson, S.D., James, P., Prat, S. and Thomas, B. (1998) Phytochrome B affects the levels of a graft-transmissible signal involved in tuberization. *Plant Physiology* 117, 29–32.
Jung, J.H., Seo, Y.H., Seo, P.J., Reyes, J.L., Yun, J., Chua, N.H., *et al.* (2007) The GIGANTEA-regulated microRNA172 mediates photoperiodic flowering independent of CONSTANS in Arabidopsis. *Plant Cell* 19, 2736–2748.
Kloosterman, B., Vorst, O., Hall, R.D., Visser, R.G. and Bachem, C.W. (2005) Tuber on a chip: differential gene expression during potato tuber development. *Plant Biotechnology Journal* 3, 505–519.
Kloosterman, B., Navarro, C., Bijsterbosch, G., Lange, T., Prat, S., Visser, R.G., *et al.* (2007) StGA2ox1 is induced prior to stolon swelling and controls GA levels during potato tuber development. *Plant Journal* 52, 362–373.
Kloosterman, B., De Koeyer, D., Griffiths, R., Flinn, B., Steuernagel, B., Scholz, U., *et al.* (2008) Genes driving potato tuber initiation and growth: identification based on transcriptional changes using the POCI array. *Functional and Integrative Genomics* 8, 329–340.
Kloosterman, B., Oortwijn, M., Uitdewilligen, J., America, T., de Vos, R., Visser, R.G., *et al.* (2010) From QTL to candidate gene: genetical genomics of simple and complex traits in potato using a pooling strategy. *BMC Genomics* 11, 158, 1–16.
Kloosterman, B., Abelenda, J.A., Gomez Mdel, M., Oortwijn, M., de Boer, J.M., Kowitwanich, K., *et al.* (2013) Naturally occurring allele diversity allows potato cultivation in northern latitudes. *Nature* 495, 246–250.
Kolomiets, M.V., Hannapel, D.J., Chen, H., Tymeson, M. and Gladon, R.J. (2001) Lipoxygenase is involved in the control of potato tuber development. *Plant Cell* 13, 613–626.

Kuhn, C., Hajirezaei, M.R., Fernie, A.R., Roessner-Tunali, U., Czechowski, T., Hirner, B., *et al.* (2003) The sucrose transporter StSUT1 localizes to sieve elements in potato tuber phloem and influences tuber physiology and development. *Plant Physiology* 131, 102–113.
Kumar, D. and Wareing, P.F. (1974) Studies on tuberization of *Solanum andigena*. II. Growth hormones and tuberization. *New Phytologist* 73, 833–840.
Lange, T. (1998) Molecular biology of gibberellin synthesis. *Planta* 204, 409–419.
Mahajan, A., Bhogale, S., Kang, I.H., Hannapel, D.J. and Banerjee, A.K. (2012) The mRNA of a Knotted1-like transcription factor of potato is phloem mobile. *Plant Molecular Biology* 79, 595–608.
Martin, A., Adam, H., Diaz-Mendoza, M., Zurczak, M., Gonzalez-Schain, N.D. and Suarez-Lopez, P. (2009) Graft-transmissible induction of potato tuberization by the microRNA miR172. *Development* 136, 2873–2881.
Martinez-Garcia, J.F., Virgos-Soler, A. and Prat, S. (2002) Control of photoperiod-regulated tuberization in potato by the Arabidopsis flowering-time gene CONSTANS. *Proceedings of the National Academy of Sciences* 99, 15211–15216.
Mignery, G.A., Pikaard, C.S. and Park, W.D. (1988) Molecular characterization of the patatin multigene family of potato. *Gene* 62, 27–44.
Navarro, C., Abelenda, J.A., Cruz-Oro, E., Cuellar, C.A., Tamaki, S., Silva, J., *et al.* (2011) Control of flowering and storage organ formation in potato by FLOWERING LOCUS T. *Nature* 478, 119–122.
Palmer, C.E. and Smith, O.E. (1969) Effect of abscisic acid on elongation and kinetin-induced tuberization of isolated stolons of *Solanum tuberosum* L. *Plant and Cell Physiology* 10, 657–664.
Park, D.H., Somers, D.E., Kim, Y.S., Choy, Y.H., Lim, H.K., Soh, M.S., *et al.* (1999) Control of circadian rhythms and photoperiodic flowering by the Arabidopsis GIGANTEA gene. *Science* 285, 1579–1582.
Peterson, R.L., Barker, W.G. and Howarth, M.J. (1985) Development and structure of tubers. In: Li, P.H. (ed.) *Potato Physiology*. Academic Press, London, pp. 123–147.
PGSC (Potato Genome Sequencing Consortium) (2011) Genome sequence and analysis of the tuber crop potato. *Nature* 479,189–195.
Pin, P.A., Benlloch, R., Bonnet, D., Wremerth-Weich, E., Kraft, T., Gielen, J.J., *et al.* (2010) An antagonistic pair of FT homologs mediates the control of flowering time in sugar beet. *Science* 330, 1397–1400.
Prat, S. (2010) Hormonal and daylength control of potato tuberization. In: Davies, P.J. (ed.) *Plant Hormones*. Springer, the Netherlands, pp. 574–596.
Proebsting, W.M., Hedden, P., Lewis, M.J., Croker, S.J. and Proebsting, L.N. (1992) Gibberellin concentration and transport in genetic lines of pea: effects of grafting. *Plant Physiology* 100, 1354–1360.
Prusinkiewicz, P., Crawford, S., Smith, R.S., Ljung, K., Bennett, T., Ongaro, V., *et al.* (2009) Control of bud activation by an auxin transport switch. *Proceedings of the National Academy of Sciences* 106, 17431–17436.
Reeve, R.M., Hautala, E. and Weaver, M.L. (1969) Anatomy and compositional variation withing potatoes I. Development histology of the tuber. *American Potato Journal* 46, 361–373.
Rensink, W.A., Iobst, S., Hart, A., Stegalkina, S., Liu, J. and Buell, C.R. (2005) Gene expression profiling of potato responses to cold, heat, and salt stress. *Functional and Integrative Genomics* 5, 201–207.
Riesmeier, J.W., Willmitzer, L. and Frommer, W.B. (1994) Evidence for an essential role of the sucrose transporter in phloem loading and assimilate partitioning. *EMBO Journal* 13, 1–7.
Rodriguez-Falcon, M., Bou, J. and Prat, S. (2006) Seasonal control of tuberization in potato: conserved elements with the flowering response. *Annual Review of Plant Biology* 57, 151–180.
Roitsch, T. and Ehneß, R. (2000) Regulation of source/sink relations by cytokinins. *Plant Growth Regulation* 32, 359–367.
Rosin, F.M., Hart, J.K., Horner, H.T., Davies, P.J. and Hannapel, D.J. (2003a) Overexpression of a knotted-like homeobox gene of potato alters vegetative development by decreasing gibberellin accumulation. *Plant Physiology* 132, 106–117.
Rosin, F.M., Hart, J.K., Van Onckelen, H. and Hannapel, D.J. (2003b) Suppression of a vegetative MADS box gene of potato activates axillary meristem development. *Plant Physiology* 131, 1613–1622.
Roumeliotis, E., Kloosterman, B., Oortwijn, M., Kohlen, W., Bouwmeester, H.J., Visser, R.G., *et al.* (2012) The effects of auxin and strigolactones on tuber initiation and stolon architecture in potato. *Journal of Experimental Botany* 63, 4539–4547.
Sanz, M.J., Mingo-Castel, A., Lammeren, A.A.M.v. and Vreugdenhil, D. (1996) Changes in the microtubular cytoskeleton precede *in vitro* tuber formation in potato. *Protoplasma* 191, 46–54.
Sergeeva, L.I., De Bruijn, S.M., Koot-Gronsveld, E.A.M., Navratil, O. and Vreugdenhil, D. (2000) Tuber morphology and starch accumulation are independent phenomena: evidence from ipt-transgenic potato lines. *Physiologia Plantarum* 108, 435–443.

Shah, S., Lee, Y.J., Hannapel, D.J. and Rao, A.G. (2011) Protein profiling of the potato petiole under short day and long day photoperiods. *Journal of Proteomics* 74, 212–230.
Shibaoka, H. (1993) Regulation by gibberellins of the orientation of cortical microtubules in plant cells. *Australian Journal of Plant Physiology* 20, 461–470.
Struik, P., Haverkort, A., Vreugdenhil, D., Bus, C. and Dankert, R. (1990) Manipulation of tuber-size distribution of a potato crop. *Potato Research* 33, 417–432.
Stupar, R.M., Bhaskar, P.B., Yandell, B.S., Rensink, W.A., Hart, A.L., Ouyang, S., *et al.* (2007) Phenotypic and transcriptomic changes associated with potato autopolyploidization. *Genetics* 176, 2055–2067.
Suttle, J. (2004) Physiological regulation of potato tuber dormancy. *American Journal of Potato Research* 81, 253–262.
Takahashi, K., Fujino, K., Kikuta, Y. and Koda, Y. (1994) Expansion of potato cells in response to jasmonic acid. *Plant Science* 100, 3–8.
Tauberger, E., Fernie, A.R., Emmermann, M., Renz, A., Kossmann, J., Willmitzer, L., *et al.* (2000) Antisense inhibition of plastidial phosphoglucomutase provides compelling evidence that potato tuber amyloplasts import carbon from the cytosol in the form of glucose-6-phosphate. *Plant Journal* 23, 43–53.
Tjaden, J., Schwoppe, C., Mohlmann, T., Quick, P.W. and Neuhaus, H.E. (1998) Expression of a plastidic ATP/ADP transporter gene in Escherichia coli leads to a functional adenine nucleotide transport system in the bacterial cytoplasmic membrane. *Journal of Biological Chemistry* 273, 9630–9636.
Turck, F., Fornara, F. and Coupland, G. (2008) Regulation and identity of florigen: FLOWERING LOCUS T moves center stage. *Annual Review of Plant Biology* 59, 573–594.
Vasquez-Robinet, C., Mane, S.P., Ulanov, A.V., Watkinson, J.I., Stromberg, V.K., De Koeyer, D., *et al.* (2008) Physiological and molecular adaptations to drought in Andean potato genotypes. *Journal of Experimental Botany* 59, 2109–2123.
Viola, R., Roberts, A.G., Haupt, S., Gazzani, S., Hancock, R.D., Marmiroli, N., *et al.* (2001) Tuberization in potato involves a switch from apoplastic to symplastic phloem unloading. *Plant Cell* 13, 385–398.
Viola, R., Pelloux, J., van der Ploeg, A., Gillespie, T., Marquis, N., Roberts, A.G., *et al.* (2007) Symplastic connection is required for bud outgrowth following dormancy in potato (*Solanum tuberosum* L.) tubers. *Plant, Cell and Environment* 30, 973–983.
Vreugdenhil, D. and Struik, P.C. (1989) An integrated view of the hormonal regulation of tuber formation in potato (*Solanum tubersosum*). *Physiologia Plantarum* 75, 525–531.
Walworth, J.L. and Carling, D.E. (2002) Tuber initiation and development in irrigated and non-irrigated potatoes. *American Journal of Potato Research* 79, 387–395.
Xu, X., van Lammeren, A.A., Vermeer, E. and Vreugdenhil, D. (1998a) The role of gibberellin, abscisic acid, and sucrose in the regulation of potato tuber formation *in vitro*. *Plant Physiology* 117, 575–584.
Xu, X., Vreugdenhil, D. and van Lammeren, A.A.M. (1998b) Cell division and cell enlargement during potato tuber formation. *Journal of Experimental Botany* 49, 573–582.
Yanovsky, M.J. and Kay, S.A. (2002) Molecular basis of seasonal time measurement in Arabidopsis. *Nature* 419, 308–312.
Yanovsky, M.J., Izaguirre, M., Wagmaister, J.A., Gatz, C., Jackson, S.D., Thomas, B., *et al.* (2000) Phytochrome A resets the circadian clock and delays tuber formation under long days in potato. *Plant Journal* 23, 223–232.
Zrenner, R., Salanoubat, M., Willmitzer, L. and Sonnewald, U. (1995) Evidence of the crucial role of sucrose synthase for sink strength using transgenic potato plants (*Solanum tuberosum* L.). *Plant Journal* 7, 97–107.
Zubko, E., Machackova, I., Malbeck, J. and Meyer, P. (2005) Modification of cytokinin levels in potato via expression of the Petunia hybrida *Sho* gene. *Transgenic Research* 14, 615–618.

5 Plant Growth and Development

Carrie H. Wohleb,[1]* N. Richard Knowles[2] and Mark J. Pavek[2]
[1]*Washington State University, Ephrata, USA;* [2]*Department of Horticulture, Washington State University, Pullman, USA*

Potatoes are classified as a C_3 plant and are members of the nightshade family (Solanaceae). The cultivated species most commonly grown for human consumption is *Solanum tuberosum*. Botanically, potatoes are considered perennials, because clones of the mother plant can develop via daughter tubers that are left unharvested. Depending on the cultivar and environment, daughter tubers may sprout while the mother plant is still alive; however, the daughter tubers typically transition through a dormancy phase prior to initiating new growth. The dormancy phase usually extends beyond the natural life of the mother plant. When potatoes are grown commercially, the vines of the mother plant senesce naturally or are killed and the daughter tubers harvested; therefore, potatoes are an annual crop.

This chapter focuses on the growth and development of *S. tuberosum* under commercial production and describes how potato plant morphology changes as plants grow and develop. To simplify the discussion on potato development over time, we separated this chapter into a logical series of growth and developmental stages including propagation, seed tuber dormancy and physiological age, tuber anatomy and sprout development, tuber initiation, canopy development, tuber bulking, and plant and tuber maturation.

Plant and tuber development are influenced largely by environment, particularly at key growth stages. To produce high-quality tubers with ideal yields, tuber size profile, shape, storability, and palatability, commercial agronomists routinely manipulate soil moisture and fertility, planting date and depth, plant population and intra- and inter-row plant spacing, source–sink relationships, vine kill date, and harvest date. Additionally, producing potatoes in the appropriate region and climate zone reduces the chances of a nature-induced crop failure. Ideal production regions have well-drained, fertile soils, 100–180 consecutive frost-free days, between 50 and 60 cm of rainfall or irrigation between plant establishment and maturation, a broad day/night temperature differential, absence of extreme heat and cold during the growing season, and proximity to market, temperature-controlled storage, and mass transportation.

5.1 Propagation: True Seed versus Seed Tubers

Potatoes are most often propagated from vegetative tissue, either whole tubers or cut pieces of tubers. Tubers used for propagation are typically called seed tubers, seed potatoes, or seed pieces. In some regions, tubers left uncut and planted whole are referred to as "single-drop" seed potatoes. Potato plants can be produced from botanical or sexual seeds, but these "true" potato seeds (TPS)

* E-mail: cwohleb@wsu.edu

are seldom used, because they are genotypically variable and give rise to plants with traits that may be completely different from those of the parent plants (Burton, 1989; Allen *et al.*, 1992). Moreover, production from TPS often requires several extra months of initial growth in a greenhouse or seedbed, followed by labor-intensive transplantation into the field. The International Potato Center (CIP), however, has been making efforts to improve TPS by selecting parental lines that produce more uniform progeny with good agronomic characters, especially for use in tropical regions or in developing countries where seed potato storage and transport is not practical (Golmirzaie *et al.*, 1994).

Plant breeders use TPS as a means of developing new potato cultivars. Potato cultivars are clonal selections with specific characteristics that are fixed only when propagated with vegetative tissue. Major characteristics important to commercial production include yield, tuber shape, size, solids content and quality, skin texture and color, and storability. It is important to understand tuber physiology and morphology when tubers are used for propagation because they can impact how the plant will develop.

5.2 Seed Tuber Dormancy

Potato tubers are usually dormant at harvest and will not sprout even under optimum conditions for growth. Dictated largely by genetic cues, weeks or months must pass before tuber dormancy is broken and sprout growth can begin. This is thought to be a natural adaptation that prevents potatoes from sprouting in the winter months, when growing conditions are adverse in the temperate zones where they originated. The first phase of tuber dormancy is referred to as “rest” and is regulated by internal physiological processes. During this period, it is thought that either the genome of dormant buds is repressed (Tuan and Bonner, 1964) or the tuber is unable to supply metabolites for bud growth (Cutter, 1992). Following the rest period, tubers may continue to be dormant due to external factors, particularly low temperatures (Hemberg, 1985; Burton *et al.*, 1992).

A number of endogenous plant growth regulators, especially abscisic acid (ABA), ethylene, cytokinin, and gibberellin, have been shown to play critical roles in the control of tuber dormancy and subsequent sprout growth. It is probably a balance of these inhibiting and promoting substances that controls dormancy and sprouting. Evidence indicates that ABA, a growth-inhibiting substance, is largely responsible for the induction and maintenance of dormancy in tubers. It has an analogous role in inhibiting the germination of true seeds (Karssen and Van Loon, 1992). ABA levels in tuber tissues are highest in dormant tubers and decline as dormancy weakens (Coleman and King, 1984; Suttle, 1995). However, there does not appear to be a threshold concentration of endogenous ABA below which dormancy will cease and sprouting occurs (Coleman and King, 1984; Suttle, 1995; Classens and Vreugdenhil, 2000). Furthermore, it has been demonstrated that application of an ABA synthesis inhibitor to dormant tubers results in premature sprouting, while subsequent application of ABA restores dormancy (Suttle and Hulstrand, 1994).

Ethylene, another growth inhibitor, probably has a role in inducing tuber dormancy, though it is not thought to maintain dormancy (Suttle, 2004a). Treatment with an ethylene inhibitor was shown to hasten sprouting in developing microtubers when applied shortly after tuberization, but later application of the inhibitor had no effect on dormancy (Suttle, 1998). Ethylene levels in tubers vary throughout the life of a tuber; ethylene levels are highest at harvest and decline in storage (Cvikrova *et al.*, 1994), but ethylene production increases again as tubers begin to sprout (Poapst *et al.*, 1968; Suttle, 2003). The increase in ethylene as sprouting begins suggests that ethylene may also play a role in the termination of dormancy, or may be involved with events occurring during early sprout growth (Suttle, 2004a).

Dormancy break is associated with cytokinins, plant growth regulators known to stimulate cell division. Cytokinin levels in dormant tubers are relatively low, and an increase in the cytokinin in tubers has been shown to immediately precede the onset of sprouting. Moreover, the application of cytokinin to dormant tubers has been shown to hasten sprouting, though sensitivity to cytokinins appears to be time dependent; for example, tubers are insensitive to cytokinins applied shortly after harvest, but become increasingly sensitive to them in storage

(Turnbull and Hanke, 1985; Suttle, 2002). The growth inhibitors that promote dormancy may also be inhibiting the effect of cytokinins during the early phase of rest (Hemberg, 1985).

Another growth-stimulating substance, gibberellin (GA), is thought to have a role in regulating either tuber dormancy and/or sprout growth immediately after dormancy break (Suttle, 2004b). GA is found at low concentrations in dormant tubers, but increases during early sprout growth. It has been shown that endogenous GA levels do not begin to increase until after the end of dormancy, suggesting GA is not a regulator of dormancy per se, but is important for regulating sprout growth immediately after dormancy break (Suttle, 2004b). Seed potatoes are sometimes treated with GA to accelerate sprout growth, and it is known that the concentration needed to accelerate tuber sprouting in the early stages of dormancy is greater than during the latter stages. Seed certification agencies commonly treat seed potatoes with GA to promote early sprouting for disease assessments in the off season. Much remains to be learned about the synthesis of plant growth regulators and the significance of the interaction between different plant growth regulators during the rest phase of tuber dormancy.

The second phase of tuber dormancy, sometimes referred to as the quiescent phase, depends on the environment surrounding a tuber. In this case, sprouting is prevented by factors outside the tuber. Temperature is critical. Tuber sprouting is greatly suppressed below 4°C. This is one of the reasons why potatoes are often stored at cool temperatures. Generally, the warmer the storage temperature, the sooner the tubers will sprout. For instance, Russet Burbank tubers stored at 8.9°C were found to sprout approximately 135 days after harvest, compared to 155 days at 7.2°C and 175 days at 5.6°C (Brandt *et al.*, 2003).

Seed potatoes for commercial production are usually stored for 6–7 months at 3–4°C in temperate zones and are later warmed for 7–10 days before handling, to promote sprouting and minimize bruising. As a general rule, seed tubers warmed (>7°C) prior to planting emerge faster from the soil than those not warmed. Most potato varieties will likely tolerate storage temperatures down to 2°C; however, storage below 2°C is not recommended (Burton *et al.*, 1992). After planting, low soil temperatures may hinder sprout growth. In general, it is recommended that soil temperatures exceed 7°C before planting.

The length of the dormant period varies among cultivars; therefore, some cultivars may require unique approaches to sprout management. For example, cultivars with longer dormancy may require a longer period of warming before planting as seed. Short-dormancy cultivars being stored for consumption may require earlier applications of chemical sprout inhibitors. When tubers were stored at 5.6°C in Idaho, USA, cultivar dormancy ranged widely; 200 days for Summit Russet, 175 days for Russet Burbank, 145 days for Umatilla Russet, and 100 days for Alturas (Brandt *et al.*, 2003, 2004, 2006).

5.3 Physiological Age of Tubers

The physiological age of tubers has a significant impact on dormancy and sprouting. Physiological age accounts for time and environmental effects. In-season stress and development, plant maturity at harvest, postharvest stress and storage temperature, humidity, and time all influence the physiological age of a tuber. Physiological age is an important consideration for seed potatoes and is often characterized as the sprouting potential of seed. Sprouts of seed tubers with advanced physiological age typically emerge faster than those from younger seed. Apical dominance, the inhibition of axillary buds by the dominant, actively growing apical bud (Michener, 1942), decreases with advancing physiological age, which results in more stems per plant, increased tuber set, and shifts in tuber size distribution toward smaller tubers. In fact, the degree of apical dominance from seed tubers is diagnostic of their physiological age. According to Krijthe (1962), the sequence of sprouting from physiologically young to physiologically old tubers is: (i) single sprout stage; (ii) multiple sprout stage; (iii) multiple and branching sprout stage; and (iv) small tuber formation stage.

Generally, any stresses encountered during growth, handling, and/or storage can accelerate the aging process in tubers, though research has demonstrated that growing conditions have a smaller effect than postharvest conditions (Knowles and Knowles, 2006). Physiologically old tubers often have higher rates of respiration (Kumar and Knowles, 1996a,b; Zabrouskov *et al.*, 2002).

Conditions that increase the rate of respiration, even for a short period, can accelerate physiological aging. For example, high temperatures during the maturation of tubers in the field, or exposure to high or fluctuating temperatures in storage, can result in higher basal respiration rates during storage, which accelerates the aging process, resulting in physiologically older tubers at planting (Blauer *et al.*, 2013b). Tuber respiration appears to be a pacemaker of aging; the longer the respiration rate remains high, the older the seed tubers will be (Blauer *et al.*, 2013b). Exposure to low temperatures, below 2.8°C, also increases respiration and aging of tubers. Factors other than temperature can increase respiration and aging, including moisture stress, inadequate fertility, and disease pressure during the growing season; bruising at harvest; and harvest of immature or overmature tubers.

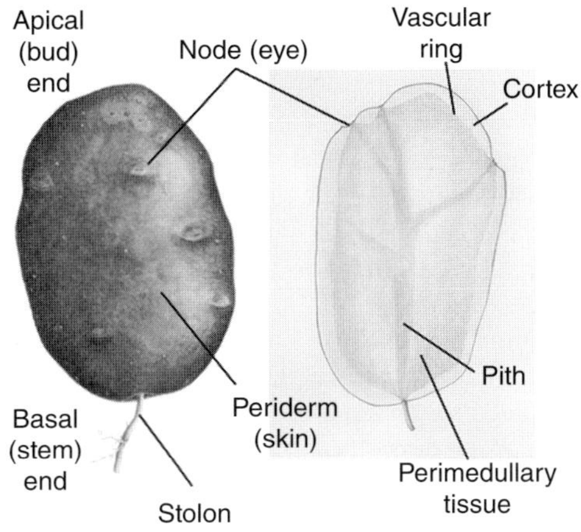

Fig. 5.1. The external and internal anatomy of a potato tuber. (Drawing courtesy of Jeanne Debons, Bend, Oregon, USA.)

5.4 Tuber Anatomy and Sprout Development

According to Cutter (1992), once one understands the anatomy of a potato eye, it becomes clear that the structure of a potato tuber is essentially an enlarged, modified, belowground stem. She explains the potato eye is a complete scale leaf with axillary buds similar to that of many plant species, containing a primary axillary bud along with second-order axillary buds that form into the first two leaf primordia. Although technically a stem, the potato tuber functions as a storage organ for carbohydrates and nutrients to facilitate its role as a vegetative propagule to perenniate the plant.

The external morphology of a tuber is similar to that of an aerial stem, and includes internodes, nodes, and scale leaves. The tuber surface is covered by an outer protective tissue called the native periderm, which includes many small pores, or lenticels, for gas exchange. Directly beneath the periderm lies an inner region of storage parenchyma tissue called the cortex, followed by a ring of vascular tissue, and an inner pith region composed largely of parenchyma tissue and interxylary phloem (Fig. 5.1).

The nodes (eyes) on tubers contain the axillary meristems (buds), which are subtended by a scale leaf, as in a leaf axil of an aerial stem. The eyes are arranged spirally on the tuber, just as axillary buds are arranged on an aboveground stem (Cutter, 1992). The meristems within each eye can sprout to form branch stems from the tuber. Sprouts that develop below the ground and in the dark are etiolated and characterized by extensive elongation of internodes.

Exposure to light induces rapid leaf development, resulting in an initial rosette or whorl of leaves around the apical meristem soon after emergence. Further growth produces leaves, branch stems, flowers, roots, stolons, and tubers (Fig. 5.2).

Generally, and depending on tuber physiological age and cultivar, only the primary (apical) bud in an eye develops to form a sprout, but if it is removed or injured, the secondary buds will become active. Moreover, apical dominance dictates that eyes on the apical end of whole tubers sprout before eyes on the basal end. Apical dominance is controlled in part by auxin, a growth-regulating substance that is produced in the apical meristem and translocated basipetally buds to growth of axillary buds (Michener, 1942 Young *et al.*, 2014). Apical dominance is reduced as tubers age; therefore, it is possible for multiple buds in one eye to sprout concurrently, resulting in multiple stems per eye from older tubers. Apical dominance is usually attenuated when tubers are cut into pieces, but there can be a noticeable difference in the rate of sprout growth from seed pieces originating from the apical versus the basal end of a tuber. Sprouts developing from

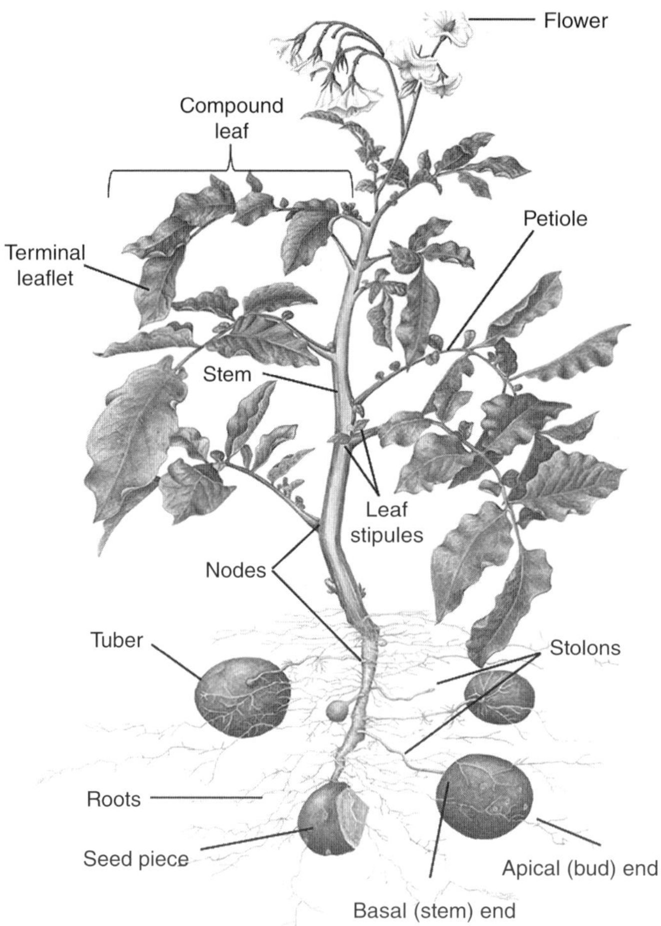

Fig. 5.2. A typical potato (*Solanum tuberosum* L.) plant and its components during early to mid season. (Drawing courtesy of Jeanne Debons, Bend, Oregon, USA.)

apically derived seed pieces usually grow more rapidly, resulting in earlier plant emergence than sprouts from basal end seed pieces. The degree of apical dominance is affected by genotype, tuber physiological age, environment, wounding, and disease. Some potato cultivars exhibit stronger apical dominance than others. The degree of apical dominance dictates the number of stems produced per seed piece, which in turn affects tuber set, number of tubers per plant, and tuber size distribution (Knowles and Knowles, 2006).

When tubers are cut for seed, an economically acceptable size range for a commercial producer is 43–71 g (Schotzko *et al.*, 1983). Regardless of shape, the number of eyes increases only slightly as tuber size increases. The number of eyes depends on genotype, and thus some potato cultivars produce more eyes than others. Seed pieces from cultivars with relatively few eyes must often be cut larger to prevent blind seed pieces.

5.5 Plant Morphology and Stand Establishment

Stand establishment refers to the period of growth from planting to the initiation of tubers. Seed potatoes may be sprouted prior to planting, or planted before sprouting occurs. Pre-sprouting, or chitting, is a practice used to promote early and more uniform emergence and earlier harvest. It is more common in regions with shorter growing seasons, especially areas in northern Europe, and for production of early potato crops. Pre-sprouting on a large scale requires specialized equipment and facilities, such as chitting trays or pre-sprouting bags, and is often labor-intensive. It is usually done under low light conditions that suppress sprout elongation, resulting in short and stocky green sprouts with greatly compressed internodes. Some of the sprouts may be removed prior to planting, to control stem numbers. A disadvantage of pre-sprouting seed is that sprouts may be wounded or broken during planting, which could compromise growth and emergence and increase disease issues.

After planting, sprouts grow upward and eventually emerge from the soil. In the dark, sprouts are etiolated. When exposed to light, chlorophyll and anthocyanins are produced. Stem color is typically green, but can be purple or a combination of dark colors like red, purple, and green, depending on the cultivar. Rapid emergence is important for early-season disease resistance in sprouts and stems (Van Emden, 1965). Reserves in the seed tuber or seed piece are the only source of energy for the developing plant until it emerges, forms leaves, and commences photosynthesis. Quick emergence also leads to rapid establishment of leaf area for the early capture of solar radiation and conversion from heterotrophic to autotrophic growth, which is essential for optimizing final tuber yield and dry matter content (Firman and Allen, 1989). The volume and mechanical resistance of soil can affect sprout growth (Taylor and Ten Broeck, 1988); however, soil moisture and temperature differences are more commonly cited as factors limiting the rate of potato emergence (Firman *et al.*, 1992). The rate of sprout growth increases greatly as soil temperatures increase from 10 to 20°C. Sprout growth is also greater when soil moisture levels are closer to field capacity than in drier soils. Additional factors that affect the rate of emergence include seed size (Iritani *et al.*, 1972) and health (Hide and Lapwood, 1982), sprout health (Carling and Leiner, 1990), sprout/eye location on the mother seed tuber (Knowles *et al.*, 1985), soil fertility and seed tuber physiological age (Firman *et al.*, 1992), and dormancy (Lai Cho *et al.*, 1983) and cultivar (Jones and Allen, 1983). It is important to note, however, that early emergence does not always result in increased yield and can increase the vulnerability of plants to early-season frost. Moreover, a balance must be struck between rapid emergence and an optimum quantity of soil covering developing tubers throughout the growing season. During the growing season, the average soil temperature is often higher nearest the soil surface. Because of this, shoots from a shallow-planted seed piece may emerge more rapidly than a deeper-planted one. However, planting seed pieces too shallow can lead to tuber greening, moisture stress due to soil surface heat and evaporation, and yield loss. Tubers should be planted deep enough to accommodate and cover the final yield appropriately, yet shallow enough for rapid, uniform emergence (Pavek and Thornton, 2009).

Roots and stolons develop from belowground nodes on the stem (Fig. 5.2). Root development begins early and before plants emerge from the soil. Adventitious roots begin to grow and develop at the base of developing sprouts. The potato root system is fibrous and highly branched, and is concentrated in the upper soil profile. However, the roots may extend to a depth of 1 m or more, depending on soil conditions and cultivar. Stolons branch off the belowground stem and typically begin to develop within 1–2 weeks following plant emergence. Potato stolons are essentially lateral stems that grow from nodes located on a belowground stem. The first stolons form at the basal nodes on the stem (near the mother tuber or seed piece) and then progress upward (Plaisted, 1957). Nodes on the belowground stem can produce both primary and secondary (lateral) stolons. Secondary stolons form in the axils of primary stolons. Moreover, branch stolons may develop from nodes on the primary or lateral stolons.

Planting depth has considerable influence on the belowground morphology of a potato plant. Seed pieces that are deep planted produce elongated belowground stems with longer internodes, more nodes and more stolons compared to those that are shallow planted. Shallow planting, on the other hand, results in plants with shortened internodes and fewer but longer stolons (Pavek and Thornton, 2009). Other factors influencing stolon and tuber growth include photoperiod, irradiance, temperature, moisture, mineral nutrient supply, seed-tuber physiological age, and date of planting (O'Brien *et al.*, 1998).

Above the ground, the stem produces a fixed number of leaves before terminating in an inflorescence (primary flower) (Allen and Scott, 2001). Leaves are compound and consist of a midrib, several leaflets, and a petiole, from which they attach to the stem (Fig. 5.2). Further leaf growth can occur via a bud located just below the flower. The bud elongates to form a new branch stem with a fixed number of leaves, and terminates in another flower (secondary flower). This process, known as sympodial branching, can be repeated several more times to result in many tiers of branch stems and leaves when growing conditions are favorable (Allen and Scott, 2001). Additional branches may arise from buds in leaf axils along the stem, but these axillary branches do not terminate in flowers. The tendency for potatoes to produce long sympodial branches explains why potato leaves and branches together are commonly referred to as "vines".

Axillary branches give the plant a bushy, upright appearance, whereas sympodial branches are spreading and prostrate. As an herbaceous perennial, potatoes are inherently indeterminate; however, cultivars are often classified, based on the amount of foliage they produce, as "determinant", "semi-determinant", or "indeterminant", depending on their growth habit. Determinant cultivars, also known as early cultivars, produce only primary flowers and have limited foliar growth. Examples of determinant cultivars are Estima, Norland, and Russet Norkotah. Indeterminant cultivars, or late cultivars, continue to produce foliage and flowers throughout the season. Examples of indeterminant cultivars are Cara and Russet Burbank.

5.6 Tuber Initiation

Soon after plants emerge, leaves and stolons begin to develop at aboveground and belowground nodes, respectively. Several stolons may form at each node, but the larger, primary stolon is the one most likely to terminate in a tuber. Lateral and branch stolons may terminate in tubers, but tubers from these stolons do not usually become as large as tubers produced on primary stolons (O'Brien *et al.*, 1998). Stolon morphology is characterized by elongated internodes and a hooked stolon tip prior to tuber initiation (Viola *et al.*, 2001).

Tuber initiation, also called tuberization, begins when stolons cease to elongate and the stolon tips begin to swell. During tuberization, the tuber first appears after cells in the subapical region of a hooked stolon begin to expand, the hook opens, and growth is marked by much radial expansion relative to elongation (Fig. 5.3a) (Viola *et al.*, 2001; Blauer *et al.*, 2013c). The result is displacement of axillary buds, as they are formed from the apical meristem along the long axis of the growing tuber (Fig. 5.3b) (Xu *et al.*, 1998).

A mature tuber thus has a distinctly recognizable apical or bud end, and a basal or stem end where the tuber attaches to the stolon. The eyes are typically concentrated at the apical end of the tuber and are fewer in number and farther apart on and near the stem end. As a result, an inconsistent number of eyes will be found on cut seed pieces, which may give rise to plants with more variable stem numbers compared to those produced from whole tubers. Stem-end seed pieces produce fewer main stems than seed pieces derived from the bud end, which, given the oftentimes direct relationship between tuber set and stem number per seed piece (Knowles and Knowles, 2006), ultimately contributes to variation in tuber size distribution in commercial production.

Tubers typically begin to form on the lower stolons first (Plaisted, 1957). The majority of tubers develop on stolons originating from the lower one-third of the belowground stem, and these tubers are usually larger than tubers that form on stolons higher up the stem (Plaisted, 1957; Pavek and Thornton, 2009). Stolons do not always produce a tuber, nor do they all form tubers simultaneously. On occasion, excessive heat or intruding light may cause a belowground stolon near the soil surface to emerge and form a leafy shoot. Stolons that develop into aboveground shoots are called secondary stems or heat runners. The term "tuber initiation" is more commonly thought of as a process that occurs over a period of time rather than the mere beginning of tuber development. Tuber initiation may continue for up to 2 weeks, but is often complete within 3–4 days of onset (O'Brien *et al.*, 1998).

Tuberization is a short-day response that can be modified by temperature and stress. Phytochrome in the leaves perceives day length. In response to environmental cues (day length and day/night temperatures), a tuberization stimulus is produced that can cross graft unions (Rodriguez-Falcon *et al.*, 2006). The nature of the tuberization stimulus is not fully understood; however, endogenous plant-growth regulators, especially GA, cytokinin, jasmonic acid, and ABA, likely play a role (Krauss, 1985). The timing of tuber initiation is controlled primarily by the ratio of GA to

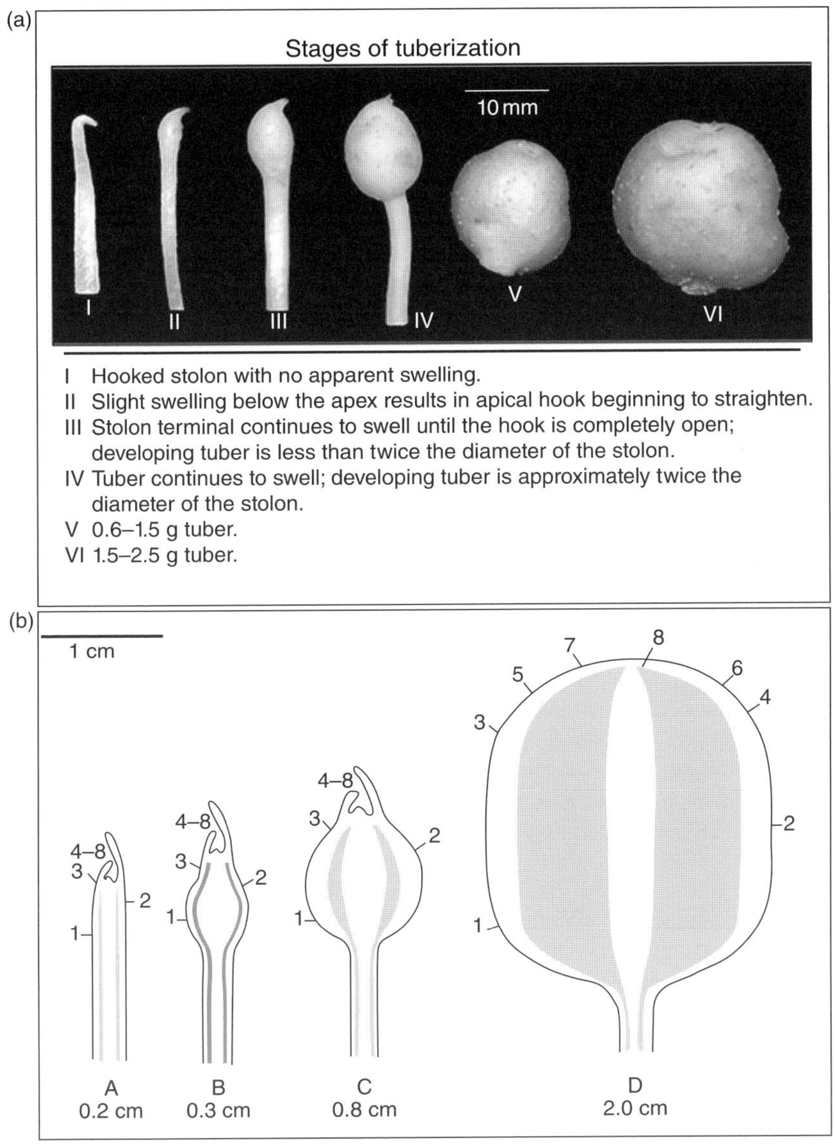

Fig. 5.3. Stages of tuberization (a and b) showing stolon and tuber morphology and thickening of the perimedullary zone (dark-shaded area) (b). The numbers 1–8 indicate the nodes. Dimensions below the letters A–D indicate stolon (A only) and tuber diameter. (Figures reproduced with permission from *Postharvest Biology and Technology*, Blauer *et al.*, 2013 (a), and *Journal of Experimental Botany*, Xu *et al.*, 1998 (b).)

ABA. A higher ratio of GA to ABA in the plant inhibits tuber production and growth in favor of foliage growth. The reverse ratio leads to tuber initiation. Excessive nitrogen fertilizer, long day length, and warm night temperatures can delay tuber initiation by increasing GA relative to ABA (Krauss, 1985).

Excluding disease, once a tuber is initiated, it either continues to expand and develop or is resorbed by the plant (Walworth and Carling, 2002). Those that remain are "set" to grow and expand. This stage is typically referred to as "tuber set"; however, tuber set may also be used to describe tuber number per plant. Despite the wording and description, tuber number per plant may change between tuber set and tuber harvest, especially when the plant is stressed (MacKerron and Jefferies, 1986). The final tuber number per plant can be influenced by temperature (Sale, 1979), soil moisture (Struik and Van Voorst, 1986; Walworth and Carling, 2002), plant health and disease (Cother and Cullis, 1985), planting depth and cultivar (Pavek and Thornton, 2009), plant spacing (Pavek and Thornton, 2006), seed tuber physiological age (Knowles and Knowles, 2006), soil fertility (Rosen and Bierman, 2012), and irradiance (Sale, 1976).

5.7 Canopy Development

As tuber initiation slows and nears completion, plants under ideal conditions focus their energy on maintaining sufficient canopy to feed the new tubers. The rate at which the tubers and canopy develop is dependent largely on environment interacting with genotype. Plants growing in favorable environments free of pests and disease, with sufficient water, nutrients, and heat units are likely to produce a larger canopy and greater tuber yields than those grown under stress. Tubers under stressed plants may actually bulk sooner and faster than those under healthy, unstressed plants; however, this usually occurs at the expense of canopy development. To maximize tuber yield or economic return, a balance must be struck between canopy size and duration and the number of growing days in a particular region. For some cultivars, excessive canopy growth late in the season may extend tuber bulking and delay plant maturity, which could result in reduced tuber yield and quality. On the other hand, an undersized canopy that senesces prematurely is likely to result in a loss of potential yield. Modern cropping systems with state-of-the-art irrigation/fertigation systems and pest and disease control allow growers to tailor potato crop growth for a particular region, cultivar, and intended market in order to maximize economic return.

Through photosynthesis, potato plants use light energy to assimilate carbon dioxide into sucrose, which is translocated from source to sink. In general, sources are net carbon producers, while sinks are net carbon users. Mature green leaves are the main source for the potato plant, while the tubers are the major sink. Conversely, seed tubers become the source of carbohydrates and nutrients during the early stages of plant establishment until plants become autotrophic.

Management of plant growth to maximize yield and economic return while minimizing water, fertilizer, pesticides, and other inputs is a primary goal in commercial crop production. Attention must be given to the partitioning of assimilates between canopy and tubers, as too much to one or the other can lead to an imbalance and a reduction in tuber yield and/or quality. Maximizing the interception of solar radiation by the potato canopy during key growth stages, and in turn the efficiency of conversion of intercepted light to photosynthetic assimilates, is essential for maximizing yield or economic return (Firman and Allen, 1988); provided the plant is healthy and has unlimited access to nutrients, water, and moderate temperatures. Often, solar radiation interception is limited due to seasonal changes associated with a particular geographical location. Moreover, in an effort to use land efficiently and to influence tuber size and yield, commercially grown potato plants are usually positioned in the field so their canopies collectively cover 100% of the soil surface (100% ground cover). As a result, the canopy of one plant often restricts the maximum solar radiation interception of its neighbor. The underlying theme is maximizing economic return on a given piece of land by producing as much of the right-sized, high-quality tubers as is biologically possible. Modern production systems manage plant population, inter- and intra-row plant spacing, and agronomic inputs to produce the ideal canopy size and duration for maximum yield of the highest-valued tuber sizes.

The ideal canopy architecture maximizes light interception during key growth stages and allows for sufficient assimilate translocation during plant maturity, leading to maximum yield or economic return. Because potato fields typically have 100% ground cover for much of the season, keeping the canopy healthy and alive for as long as possible becomes the limiting factor for optimizing output. This is not to say that the canopy should stay green and vigorous up until harvest; rather, a healthy canopy should be maintained as long as possible while still allowing for proper plant maturity prior to harvest. The ideal canopy size and duration varies by cultivar and season length.

Keeping the plant's source to sink ratio in balance is important when trying to maximize profits from crop production. Therefore, it is essential that the relationship between foliar and tuber growth be maintained in a manner that will maximize tuber yield and quality.

Donald (1962) coined the term "harvest index" (HI) to describe the source–sink relationship of wheat varieties as a ratio indicating the percentage of the plant's total weight coming from grain yield versus straw yield. He believed wheat varieties that produced a relatively high grain yield compared to straw yield were more photosynthetically efficient than those where the straw yield was dominant. The HI can be used as a tool for any plant to describe the relationship between source and sink. For potatoes, HI expresses tuber growth (yield) as a percentage of total plant biomass (tubers plus aboveground foliar growth):

$$\text{Harvest index (\%)} = (\text{tuber fresh wt}/(\text{foliar} + \text{tuber fresh wt})) \times 100$$

Hence, a HI of 40% indicates that 40% of total plant fresh weight is tubers. Following tuber set, HI increases throughout the season, eventually reaching 100 as vines fully senesce. The HI, source–sink relationship, tuber bulking, and other crop growth milestones are discussed later in the chapter, in the section "Stages of Crop Growth and Development: A Long-Season Cultivar Case Study".

5.8 Tuber Bulking

Once the canopy is well established and tuberization finished, tuber bulking begins. Final yield is ultimately determined by the rate and duration of tuber bulking. The rate and duration of bulking depends largely on canopy health and size, genotype, soil and air temperatures, the night to daytime temperature differential, day length, and incident radiation (Bodlaender, 1963). As discussed below in the long-season cultivar case study, agronomic inputs also play a major role in tuber bulking and are often manipulated to maximize economic and/or biological yield. Soon after tuber set, source leaves begin partitioning an increasing amount of dry matter to tubers. The rapidly growing tubers soon become the dominant sink and largely drive the partitioning activity of the foliage through their increasing demand for carbon, water, and nutrients (Engels and Marschner, 1986). The GA to ABA ratio in the plant also regulates carbon partitioning between the source and sink. Excessive GA due to high temperatures or long photoperiods can slow tuber growth as a result of a reduction in the activity of ADP-glucose (ADPG)-pyrophosphorylase in the tuber (Dwelle, 1985). Alternatively, low GA levels favor tuber bulking.

Tuber growth is a function of both cell division and expansion; however, cell division plays a larger role in determining final tuber size. According to Plaisted (1957), 200-g Cobbler tubers had approximately 500-fold more cells than their 37-g counterparts, but only tenfold more cell volume. Over the life of the growing tubers, from tuberization to plant maturity, tuber growth is typically sigmoidal; growth is initially slow as tubers begin to swell, but is rapid and linear soon after tuberization is complete. As an example, tuber growth of Alpine Russet is linear between 70 and 120 days after planting (DAP), when grown in the Columbia Basin of Washington, USA. During this period, tuber weight is increasing an average of 1.25 t ha^{-1} day^{-1} (Fig. 5.4). As the canopy begins to senesce, however, tuber growth slows and eventually ceases prior to complete foliar senescence.

Final tuber size is determined largely by genotype, inter- and intra-row plant spacing (plant population), management inputs, environment, canopy size, and duration. Commercial producers are often able to increase average tuber size by increasing the inter- and intra-row distance between plants (e.g. reducing plant competition), and vice versa (Pavek and Thornton, 2006). However, decreases in plant population can lead to yield reduction, even though average tuber size may increase. Because growers are often paid premiums for certain tuber sizes, yield reductions may be acceptable if associated with an increase in average tuber size. The commercial production goal is to produce the largest yield of the most valuable tubers, regardless of total yield per hectare. To accomplish this, growers must first identify what is the most valuable tuber size profile for the intended market. Once this is defined, the grower must find the right combination of plant population and spatial arrangement to produce the ideal tuber size profile. Final tuber size may also be altered through a deliberate change in stem number per plant.

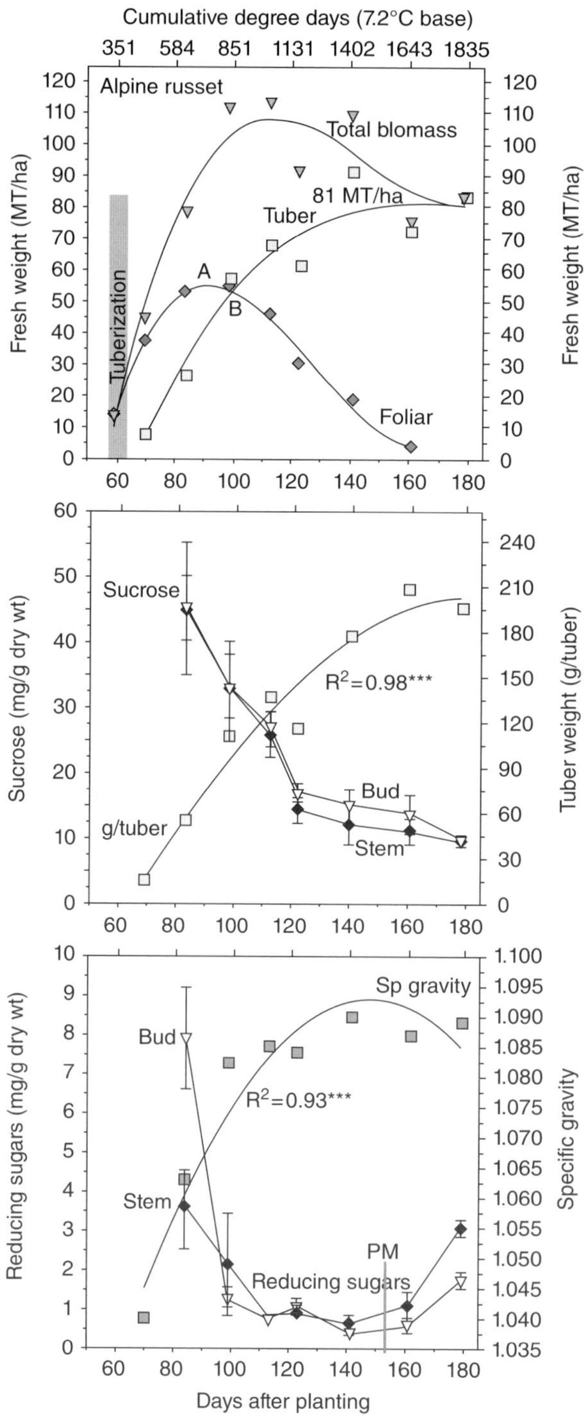

Fig. 5.4. Foliar and tuber growth responses of Alpine Russet at Othello, Washington, USA, in 2012 (planted 9 April). Cumulative degree days at the corresponding days after planting (DAP) are shown (top row). Harvest index (HI) equals tuber fresh weight as percent of total plant (tubers + foliage) fresh weight at maximum foliar growth. At maximum foliar development (92 DAP, point A, top graph), the HI was 44%. The DAP to 50% HI (where foliar and tuber growth curves cross, point B) is also indicated. Changes in tuber sucrose concentrations, (bud and stem halves of tubers) average tuber weights (middle row), reducing sugars (glucose and fructose), and specific gravity (bottom row) are also shown as components of physiological maturity (PM). PM was estimated at 153 DAP.

An increase in stem number per plant is often followed by an increase in tuber number per plant, which ultimately leads to a decrease in average tuber size, and vice versa (Knowles and Knowles, 2006). Stem number per plant can be changed by manipulating the physiological age of seed potatoes (Knowles and Knowles, 2006), and through the use of plant growth regulators

(Mikitzel and Knowles, 1990; Blauer *et al.*, 2013a). For each variety, growing region, and market, the proper combination of cultural management, plant population, and spatial positioning is needed to produce the highest-valued tuber size profile. Because tuber bulking is dependent on canopy health and duration, preventing foliage damage from pests is essential.

5.9 Stages of Crop Growth and Development: A Long-Season Cultivar Case Study

The principal stages of crop growth and development include: (i) sprouting; (ii) plant emergence and establishment; (iii) vine growth and foliar canopy development; (iv) flowering and tuberization; (v) tuber bulking; (vi) foliar senescence; and (vii) tuber maturation. The timing and overlap among these stages can vary considerably and depend on cultivar, management, climatic, and edaphic variables, and stresses experienced by the crop throughout the season. Managing a crop to coordinate and synchronize the onset and duration of these stages within the environmental constraints of a particular growing area is the key to optimizing yield and quality for maximum economic return.

Potato inherently has an indeterminate growth habit, and thus vegetative growth continues during flowering and throughout the season. However, cultivars vary in their expression of indeterminism, which affects how the crop uses and partitions water and nutrients to foliage and tubers. Tuber yield and quality can be affected negatively if the seven stages of development are not synchronized correctly for a particular cultivar and area of production.

The relative and optimal timing of each developmental stage depends on many factors, including cultivar, length of the growing season, management inputs, growing environment (e.g. light intensity, photoperiod, day/night temperature), and market expectations for yield and tuber size distribution. Achieving optimum tuber yield and quality involves managing source–sink (foliar/tuber growth) relationships at critical periods during the growing season. In the highly productive areas of the Pacific Northwest USA (e.g. Columbia Basin of Washington and Oregon), crop growth profiling studies have revealed key milestones and measurable indices of development for late-season cultivars that relate to yield potential and quality (Fig. 5.4) (Knowles *et al.*, 2013). Regardless of production area, these milestones can be used to gauge the effects of agronomic inputs on season-long crop development to facilitate formulating best management practices for a particular cultivar and end use.

5.10 Milestones of Crop Development

- Harvest index (HI) at maximum foliar growth.
- DAP and/or cumulative degree days to 50% HI and yield at 50% HI (foliar biomass = tuber biomass at 50% HI).
- DAP and/or cumulative degree days to:
  - – maximum foliar fresh weight;
  - – maximum tuber yield;
  - – maximum tuber specific gravity;
  - – minimum tuber sucrose concentration;
  - – minimum tuber reducing sugar concentration; and
  - – tuber physiological maturity.

  } These milestones define tuber physiological maturity (PM) [applies to the last four items]
- Maximum foliar biomass.
- Maximum tuber yield.
- Specific gravity at harvest.

Source–sink relationships are indicated by the HI. The HI is dynamic, increasing rapidly at the beginning of the season during early to mid-bulking, and coincident with the period of rapid canopy (foliar) and tuber development. At maximum foliar development (~92 DAP in Fig. 5.4), the HI should fall within the range of 40–50%, favoring foliar growth over tuber growth, for most (but not all) indeterminate late-maturing cultivars in long-season growing areas like the Columbia Basin. The HI at point (A) in Fig. 5.4 is 44% (e.g. tubers account for 44% of plant fresh weight at this point in the growing season). This index can inform management; for maximum yield potential, early-season fertility and irrigation should be managed to promote rapid foliar development during plant establishment, tuberization, and early bulking to produce sufficient leaf area to support continued bulking over the prolonged growing season (150+ days). If foliar growth is restricted early in the season (e.g. by lack of nutrients, water, or stress), HI at maximum foliar development will often exceed 50%, favoring tuber growth over foliar growth, which, for many cultivars, will limit the total amount and duration of subsequent foliar growth. The resulting source–sink imbalance can result in early vine senescence, which in turn can compromise yield potential in a long-season area. Conversely, too much nitrogen early in the season can promote excessive foliar growth and delay tuberization (e.g. HI <40% at maximum foliar development), limiting the available window for bulking and resulting in reduced yield over the available growing season.

By definition, tuber growth equals foliar growth when the HI is 50%. This index is thus marked by the intersection of tuber and foliar yield curves (point (B) in Fig. 5.4). If 50% HI occurs prior to maximum foliar growth, the HI at maximum foliar growth will exceed 50% (e.g. tubers will account for more than 50% of plant biomass). Hence, the timing of 50% HI relative to maximum foliar growth can also indicate a pending source–sink imbalance that will potentially compromise maximum yield, due to limiting foliar growth and premature vine senescence later in the season. The example given is for a late-season, indeterminate cultivar (e.g. Alpine Russet, Ranger Russet, and Alturas) grown in a long-season area. The relationships among these indices and specific index recommendations to maximize yield and quality will differ for production areas with shorter seasons, production of late-season cultivars for early markets, short-season cultivars, and cultivars with a more determinate growth habit. The goal is to build a robust canopy as early as possible and to maintain it for as long as possible, within reason. The potato plant needs time to mature naturally for optimum yield, storability, and economic return. Plants with ideal foliage growth early to mid season achieve spectacular yields if allowed enough time to mature naturally (Fig. 5.4). This means cutting off and depleting the soil of nitrogen after vines have peaked, about late July or early August in the Columbia Basin, to stimulate foliar senescence and hasten crop maturation.

In addition to yield, optimizing source–sink relationships to the growing season by synchronizing critical growth stages properly is important for achieving tuber physiological maturity (PM) at season end, which in turn affects tuber quality at harvest and the retention of tuber quality during storage. PM coincides with maximum dry matter (specific gravity) and minimum concentrations of sucrose and reducing sugars in tubers. It equates to the average DAP to achieve maximum tuber yield, maximum specific gravity, and minimum concentrations of sucrose and reducing sugars in tubers (e.g. PM = 153 DAP in Fig. 5.4). Also, PM informs end-of-season management and harvest timing. Healthy tubers harvested within 7–10 days of PM retain processing quality the longest in storage (Knowles *et al.*, 2009, 2011). Delaying harvest well beyond PM exposes tubers to fluctuations in soil temperature, which can accelerate tuber aging and ultimately compromise retention of processing quality during storage, but these effects are cultivar dependent. For many long-season cultivars, reducing sugars (glucose and fructose) often increase following PM and prior to harvest, particularly in the stem ends of tubers (Fig. 5.4). This response reflects overmaturation and signals the beginning of a progressive loss in quality. The result can be sugar ends at harvest, increased propensity to develop sugar ends during storage, earlier than normal increase in reducing sugars throughout the entire tuber during storage, and/or early sprouting. Delaying harvest significantly beyond PM can thus affect retention of postharvest quality for many frozen processing or chipping cultivars. In practice, the approximate timing of PM should be part of the management recommendations for new cultivars. It is best specified by three criteria: DAP to PM, cumulative degree days (7.2°C base) from planting to PM, and the degree of vine senescence at PM. For example, PM for Alpine Russet in the central Columbia Basin is achieved at approximately 150–160 DAP, 1523–1643 cumulative degree days (°C), and 70–80% vine senescence.

Agronomic inputs such as nitrogen rate and scheduling can effectively shift the relative timing and duration of growth stages and the various milestones of crop development to affect final yield and quality. For example, as the nitrogen rate increases, foliar growth, duration, and tuber bulking are prolonged, and the attainment of tuber PM is delayed. In the highly productive long-season growing areas of the Pacific Northwest, the result is usually high yields and improved quality and storability, as harvest most often occurs closer to PM. Conversely, a crop grown with a relatively low rate of seasonal nitrogen will generally set tubers early and produce less foliar growth; also, foliar senescence, tuber maturation, and PM will occur before the season has ended. Yield will be lower and the ability to retain processing quality during storage can be compromised if the maturation period under dead vines is prolonged and harvest delayed until late in the season (beyond PM) (Fig. 5.5). Nitrogen rate and duration of application can thus affect the retention of processing quality significantly by influencing the timing of PM in relation to harvest.

5.11 Plant and Tuber Maturation

As potato plants mature, chlorophyll levels in the leaves decline and the rate of photosynthesis decreases. Carbohydrates and phloem-mobile metabolites translocate from senescent foliage to the tubers, and tuber dry matter content reaches a maximum. The senescing leaves turn yellow and abscise from stems, which also eventually senesce. At this time, the tuber periderm (skin) thickens and becomes stronger. Potato tuber maturity is often discussed in terms of physiological and physical maturity, but these two aspects of maturity do not always happen simultaneously (Sabba *et al.*, 2007).

As discussed earlier, tubers reach PM when tuber yield and dry matter accumulation are at maximum levels and tuber sucrose and reducing sugar levels are at a minimum. Although not always possible, it is ideal to harvest healthy tubers as near to PM as possible, to achieve high yields and to maximize processing quality at harvest and during long-term storage. Dry matter content is important to processors, because product recovery and oil absorption are directly affected. Because tubers with high dry matter contain less water, they require less energy during frying or dehydration and absorb less oil during frying.

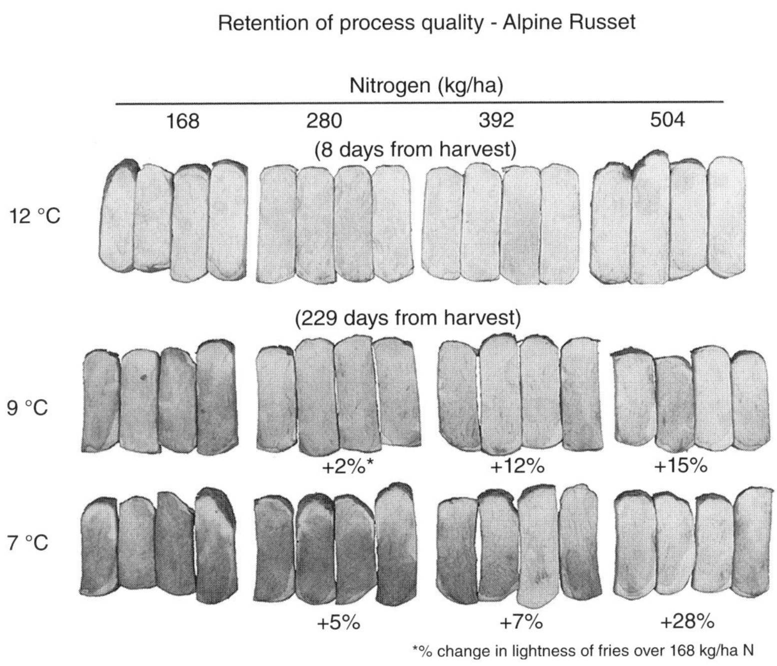

Fig. 5.5. Effects of nitrogen rate and storage temperature on retention of process quality of Alpine Russet tubers following wound healing (8 days at 12°C) and 221 days of storage at 9 and 7°C (229 days total). The percent values indicate the degree to which tubers fried lighter than those grown with 168 kg ha^{-1} nitrogen (N). Physiological maturity (PM) (see Fig. 5.4) was delayed as N rate increased, resulting in the high N crop being harvested closer to PM than the low N crop. Each fry plank is from a different tuber selected to represent the average fry color in a 12-tuber sample.

Tubers harvested when immature or overmature can contain high levels of reducing sugars (Driskill *et al.*, 2007; Sabba *et al.*, 2007), which can increase in storage and result in processed products that are excessively dark and bitter tasting (Kumar *et al.*, 2004).

Physical maturity, which is the strengthening of the periderm during tuber maturation, is commonly referred to as "skin set". The periderm is composed of three cell types: phellem (or cork), phellogen (or cork cambium), and phelloderm. The phellem, which is derived from the phellogen cells underneath, forms a protective layer on the outside of the tuber when its cells become suberized and then die. Suberization involves deposition of a waxy, fatty material, called suberin, within the cell wall (Kolattukudy, 1984). This suberized layer of cells plays an important role in resistance to infection by bacterial and fungal pathogens (Lulai and Corsini, 1998) and resistance to water vapor loss (Kolattukudy, 1984). The phelloderm, the innermost part of the periderm, is also derived from the phellogen and is there to provide energy and biochemical building materials for periderm growth. The phellogen, in the middle, is a single layer of meristematic cells, with cell walls that are very thin and elastic during tuber bulking to allow for rapid cell division and expansion as tubers increase in size. Immature phellogen fractures easily, which causes the phellem to slough off when immature tubers are abraded during harvest and handling (Lulai and Freeman, 2001). When tuber growth stops, the phellogen becomes meristematically inactive and its cell walls thicken and become rigid (Lulai and Freeman, 2001). The width of radial cell walls in the phellogen can more than double in size between the mid-bulking period and skin set (Sabba and Bussan, 2012). This makes the tuber highly resistant to skinning injury. In fact, the progress of periderm maturation can be monitored easily by digging a sample of tubers and then assessing how easily the skin "slips" when pressure is applied with a sideways motion of the thumb. Skin set is accomplished when the skin no longer slips with applied pressure.

Skin set is influenced greatly by cultivar and vine condition, and is also affected by soil moisture and temperature. The rate of skin set varies significantly among cultivars. This is due mainly to genetic differences, but can also depend on how cultivars respond to conditions in the field (Lulai and Orr, 1993). It is generally recognized that smooth-skinned cultivars set skin more slowly than russet-skinned varieties, such as Russet Burbank. Soil moisture and soil temperature affect the rate of skin set, but cultural practices and growing conditions often modify their effects in the field (Lulai and Orr, 1993; Sabba and Bussan, 2012). In general, a moderately dry soil, with 70–75% available soil moisture, is recommended to promote skin set. Soil temperatures in the range of 21–24°C are ideal for skin set, but temperatures below 7°C or above 32°C will delay skin maturation. Therefore, tubers that encounter cool, wet weather at the end of the season usually take longer to mature properly. Periderm maturation usually begins in the field, but can and does continue after harvest, depending on temperature and the humidity conditions in storage (Lulai and Orr, 1993).

The process of skin set is initiated after tuber growth ceases, usually at the time the vines senesce in the field. Therefore, tubers harvested when the vines are still green and active are highly susceptible to skinning injury. In some regions, immature tubers are harvested while the vines are still green, but these potatoes are usually marketed or processed immediately. Depending on the cultivar and soil conditions, tubers require 10–21 days after vines are killed, removed, or have senesced completely for the skin to set sufficiently to resist skinning damage at harvest. It should also be noted, however, that delaying harvest for many weeks can increase problems with soilborne diseases, like silver scurf and Rhizoctonia black scurf, and can result in tubers that are physiologically overmature.

A fertility management plan that allows soil nitrogen levels to deplete towards the end of the growing season encourages natural vine senescence and time for proper skin set before harvest. Short periods of freezing or near-freezing temperatures that kill vines (killing frost) a few weeks before harvest can also promote skin set. However, it is not always practical to allow vines to senesce naturally before harvest; therefore, many potato producers remove or kill vines via mechanical defoliation or chemical desiccation. Mechanical defoliation methods include flaming, pulling, undercutting, rolling, flailing, and beating vines. Chemical vine desiccation products include sulfuric acid and non-selective herbicides such as diquat dibromide, paraquat dichlor, or endothall. A combination of mechanical and chemical methods may be used to kill vines. Each method has pros and cons that should be considered carefully before use. Methods that kill the vines slowly are preferred, as rapid vine kill may cause discoloration of the tuber vascular ring on occasion, called stem-end browning (Hiller *et al.*, 1985).

Skin set typically improves if vines are killed after natural vine senescence has begun (Wiltshire *et al.*, 2005). Because of this, it is often recommended that vine kill operations be delayed until at least half the leaves have started to turn yellow. In addition to killing vines to promote skin set, minimize postharvest injury, bruising, weight loss, and disease susceptibility, vines may be terminated to control tuber size, avoid insect or disease infestations, and facilitate vine removal during harvesting.

5.12 Summary

From plant propagation to plant and tuber maturity, all potato varieties share commonalities; however, the influences of growing region, cultural practices, and environment cannot be overstated. Differences in season length, radiation intensity and duration, soil type and quality, temperature, wind, irrigation, rainfall, seed health and physiology, nutrient availability, intra- and inter-row plant spacing, plant population, cultivation, pest load and management influence potato growth. Because of this, it is absolutely essential that cultural management recommendations be developed locally through research conducted on cultivars common to the local markets. Cultural management standards set forth by private and public researchers, on-farm experimentation, experience, education, and intuition enable growers to identify and implement practices designed to keep their operation profitable and competitive. To produce profitable, high-quality yields with the ideal tuber size profile, shape, storability, and palatability for each cultivar of interest, local researchers must identify the ideal amounts of soil moisture and fertility, plus the proper planting date and depth, plant population, intra- and inter-row plant spacing, and integrated pest management for each locally grown variety.

References

Allen, E.J. and Scott, R.K. (2001) *The Agronomy of Effective Potato Production*. British Potato Council, Oxford, UK.

Allen, E.J., O'Brien, P.J. and Firman, D. (1992) Seed tuber production and management. In: Harris, P.M. (ed.) *The Potato Crop: The Scientific Basis for Improvement*. Chapman and Hall, London, pp. 247–291.

Blauer, J.M., Knowles, L.O. and Knowles, N.R. (2013a) Manipulating stem number, tuber set and size distribution in specialty potato cultivars. *American Journal of Potato Research* 90, 470–496.

Blauer, J.M., Knowles, L.O. and Knowles, N.R. (2013b) Evidence that tuber respiration is the pacemaker of physiological aging in seed potatoes (*Solanum tuberosum* L.). *Journal of Plant Growth Regulation* 32, 708–720.

Blauer, J.M., Kumar, G.N.M., Knowles, L.O., Dhingra, A. and Knowles, N.R. (2013c) Changes in ascorbate and associated gene expression during development and storage of potato tubers (*Solanum tuberosum* L.). *Postharvest Biology and Technology* 78, 76–91.

Bodlaender, K.B.A. (1963) Influence of temperature, radiation and photoperiod on development and yield. In: Ivins, J.D. and Milthorpe, F.L. (eds) *The Growth of The Potato*. Butterworths, London, pp. 199–210.

Brandt, T.L., Kleinkopf, G., Olson, N. and Love, S.L. (2003) *Storage Management for Umatilla Russet Potatoes.* BUL 839. University of Idaho Extension, Moscow, Idaho.

Brandt, T.L., Kleinkopf, G., Olson, N. and Love, S.L. (2004) *Storage Management for Summit Russet Potatoes.* CIS 1123. University of Idaho Extension, Moscow, Idaho.

Brandt, T.L., Olson, N., Kleinkopf, G. and Novy, R. (2006) *Storage Management of Alturas Potatoes.* CIS 1134. University of Idaho Extension, Moscow, Idaho.

Burton, W.G. (1989) *The Potato*, 3rd edn. John Wiley and Sons, New York.

Burton, W.G., van Es, A. and Hartmans, K.J. (1992) The physics and physiology of storage. In: Harris, P.M. (ed.) *The Potato Crop: The Scientific Basis for Improvement*. Chapman and Hall, London, pp. 608–727.

Carling, D.E. and Leiner, R.H. (1990) Effect of temperature on virulence of *Rhizoctonia solani* and other *Rhizoctonia* on potato. *Phytopathology* 80, 930–934.
Classens, M.M.J. and Vreugdenhil, D. (2000) Is dormancy breaking of potato tubers the reverse of tuber initiation? *Potato Research* 43, 347–369.
Coleman, W.K. and King, R.R. (1984) Changes in endogenous abscisic acid, soluble sugars and proline levels during tuber dormancy in *Solanum tuberosum* L. *American Journal of Potato Research* 61, 437–449.
Cother, E.J. and Cullis, B.R. (1985) Tuber size distribution in cv. Sebago and quantitative effects of *Rhizoctonia solani* on yield. *Potato Research* 28, 1–14.
Cutter, E.G. (1992) Structure and development of the potato plant. In: Harris, P.M. (ed.) *The Potato Crop*. Chapman and Hall, London, pp. 65–161.
Cvikrova, M., Sukhova, L.S., Eder, J. and Korableva, N.P. (1994) Possible involvement of abscisic acid, ethylene and phenolic acids in potato tuber dormancy. *Plant Physiology and Biochemistry* 32, 685–691.
Donald, C.M. (1962) In search of yield. *Journal of the Australian Institute of Agricultural Science* 28, 171–178.
Driskill, E.P., Knowles, L.O. and Knowles, N.R. (2007) Temperature-induced changes in potato processing quality during storage are modulated by tuber maturity. *American Journal of Potato Research* 84, 367–383.
Dwelle, R.B. (1985) Photosynthesis and assimilate partitioning. In: Li, P.H. (ed.) *Potato Physiology*. Academic Press, Orlando, Florida, pp. 35–58.
Engels, C. and Marschner, H. (1986) Allocation of photosynthate to individual tubers of *Solanum tuberosum* L. I. Relationship between tuber growth rate and enzyme activities of the starch metabolism. *Journal of Experimental Botany* 37, 1795–1803.
Firman, D.M. and Allen, E.J. (1988) Field measurements of the photosynthetic rate of potatoes grown under different amounts of nitrogen fertilizer. *Journal of Agricultural Science* 11, 151–163.
Firman, D.M. and Allen, E.J. (1989) Relationship between light interception, ground cover and leaf area index in potatoes. *Journal of Agricultural Science* 113, 355–359.
Firman, D.M., O'Brien, P.J. and Allen, E.J. (1992) Predicting the emergence of potato sprouts. *Journal of Agricultural Science* 118, 55–61.
Golmirzaie, A.M., Malagamba, P. and Pallais, N. (1994) Breeding potatoes based on true seed propagation. In: Bradshaw, J.E. and Mackay, G.E. (eds) *Potato Genetics*. CAB International, Wallingford, UK, pp. 499–513.
Hemberg, T. (1985) Potato rest. In: Li, P.H. (ed.) *Potato Physiology*. Academic Press, Orlando, Florida, pp. 353–388.
Hide, G.A. and Lapwood, D.H. (1982) Disease aspects of potato production. In: Harris, P.M. (ed.) *The Potato Crop: The Scientific Basis for Improvement*, 2nd edn. Chapman and Hall, London, pp. 407–415.
Hiller, L.K., Koller, D.C. and Thornton, R.E. (1985) Physiological disorders of potato tubers. In: Li, P.H. (ed.) *Potato Physiology*. Academic Press, Orlando, Florida, pp. 389–443.
Iritani, W.M., Thornton, R., Weller, L. and O'Leary, G.O. (1972) Relationship of seed size, spacing, stem numbers, to yield of Russet Burbank potatoes. *American Potato Journal* 49, 463–469.
Jones, J.L. and Allen, E.J. (1983) Effects of date of planting on plant emergence, leaf growth, and yield in contrasting potato cultivars. *Journal of Agricultural Science* 101, 81–95.
Karssen, C.M. and Van Loon, L.C. (1992) Probing hormone action in developing seeds by ABA-deficient and -insensitive mutants. In: Karssen, C.M., Van Loon, L.C. and Vreugdenhil, D. (eds) *Progress in Plant Growth Regulation*. Kluwer Academic Press, Dordrecht, the Netherlands, pp. 43–53.
Knowles, N.R. and Knowles, L.O. (2006) Manipulating stem number, tuber set, and yield relationships for northern- and southern-grown potato seed lots. *Crop Science* 46, 284–296.
Knowles, N.R., Iritani, W.M. and Weller, L.D. (1985) Plant growth response from aged potato seed-tubers as affected by meristem selection and NAA. *American Potato Journal* 62, 289–300.
Knowles, N.R., Driskill, E.P. Jr and Knowles, L.O. (2009) Sweetening responses of potato tubers of different maturity to conventional and non-conventional storage temperature regimes. *Postharvest Biology and Technology* 52, 49–61.
Knowles, N.R., Pavek, M.J., Hiles, C., Knowles, L.O. and Holden, Z.J. (2011) Nitrogen management affects tuber physiological maturity and retention of processing quality [Abstract]. The 94th Annual Meeting of the Potato Association of America. *American Journal of Potato Research* 88, 49.
Knowles, N.R., Knowles, L.O. and Pavek, M.J. (2013) In-season management and postharvest quality – lessons learned from Alpine, Sage Russet and other cultivars. In: *Proceedings of the Washington-Oregon Potato Conference*, 29–31 January, Kennewick, Washington. Washington State Potato Commission, Moses Lake, Washington, pp. 34–47.

Kolattukudy, P.E. (1984) Biochemistry and function of cutin and suberin. *Canadian Journal of Botany* 62, 2918–2933.

Krauss, A. (1985) Interaction of nutrients and tuberization. In: Li, P.H. (ed.) *Potato Physiology*. Academic Press, Orlando, Florida, pp. 209–230.

Krijthe, N.K. (1962) Observations on the sprouting of seed. *European Potato Journal* 5, 316–332.

Kumar, D., Singh, B. and Kumar, P. (2004) An overview of the factors affecting sugar content of potatoes. *Annals of Applied Biology* 145, 247–256.

Kumar, G.N.M. and Knowles, N.R. (1996a) Nature of enhanced respiration during sprouting of aged potato seed tubers. *Physiologia Plantarum* 97, 228–236.

Kumar, G.N.M. and Knowles, N.R. (1996b) Oxidative stress results in increased sinks for metabolic energy during aging and sprouting of potato seed-tubers. *Plant Physiology* 112, 1301–1313.

Lai Cho, J., Iritani, W.M. and Martin, M.W. (1983) Comparison of methods for measuring dormancy of potatoes. *American Potato Journal* 60, 169–177.

Lulai, E.C. and Corsini, D.L. (1998) Differential depositions of suberin phenolic and aliphatic domains and their roles in resistance to infection during potato tuber (*Solanum tuberosum* L.) periderm. *Physiological and Molecular Plant Pathology* 53, 209–222.

Lulai, E.C. and Freeman, T.P. (2001) The importance of phellogen cells and their structural characteristics in susceptibility and resistance to excoriation in immature and mature potato tuber (*Solanum tuberosum* L.) periderm. *Annals of Botany* 88, 555–561.

Lulai, E.C. and Orr, P.H. (1993) Determining the feasibility of measuring genotypic differences in skin-set. *American Potato Journal* 70, 599–609.

MacKerron, D.K.L. and Jefferies, R.A. (1986) The influence of early soil moisture on tuber numbers in potatoes. *Potato Research* 29, 299–312.

Michener, H.D. (1942) Dormancy and apical dominance in potato tubers. *American Journal of Botany* 29, 558–568.

Mikitzel, L.J. and Knowles, N.R. (1990) Effect of potato seed-tuber age on plant establishment and amelioration of age-linked effects with auxin. *Plant Physiology* 93(3), 967–975.

O'Brien, P.J., Allen, E.J. and Firman, D.M. (1998) A review of some studies into tuber initiation in potato (*Solanum tuberosum*) crops. *Journal of Agricultural Science* 130, 251–270.

Pavek, M.J. and Thornton, R.E. (2006) Agronomic and economic impact of missing and irregularly spaced potato plants. *American Journal of Potato Research* 83, 55–66.

Pavek, M.J. and Thornton, R.E. (2009) Planting depth influences potato plant morphology and economic value. *American Journal of Potato Research* 86, 56–67.

Plaisted, P.H. (1957) Growth of the potato tuber. *Plant Physiology* 32(5), 445–453.

Poapst, P.A., Durkee, A.B., McGugan, W.A. and Johnston, F.B. (1968) Identification of ethylene in gibberellic-acid-treated potatoes. *Journal of the Science of Food and Agriculture* 19, 325–327.

Rodríguez-Falcón, M., Bou, J. and Prat, S. (2006) Seasonal control of tuberization in potato: conserved elements with the flowering response. *Annual Review of Plant Biology* 57, 151–180.

Rosen, C.J. and Bierman, P.M. (2012) Potato yield and tuber set as affected by phosphorus fertilization. *American Journal of Potato Research* 85, 110–120.

Sabba, R.P. and Bussan, A.J. (2012) Comparison of skin-set and periderm maturation in 'Red Norland' potatoes grown in two soil types in Wisconsin. *American Journal of Potato Research* 89, 508–511.

Sabba, R.P., Bussan, A.J., Michaelis, B.A., Hughes, A., Drilias, M.J. and Glynn, M.T. (2007) Effect of planting and vine-kill timing on sugars, specific gravity and skin set in processing potato cultivars. *American Journal of Potato Research* 84, 205–215.

Sale, P.J.M. (1976) Effect of shading at different times on the growth and yield of the potato. *Australian Journal of Agricultural Research* 27, 557–566.

Sale, P.J.M. (1979) Growth of potatoes (*Solanum tuberosum* L.) to the small tuber stage as related to soil temperature. *Australian Journal of Agricultural Research* 30, 667–675.

Schotzko, R.T., Hyde, G.M. and Thornton, R.E. (1983) The dollars and cents of the 1982 potato seed size and spacing survey. *Proceedings of the Washington State Potato Conference* 22, 23–29.

Struik, P.C. and Van Voorst, G. (1986) Effects of drought on the initiation, yield, and size distribution of *Solanum tuberosum* L. *American Potato Journal* 29, 487–500.

Suttle, J.C. (1995) Postharvest changes in endogenous ABA levels and ABA metabolism in relation to dormancy in potato tubers. *Physiologia Plantarum* 95, 233–240.

Suttle, J.C. (1998) Involvement of ethylene in potato microtuber dormancy. *Plant Physiology* 118, 843–848.

Suttle, J.C. (2002) Dormancy-related changes in cytokinin efficacy and metabolism in potato tubers during postharvest storage. *Plant Growth Regulation* 35, 199–206.

Suttle, J.C. (2003) Auxin-induced sprout growth inhibition: role of endogenous ethylene. *American Journal of Potato Research* 80, 303–309.

Suttle, J.C. (2004a) Involvement of endogenous gibberellins in potato tuber dormancy and early sprout growth: a critical evaluation. *Journal of Plant Physiology* 161, 157–164.

Suttle, J.C. (2004b) Physiological regulation of potato tuber dormancy. *American Journal of Potato Research* 81, 253–263.

Suttle, J.C. and Hultstrand, J.F. (1994) Role of endogenous abscisic acid in potato microtuber dormancy. *Plant Physiology* 105, 891–896.

Taylor, A.G. and Ten Broeck, C.W. (1988) Seedling emergence forces of vegetable crops. *HortScience* 23(2), 367–369.

Tuan, D.Y. and Bonner, J. (1964) The control of seed dormancy. *Plant Physiology* 39(5), 768–772.

Turnbull, C.G.N. and Hanke, D.E. (1985) The control of bud dormancy in potato tubers. Evidence for the primary role of cytokinins and a seasonal pattern of changing sensitivity to cytokinin. *Planta* 165, 359–365.

Van Emden, J.H. (1965) *Rhizoctonia solani*: results of experiments. *European Potato Journal* 8, 188–189.

Viola, R., Roberts, A.G., Haupt, S., Grazzani, S., Hancock, R.D., Marmiroli, N., *et al.* (2001) Tuberization in potato involves a switch from apoplastic to symplastic phloem unloading. *Plant Cell* 13, 385–398.

Walworth, J.L. and Carling, D.E. (2002) Tuber initiation and development in irrigated and non-irrigated potatoes. *American Journal of Potato Research* 79, 387–395.

Wiltshire, J., Milne, F. and Peters, J. (2005) Improving the understanding and management of skin set and bloom in potatoes. British Potato Council Project Report 807/222. British Potato Council, Oxford, UK.

Xu, X., Vreugdenhil, D. and van Lammeren, A.A.M. (1998) Cell division and cell enlargement during potato tuber formation. *Journal of Experimental Botany* 49 (320), 573–582.

Young, N.F., Ferguson, B.J., Antoniadi, I., Bennett, M.H., Beveridge, C.A., and Turnbull, C.G.N. (2014) Conditional auxin response and differential cytokinin profiles in shoot branching mutants. *Plant Physiology* 165, 1723–1736.

Zabrouskov, V., Kumar, G.N.M., Spychalla, J.P. and Knowles, N.R. (2002) Oxidative metabolism and the physiological age of seed potatoes are affected by increased α-linolenate content. *Physiologia Plantarum* 116, 172–185.

6 Commercial Potato Production and Cultural Management

Mark J. Pavek*
Department of Horticulture, Washington State University, Pullman, USA

Potatoes are grown throughout the world; however, most commercial production is concentrated in geographic regions that have more than 100 consecutive frost-free days, well-drained, fertile soils, approximately 50–60 cm of rainfall or irrigation between planting and harvest, a broad day/night temperature differential, absence of extreme temperatures during the growing season, and close proximity to markets, temperature-controlled storages and mass transportation. Ninety-three percent of the world's potatoes are grown in the northern hemisphere (Hijmans, 2001). The two major global potato production zones are temperate climates with a latitude between 45°N and 57°N, where potatoes are grown as a summer crop, such as Western and Eastern Europe, northern China, northern USA, and southern Canada, and subtropic lowlands with a latitude between 23°N and 34°N, where potatoes are grown as a winter crop, such as the Ganges plain, southern China, southern USA, northern Mexico, and Egypt (Hijmans, 2001).

The goal of most commercial producers is to maximize profits and maintain or expand the farm's productivity long term. As with most businesses, the commercial production manager must coordinate all the activities necessary to produce a quality product. This means the manager must have a thorough understanding of the intended market(s) and buyer expectations. Because potatoes are typically bought and sold by size (Table 6.1), commercial producers often manipulate in-field production such that the yield of the most valuable tuber sizes is maximized. The tuber size profile can be reduced or expanded by altering inter- and intra-row seed spacing, controlling days of growth by planting late or killing vines early, regulating inputs like fertilizer and water, applying growth regulators, and manipulating seed tuber physiological age.

In developed countries, cultural management recommendations for various potato varieties are often available through publicly or privately funded research organizations such as state and federal government agencies, potato commissions, consultants, and international cooperatives. Many aspects must be considered when producing potatoes for profit. This chapter focuses on the principles and components key to successful commercial potato production. Readers should also consider information found in other chapters of this volume, as many contain detailed accounts of topics such as whole-plant growth and development, integrated pest and weed management, postharvest physiology, variety development, seed production and certification, etc. For variety and region-specific cultural management recommendations, growers should seek advice locally from public or private agronomists.

* E-mail: mjpavek@wsu.edu

Table 6.1. US fresh market size guide for Russet-type potatoes and 4-year average prices per size category.

Carton size (22.68 kg cartons)	Actual number of tubers per carton	US Number 1 grade tuber size range[a] per carton (g)	Four-year average prices (US$/45 kg)[b]
		<170	8.60
110-Count	105–116	170–227	13.54
100-Count	95–105	199–255	13.54
90-Count	86–95	227–312	14.71
80-Count	76–84	255–340	17.02
70-Count	67–74	284–369	17.62
60-Count	57–63	312–397	17.61
50-Count	48–53	284–482	17.06
		>482	8.60

Notes: [a]Typical size ranges, US Department of Agriculture allows wider ranges than shown. Assumes tubers <170 g and >482 g are sold in bulk to processors; [b]sales FOB shipping point, Columbia Basin, Washington, USA, USDA Federal-State Market News Service 2007–2010 (http://www.ams.usda.gov/). Tubers <170 g and >482 g priced as process market culls using regional process-cull market values.

6.1 Market and Variety Choice

Two major marketing avenues exist for potatoes intended for human consumption: fresh or processing. Fresh market potatoes are those typically purchased as a fresh, raw commodity by consumers or restaurants. Processing converts potatoes into items such as frozen fries (chips in the UK), chips (crisps in the UK), vodka, schnapps, starch, flour, dehydrated flakes and pearls, potato salad, etc. Prior to the growing season, potato producers must decide if they want to sell their potatoes to fresh or process market buyers. They must also decide whether or not to forward contract their potatoes or to sell them on the open market. Both marketing strategies have their advantages and disadvantages, but many buyers prefer to obtain potatoes through a forward contract. The forward contract removes significant uncertainty by specifying the quantity, quality, size range, and variety the producer must supply at a certain time to the buyer at a previously arranged price. On the other hand, the open market follows the laws of supply and demand. Prices paid to the grower are excellent when demand is high and supplies limited; however, the open market gamble may result in financial loss to the grower if supplies are abundant and demand low. Alternatively, the buyer benefits when prices fall due to excessive supply.

Within the fresh or processing sectors, growers can choose to market their potatoes as organically or conventionally grown. In most developed countries, potatoes marketed as organic must be certified as such by a governing body. Organic certification assures consumers the commodity has been grown without the use of synthetic inputs such as synthesized pesticides or fertilizers and all requirements set forth by the governing body (Johnson and Sideman, 2006) have been followed. Ultimately, the decision to grow potatoes organically is a business decision, but some producers choose organic over conventional production because they believe it is better for the environment (Delate and Hartzler, 2003) and/or healthier for consumption (Davies *et al.*, 1995).

Once the market choice is decided, the selection of varieties is typically limited to the demands of the marketplace. As an example, only a handful of potato varieties meet the stringent requirements set forth by McDonalds® for production of their fries. Moreover, approximately 80% of potato varieties used by Frito-Lay® to make chips in the USA are owned by Frito-Lay® (Kirkman, 2007). In addition to having unique post-frying attributes like mouth-feel, flavor, aftertaste, crispiness, interior color and texture, varieties intended for the fry or chip market must be low in defect level and sugar content, have easy-to-peel skin and storability, and be profitable for everyone in the marketing chain—from grower to franchise owner. In addition to marketability, seed must be available in sufficient quantities for planting, and the variety must fit the growth requirements of the production region, such as number of days to maturity, disease resistance, ability to withstand weather extremes, etc. Unless a new variety is requested by a buyer, growers should assess the potential market outlets thoroughly prior to planting a variety that is unfamiliar to buyers. On the other hand, if a new variety gets ample exposure and is promoted prior to hitting the market, early adopters may find new varieties a financial boon, especially when supplies of the new variety are limited. Similarly, marketers and growers can use exclusive-rights ownership to control the supply of privately owned varieties, with the hope of reaping financial rewards greater than those coming from public varieties (Guenthner, 2003).

6.2 Pre-Plant Preparation

Many decisions must be made prior to and during the planting process. Pre-planting considerations include field selection and preparation, whole-farm crop rotation, soil testing for nutrients and pests, pre-plant fertilizer and pesticide applications, time of planting, equipment and irrigation needs and capabilities, potato variety and market, postharvest storage, crop and personnel management, and financial considerations.

Field and soil considerations

Established commercial farms typically engage in a whole-farm crop rotation strategy where certain crops proceed or follow others in the same field across production years. The exception is when growers choose to lease fields randomly rather than farm within a set series of fields. Crop rotations vary by region and farms, but the underlying theme is to prevent build-up of insect pests, weeds, and diseases, and to maintain or improve the soil health by cycling particular crops in and out of a particular field (Magdoff and van Es, 2000). When growing conditions allow, green manure crops are often planted in between rotation crops to help build the soil organic matter, capture and hold leachable nutrients, prevent soil erosion, reduce weeds (by allowing germination but not seed development), and suppress nematode and disease activity with natural plant compounds like glucosinolates found in most *Brassica* species (McGuire, 2003).

Ideal potato fields have limited slope, well-drained, fertile, and non-compacted soil with limited disease and pest pressure, overhead irrigation and/or sufficient rainfall, and are free from damaging or quarantined pests, rocks, clods, and excessive debris from previous crops. Other ideal soil properties include low concentrations of salt and sodium and a soil pH of 6.0–7.5 (Magdoff and van Es, 2000). Medium- to coarse-textured soils are often favored over fine-textured soils because they drain more effectively, thereby preventing anaerobic conditions and uneven soil moisture issues. In addition, coarser soils are desirable because they wash easily from tubers, especially when compared to soils with high clay content. Examples of medium and coarse soils include silt loams, loams, sandy loams, and sand (McDole *et al.*, 1974).

Weeks, or months, prior to planting, fields with high levels of disease (verticillium wilt), weed seeds, wireworm, and nematodes are typically fumigated with a pesticide. Unless the grower knows the field history well, soil samples should be collected and tested to determine disease and pest load. Soil samples are also recommended for determining soil properties like pH, nutrient levels, and organic matter.

Prior to planting, and after soil samples have been analyzed, decisions regarding nutrient source and delivery to the plant must be made. Overhead irrigation systems allow the grower to apply fertilizer, water, and plant protection products (also referred to as fertigation or chemigation) simultaneously. This offers flexibility throughout the season to match nutrient applications with plant needs. Where overhead irrigation is not available, however, growers must apply fertilizer prior to planting, during planting, and/or during post-planting tillage operations. Moreover, once foliage approaches 100% ground cover, tillage operations typically cease for the growing season. Because early spring soils are often saturated with water, operations with overhead irrigation systems may choose to apply and incorporate some or all fertilizer with ground applicators and tillage equipment prior to 100% ground cover. This ensures that the plants receive an adequate supply of nutrients if early irrigation and fertigation are not possible. Once fertilizer application timing and amount is determined, the fields can be tilled accordingly. Specific nutrient information is provided later in the chapter.

6.3 Pre-Plant Pest Management Considerations

Pest management must be considered prior to planting, as there are numerous options, many of which involve chemical use. Fertilizers, fungicides, insecticides, miticides, and herbicides can all be applied through the overhead irrigation systems. Because some products must be mixed with or watered into the soil, growers without overhead irrigation may have to incorporate them via tillage or rainfall. To reduce tillage operations, plant protection products like herbicides may be incorporated with fertilizer during the same tillage pass. Overhead irrigation provides more flexibility for pesticide application timing than ground application because it does not rely on tractors, which may be busy elsewhere or restricted due to poor field and weather conditions.

6.4 Integrated Pest Management

Potato production requires ample preparation to prevent and manage season-long pest-related issues. Together, canopy quality and duration, and plant health are essential in obtaining high yields of superior quality. Diseases, insects, weeds, mites, and nematodes can all affect plant health and reduce economic return. The idea behind integrated pest management, or IPM, is to prevent economically significant pest damage holistically by utilizing and incorporating production inputs and practices synergistic for optimizing plant health. As an example, herbicides, tillage, and proper crop rotation can all be used together for weed prevention (Liebman *et al.*, 1996). By choosing a potato variety with a large canopy and verticillium wilt resistance, weed control via crop competition is likely farther into the season when compared to a variety with small, verticillium-susceptible vines. Components and inputs key to potato IPM include crop rotation, green manure crops, beneficial organisms (insects, spiders, fungi, etc.), crop protection chemicals, disease-free certified seed, sanitized equipment and potato storage, disease-resistant varieties, neighboring crops that will not harbor insect pests or mites, maintaining healthy potato plants, the timing of planting, vine kill and harvest, and more.

Aside from major weeds like nightshade (*Solanum* spp.) and common lambsquarters (*Chenopodium album*), some economically important potato pests include various nematode species, spider mites (*Tetranychus urticae*), aphids (*Myzus persicae* and *Macrosiphum euphorbiae*), wireworms (*Agriotes* spp.), Colorado potato beetle (*Leptinotarsa decemlineata*), slugs (*Arion* spp.), larvae of various species within the Lepidoptera family, black scurf (*Rhizoctonia solani*), late blight (*Phytophthora infestans*), early blight (*Alternaria solani*), powdery scab (*Spongospora subterranea*), common scab (*Streptomyces scabies* and *Streptomyces turgidiscabies*), silver scurf (*Helminthosporium solani*), bacterial ring rot (*Clavibacter michiganense* subsp. *sepedonicus*), and various viruses like *Potato virus Y* (PVY), X (PVX), A (PVA), S (PVS), *Potato leaf roll virus* (PLRV), corky ringspot (*Tobacco rattle virus*), and *Potato mop-top virus* (PMTV). Insect pests like aphids not only damage plants via feeding but also by vectoring viruses like PVY and PVA, which may compromise yield (Nolte *et al.*, 2003).

Beneficial organisms include parasitoid wasps (*Aphidius* spp.), big-eyed bugs (*Geocoris* spp.), ladybird beetles (*Hippodamia convergens*, *Harmonia axyridis*, and *Coleomegilla maculate*), lacewings (*Chrysopa* spp. and *Chrysoperia* spp.), and bacterial endotoxins like *Bacillus thuringiensis* var. *israelensis* (Dreves *et al.*, 2009).

Major pests and management strategies are not discussed in this chapter as they are covered in detail throughout other chapters in this volume.

6.5 Field and Seedbed Preparation

Field and seedbed preparations are often specific to the growing region; what works in one area may not work in another. There are considerable regional differences such as expected rainfall at various intervals during the season, rock and clod removal or placement, potential wind and water erosion prior to planting, planting and harvest dates, irrigation type, and grower preferences. Similar to most cropping systems, fields destined for potato production must be tilled to loosen the soil and incorporate organic matter from the previous crop prior to planting.

Depending on the previous crop and the types of tillage performed, reduced tillage may be an option for improving the soil (Grant and Epstein, 1973). Tillage is also used to incorporate pre-plant fertilizer and other soil amendments. If planting and harvest logistics allow, one goal of field preparation should be to reduce compaction from previous field operations. This can be accomplished by using deep-tillage implements such as subsoilers, rippers, and para-plows (Magdoff and van Es, 2000). Soil compaction is reduced when steel shanks from these implements are pulled deeply (>30 cm) through the field. Despite the benefits of deep tillage, there are production regions where tractors and harvesters will sink if the soil is significantly loosened (especially within the furrow) and rain abundant. Although the fields in these regions are used for production, yields likely suffer from soil compaction (Stark and King, 1997).

The top 30–60 cm of soil in the potato seedbed should be loose and moist prior to planting, and relatively free from rocks, clods, and excessive crop debris. Seedbeds are shaped prior to, during, and/or after planting. If conditions allow, seedbeds should be shaped with a minimal amount of tractor passes. None the less, some growers must perform multiple operations to disrupt weeds, incorporate fertilizer and herbicide, relocate or reduce stones and clods, and add soil to the seedbed.

6.6 Planting the Crop

Planting time varies by region, depending on local climatic conditions, potato variety, and intended markets. Most potatoes grown in the temperate climates are planted from April to early June. Winter crops are typically planted from November to February. Some regions have weather that is conducive to year-round production. Ideally, potatoes should be planted into soils with temperatures conducive for rapid growth and wound healing (especially when using cut seed), which is typically above 7°C (Isleib and Thompson, 1959). However, soils warmer than 19°C may reduce tuber number and early-season growth (Van Dam *et al.*, 1996). At the time of planting, soils should be moist but not excessively wet. In irrigated desert areas, it is common to pre-irrigate prior to planting. Planting into soils that are excessively cold, hot, wet, or dry may increase the potential for seed-piece decay and delayed emergence. Within reasonable limits, the earlier a crop is established, the more potential there is to maximize tuber yield and dry matter content. In regions where the growing season does not limit the plant's ability to mature fully, planting dates are often selected in an effort to provide the crop with a growing environment that will produce the highest economic yield for a desired market.

6.7 Potato Propagation and Seed Preparation

Vegetative propagation using whole or cut tubers is the most common method for starting a potato crop. Depending on the size of the seed tubers, growers may leave the tubers whole or cut them into smaller pieces. Tubers left uncut and planted whole are often referred to as "whole seed", "seed", "seed potatoes/tubers", or "single drop". Cut tubers are also referred to as "seed" or "seed potatoes/tubers", but are most often called "seed pieces". Most potato plants produce "true" botanical seed, or true potato seed (TPS); however, TPS is seldom used because each seed develops into a unique plant with traits unlike those of the parent plants (Burton, 1989). In addition, TPS-planted crops require several extra months of initial growth compared with those grown from seed tubers; often, the initial growth takes place in a greenhouse or seedbed and is followed by field transplantation. Compared with TPS, seed tubers allow growers to produce plants and tubers with traits identical (clone) to the original variety of interest, typically with rapid and uniform emergence. TPS is occasionally used in tropical or developing countries because seed potato storage and transport is often expensive and unavailable (Golmirzaie *et al.*, 1994).

According to Struik and Wiersema (1999), high potato yields are correlated strongly with the use of high-quality, certified seed tubers. Planting high-quality seed is perhaps the most important step in successful potato production and cannot be neglected. Seed certification programs are relied upon to regulate seed quality in most developed countries and an increasing number of less-developed countries (UNECE, 2013).

Within strict protocols and predefined conditions, seed lots that receive a "certified" classification have been tested for quality and approved for production by the certification agency. In order for a seed lot to be approved, it must have a high level of varietal purity and be free from, or have minimally tolerable levels of, certain seedborne pathogens and a minimal amount of other issues, like seedborne herbicide carry-over. When possible, growers should only buy and plant certified potato seed.

To promote early and uniform emergence in regions with short growing seasons, seed tubers may be pre-sprouted prior to planting. Pre-sprouting, or chitting, requires specialized equipment and facilities, and is often labor-intensive. The presence of low light suppresses sprout elongation during chitting and results in short green sprouts. Some of the sprouts may be removed prior to planting to reduce stem number per plant. Disadvantages of using pre-sprouted seed include additional expense, equipment and labor, planter performance issues (sprouts getting caught on equipment), and emergence issues due to sprout injuries during planting.

Some growers cut seed potatoes to reduce costs, extend limited seed stocks, break apical dominance, and reduce tuber size for improved planter performance. Schotzko *et al.* (1983) determined that the economically acceptable seed tuber and seed piece size range for commercial producers was 43–71 g. Seed pieces above that range were found to be wasteful, and those below compromised emergence. One size range does not work for all varieties, however. The Schotzko *et al.* (1983) seed piece weight range was specific to Russet Burbank, which had eyes well distributed across the tuber. Some varieties have few eyes, which are often concentrated near one tuber end (bud end). Seed pieces without eyes (blind seed pieces) will not sprout; therefore, seed pieces of varieties with few eyes must be cut large enough to ensure that at least one eye is found on every piece.

Cut seed is either planted immediately or allowed to wound heal (suberize). Suberized seed pieces typically fare better than freshly cut seed in adverse germination conditions such as cool, wet soils (Secor and Gudmestad, 1993). Moreover, properly suberized seed may be stored for as long as whole seed, which allows seed to be cut and stored well before planting, freeing up resources and management in preparation for planting. To prevent shrinkage and rot, humid air must flow continuously through the seed piece pile during suberization (Secor and Gudmestad, 1993). Whether suberized or not, growers often treat seed tubers and pieces during or prior to planting with a fungicide and/or systemic insecticide to protect against soil- and seedborne fungi and in-season insect pests. To promote rapid emergence, seed is often removed from refrigerated storage and "warmed" several weeks prior to planting. Sanitary seed-cutting conditions are essential for preventing disease and bacteria transmission across or within seed lots. Additional information on seed tuber physiology, seed potato production and certification, and insect and disease prevention may be found throughout this volume.

6.8 In-Row Seed Piece Spacing and Plant Population

The ideal combination of plant population, row width, and in-row seed piece spacing for a particular variety and region is the one that produces the optimum tuber size profile (see Table 6.1), and in turn, maximum grower revenue. The spatial configuration of seed pieces, and ultimately healthy plants, plays a major role in determining the tuber size profile. Other important and often growth-limiting factors include season length, variety, irrigation and nutrient availability, and soil type. Research has shown that stem number per plant also affects tuber size profile (Knowles and Knowles, 2006). In 2006, Knowles and Knowles reported that physiologically older seed tubers produced higher stem and tuber numbers per plant and smaller tuber size profiles compared with physiologically younger seed. In addition to seed age, stem number is correlated highly with planted seed tuber weight (Wurr *et al.*, 1992). Wurr *et al.* (1992) found that as planted seed tuber weight increased, stem number increased. Using planted seed tuber weight to boost stem number might provide the desired results; however, the economics of planting larger seed is often prohibitive (Schotzko *et al.*, 1983). Because the ability to predict stem number is difficult, growers typically opt for spatial configurations determined via research, or those that historically have provided the best returns across years and seed lots.

Generally, the average harvested tuber weight per plant or unit area decreases as plants are crowded together and increases when plants have less competition from their neighbor (Pavek and Thornton, 2006). Without proper uniform spatial arrangement, plant population by itself is of marginal importance in optimizing economic return. Pavek and Thornton (2006) demonstrated this principle by altering the in-row spacing of identical plant populations within the same unit of area. In this case, the uniform, properly spaced treatment produced a higher-valued tuber size profile than the irregular spacing treatment, which contained the same number of potato plants. Once the optimum inter- and intra-row spacing is determined, the accompanying plant population can be calculated.

In-row spacing values in the range of 15–30 cm are common. Variety characteristics like average stem and tuber number per plant, average tuber size, and days to reach maturity need to be defined prior to selecting the appropriate seed piece spacing. To minimize oversized tubers, varieties with low tuber numbers per plant should be spaced closer together than those with higher tuber sets, and vice versa. Wider in-row spacing may be used to increase average tuber size when growing late-maturing varieties in regions with short seasons. Closer in-row spacing may be more desirable where season length is long and moisture and nutrients are not expected to be limiting. Rapidly growing and excessively large tubers are more prone to internal defects such as hollow heart and brown center; therefore, a simple in-row spacing reduction could lead to higher-quality tubers, simply by regulating tuber growth and preventing the production of oversized tubers (Rex and Mazza, 1989).

Because growing conditions, varieties, and market needs vary from region to region, seed piece spacing requirements for particular varieties are not consistent across regions. The proper inter- and intra-row spacing for each variety and region must be determined through local research.

6.9 Conventional Rows versus Beds

Potatoes are typically planted into some type of row configuration. Furrows between conventionally-ridged potato rows help prevent tuber damage from tractor tires and allow for inter-row post-planting weed disruption, incorporation of fertilizer and herbicides, water drainage, furrow irrigation, potato hill enlargement, increased seed piece depth, erosion prevention, and soil compaction mitigation. Conventional rows are essentially raised mounds of soil (ridges/hills), commonly 76–91 cm wide, that run parallel to each other the length of the field; each ridge contains one row of linearly spaced potatoes. Some growers forego conventional row systems and plant their potatoes into wide beds. Beds are essentially wide ridges that contain multiple closely spaced rows. In this system, furrows are only located at intervals consistent with the tractor, planter, and harvester track width. Furthermore, furrows are absent between all rows contained within the bed. Beds are more conducive to high planting densities intended to minimize average tuber size (Mundy *et al.*, 1999). In addition, bed systems are occasionally more efficient at capturing irrigation water or rainfall than conventional ridge systems (Robinson, 1999). In areas with excessive rainfall, however, capturing and holding rainfall during certain times in the season may prove undesirable. On average, soil temperatures throughout the bed fluctuate less than those in ridges (Mundy *et al.*, 1999), presumably due to the insulating factor associated with the extra soil found between each bed-planted row, which is normally removed when conventional ridges are formed. Prior to emergence, the insulating factor in beds may actually inhibit the accumulation of early-season soil heat units and delay emergence.

Bed systems do present some disadvantages when compared with conventional rows. Inter-row post-planting operations commonly conducted in ridge systems (Fig. 6.1) are not possible in bed systems due to the lack of furrows between every row. The conventional row system also allows for furrow irrigation. In addition, tractor horsepower requirements might increase when harvesting from beds, as harvesters must sift through more soil than with conventional rows.

Fig. 6.1. Tillage of conventional rows to disrupt weed growth. (Photo courtesy of M.J. Pavek.)

6.10 Depth of Planting

As with seed piece spacing, the ideal planting depth is the one that helps to optimize grower revenue. The ideal planting depth positions seed pieces, stems, roots, and tubers in the most favorable season-long growth environment. It also provides ample room for tubers, like those seen in Fig. 6.2, to expand and mature under a protective layer of soil.

Moisture and temperature differences throughout a soil profile can influence emergence rate strongly (Pavek and Thornton, 2009). If the soil is adequately moist and warm (>7°C), sprouts typically emerge more rapidly from shallower-planted potato seed pieces. This is largely because the shallower-planted seed pieces accumulate heat units quicker than deeper-planted ones. Moreover, sprouts have less soil to travel through prior to emerging from the soil surface. In relatively hot, dry regions, deeper planting depths may provide more soil moisture and less extreme temperatures throughout the season.

Rapid emergence reduces the severity of early season diseases like *R. solani* (Van Emden, 1965). Rapid emergence also hastens the plant's ability to capture early solar radiation, which is essential for optimizing tuber yield and dry matter (Firman and Allen, 1989). At the time of planting, the soil temperatures at seed piece depth should be at or above 7°C (Bohl *et al.*, 2003). Soils should be moist but not excessively wet or dry. If possible, pre-plant irrigation should be used to add moisture to excessively dry soils. Planting depths typically range from 5 to 20 cm (measured from the top of the seed piece to the top of the ridge/hill). Due to post-planting tillage operations, however, final seed piece depth typically ranges from to 10 to 20 cm. In many regions, it is common to plant shallow to encourage early emergence. Many of the growers within these regions build and shape the final hill with one or more post-planting tillage operations. In addition to hastened sprout emergence, shallow planting depths may reduce at-harvest tractor horsepower needs by minimizing harvester blade depth. The drawbacks of shallow planting may include lack of adequate soil moisture compared to deeper soil, along with higher yields of undersized, green, and surface-exposed tubers (Pavek and Thornton, 2009). Although deeper planting depths may provide better soil moisture and less tuber issues, the drawbacks may include yield reduction and disease issues due to delayed plant emergence. The ideal planting depth balances rapid sprout emergence with the space requirements of expanding and mature tubers (Pavek and Thornton, 2009). Within limits, varieties that produce large tubers and high yields may need to be planted deeper and into a wider hill than those that do not. Mature and rapidly bulking tubers require a protective layer of soil

Fig. 6.2. Large tubers and yields like these typically require a large, broad ridge (hill) to protect against tuber greening and quality loss. (Photo courtesy of M.J. Pavek.)

that limits tuber greening and protects the tubers and roots against temperature and moisture extremes throughout the season.

Planting depth should be adjusted for each growing region, variety, and anticipated harvest date. Seed pieces should be planted deep enough to avoid soil moisture deficits, tuber greening, and to accommodate the predicted yield. Potatoes should also be planted shallow enough to enable rapid, uniform emergence, thereby reducing disease and allowing for early growth. Other important planting depth considerations include cultivar physiology and growth characteristics, post-planting tillage and weed control needs, harvester capabilities, harvest date (early versus late), and the requirements of the intended market.

6.11 Planter Types and Common Issues

Commercial producers typically rely on cup- or pick-type planters. As the name infers, cup planters use small cups to move seed tubers or pieces from a planter hopper into a planter-created furrow. There are various cup-type models, and some incorporate a vacuum to aid in cup filling and seed dropping. Instead of a cup, the pick planter uses a small nail-like pick to pierce and move the seed from a large bowl or the hopper into the planter furrow. Pick planters may facilitate seed decay and disease spread by creating an entry wound on each seed piece. In addition, the delicate picks must be inspected routinely for damage from foreign material. Other planter types exist, but are less common in large-scale production.

Planting seed pieces of different shapes and sizes can prove difficult. A 2005 survey indicated that an average of 6% (2700 plants ha^{-1}) of the intended plant population was missing from Washington State (USA) potato fields due to planter skips (Pavek and Thornton, 2005). Moreover, the other 94% was non-uniformly spaced. A related study determined these issues collectively were costing growers US\$200–250 ha^{-1} (Pavek and Thornton, 2006). Pavek and Thornton (2005) postulated that the combination of poor planter design and irregularly shaped seed was largely to blame. In 1983, Entz and LaCroix reported that whole seed tubers, of uniform shape and size, planted more uniformly than cut, irregularly shaped seed pieces; this finding was confirmed in 2005 by Pavek and Thornton. Other studies suggest planter performance is compromised by insufficient management and maintenance, improper seed-bowl levels, improper ground speed, and planter design (Hyde *et al.*, 1979; Sieczka *et al.*, 1986).

In an effort to prevent planter skips and to optimize planter performance, growers should routinely maintain the planter, follow the planter, uncover planted potatoes, and assess accuracy. Through a simple visual assessment, one can determine whether or not a planter is functioning as it should. If operators experience poor planter performance, they should adjust tractor speed, check for mechanical issues, make sure hopper seed levels are adequate, etc., until performance improves. If performance does not improve, growers should contact the manufacturer for assistance or assess different planter makes and models.

6.12 Planting Operations and Post-Planting Tillage

To minimize tractor passes and optimize the placement and effectiveness of fertilizers, pesticides, and fungicides, growers may utilize the planting operation to apply liquid or granular products in-furrow and/or in a band within or on the ridge. During planting, growers have the option to place seed tubers at their final depth into a ridge of final shape, or to plant tubers shallow with the intention of enlarging and shaping the ridge later in the season. Because sprouts are likely to emerge more rapidly from shallow-planted seed, growers in some regions may choose initially to plant shallow and then to enlarge the ridge, after the plant emerges, with one or more tillage or ridging (hilling) operations. Ridging or re-ridging is conducted to hasten plant growth, incorporate herbicides and fertilizers, disrupt weeds, and provide enough soil for expanding tubers. In wet regions, seed tubers are often positioned high in the ridge, well above the furrow, to reduce the rot potential from standing or draining water. Some commercial operations choose to "drag off" the top of the ridge after planting and before emergence to reduce seed piece depth and disrupt weeds. Once sprouts emerge from the soil, the dragged-off rows are shaped and enlarged via a hilling operation.

6.13 Growing and Harvesting the Crop

Once the potato sprouts emerge from the soil, cultural management decisions must be made and plant growth and development managed to optimize tuber yield, size profile, and postharvest quality.

Potato water use

Irrigation is used extensively to optimize potato growth. Proper irrigation timing and rate are important to tuber yield and quality (Ojala *et al.*, 1990). Lack of soil moisture may reduce tuber yield and quality (Shock *et al.*, 1992), as may excessive moisture (Holder and Cary, 1984). Nelson and Hwang (1975) characterized plant water use at four individual potato growth stages: emergence to tuber initiation; the transition from tuber initiation to tuber bulking; tuber bulking; and senescence/tuber ripening. From emergence to tuber initiation, they indicated that young plants retained a high percentage of moisture and that water use was high; foliage growth and water use were significantly correlated. As plants transitioned out of tuber initiation, they held less moisture and their water use peaked and leveled off. As growth shifted from canopy to tubers (tuber bulking), both water use and moisture retention remained constant, but declined as plants senesced (Nelson and Hwang, 1975).

Water stress and overabundance

Water stress before tuber initiation may lead to a reduction in tuber number (van Loon, 1981). It may also affect fry and chip color. Shock *et al.* (1992) found that drought stress during tuber initiation improved postharvest fry/chip color; Iritani and Weller (1973) suggested the opposite. Shock *et al.* (1992) indicated the difference between the two studies might have more to do with the definition of "tuber initiation" than with differences in potato physiology. They further explained that if they widened the tuber initiation window to include late-developing tubers (end of June), then indeed, water stress might cause dark (stem-end) fries. Shock *et al.* (1993) supported this claim by imposing water stress on two varieties during late June and early July. Following harvest, both cultivars produced dark stem-end fries (also known as "sugar ends") due to an increase in reducing sugars. Eldredge *et al.* (1996) indicated that water stress during the early bulking stage also resulted in sugar ends on the variety Russet Burbank. For some varieties, water stress, whether from too much or not enough water, can lead to tuber malformations, such as knobs, growth cracks, and irregular shape (Sparks, 1958; Kleinkopf, 1982; Holder and Cary, 1984).

Excessive soil moisture may exacerbate the development of diseases such as late blight (*P. infestans*), blackleg (*Erwinia carotovora* var. *atroseptica*), softrot (*Erwinia* spp.), and pinkrot (*Phytophthora erythroseptica*), among others (Adams and Stevenson, 1990).

Irrigation and available soil water recommendations

Maintaining moist soils in the plant's root zone (~75% available soil water (ASW)) from 2 to 7 weeks after sprouts emerge significantly reduces common scab (*S. scabies*) (Davis *et al.*, 1974). Because plants may not be able to remove all soil water, ASW is defined as the water within the plant-extractable range, sufficient enough to prevent plant wilt and death. The upper limit is referred to as field capacity, while the lower limit is called the wilting point. Stark and King (1997) recommend maintaining ASW in the primary root region (0–45 cm) between 65% and 85% during the active growth stages of canopy establishment, tuberization, and rapid tuber bulk. They also suggest an ASW of 70–80% during planting to hasten plant emergence. When vines have senesced or are killed, they recommend decreasing ASW to 60–65% to allow for proper skin set and prevent tuber rot. Just prior to harvest, they suggest an ASW above 60% to reduce bruising, which is exacerbated by tuber dehydration and cloddy soils.

Various irrigation scheduling instruments, like tensiometers, use units of measure called bars or centibars (cbar) (1 bar = 100 cbar). Tensiometers measure soil water tension (suction), which mimics the energy plants must exert to extract water from soil. The relationship between ASW and soil water tension depends on the soil texture. For coarse-textured soils, soil water tension at field capacity or 100% ASW may measure 10 cbar. In a fine-textured soil, the field capacity may equal 30 cbar (Ley *et al.*, 1994). Sandy loam soil ASW values of 65–85% should measure close to 25 to 15 cbar, respectively, while a loam soil may measure 55 to 20 cbar, respectively (Ley *et al.*, 1994).

Common irrigation methods for potatoes include sprinkler systems such as center pivots, lateral/linear move systems, wheel- and hand-lines/solid set, and big gun or boom hose reels, as well as furrow/gravity systems, and drip-tape or drip-hose systems. Most of these systems are designed to be used on a regular basis as the primary water source for the potato crop; however, the big gun and boom hose reel systems are typically used to supplement fields that are otherwise rain fed.

Soil water monitoring and measurement

Soil water is typically measured and monitored with tensiometers, neutron probes, soil moisture blocks, gravimetric sampling, the balance sheet method, remote sensing, and soil feel and appearance.

Tensiometers are constructed of airtight, water-filled tubes that house porous ceramic tips on one end. The ceramic tip is placed in the soil at the desired depth and an aboveground vacuum gauge displays water tension measurements in the range of 0–80 cbar (Morris, 2006). Tensiometers are relatively inexpensive, common, and useful; however, they must be maintained and checked often. A common issue with tensiometers is the loss of tension due to drying periods and poor soil contact.

Neutron probes measure the hydrogen associated with water (H_2O) (Ley *et al.*, 1994). They can be very accurate, but are also expensive and require training, radioactive material licensing, and calibration to use. Growers who rely on neutron probes often contract with a commercial irrigation monitoring company that owns and operates the probes.

Electrical resistance blocks or capacitance probes measure the capacity of the soil to store electricity (electrical resistance). Soil moisture and electricity are correlated; wet soils hold more electricity than dry soils. Electrical resistance is low in wet soils, which make it easier for the blocks to conduct electricity (Morris, 2006).

Two common types of blocks are the granular matrix sensors and gypsum blocks. The blocks are buried and absorb water from the soil. One common issue is the lack of good soil-to-block contact; air pockets can provide erroneous measurements. The blocks are connected to wires, which plug into a portable data logger or meter (Morris, 2006).

Gravimetric sampling is one of the more accurate soil water measuring techniques and involves soil sampling, weighing wet and dried soil, and calculations involving soil bulk density. Due to the labor and time involved, this sampling is usually not used on a large scale.

The balance sheet method relies on known or predicted daily evapotranspiration (ET) measurements and real-time and forecasted weather data. Irrigation is scheduled to replace all or some of the soil water used. Computer models are available on the Internet, or through companies that have created values for specific varieties grown in specific regions. This method of irrigation scheduling can be rather accurate, provided correct model input values are used during calculations.

The most traditional and perhaps most reliable method is digging, feeling, and viewing soil from the rooting zone of the potato. Soil is often collected using soil probes pushed to the desired testing depth. Although this method relies heavily on experience, Ley *et al.* (1994) have produced a guide that matches soil feeling and appearance to percent ASW. Drawbacks include reliance on judgment, which could lead to differences among multiple testers, and inability to probe at the depth desired due to rocks and compaction.

Modern soil water measurement and monitoring methods incorporate cellular and wireless data transfer directly from the instrument or field to provide real-time information to growers, enabling them to adjust irrigation systems quickly from their computers or phones. Remote sensing systems use buried sensors that detect soil moisture changes, which are fed into a data logger (Morris, 2006). The data logger sends wireless or cellular signals to the user's phone or computer, allowing for real-time monitoring of multiple fields. The disadvantages of remote sensing are the relatively high cost for each unit (>US$1000) and the lack of field-monitoring visitations. As with most technology, these units should become more affordable over time; however, it is difficult or impossible to replace actual field visitations by an agronomist. Although many of the soil water measuring instruments provide good information, growers should incorporate a mixture of instrument use and hands-on measuring to ensure successful production.

Plant nutrition and fertilizer

Mineral nutrients are essential for potato plant growth and development (Kolbe and Stephan-Beckmann, 1997a,b). Mineral nutrients found in the soil are typically classified either as a macronutrient or a micronutrient. As their name indicates, plants require and use more macronutrients than micronutrients, but nutrients from both categories are important to plant health. Macronutrients important for potato production include nitrogen (N), phosphorus (P), potassium (K), calcium (Ca), magnesium (Mg), and sulfur (S); however, potato plants and tubers extract far more N, P, and K from the soil than other nutrients. Micronutrients essential to potato growth include boron (B), copper (Cu), iron (Fe), chloride (Cl), manganese (Mn), and zinc (Zn).

In most situations, potatoes grown for profit require fertilizer applications. Unless growers are extremely familiar with their fields and rotations, a pre-plant soil analysis should be used to determine nutrient levels, organic matter, pH, cation-exchange capacity (CEC), and other soil characteristics. Whether manure or synthesized fertilizer is used, crop nutrient needs are determined largely by the region, season length and anticipated harvest date, and market choice. Recommended application timing and amounts are usually based on research (Lang *et al.*, 1999; Stark and Westermann, 2003; Dampney *et al.*, 2013) or local experience.

Fertilizer recommendations typically exceed the actual quantity needed by the plants due to tie-up in the soil by microorganisms (immobilization) and chemical reactions, soil leaching and volatilization, excessively large gaps between plants/poor stands, lack of in-season monitoring and experience, and a favorable benefit–cost ratio (excessive fertilizer applications may be less expensive than yield losses caused by nutrient deficiencies). Excessive fertilizer applications can damage the environment and lead to human health issues.

As a result, research has focused on improving nutrient application efficiency (Westermann *et al.*, 1988; Waddell *et al.*, 2000; Davenport *et al.*, 2005). Excessive soil nutrients, especially nitrogen, may promote a source–sink imbalance, which can delay tuber initiation, plant maturity and tuber bulking, and promote excessive foliage growth at the expense of tuber yield.

An important concept to remember when applying fertilizers is that maximum biological yield does not always result in maximum economic return (Fig. 6.3).

Figure 6.3 demonstrates this principle with the potato variety Alturas, grown at different rates of full-season N in the Columbia Basin of Washington during 2009–2011. The maximum biological yield of 86.3 t ha^{-1} was realized with 485 kg N ha^{-1}; however, the maximum economic return of US$12,042 ha^{-1} came from a yield of 85 t ha^{-1}, which was produced with only 410 kg N ha^{-1}. Moreover, the economic return from potatoes produced with 333 kg N ha^{-1} was identical to that of potatoes produced with 485 kg N ha^{-1} (US$11,720 ha^{-1}) (Fig. 6.3). Although the extra 152 kg N ha^{-1} increased the yield from 83 to 86.3 t ha^{-1}, the expense of the additional nitrogen offset any economic gain that would have come from the extra 3.3 t ha^{-1} of yield.

Based on data from Kolbe and Stephan-Beckmann (1997b), whole potato plants may remove 3.6, 0.7, and 6.4 kg N, P_2O_5, and K_2O, respectively, from the soil per tonne of fresh plant and tuber weight. Considering those values, fresh-weight yields exceeding 79 t ha^{-1}, similar to those found in the Columbia Basin of Washington State may extract more than 286, 56, and 509 kg ha^{-1} of N, P_2O_5, and K_2O, respectively, from the soil during one growing season (Kolbe and Stephan-Beckmann, 1997b). In contrast, a 1980 study by Dow and Cline calculated that a 79 t ha^{-1} yield would remove 422, 47, and 520 kg ha^{-1} of N, P_2O_5, and K_2O, respectively (Dow and Cline, 1980). Presumably, the differences between the two studies were due to growing environment and variety differences.

To achieve a 79 t ha^{-1} tuber yield for Russet Burbank grown for approximately 150 days in the Columbia Basin of Washington State, researchers recommend depleted soils receive approximately 391, 330, and 537 kg ha^{-1} of N, P_2O_5, and K_2O fertilizer, respectively (Lang *et al.*, 1999). Alternatively, growers in shorter-season regions like the UK may need to apply rates of 270, 250, and 360 kg ha^{-1} of N, P_2O_5, and K_2O, respectively, for Russet Burbank when grown for 120 days (Dampney *et al.*, 2013). It is interesting

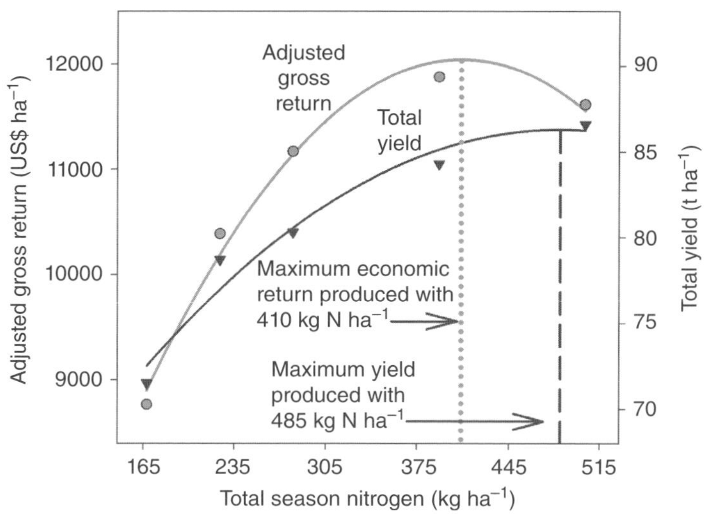

Fig. 6.3. Alturas total yield (R^2 = 0.96) and nitrogen-cost adjusted gross return (R^2 = 0.99) when grown with five total season nitrogen rates (168, 224, 280, 392, and 504 kg ha^{-1}) in the Columbia Basin of Washington, USA, 2009–2011. Maximum economic return of US$12,042 ha^{-1} (gray dotted line) came from potatoes produced with 410 kg N ha^{-1}, while the maximum biological yield (black dashed line) of 86.3 t ha^{-1} was produced with 485 kg N ha^{-1}. (Unpublished data from M.J. Pavek.)

to note that the recommended rates from these two regions are remarkably similar when divided by the anticipated days of growth (120 versus 150). The resulting recommended NPK rates per day of growing season in the UK versus the Columbia Basin are 2.3 versus 2.6 kg N ha^{-1}, 2.1 versus 2.2 kg P_2O_5 ha^{-1}, and 3.0 versus 3.6 kg K_2O ha^{-1}, respectively. Since both research teams used a common variety, this observation might strengthen their recommendations as being universally acceptable for Russet Burbank.

Plant and soil nutrient status

In-season nutrient sampling is usually more intense for overhead-irrigated farms than for rainfed systems. Real-time data are essential for overhead-irrigated farms, as fertilizer can and may need to be applied at any point during the season via the irrigation water (fertigation). None the less, all farms can benefit from an in-season nutrient monitoring program. For efficient and effective nutrient management, plants and soils should be sampled, analyzed, and monitored throughout the season. Sampling soils where most of the potato roots are found (0–45 cm) is ideal. Deeper samples may be collected and are especially helpful for tracking and monitoring nutrient movement through the soil profile to prevent leaching.

Petioles are commonly used for plant nutrient analysis. Typically, the fourth leaf (newest fully expanded leaf) from the top of the plant is removed and the leaves stripped to isolate the petiole (Stark and Westermann, 2003). Depending on field size, a representative sample—perhaps 50 or more petioles—should be collected randomly throughout each field or field section and combined for analysis. Due to natural field variability, larger fields may be divided into sampling grids to allow for sectional management. Base soil and petiole levels exist for many regions (Dow, 1980; Rykbost *et al.*, 1993; Walworth and Muniz, 1993; Lang *et al.*, 1999; Stark and Westermann, 2003) and may be extrapolated for regions without recommendations. Table 6.2 displays petiole values common to the Columbia Basin of Washington. Petioles should be collected from healthy, unstressed plants. McDole and McMaster (1978) reported that moisture-stressed plants exhibited higher petiole nitrate levels than those being grown with optimum irrigation levels.

Table 6.2. Typical N, P, K, and S petiole values for three commonly grown cultivars during June and July[a] in the Columbia Basin of Washington, USA. Nitrogen petiole values are displayed in a range developed to maximize economic return for each cultivar. (Based on unpublished data from M.J. Pavek, 2006–2012.)

		End of tuber initiation	100% ground cover	Early tuber bulk	Mid tuber bulk
Variety	Nutrient[b]	15 June	30 June	15 July	30 July
		%			
Ranger Russet	N	2.6–2.9	2.4–2.7	2.2–2.5	2.1–2.4
	P	0.49	0.36	0.26	0.25
	K	10.80	10.30	7.20	6.70
	S	0.21	0.20	0.16	0.20
Russet Burbank	N	2.4–2.7	2.0–2.3	1.8–2.1	1.5–1.8
	P	0.44	0.30	0.20	0.14
	K	10.90	9.30	7.40	6.50
	S	0.22	0.18	0.16	0.20
Umatilla Russet	N	2.8–3.1	2.6–2.9	2.3–2.6	1.9–2.2
	P	0.44	0.41	0.25	0.18
	K	10.60	8.90	7.60	6.70
	S	0.29	0.24	0.25	0.28

Notes: [a]Based on ≥150 days of growth from planting to harvest. Typical dates: planting = 15 April; emergence = 15 May; harvest = 15 September. Typical potato fertilizer rates for depleted soils in the Columbia Basin: N = 400 kg ha^{-1}; P_2O_5 = 280 kg ha^{-1}; K_2O = 540 kg ha^{-1}; and SO_4^{2-} = 45 kg ha^{-1}. [b]To convert values to ppm, multiply by 10,000; e.g. 2.6% N = 26,000 ppm N.

Fertilizer application options

Various methods exist to deliver fertilizer to potatoes. Growers with overhead irrigation, like that from center pivots, have the ability to fertigate as needed throughout the season. Applying fertilizers like nitrogen through an overhead irrigation system allows the plants to receive the nutrient as needed with extreme precision and uniform distribution. Growers without overhead irrigation have fewer options, and their fertilizer is typically applied and incorporated with tillage equipment prior to planting, during planting, and/or prior to row closure (100% ground cover); however, foliar nutrients are occasionally applied after row closure via ground or air applicators. Foliar applications are typically used for smaller, supplemental nutrient applications. By using nutrient applications via the overhead-irrigation system, losses due to nitrogen leaching and volatilization are minimized because the application is matched closely to real-time plant growth during the season (Waddell *et al.*, 2000). Moreover, applying all fertilizer preplant or early in the season is inefficient, as plants may recover only 60% of that applied (Westermann *et al.*, 1988). To maximize efficiency and minimize leaching and volatilization, many growers apply nutrients via fertigation in small quantities twice or more each week during the growing season. The ability to "spoon feed" nutrients to the plants through the irrigation system enables growers to tweak plant growth during crucial plant development stages in an effort to optimize yield, postharvest quality and storability, and in turn, economic return.

Nitrogen fertigation

Compared with other macronutrients, nitrogen has the largest effect on plant growth and final tuber yield (Rykbost *et al.*, 1993). Nitrogen fertigation quantities and timing should match plant growth. This can be accomplished by maintaining petiole and soil levels in accordance with recommendations based on previous research (as indicated previously), and by having a thorough understanding of plant growth and development, especially as it relates to a particular variety and region.

For a predicted yield of 79 t ha^{-1} in the Columbia Basin of Washington State, Lang *et al.* (1999) recommended a nitrogen fertilizer rate of 392 kg ha^{-1}, after accounting for residual soil nitrogen and potential microbial tie-up. They also suggested applying approximately 17–35% (67–134 kg N ha^{-1} for a 79 t ha^{-1} yield) in the top 30 cm of soil prior to or at planting, with the remainder applied in-season. They estimated 30–40% of total nitrogen uptake occurred by the end of tuber initiation, and the largest daily nitrogen demand came during rapid tuber bulk (~early July), when plants were extracting upwards of 4.5 kg N ha^{-1} day^{-1}.

It is common for Columbia Basin growers to fertigate 22–34 kg N ha^{-1} during one irrigation cycle, with weekly totals of up to 84 kg N ha^{-1}, especially during early, rapid tuber bulk. Due to differences in seasons, varieties, and mineralization rates, it is important to track petiole and soil nitrogen and to adjust in-season application rates accordingly. To allow for proper plant and tuber maturity, soil nitrogen should be depleted and in-season applications should cease approximately 30–45 days prior to harvest.

Potato source–sink relationships

One relationship that is crucial to profitable potato production is the source–sink relationship. The term "source–sink" is somewhat self-descriptive in that plant organs that produce synthesized food material are referred to as the "source" and those that store them are called the "sink". Sources are net carbon producers, while sinks are net carbon users (Dwelle, 1990). Mature green potato leaves act as the main source; tubers become the major sink when they start to bulk. Keeping the source–sink ratio in balance is thought to be important when trying to maximize profits via crop production. The sink provides the income for potato growers, and it is therefore essential that the relationship between the leaves and tubers is maintained in a manner that will maximize final tuber weight and quality. Through the use of in-season nitrogen fertigation, growers can manipulate plant growth at key stages to maximize their profits. For an in-depth discussion on this concept and potato growth, see Chapter 5 of this volume.

Plant maturity

Cultural management decisions are typically based on a predetermined harvest date or window. With a harvest target date, growers can adjust input applications and quantities to allow for proper plant and tuber maturity. Having knowledge of the intended harvest date is also important for pesticide applications, as most plant protection products require a post-application preharvest interval after each application. Tuber maturity is an important factor during harvest. Whether the vines senesce naturally or prematurely, the process sets the stage for tuber skin set and physiological maturity. Proper skin set at harvest is important for crop storage and marketing (especially when marketed fresh). Immature tubers are highly susceptible to shatter bruise and skinning during harvest (Thornton and Bohl, 1998). Mature tubers may wound heal faster and resist weight loss associated with dehydration (McGee *et al.*, 1985). Soil type, temperature, moisture, variety, and management of fertilizer, irrigation, desiccation timing, and season-long cultural management practices contribute to the speed of skin set. Poor skin set results in tuber skinning and wounding, which may lead to pathogen infection and premature tuber dehydration. Proper skin set may take up to 22 days or longer following vine death (Bowen *et al.*, 1996).

Defoliation

Due to variety maturity differences and early harvest dates, potato foliage may need to be killed mechanically or chemically to induce tuber maturity and skin set prior to harvest and storage. Mechanical defoliation methods include flaming, pulling, cutting, undercutting, rolling, flailing, and beating vines. Chemical vine desiccation products include sulfuric acid and non-selective herbicides such as diquat dibromide, paraquat dichlor, or endothall. Growers may incorporate mechanical and chemical methods, especially if vines are large and immature and/or weeds are prevalent.

In addition to skin set and tuber maturity, defoliation is also used to control tuber size, avoid insect or disease infestations, and facilitate harvest. Vine desiccation and removal is effective in stopping late blight spore production on foliage, which may reduce contamination of healthy tubers during harvesting operations and storage. In addition, seed growers may remove foliage from healthy plants prematurely to prevent virus spread by late-season aphid infestations.

Harvest

Skinning and bruising are perhaps the most common impact-related defects occurring during and after harvest. In 1996, Brook estimated that a 1% decline in tuber impact-related defects was valued at US$7.5 million (Brook, 1996). Mature tubers resist skinning, but may be more susceptible to blackspot bruise than immature tubers (Corsini *et al.*, 1999). Moreover, a variety's genetic background can impact its susceptibility to bruising greatly (Pavek *et al.*, 1993), as can soil temperature and moisture, and tractor and harvester chain speed (Smittle *et al.*, 1974). To minimize bruising, available soil water at harvest should be 60–80% to allow for good soil and tuber separation, and harvester soil load (Thornton *et al.*, 1973).

In general, bruising increases as tuber pulp temperature decreases, but tuber hydration also plays a role. Dehydrated tubers are more susceptible to blackspot bruise, and fully hydrated tubers more susceptible to shatter bruise (Thornton *et al.*, 1973). Depending on the variety's susceptibility to either bruise type, tuber hydration must be balanced to reduce both. For moderately hydrated tubers (60–80% ASW), the ideal tuber pulp temperature range at harvest is approximately 15–18°C. Tuber decay may increase in storage when at-harvest pulp temperatures are above 18°C. Tubers with pulp temperatures below 7°C during harvest and handling are prone to tuber bruising. Harvester chains should be padded and potential impact zones modified to prevent tuber damage. Harvester chain conveyor and ground speed should be adjusted to allow for a full, uniform flow of potatoes through the whole machine (Thornton and Bohl, 1998). Blades on harvesters and windrowers should be adjusted to prevent tubers from colliding with the primary chain on pickup. In addition, conveyor drop heights should be minimized (<20 cm) to reduce bruising.

6.14 Summary

The commercial potato production cycle is essentially an exercise in converting solar radiation, CO_2, water, and nutrients into tuber dry matter—profitably. Practices implemented throughout the production cycle must be adjusted to produce potatoes of marketable size, quantity, and quality. The production journey begins by identifying a market avenue for the harvested tubers. This is followed by selecting a variety that is marketable and, if possible, efficient with inputs and resistant to major diseases. Because all varieties do not grow the same, it is essential that agronomic practices be adjusted to fit each variety and growing region. Specific variety recommendations are often available from the organization that developed the variety and/or government agencies tasked with providing grower outreach.

The soil must be conducive to potato production, free from clods and rocks, well drained and fertile, and properly prepared to sustain growth season-long. This includes reducing compaction and excessive crop residue, addressing pests, and testing for and adding nutrients and water as needed. Proper planting depth and spatial arrangement of seed potatoes are critical in optimizing the production of marketable-sized tubers. An anticipated harvest date must be established so input level and cultural practices can be tailored to optimize economic return within the harvest window.

Canopy and plant health must be maintained by practicing integrated pest management and monitoring and measuring irrigation and plant and soil nutrient status throughout the season. Production inputs should be used to maximize economic return, not necessarily yield. Growers must embrace the idea that maximum biological yield is not necessarily the same as the maximum economic yield; at some level of plant growth, the marginal gain in yield that may come from adding more inputs is offset by the cost of those inputs (the law of diminishing returns). Proper input levels are typically determined and reported by public and private research organizations (Lang *et al.*, 1999; Stark and Westermann, 2003; Dampney *et al.*, 2013). A balance of plant maturity, ideal soil moisture and temperature, and proper harvester operation is essential in preventing tuber skinning and bruising, and in turn optimizing quality and economic return.

References

Adams, S.S. and Stevenson, W.R. (1990) Water management, disease development, and potato production. *American Potato Journal* 67, 3–11.

Bohl, W.H., Olsen, N., Love, S.L. and Nolte, P. (2003) Seed planting and management. In: Stark, J.C. and Love, S.L. (eds) *Potato Production Systems*. University of Idaho Agricultural Communications, Moscow, Idaho, pp. 91–113.

Bowen, S.A., Muir, A.Y. and Dewar, C.T. (1996) Investigations into skin strength in potatoes: factors affecting skin adhesion strength. *Potato Research* 39, 313–321.

Brook, R.C. (1996) *Potato Bruising; How and Why; Emphasizing Blackspot Bruise*. Running Water Publishing, Haslett, Michigan.

Burton, W.G. (1989) *The Potato*, 3rd edn. John Wiley and Sons, Inc., New York.

Corsini, D., Stark, J. and Thornton, M. (1999) Factors contributing to the blackspot bruise potential of Idaho potato fields. *American Journal of Potato Research* 76, 221–226.

Dampney, P., Collins, G., Sagoo, L. and Allison, M. (2013) Crop nutrition for potatoes. British Potato Council, Agriculture and Horticulture Development Board (www.potato.org.uk, accessed 24 November 2013).

Davenport, J.R., Milburn, P.H., Rosen, C.J. and Thornton, R.E. (2005) Environmental impacts of potato nutrition management. *American Journal of Potato Research* 82, 321–328.

Davies, A., Titterington, A. and Cochrane, C. (1995) Who buys organic food: a profile of the purchases of organic food in Northern Ireland. *British Food Journal* 10, 17–23.

Davis, J.R., McMaster, G.M., Callihan, R.H., Garner, J.G. and McDole, R.E. (1974) The relationship of irrigation timing and soil treatments to control potato scab. *Phytopathology* 64, 1404–1410.

Delate, K. and Hartzler, R. (2003) *Weed Management for Organic Farms.*, Publication pm 1883, Iowa State University Extension, Ames, Iowa.
Dow, A.I. (1980) *Critical Nutrient Ranges in Northwest Crops*. Western Regional Extension Publication WREP43, Washington State University, Pullman, Washington.
Dow, A.I. and Cline, T.A. (1980) Growth curve, yield, nutrient removal and petiole nutrient levels of potatoes grown on sandy soils under three center pivot irrigation systems. In: *Proceedings of the 31st Annual Northwest Fertilizer Conference,* Salt Lake City, UT, USA, 14–16 July 1980. Northwest Plant Food Association, Portland, Oregon, pp. 61–74.
Dreves, A.J., Coop, L. and Ambrosino, M. (2009) Biological control. In: Hollingsworth, C.S. (ed.) *2009 Pacific Northwest Insect Management Handbook*. Oregon State University Extension and Station Communications, Corvallis, Oregon, pp. 614–619.
Dwelle, R.B. (1990) Source/sink relationship during tuber growth. *American Potato Journal* 67, 829–834.
Eldredge, E.P., Holmes, Z.A., Mosley, A.R., Shock, C.C. and Stieber, T.D. (1996) Effects of transitory water stress on potato tuber stem-end reducing sugar and fry color. *American Potato Journal* 73, 517–530.
Entz, M.H. and LaCroix, L.J. (1983) A survey of planting accuracy of commercial potato planters. *American Potato Journal* 60, 617–623.
Firman, D.M. and Allen, E.J. (1989) Relationship between light interception, ground cover and leaf area index in potatoes. *Journal of Agricultural Science* 113, 355–359.
Golmirzaie, A.M., Malagamba, P. and Pallais, N. (1994) Breeding potatoes based on true seed propagation. In: Bradshaw, J.E. and Mackay, G.R. (eds) *Potato Genetics*. CAB International, Wallingford, UK, pp. 499–513.
Grant, W. J. and Epstein, E. (1973) Minimum tillage for potatoes. *American Potato Journal* 50, 193–203.
Guenthner, J.F. (2003) Principles of economics and marketing. In: Stark, J.C. and Love, S.L. (eds) *Potato Production Systems*. University of Idaho Agricultural Communications, Moscow, Idaho, pp. 383–397.
Hijmans, R.J. (2001) Global distribution of the potato crop. *American Journal of Potato Research* 78, 403–412.
Holder, C.B. and Cary, J.W. (1984) Soil oxygen and moisture in relation to Russet Burbank potato yield and quality. *American Potato Journal* 61, 67–75.
Hyde, G.M., Thornton, R.E. and Kunkel, R. (1979) Potato planter mechanism performance. In: *Proceedings of the 18th Annual Washington Potato Conference, Moses Lake, Washington*. pp 55–59.
Iritani, W.M. and Weller, L. (1973) The development of translucent end tubers. *American Potato Journal* 50, 223–233.
Isleib, D.R. and Thompson, N.R. (1959) The influence of temperature on the rate of root and sprout growth of potatoes. *American Potato Journal* 36, 173–178.
Johnson, S.B. and Sideman, E. (2006) *Producing Potatoes Organically in Maine*. Bulletin No 2419, University of Maine Extension, Orono, Maine.
Kirkman, M.A. (2007) Global markets for processed potato products. In: Vreugdenhil, D. (ed.) *Potato Biology and Biotechnology, Advances and Perspectives*. Elsevier Ltd, Oxford, UK, pp. 27–44.
Kleinkopf, G.E. (1982) Potato. In: Teare, I. and Peet, M. (eds) *Crop–Water Relations*. John Wiley and Sons, New York, pp. 287–305.
Knowles, N.R. and Knowles, L.O. (2006) Manipulating stem number, tuber set, and yield relationships for northern- and southern-grown potato seed lots. *Crop Science* 46, 284–296.
Kolbe, H. and Stephan-Beckmann, S. (1997a) Development, growth, and chemical composition of the potato crop (*Solanum tuberosum* L.). I. Leaf and stem. *Potato Research* 40, 111–129.
Kolbe, H. and Stephan-Beckmann, S. (1997b) Development, growth, and chemical composition of the potato crop (*Solanum tuberosum* L.). II. Tuber and whole plant. *Potato Research* 40, 135–153.
Lang, N.S., Stevens, R.G., Thornton, R.E., Pan, W.L. and Victory S. (1999) *Potato Nutrient Management Guide for Central Washington*. Bulletin No 1871, Washington State University Extension, Pullman, Washington.
Ley, T.W., Stevens, R.G., Topielec, R.R. and Neibling, W.H. (1994) Soil water monitoring and measurement. *Pacific Northwest Publication PNW0475*. Washington State University, Pullman, Washington.
Liebman, M., Drummond, F.A., Corson, S. and Zhang, J. (1996) Tillage and rotation crop effects on weed dynamics in potato production systems. *Agronomy Journal* 88, 18–26.
McDole, R.E. and McMaster, G.M. (1978) Effects of moisture stress and nitrogen fertilization on tuber nitrate–nitrogen content. *American Potato Journal* 55, 611–619.
McDole, R.E., McMaster, G.M. and Larsen, D.C. (1974) *Available Water-Holding Capacities of Soils in Southern Idaho*. Bulletin No 236, University of Idaho Extension, Moscow, Idaho.
McGee, E., Jarvis, C. and Duncan, H.J. (1985) Wound healing in potato tuber tissue. I. Effects of maturity at harvest. *Potato Research* 28, 91–99.

McGuire, A. (2003) *Using Green Manures in Potato Cropping Systems*. Bulletin EB1951E, Washington State University Cooperative Extension, Pullman, Washington.

Magdoff, F. and van Es, H. (2000) *Building Soils for Better Crops*, 2nd edn. Sustainable Agricultural Research and Education Program, Beltsville, Maryland, 230 pp.

Morris, M. (2006) *Soil Moisture Monitoring: Low Cost Tools and Methods*. A publication of ATTRA – National Sustainable Agricultural Information Service. National Center for Appropriate Technology, Butte, Montana, 11 pp.

Mundy, C., Creamer, N.G., Crozier, C.R. and Wilson, G.L. (1999) Potato production on wide beds: impact on yield and selected soil physical characteristics. *American Potato Journal* 76, 323–330.

Nelson, S.H. and Hwang, K.E. (1975) Water usage by potato plants at different stages of growth. *American Potato Journal* 52, 331–339.

Nolte, P., Miller, J.S., Geary, B.D. and Corsini, D.L. (2003) Disease management. In: Stark, J.C. and Love S.L. (eds) *Potato Production Systems*. University of Idaho Agricultural Communications, Moscow, Idaho, pp. 153–183.

Ojala, J.C., Stark, J.C. and Kleinkopf, G.E. (1990) Influence of irrigation and nitrogen management on potato yield and quality. *American Potato Journal* 67, 29–43.

Pavek, M.J. and Thornton, R.E. (2005) A survey of stand establishment and in-row spacing uniformity in Washington potato fields. *American Journal of Potato Research* 82, 463–469.

Pavek, M.J. and Thornton, R.E. (2006) Agronomic and economic impact of missing and irregularly spaced potato plants. *American Journal of Potato Research* 83, 55–66.

Pavek, M.J. and Thornton, R.E. (2009) Planting depth influences potato plant morphology and economic value. *American Journal of Potato Research* 86, 56–67.

Pavek, J.J., Brown, C.R., Martin, M.W. and Corsini, D.L. (1993) Inheritance of blackspot bruise resistance in potato. *American Potato Journal* 70, 43–48.

Rex, B.L. and Mazza, G. (1989) Cause, control and detection of hollow heart in potatoes: a review. *American Potato Journal* 66, 165–183.

Robinson, D. (1999) A comparison of soil–water distribution under ridge and bed cultivated potatoes. *Agricultural Water Management* 42, 189–204.

Rykbost, K.A., Christensen, N.W. and Maxwell, J. (1993) Fertilization of Russet Burbank in short-season environment. *American Potato Journal* 70, 699–710.

Schotzko, R.T., Hyde, G.M. and Thornton, R.E. (1983) The dollars and cents of the 1982 potato seed size and spacing survey. *Proceedings of the Washington State Potato Conference* 22, 23–29.

Secor, G.A. and Gudmestad, N.C. (1993) Handling and planting seed tubers. In: Rowe, R.C. (ed.) *Potato Health Management*. The American Phytopathological Society, St Paul, Minnesota, pp. 27–34.

Shock, C.C., Zalewski, J.C., Steiber, T.D. and Burnett, D.S. (1992) Impact of early-season water deficits on Russet Burbank plant development, tuber yield and quality. *American Potato Journal* 69, 793–803.

Shock, C.C., Holmes, Z.A., Stieber, T.D., Eldredge, E.P. and Zhang, P. (1993) The effect of timed water stress on quality, total solids, and reducing sugar content of potatoes. *American Potato Journal* 70, 227–241.

Sieczka, J.B., Ewing, E.E. and Markwardt, E.D. (1986) Potato planter performance and effects of non-uniform spacing. *American Potato Journal* 63, 25–37.

Smittle, D.A., Thornton, R.E., Peterson, C.L. and Dean, B.B. (1974) Harvesting potatoes with minimum damage. *American Potato Journal* 51,152–164.

Sparks, W.C. (1958) Abnormalities in the potato due to water uptake and translocation. *American Potato Journal* 35, 430–436.

Stark, J.C. and King, B.A. (1997) *Potato Irrigation Management*. Bulletin No 789, University of Idaho Extension, Moscow, Idaho.

Stark, J.C. and Westermann, D.T. (2003) Nutrition management. In: Stark, J.C. and Love, S.L. (eds) *Potato Production Systems*. University of Idaho Agricultural Communications, Moscow, Idaho, pp. 115–135.

Struik, P.C. and Wiersema, S.G. (1999) *Seed Potato Technology*. Wageningen Press, Wageningen, the Netherlands.

Thornton, M. and Bohl, W. (1998) *Preventing Potato Bruise Damage*. Bulletin 725, University of Idaho Extension, Moscow, Idaho.

Thornton, R.E., Smittle, D.A. and Peterson, C.L. (1973) *Reducing Potato Damage During Harvest*. Bulletin 646, Washington State University Extension, Pullman, Washington.

UNECE Specialized Section on Standardization of Seed Potatoes (2013) International Standard for Certification and Marketing of Seed Potatoes (http://www.unece.org/trade/agr/standard/potatoes/pot_e.html, accessed 2 May 2014).

Van Dam, J., Kooman, P.L. and Struik, P.C. (1996) Effects of temperature and photoperiod on early growth and final number of tubers in potato (*Solanum tuberosum* L.). *Potato Research* 39, 51–62.
Van Emden, J.H. (1965) *Rhizoctonia solani*: results of experiments. *European Potato Journal* 8,188–189.
van Loon, C.D. (1981) The effects of water stress on potato growth, development, and yield. *American Potato Journal* 58, 51–69.
Waddell, J.T., Gupta, S.C., Moncrief, J.F., Rosen, C.J. and Steele, D.D. (2000). Irrigation- and nitrogen-management impacts on nitrate leaching under potato. *Journal of Environmental Quality* 29, 251–261.
Walworth, J.L. and Muniz, J.E. (1993) A compendium of tissue nutrient concentrations for field-grown potatoes. *American Potato Journal* 70, 579–597.
Westermann, D.T., Kleinkopf, G.E. and Porter, L.K. (1988) Nitrogen fertilizer efficiencies on potatoes. *American Potato Journal* 65, 377–386.
Wurr, D.C.E., Fellows, J.R., Hogge, M.C., Booth, D.M. and Allen, E.J. (1992) The feasibility of predicting the number of stems per tuber produced by potato seed stocks from measurements made before planting. *Potato Research* 35, 149–155.

7 Plant–Water Relations and Irrigation Management of Potato

David Levy[1]* and Warren K. Coleman[2]
[1]Institute of Plant Sciences and Genetics in Agriculture, The Hebrew University of Jerusalem, Rehovot, Israel; [2]Potato Research Centre, Agriculture and Agri-Food Canada, Fredericton, Canada (retired)

At the start of the 21st century, the world's population used over 50% of the land's photosynthetic products and no less than 75% of the world's fresh water for agriculture (Rojstazer *et al.*, 2001; Araus *et al.*, 2008; Krausmann *et al.*, 2013). This ability is not sustainable, due to growing population pressures, water resources degradation, and climate change (Schmidhuber and Tubiello, 2007; Foley *et al.*, 2011).

Climate change will affect the production of crops like potato, which are sensitive to water shortages. Predicted fluctuations in weather during the next 50 years include frequent extreme temperatures and increasing drought in a number of important crop production regions (Gornall *et al.*, 2010; Jaggard *et al.*, 2010). These changes will adversely affect potato production in most countries, and will lead to significant declines in tuber production during the 21st century (Hijmans, 2003; Schafleitner *et al.*, 2011).

The potato is the third most important food crop globally (after rice and wheat) in terms of human consumption and has a high nutritional value (CIP, 2013). Under rain-fed conditions, potato productivity of dietary energy is greater than that of wheat, maize, and rice (Renault and Wallender, 2000). However, the potato plant is very susceptible to drought. Drought is an environmental stress imposed by moisture deficit, leading to the reduction of water potential and cell turgor below their optimum values, which leads to disturbances of physiological processes (Turner and Kramer, 1980). The potato plant requires appropriate water management practices to secure yield and quality.

This chapter examines how the potato plant responds to different moisture levels, and outlines advances in dealing with drought and water management. Additional information on plant–water relations can be found in: Kramer (1983); Taylor *et al.* (1983); Harris (1992); Kramer and Boyer (1995); Chaves *et al.* (2003); Gopal and Paul Khurana (2006); Haverkort and MacKerron (2006); Vreugdenhil (2007); Blum (2010); Turkan (2011); Venkateswarlu *et al.* (2011); Yadav *et al.* (2011); and Aroca (2012).

7.1 Potatoes and Drought

Drought continues to be viewed as the "most important physiological stress" affecting potatoes (Burton, 1981, p. 3). Understanding the potato plant's response to this stress is essential for developing appropriate solutions (Moffat, 2002; Araus *et al.*, 2008; Blum, 2011). Several mechanisms involved in coping with drought stress have been described, such as drought avoidance, or drought escape, and drought tolerance (Turner and Kramer, 1980; Chaves *et al.*, 2003; Schafleitner, 2009; Harb *et al.*, 2010; Monneveux *et al.*, 2013).

* E-mail: levy.david@mail.huji.ac.il

Drought avoidance

The synchronization of crop growth and development with water supply is a successful strategy for escaping from the adverse effects of drought (Rockström, 2003; Blum, 2009). Plants that complete their growing cycle before or at the onset of drought are drought-avoiding or drought-escaping plants, since they are seldom severely stressed (Turner and Kramer, 1980). Consequently, early crop maturation may be better suited to regions where growth is limited by environmental constraints such as high temperatures and limited water availability.

A potato model (Haverkort and Kooman, 1997) suggested that early-maturing potatoes are sensitive to early-season transient drought but escape late-season drought and have a better harvest index when compared to late-maturing cultivars. Stark *et al.* (2013) noted that the early-maturing cultivar, Russet Norkotah, was the least sensitive to drought, particularly to late-season drought, as compared to later-maturing cultivars. Under Mediterranean growing conditions, early-maturing potato cultivars had smaller yield losses when compared to later-maturing cultivars that experienced increasing temperatures and evaporation demand during late spring and early summer (Levy, 1986). Drought avoidance requires breeding and selecting potato plants for specific growing sites that are based on the harmonization of plant development with growing conditions and water supply (Levy *et al.*, 2013).

Drought tolerance

Potatoes require a well-managed water supply to secure yield and quality. However, in many major potato production regions, potatoes are exposed to sporadic droughts due to short-term hot and dry spells, an irregular rainy season, or improperly delivered irrigation. Hence, moderate drought tolerance in potato is useful to help cope with seasonal droughts. For example, Stark *et al.* (2013) reported a differential response of six potato cultivars to seasonal drought patterns, while Lynch and Tai (1989) observed significant differences in tuber yields among eight cultivars exposed to water deficit. Both reports showed that potato cultivars varied in their response to drought at different growth stages. These effects indicate that water management should be cultivar dependent when water supply is limited. Due to the importance of relative drought tolerance to growers, many national organizations note the cultivars that are more tolerant to drought than other cultivars. In the UK, potato cultivars are ranked on a scale of drought tolerance that is based on ware yield under field conditions (Harris, 1992).

Several plant traits can help in dealing with drought. For example, cultivars with deep root systems allow water absorption from lower soil layers. In addition, stomata sensitive to water deficits reduce transpiration, osmotic adaptation helps to maintain turgor and assists recovery when potato plants are exposed to an adequate water supply.

7.2 Potato Roots

Potato plants growing from tubers produce branching fibrous roots from adventitious roots associated with above- and belowground shoots (Cutter, 1978; Kratzke and Palta, 1985). The potato plant is shallow rooted and exhibits inhibition of root growth by soil strength levels above 1 MPa.

Currently, a major cause of drought sensitivity is a result of the limited ability of potato roots to transport water efficiently to stems and leaves (Chaves *et al.*, 2003). Essential factors for coping with drought stress include superior root architecture that allows potato roots to grow in high-strength soils with efficient water extraction and transport (Taylor *et al.*, 1983; Iwama and Yamaguchi, 2006; Iwama, 2008; Tardieu, 2012). Increased efficiency for extracting soil water will require major changes to features such as root growth rate, external morphology, and internal structure (Clarke *et al.*, 2003; Boyer *et al.*, 2010; Jaramillo *et al.*, 2013). The root system will also need to possess increased water transport efficiency in order to avoid, or significantly delay, the effects of drought stress (Burton, 1981; Levy *et al.*, 2013).

Stolons and tubers

Lateral shoot buds under the soil surface often develop into diageotrophic stems called stolons.

In response to cues from the environment such as photoperiod, the subapical region develops into a tuber, where over 70% of the dry matter produced by the plant will accumulate (Cutter, 1978). When tuberizing potato plants are exposed to drought stress, an alteration in partitioning occurs that results in reduced starch synthesis, increased sucrose levels, and reduced tuber dry matter (Cutter, 1978; Geigenberger *et al.*, 1997; Levy *et al.*, 2013).

Tuber growth is reduced when tuber water potential falls below –0.4 MPa (Gandar and Tanner, 1976), although it will resume when the plant is rehydrated. However, the effects on total and marketable yield are complex since they depend, inter alia, on such variables as timing and severity of the drought stress, soil characteristics, tuber number, and the simultaneous occurrence of other abiotic stresses (Iwama and Yamaguchi, 2006; Schafleitner, 2009). In addition, drought stress will affect the interactions between developing tubers and metabolic events in aboveground potato shoots, such as photosynthesis (Basu *et al.*, 1999).

7.3 Leaf Growth and Development

There is a close correspondence between leaf and tuber water relations, with inhibition of growth at tissue water potentials below –0.4 MPa (Gandar and Tanner, 1976; Bethke *et al.*, 2009). Maintenance of leaf turgor under mild to moderate drought may be due to one or more changes in such features as osmotic adjustment, altered cell wall elasticity, and reduced cell size (Radin, 1983; Heuer and Nadler, 1998). However, under extreme drought, when water potentials fell below –2.0 MPa, researchers observed a loss of potato leaf viability (Ackerson *et al.*, 1977; Shimshi *et al.*, 1983).

In potato plants, the earliest observed response to drought is stomatal closure. However, with increasing drought stress, the first morphological effect observed is reduced leaf size (Jefferies, 1993; Jefferies and MacKerron, 1993). Other effects include reduced leaf expansion rate, reduced leaf formation, and increased rates of leaf senescence (Fleisher *et al.*, 2008a,b).

The production of smaller leaves is a positive adaptation to an arid environment. For example, small leaves are able to dissipate sensible heat better than larger leaves (Nobel, 1999). In addition, small leaves may possess a greater vein density, and this feature has been associated with increased drought tolerance in a range of plant species (Brodribb and Holbrook, 2003; Scoffoni *et al.*, 2011). Increased drought tolerance may be related to the observation that vein density has been correlated with hydraulic conductivity (Brodribb *et al.*, 2010) and maximum rate of photosynthesis. Whether vein density is optimized at specific levels of drought stress is unknown (Noblin *et al.*, 2008), as is the role played by stomata in the potato plant.

7.4 Stomata and the Cuticle

Stomata serve as major control points for uptake of carbon dioxide and loss of water. For example, on days of high evaporative demand, stomata begin to close when the leaf water potential falls below –0.1 MPa (Levy *et al.*, 2013). As stomata close (e.g. midday depression of stomatal conductance), water loss by transpiration is affected more than carbon assimilation (Yoo *et al.*, 2009). This optimization of carbon gain in relation to water loss may be affected by such diverse factors as xylem sap pH, decreased hydraulic conductance, stomatal heterogeneity, circadian rhythms, xylem-borne abscisic acid (ABA), or may represent a stomatal response to excessive solar radiation (Mott and Buckley, 1998; Brodribb and Holbrook, 2003; Chaves *et al.*, 2003; Buckley, 2005; Liu *et al.*, 2005). Understanding the roles and timing of these factors in the potato plant will be important research goals. However, potato cultivars developed in the future will have to exhibit increased yield per unit of water used, such as increased water-use efficiency (WUE), in conjunction with no significant decrease in harvest index (Yoo *et al.*, 2009; Levy *et al.*, 2013).

Potato cultivars and *Solanum* species exhibit a wide range of stomatal conductance responses that may be due to the size and frequency of stomata as well as their distribution on adaxial and abaxial leaf surfaces (Dwelle *et al.*, 1983; Ekanayake and De Jong, 1992; Chaves *et al.*, 2003; Coleman, 2008). As a cautionary note, increased CO_2 levels associated with climate change may have pronounced effects on stomatal density and patterning (Brownlee, 2001;

Lawson *et al.*, 2002). In addition, increased CO_2 levels may partially alleviate the detrimental effects of drought (Fleisher *et al.*, 2008a,b). The rise in atmospheric CO_2 levels associated with climate change has the potential to increase the photosynthesis of plants, growth of crops, and WUE (Polley, 2002).

Stomatal closure is the dominant factor affecting photosynthesis during moderate drought. Severe drought can inhibit critical metabolic processes (Flexas and Medrano, 2002). With complete stomatal closure, gas exchange shifts from stomatal to cuticular control (Boyer *et al.*, 1997). Cuticular wax may be an important factor in determining drought tolerance (Coleman, 1986; Kosma *et al.*, 2009) and is amenable to genetic analysis (e.g. Lu *et al.*, 2012).

Leaf senescence

Accelerated senescence of leaves is a common response to abiotic stresses such as drought. Because of the loss of major photosynthetic organs, potato plants will often exhibit reduced yields. To optimize the plant's response to drought, manipulation of leaf senescence may be an important goal for crop management (Rivero *et al.*, 2007). However, resistance to drought-induced leaf senescence did not provide a consistent response in soybean and sorghum when their "stay green" lines were studied. An analysis of the benefits of increased functional leaf life for potato production would be a useful goal.

7.5 Osmotic Adjustment

Osmotic adjustment (OA) in potato may be important in dealing with drought stress (Levy, 1983, 1992; Coleman, 1986; Jefferies, 1993; Heuer and Nadler, 1998). In arid climates, potato crops can be exposed to transient drought and heat stress resulting from high ambient temperatures and low air humidity. Leaf wilting is often observed, and may lead to a loss of tuber yield and quality. Leaf turgor may be preserved by OA, and this situation can promote faster recovery on alleviation of drought stress (Levy, 1992).

Solute accumulation during a decline in leaf water potential is important for maintaining leaf hydration, avoiding cellular desiccation damage and promoting drought resistance in crop plants (Jones, 1998; Chaves and Oliveira, 2004; Blum, 2011). In wheat, potassium accumulation under drought stress (controlled by a recessive osmoregulation gene, *or*) has been linked to greater yield (Richards, 2006). Many plant species naturally accumulate glycine, betaine, and proline when subjected to different abiotic stresses. These compounds are thought to be involved in mediating OA and in protecting subcellular structures in stressed plants (Ashraf and Foolad, 2007).

Factors other than OA may play important roles in potato production. In conjunction with water status analysis, the protective roles of small proteins such as dehydrins need to be studied in potato (Cellier *et al.*, 1998, 2000; Jones, 2007).

7.6 Irrigation Management

In view of the great susceptibility of potato to drought, as well as a decrease in water resources, farmers are challenged to save water without sacrificing yield or quality. Optimized water use, avoiding wastage and leakage, and employment of efficient irrigation protocols have been recommended. For example, the introduction of efficient irrigation technologies such as micro- and trickle irrigation systems has been suggested (Lamm *et al.*, 2007). The low-pressure (LPS) drip irrigation system designed to operate at low pressures (0.14–0.21 kg cm^{-2}), and with drippers of low flow rates, was shown to conserve 40% of the irrigation water in potatoes grown in Mexico compared to the common flooding irrigation system (Dowgert *et al.*, 2006).

Attempts to save water have prompted the assessment of deficit irrigation protocols and the development of precision irrigation for reducing inputs and increasing resource-use efficiency to improve sustainability (Fereres and Soriano, 2007; Sepaskhah and Ahmadi, 2010).

Deficit irrigation and partial root zone drying

Potato fields fully covered with green active canopy are commonly irrigated with amounts attempting to restore evapotranspiration demand of the area (ET-100%). The data presented here suggest the crop may have been over-irrigated in some cases, and employing soil-moisture sensors at the root zone may be useful for water saving.

Studies of deficit irrigation (DI) under field conditions have become increasingly important in areas that rely heavily on fresh water supply. In Oregon, USA, Shock and Feibert (2002) studied three DI treatments. At each irrigation application, 100, 70, and 50% of the accumulated evapotranspiration (ETc) was applied. Over 3 years, the potato crop lost total yield and grade under DI, with reduced profitability. Alva (2008) working in Prosser, Washington, USA, with cultivar Ranger Russet obtained similar results. These results demonstrated that DI for potato growing in the semi-arid environment of western USA was not economically viable. Stark *et al.* (2013) working in Aberdeen, Idaho, found that the greatest reduction in total yield and in US No 1 yield occurred when potatoes were irrigated at 70% ET until the beginning of late bulking, when irrigation was terminated. They also reported a differential response to drought among the six cultivars included in their experiments.

Geerts and Raes (2009) discussed DI as a strategy to maximize crop water productivity in dry areas. They state that no single DI strategy applies to all crops, the levels of tolerance to drought differ among crops, at the various stages of plant development for each crop. Hence, scheduling irrigation for the period from tuber initiation through mid-bulking, and selecting cultivars that use less water and/or are less sensitive to water stress, are recommended.

Fabeiro *et al.* (2001), working in a semi-arid climate in Spain, also studied DI in potato and found tuber yield was correlated with the amount of water applied. The most adverse effect on yield occurred when water shortage was imposed during the bulking and maturation stages. However, Darwish *et al.* (2006) did not observe reduced tuber yield of potato cultivar Santana under DI in Lebanon using 80% of measured evapotranspiration compared to the control of 100%. Similarly, Nimah and Bashour (2010) found that irrigating 50–75% of the root zone of potato and pepper gave better yield than the control in a study of potato cultivar Spunta and pepper cultivar Andalus in Lebanon. Water saving of up to 195 mm for potato and 192 mm for pepper was obtained, without reduction in total yield for both crops, by limiting water penetration to the top 50% of the root zone.

Partial root zone drying (PRD) is a modified form of deficit irrigation. This procedure involves wetting only one part of the root zone in each irrigation event, letting the other part of the root zone dry, and then rewetting it by shifting water application to the dry side and letting the other side dry. Xie *et al.* (2012), working in a semi-arid climate in northern China, found that reducing irrigation to 50% of the control did not affect tuber yield, whereas WUE was increased significantly, regardless of the irrigation method. A partial root zone drying (PRD50) irrigation procedure, where furrows were alternately watered and dried in each irrigation cycle, gave the best results. Yield differences among cultivars and locations were indicative of a genotype–environment (G × E) interaction. These interactions probably explain the conflicting data and suggest DI needs to be assessed under local climatic and management conditions.

Muñoz-Carpena *et al.* (2005) found that irrigating tomato with a low-volume/high-frequency (LVHF) soil moisture-based drip irrigation on shallow sandy soil reduced water use while not significantly affecting tomato yield. Tensiometers at the –15 kPa setpoint resulted in 73% reduction in water use compared to a control irrigation treatment, or 50% with respect to the historical evapotranspiration demand of the area (ET-100%). These authors reported the development of an inexpensive irrigation controller that reduced water use up to 61% compared to evapotranspiration-based application and up to 79% when using switching tensiometers. Additional studies are needed to investigate opportunities to conserve water using DI and PRD under different growing conditions and with different potato cultivars.

Precision irrigation

Precision irrigation (PI) is defined as timely and accurate water application in accordance with spatial and temporal soil properties in response to plant requirements during different growth stages (Al-Kufaishi *et al.*, 2006). PI is an important component of the precision agriculture (PA) concept that aims at increased farming efficiency and profitability and enhances environmental sustainability.

Al-Kufaishi *et al.* (2006) used a simulation model to calculate the amount of water required by sugarbeets in a field experiment in Germany. Spatially variable irrigation applied by a central pivot system was compared to uniform irrigation. The comparison revealed that loss of water was greater for uniform applications than variable rate applications (VRA). PI systems applying variable rates of irrigation (VRI) are already available commercially. The variable irrigation is planned according to "management zones" defined by soil type and topography using points collected from a GPS device. The control of VRA (or VRI) using the actual special water status of the crop promises a better response to the crop needs and a better WUE. A remote thermal imagery system is being developed for this purpose.

Canopy temperature, measured by infrared thermometry, provides a non-invasive monitoring of the water status of plants (Cohen *et al.*, 2005). This system can be performed automatically and provides spatial information of surface temperature. This approach enables mapping of canopy temperature variability over large areas.

Canopy temperatures were shown to be well related to soil water within the root zone of cotton, as well as a stomatal conductance index derived from canopy temperature (Padhi *et al.*, 2012). The relatively low spatial resolution of the thermal images necessitated the use of high spatial resolution of red, green, and blue (RGB) images (Moller *et al.*, 2007). Zakaluk and Sri Ranjan (2006) investigated the feasibility of using a digital camera to determine the leaf water potential (ΨL) of potato plants by capturing RGB digital images in the visible region of the electromagnetic spectrum. Their work showed promise of using RGB to assess leaf water potential and soil nitrate status. The crop water stress index (CWSI) is used to normalize the canopy temperature from thermal images (Tc, °C) and water status of the plants (Jones *et al.*, 2002). Analysis of the ground thermal and RGB images from three dates during each of two growing seasons showed highly significant correlations between CWSI and stomatal conductance during the tuber-bulking stage ($R^2 = 0.8–0.96$, $P < 0.01$). Aerial thermal and RGB images revealed significant correlation between CWSI and stomatal conductance ($R^2 = 0.8$) (Rud *et al.*, 2014). Durigon and de Jong van Lier (2013) experimentally compared CWSI to the transpiration reduction function (FRF) using common bean grown in Brazil under full and deficit irrigation. They found canopy temperature combined with CWSI were as effective at predicting plant water stress as soil-based factors like soil water pressure head. The results presented suggest that thermal images may serve as an efficient control tool for variable irrigation rates and should be assessed in potato.

7.7 Potato Responses to Waterlogging

Abiotic stresses that are based on too much water (flooding) or too little water (drought) largely determine the global distribution of plants and agricultural productivity (Visser *et al.*, 2003; Jackson *et al.*, 2009). In a flooded environment, soils rapidly become deficient in oxygen. This condition results in the death of potato plants due to the cessation of aerobic root respiration, loss of water absorption, and loss of ion uptake and transport (Visser *et al.*, 2003). If the flooding incident is brief, for example less than a few hours, potato plants will often survive. Nevertheless, longer-term exposure will usually result in plant death, since potato plants have no or very limited ability to transport oxygen from aboveground shoots to belowground roots (Vasellati *et al.*, 2001). Although attempts have been made to characterize tolerance to waterlogging in crop plants (e.g. Dennis *et al.*, 2000), the development of flood tolerance in potato has not been successful to date.

7.8 Coping with Drought

There are currently seven cultivated potato species and 199 wild potato species found in various regions of the world (Hijmans and Spooner, 2001). Different wild potato species often demonstrate effective adaptive responses to a wide range of abiotic and biotic stresses.

Successful strategies for dealing with drought stress will require breeding for adaptation. Drought resistance traits will need to be selected carefully (Tardieu, 2012). Incorporation of drought tolerance into cultivated potato will involve both traditional breeding as well as analytical breeding with gene transfer. These breeding approaches will necessitate an in-depth knowledge of the potato's ecophysiology (Parry *et al.*, 2005). In addition, a better understanding of potato physiology and adaptation mechanisms will be critical for dealing with future challenges.

Simulation of potato growth and development could provide the necessary insights into adaptation to drought (e.g. Campbell *et al.*, 1976; Spitters and Shapendonk, 1990; Peiris *et al.*, 1996; Jones, 1998; Sinclair, 2000; Sinclair and Muchow, 2001; Passioura, 2007). However, overall success will require collaboration across several disciplines that range from the molecular to the ecological, with close links to climate research (Bazzaz, 2001; Howden *et al.*, 2007; Morison *et al.*, 2008; Weart, 2013).

7.9 Summary

Irrigation management systems are constantly improving and provide the farmer with tools for increased efficiency of water use. Implementation of efficient irrigation systems such as micro-irrigation and drip irrigation was shown to conserve water. Soil moisture sensors are useful monitors of soil water status and are used to control irrigation schedules and quantities. Potato production systems could probably benefit from an LVHF soil moisture-based drip irrigation system (Muñoz-Carpena *et al.*, 2005). The low-pressure/low-flow drip irrigation system (Dowgert *et al.*, 2006) is also a promising system that deserves additional studies.

Current VRI irrigation regimes are based on GPS data of uneven field conditions, topography, microclimate, soil gradients, etc. Precision irrigation is expected to develop further and to provide a reliable representation of the spatial water status of crops, including in the potato crop (Rud *et al.*, 2014). This technology may facilitate VRI based on variations in water requirements among potato cultivars with diverse water needs or different tolerances to drought, and respond in real time to the crop's water requirements at various growth stages. Combining an understanding of potato–water relations with recent advances in irrigation management may provide the tools necessary to conserve water while securing potato yields under challenging climate conditions.

References

Ackerson, R.C., Krieg, D.R., Miller, T.D. and Stevens, R.G. (1977) Water relations and physiological activity of potatoes. *Journal of the American Society for Horticultural Science* 102, 572–575.

Al-Kufaishi, S.A., Blackmore, B.S. and Sourell, H. (2006) The feasibility of using variable rate water application under a central pivot irrigation system. *Irrigation and Drainage Systems* 20, 317–327.

Alva, A.K. (2008) Setpoints for potato irrigation in sandy soils using real-time, continuous monitoring of soil-water content in soil profile. *Journal of Crop Improvement* 21, 117–137.

Araus, J.L., Slafer, G.A., Royo, C. and Serret, M.D. (2008) Breeding for yield potential and stress adaptation in cereals. *Critical Reviews in Plant Sciences* 27, 377–412.

Aroca, R. (2012) *Plant Responses to Drought Stress: From Morpological to Molecular Features*. Springer, New York.

Ashraf, M. and Foolad, M.R. (2007) Role of glycine betaine and proline in improving plant abiotic stress resistance. *Environmental and Experimental Botany* 59, 206–216.

Basu, P.S., Sharma, A., Garg, I.D. and Sukumaran, N.P. (1999) Tuber sink modifies photosynthetic response in potato under water stress. *Environmental and Experimental Botany* 42, 25–39.

Bazzaz, F.A. (2001) Plant biology in the future. *Proceedings of the National Academy of Sciences* 98, 5441–5445.

Bethke, P.C., Sabba, R. and Bussan, A.J. (2009) Tuber water and pressure potentials decrease and sucrose contents increase in response to moderate drought and heat stress. *American Journal of Potato Research* 86, 519–532.

Blum, A. (2009) Effective use of water (EUW) and not water-use efficiency (WUE) is the target of crop yield improvement under drought stress. *Field Crops Research* 112, 119–123.

Blum, A. (2010) *Plant Breeding for Water-Limited Environments*. Springer, New York.

Blum, A. (2011) Drought resistance – is it really a complex trait? *Functional Plant Biology* 38, 753–757.
Boyer, J.S., Wong, S.C. and Farquhar, G.D. (1997) CO_2 and water vapor exchange across leaf cuticle (epidermis) at various water potentials. *Plant Physiology* 144, 185–191.
Boyer, J.S., Silk, W.K. and Watt, M. (2010) Path of water for root growth. *Functional Plant Biology* 37, 1105–1116.
Brodribb, T.J., and Holbrook, N.M. (2003) Stomatal closure during leaf dehydration, correlation with other leaf physiological traits. *Plant Physiology* 132, 2166–2173.
Brodribb, T.J., Field, T.S. and Sack, L. (2010) Viewing leaf structure and evolution from a hydraulic perspective. *Functional Plant Biology* 37, 488–498.
Brownlee, C. (2001) The long and short of stomatal density signals. *Trends in Plant Science* 6, 441–442.
Buckley, T.N. (2005) The control of stomata by water balance. *New Phytologist* 168, 275–292.
Burton, W.G. (1981) Challenges for stress physiology in potato. *American Potato Journal* 58, 3–14.
Campbell, M.D., Campbell, G.S., Kunkel, R. and Papendick, R.I. (1976) A model describing soil–plant–water relations for potatoes. *American Potato Journal* 53, 431–441.
Cellier, F., Conéjéro, G., Breitler, J.C. and Casse, F. (1998) Molecular and physiological responses to water deficit in drought-tolerant and drought-sensitive lines of sunflower: accumulation of dehydrin transcripts correlates with tolerance. *Plant Physiology* 116, 319–328.
Cellier, F., Conéjéro, G. and Casse, F. (2000) Dehydrin transcript fluctuations during a day/night cycle in drought-stressed sunflower. *Journal of Experimental Botany* 51, 299–304.
Chaves, M.M. and Oliveira, M.M. (2004) Mechanisms underlying plant resilience to water deficits: prospects for water-saving agriculture. *Journal of Experimental Botany* 55, 2365–2384.
Chaves, M.M., Maroco, J.P. and Pereira, J.S. (2003) Understanding plant responses to drought – from genes to the whole plant. *Functional Plant Biology* 30, 239–264.
CIP (International Potato Center) (2013) Potato Facts and Figures (http://cipotato.org/potato/facts, accessed 12 September 2013).
Clarke, L.J., Whalley, W.R. and Barraclough, P.B. (2003) How do roots penetrate strong soil? *Plant and Soil* 255, 93–104.
Cohen, Y., Alchanatis, V., Meron, M., Saranga, Y. and Tsipris, J. (2005) Estimation of leaf water potential by thermal imagery and special analysis. *Journal of Experimental Botany* 56, 1843–1852.
Coleman, W.K. (1986) Water relations of two potato (*Solanum tuberosum* L.) cultivars. *American Potato Journal* 63, 263–276.
Coleman, W.K. (2008) Evaluation of wild *Solanum* species for drought resistance. 1. *Solanum gandarillasii* Cardenas. *Environmental and Experimental Botany* 62, 221–230.
Cutter, E.G. (1978) Structure and development of the potato plant. In: Harris, P.M. (ed.) *The Potato Crop*. Chapman and Hall, London, pp.70–152.
Darwish, T.M., Atallah, T.W., Hajhasan, S. and Haidar, A. (2006) Nitrogen and water use efficiency of fertigated processing potato. *Agricultural Water Management* 85, 95–104.
Dennis, E.S., Dolferus, R., Ellis, M., Rahman, M., Wu, Y., Hoeren, F.U., *et al.* (2000) Molecular strategies for improving waterlogging tolerance in plants. *Journal of Experimental Botany* 51, 89–97.
Dowgert, M., Marsh, B., Hutmacher, R., Thompson, T.L., Hannaford, D., Phene, J., *et al.* (2006) Low pressure drip irrigation lessens agricultural inputs (http://www.irrigationtoolbox.com/ReferenceDocuments/TechnicalPapers/IA/2006/012.pdf, accessed 28 May 2014).
Durigon, A. and de Jong van Lier, Q. (2013) Canopy temperature versus soil water pressure head for the prediction of crop water stress. *Agricultural Water Management* 127, 1–6.
Dwelle, R.B., Hurley, P.J. and Pavek, J.J. (1983) Photosynthesis and stomatal conductance of potato clones (*Solanum tuberosum* L.). *Plant Physiology* 72, 172–176.
Ekanayake, I.J. and De Jong, J.P. (1992) Stomatal response of some cultivated and wild tuber-bearing potatoes in warm tropics as influenced by water deficits. *Annals of Botany* 70, 53–60.
Fabeiro, C., Olalla, F.M.D. and de Juan, J.A. (2001) Yield and size of deficit irrigated potatoes. *Agricultural Water Management* 48, 255–266.
Fereres, E. and Soriano, M.A. (2007) Deficit irrigation for reducing agricultural water use. *Journal of Experimental Botany* 58, 147–159.
Fleisher, D.H., Timlin, D.J. and Reddy, V.R. (2008a) Elevated carbon dioxide and water stress effects on potato canopy gas exchange, water use, and productivity. *Agricultural and Forest Meteorology* 148, 1109–1122.
Fleisher, D.H., Timlin, D.J. and Reddy, V.R. (2008b) Interactive effects of carbon dioxide and water stress on potato canopy growth and development. *Agronomy Journal* 100, 711–719.

Flexas, J. and Medrano, H. (2002) Drought-inhibition of photosynthesis in C3 plants: stomatal and non-stomatal limitations revisited. *Annals of Botany* 89, 183–189.

Foley, J.A., Ramankutty, N., Brauman, K.A., Cassidy, E.S., Gerber, J.S., Johnston, M., *et al.* (2011) Solutions for a cultivated planet. *Nature* 478, 337–342.

Gandar, P.W. and Tanner, C.B. (1976) Leaf growth, tuber growth, and water potential in potatoes. *Crop Science* 16, 534–538.

Geerts, S. and Raes, D. (2009) Deficit irrigation as an on-farm strategy to maximize crop water productivity in dry areas. *Agricultural Water Management* 96, 1275–1284.

Geigenberger, P., Reimholz, R., Geiger, M., Merlo, L., Canale, V. and Stitt, M. (1997) Regulation of sucrose and starch metabolism in potato tubers in response to short-term water deficit. *Planta* 201, 502–518.

Gopal, J. and Paul Khurana, S.M. (2006) *Handbook of Potato Production, Improvement, and Postharvest Management*. Haworth Press, New York.

Gornall, J., Betts, R., Burke, E., Clark, R., Camp, J., Willett, K., *et al.* (2010) Implications of climate change for agricultural productivity in the early twenty-first century. *Philosophical Transactions of the Royal Society of London, Series B: Biological Sciences* 365, 2973–2989.

Harb, A., Krishnan, A., Ambavaram, M.M.R. and Pereira, A. (2010) Molecular and physiological analysis of drought stress in Arabidopsis reveals early responses leading to acclimation in plant growth. *Plant Physiology* 154, 1254–1271.

Harris, P.M. (1992) *The Potato Crop*, 2nd edn. Chapman and Hall, London.

Haverkort, A.J. and Kooman, P.L. (1997) The use of system analysis and modeling of growth and development in potato ideotyping under conditions affecting yield. *Euphytica* 94, 191–200.

Haverkort, A.J. and MacKerron, D.K. (2006) *Management of Nitrogen and Water in Potato Production.* Academic Publishers, Wageningen, the Netherlands.

Heuer, B. and Nadler, A. (1998) Physiological response of potato plants to soil salinity and water deficit. *Plant Science* 137, 43–51.

Hijmans, R.J. (2003) The effect of climate change on global potato production. *American Journal of Potato Research* 80, 271–280.

Hijmans, R.J. and Spooner, D.M. (2001) Geographic distribution of wild potato species. *American Journal of Botany* 88, 2101–2112.

Howden, S.M., Soussana, J.F., Tubiello, F.N., Chhetri, N., Dunlop, M. and Meinke, H. (2007) Adapting agriculture to climate change. *Proceedings of the National Academy of Sciences* 104, 19691–19696.

Iwama, K. (2008) Physiology of the potato: new insights into root system and repercussions for crop management. *Potato Research* 51, 333–353.

Iwama, K. and Yamaguchi, J. (2006) Abiotic stresses. In: Gopal, J. and Paul Khurana, S.M. (eds) *Handbook of Potato Production, Improvement, and Postharvest Management*. Haworth Press, New York, pp. 231–278.

Jackson, M.R., Ishizawa, K. and Ito, O. (2009) Evolution and mechanisms of plant tolerance to flooding stress. *Annals of Botany* 103, 137–142.

Jaggard, K.W., Qi, A.M. and Ober, E.S. (2010) Possible changes to arable crop yields by 2050. *Philosophical Transactions of the Royal Society of London, Series B: Biological Sciences* 365, 2835–2851.

Jaramillo, R.E., Nord, E.A., Chimungu, J.G., Brown, K.M. and Lynch, J.P. (2013) Root cortical burden influences drought tolerance in maize. *Annals of Botany* 112, 429–437.

Jefferies, R.A. (1993) Responses of potato genotypes to drought. I. Expansion of individual leaves and osmotic adjustment. *Annals of Applied Biology* 122, 93–104.

Jefferies, R.A. and MacKerron, D.K.L. (1993) Responses of potato genotypes to drought. II. Leaf area index, growth and yield. *Annals of Applied Biology* 122, 105–112.

Jones, H.G. (1998) Stomatal control of photosynthesis and transpiration. *Journal of Experimental Botany* 49, 387–398.

Jones, H.G. (2007) Monitoring plant and soil water status: established and novel methods revisited and their relevance to studies of drought tolerance. *Journal of Experimental Botany* 58, 119–130.

Jones, H.G., Stoll, M., Santos, T., de Sousa, C., Chaves, M.M. and Grant, O.M. (2002) Use of infrared thermography for monitoring stomatal closure in the field: application to grapevine. *Journal of Experimental Botany* 53, 2249–2260.

Kosma, D.K., Bourdenx, B., Bernard, A., Parsons, E.P., Lu, S., Joubes, J., *et al.* (2009) The impact of water deficiency on leaf cuticle lipids of Arabidopsis. *Plant Physiology* 151, 1918–1929.

Kramer, P.J. (1983) *Water Relations of Plants*. Academic Press, New York.

Kramer, P.J. and Boyer, J.S. (1995) *Water Relations of Plants and Soils*. Academic Press, New York.

Kratzke, M.G. and Palta, J.P. (1985) Evidence for the existence of functional roots on potato tubers and stolons: significance in water transport to the tuber. *American Potato Journal* 62, 227–236.

Krausmann, F., Erb, K.H., Gingrich, S., Haberl, H., Bondeau, A., Gaube, V., *et al.* (2013) Global human appropriation of net primary production doubled in the 20th century. *Proceedings of the National Academy of Sciences* 110, 10324–10329.

Lamm, F.R., Ayars, J.E. and Nakayama, F.S. (2007) *Microirrigation for Crop Production: Design, Operation, and Management*. Elsevier, Amsterdam, the Netherlands.

Lawson, T., Craigon, J., Black, C.R., Colls, J.J., Landon, G. and Weyers, J.D.B. (2002) Impact of elevated CO_2 and O_3 on gas exchange parameters and epidermal characteristics in potato. *Journal of Experimental Botany* 53, 737–746.

Levy, D. (1983) Varietal differences in the response of potatoes to repeated short periods of water stress in hot climates. I. Turgor maintenance and stomatal behavior. *Potato Research* 26, 303–313.

Levy, D. (1986) Genotypic variation in the response of potatoes (*Solanum tuberosum* L.) to high ambient temperatures and water deficit. *Field Crops Research* 15, 85–96.

Levy, D. (1992) Osmotic potential of potatoes subjected to a single cycle of water deficit. *Potato Research* 35, 17–24.

Levy, D., Coleman, W.K. and Veilleux, R.E. (2013) Adaptation of potato to water shortage: irrigation management and enhancement of tolerance to drought and salinity. *American Journal of Potato Research* 90, 186–206.

Liu, F., Jensen, C.R., Shahanzari, A., Andersen, M.N. and Jacobsen, S.E. (2005) ABA regulated stomatal control and photosynthetic water use efficiency of potato (*Solanum tuberosum* L.) during progressive soil drying. *Plant Science* 168, 831–836.

Lu, S., Zhao, H., Des Mariais, D.L., Parsons, E.P., Wen, X., Xu, X., *et al.* (2012) Arabidopsis ECERIFERUM9 involement in cuticle formation and maintenance of plant water status. *Plant Physiology* 159, 930–944.

Lynch, D.R. and Tai, G.C.C. (1989) Yield and yield component response of eight potato genotypes to water stress. *Crop Science* 29, 1207–1211.

Moffat, A.S. (2002) Finding new ways to protect drought-stricken plants. *Science* 296, 1226–1229.

Moller, M., Alchanatis, V., Cohen, Y., Meron, M., Tsipris, J., Naor, A., *et al.* (2007) Use of thermal and visible imagery for estimating crop water status of irrigated grapevine. *Journal of Experimental Botany* 58, 827–838.

Monneveux, P., Ramirez, D.A. and Pino, M.T. (2013) Drought tolerance in potato (*S. tuberosum* L.): can we learn from drought tolerance research in cereals? *Plant Science* 205–206, 76–86.

Morison, J.I.L., Baker, N.R., Mullineaux, P.M. and Davies, W.J. (2008) Improving water use in crop production. *Philosophical Transactions of the Royal Society Series B* 363, 639–658.

Mott, K.A. and Buckley, T.N. (1998) Stomatal heterogeneity. *Journal of Experimental Botany* 49, 407–417.

Muñoz-Carpena, R., Dukes, M.D., Li, Y.C.C. and Klassen, W. (2005) Field comparison of tensiometer and granular matrix sensor automatic drip irrigation on tomato. *HortTechnology* 15, 584–590.

Nimah, N.M. and Bashour, R. (2010) Modeling deficit irrigation water productivity as a function of crop root depth. Paper No 1008942. 2010 American Society of Agricultural and Biological Engineers (ASABE) Annual International Meeting, Pittsburgh, Pennsylvania.

Nobel, P.S. (1999) *Physicochemical & Environmental Plant Physiology*, 2nd edn. Academic Press, San Diego, California.

Noblin, X., Mahadevan, L., Coomaraswamy, I.A., Weitz, D.A., Holbrook, N.M. and Zwieniecki, M.A. (2008) Optimal vein density in artificial and real leaves. *Proceedings of the National Academy of Sciences* 105, 9140–9144.

Padhi, J., Misra, R.K. and Payero, J.O. (2012) Estimation of soil water deficit in an irrigated cotton field with infrared thermography. *Field Crops Research* 126, 45–55.

Parry, M.A.J., Flexas, J. and Medrano, H. (2005) Prospects for crop production under drought: research priorities and future directions. *Annals of Applied Biology* 147, 211–226.

Passioura, J. (2007) The drought environment: physical, biological and agricultural perspectives. *Journal of Experimental Botany* 58, 113–117.

Peiris, D.R., Crawford, J.W., Grashoff, C., Jefferies, R.A., Porter, J.R. and Marshall, B. (1996) A simulation study of crop growth and development under climate change. *Agricultural and Forest Meteorology* 79, 271–287.

Polley, H.W. (2002) Implications of atmospheric and climate change for crop yield and water use efficiency. *Crop Science* 42, 131–140.

Fig. 8.1. Mr Andrew Houser with the Colorado Potato Certification Service in the USA, conducting a second field inspection for certified seed.

During final inspection, each lot's information for the growing season is examined, and the agency makes certain there are no problems leading to de-certification prior to harvest. Additionally, it may be necessary or required to kill or remove the haulms (vines) chemically to control the potential for late-season virus spread by vectors that often peak late season (NAK, 2013).

During the inspections, a set of disease tolerances are followed that dictate where the seed lot fits regarding classification, and whether or not the seed is eligible for certified status. If a lot is deemed too high for any given tolerance, it may be downgraded to the next appropriate classification or rejected from certification. Any lot rejected from certification is still eligible to be moved into the commercial potato channels for consumption as table-stock (fresh market) or processing potatoes.

Disease testing

Disease and pathogen testing has improved dramatically in the past few years. The ELISA test (enzyme-linked immunosorbent assay) is the primary workhorse for the seed potato industry and is used to detect viruses and bacterial problems (Clark and Adams, 1977; Knutson, 1998). However, variations and technologies such as the use of polymerase chain reaction (PCR) techniques have made detecting difficult pathogens and verifying strain variations in viruses much easier. For example, Singh *et al.* (2013) have recently shown that a real-time, reverse transcription (RT)-PCR assay is quite effective in detecting *Potato virus Y* (PVY) in dormant tubers. Testing via PCR allows programs to test stocks almost any time during the year. Many areas of the world require soil testing prior to planting for verifying that nematodes are not present in the field or that powdery scab is within a reasonable level for seed production.

It is fairly common to have foliage tested during the growing season for various viruses like *Potato virus X* (PVX), PVS, and PVY, as well as other potential problems. This is true both in the field during the growing season and in post-harvest test plots (Sather *et al.*, 2013). With the explosion of different strains and their impact on crops, PVY has become a worldwide concern (Crosslin, 2013). In response, programs have been devoting more resources to testing for the pathogen, and have made changes in the manner in which seed potatoes are produced. It is also standard to have the progeny tubers or their vegetative growth tested for virus and bacterial pathogens (postharvest testing). Laboratory testing has become one of the most effective mechanisms to assist certification programs in making good science-based decisions about the health of the certified seed lots.

Disease tolerances

Disease tolerances are established for varietal mixtures and disease content. They tend to be highest in early inspections and ratchet down for later inspections. This gives the seed producers the opportunity to remove diseased plants from the field between inspections.

Roguing, or the removal of diseased plants from the field, can be quite effective for certain virus problems such as leaf roll and the mosaic viruses (PVY, PVA, PVM), but less effective to non-effective for viruses that may be latent or mild in the crop, depending on strain or cultivar characteristics, including certain PVY strains, PVX, and PVS (Hane and Hamm, 1999; Whitworth *et al.*, 2010; Davidson *et al.*, 2013). Additionally, roguing is ineffective for bacterial diseases such as blackleg/soft rot and bacterial ring rot.

During the inspections, the certification personnel make precise plant counts (plants per hectare) within the seed lot, and often have minimal count numbers like 2500 plants or 100% of the plants grown for small lots of 1 ha or less, or 2500 plants ha^{-1} for larger lots. Inspections are conducted across representative portions of the lot, during which time the number of plants that have disease symptoms, varietal mixtures, or other problems are recorded during the count. The inspector may make a percentage count on each of the problems observed and match the percentage to the tolerances by seed classification to ensure the lot qualifies as certified seed.

As part of the certification process, each country has a set of standards regarding disease and pest tolerances. For example, in the USA and Canada, the State/National Harmonization Program sets the maximum tolerances for certified seed potatoes after the final field test. The tolerance is 1.0% for leaf roll, 2.0% for mosaic, with a total virus content of 3.0%, and 0.5% for variety mixture, and zero tolerance for bacterial ring rot (USDA, 2006). In comparison, the European Union allows maximum tolerances for certified seed potatoes at 10.0% for viruses, with no more than 2.0% severe, 2.0% for blackleg, 0.5% for variety mixture, and zero tolerance for bacterial ring rot, potato cyst nematode, potato wart, brown rot, and potato spindle tuber viroid (UNECE, 2013b).

The term "zero tolerance" means none of the disease or pest of interest was found during normal visual inspections of the crop, or during routine testing of the soil and foliage. If a zero tolerance disease or pest is found at any time during a seed lot's life, the lot will be de-certified, and often the farm or certain production fields put on quarantine status. Certification, however, is not a guarantee that the lot inspected is free from any zero tolerance disease or pest (Sather *et al.*, 2013).

The Specialized Section on Standardization of Seed Potatoes within the United Nations Economic Commission for Europe (UNECE) has spent many years developing tolerances and seed certification practices to help standardize the certification process (UNECE, 2013). To quote, "It is a unique international frame of reference, covering all aspects related to seed-potato certification: (a) varietal identity and purity; (b) genealogy and traceability; (c) diseases and pests; (d) external quality; (e) sizing; and (f) labeling" (UNECE, 2013, p. 1). The UNECE section has participants from most regions of the world, and members representing all European countries, the USA, and Canada. This is an important framework within the world's seed certification community because it is a global representation.

Postharvest testing

When lots have been found to meet the applicable rules, regulations, and disease tolerances, the grower is allowed to harvest and move the potatoes into an approved storage. Certification officials will conduct a storage inspection to ensure the lots have been stored in a manner consistent with good seed practices, have appropriate isolation from non-certified stocks, can be verified by location information, and have no obvious disease issues. Also, growers will select a subset of tubers at harvest representative of each individual seed lot for additional postharvest testing.

Evaluation of tubers for additional testing may be carried out in a number of ways. Tubers may have dormancy broken and be planted in a suitable off-season location for growth and inspection. This is common in the USA and Canada, where the samples are planted during the winter months in temperate southern climates like Florida, Arizona, California, and Hawaii, or in a greenhouse.

After plant foliage reaches a 40–50 cm height, inspections are conducted by official certification personnel, who look for disease problems, varietal mixtures that may have gone unnoticed during the growing season, or viruses that were spread too late in the season to detect. Inspections are one of the main methods for determining if undetectable levels of disease have moved into the lot, if cultivar mixing took place during harvest or storage, or if any chemical carry-over has occurred. In some cases, these evaluations are the final tool to determine if a lot should be eligible for sale as certified seed.

In most other cases, postharvest testing is used to determine re-certification eligibility for the following year. If field growth is not possible or preferred, agencies will use greenhouses to grow the tubers and either visually read or use laboratory testing techniques to examine the tubers for the presence of various pathogens—especially virus-related problems (NIVAP, 2005). Finally, agencies routinely test individual tubers for the presence of specific issues (primarily viruses) in lieu of field or greenhouse growth.

Testing can be very effective and is useful in separating out lots exceeding tolerance levels for certified seed. The disadvantages are that testing must be pathogen specific (thus chemical carry-over may be difficult, if not impossible, to detect), and often costs are higher. However, testing can be performed at almost any time after the sample has been taken and stand issues, insect problems, and other normal situations of field growth are not a problem.

Once again, disease tolerances are employed to verify seed lot status. The earlier the generation or class of seed, the lower the tolerances allowed. In the USA, the State/National Harmonization Program allows for no more than 1% of the samples to be infected with mosaic-type viruses, and no more than 0.5% infected with *Potato leaf roll virus* (USDA, 2006). In the Netherlands, however, each class of seed falls within a certain range of virus allowed with class S (the basic seed), having a tolerance of 0 plants out of 200, while class C (the oldest generation) is allowed a tolerance of 10 plants out of 100 (NIVAP, 2005).

Storage inspections

In countries with certification systems, storage inspections are conducted on each facility holding seed potatoes. Inspectors check the adequacy of the storage units to be sure they are capable of holding seed at the appropriate temperatures (3–4°C), relative humidity (95% +), and that air flow into the bins is adequate to prevent unwanted sprouting. Often, seed lots are harvested and stored in cribs (typically holding around 0.5 to 1 t per unit), which may be stacked to fill the storage area or are stored in bins as a bulk unit. Bins for bulk storage are normally in the 450–1200 t range, to maintain adequate isolation and manage the environment appropriately.

Inspectors also check the location of each lot within the structure, verify the number of metric tonnes stored, and verify each seed lot is adequately identified by a label or tag. Seed lots within the storage unit are also examined for the presence of tuber issues (rots, cultivar mixing, etc.), and maps of the storage unit with seed lot locations are produced. These steps allow each seed lot to have complete traceability from the application process, to the field growth, and finally to the storage unit. Once in storage, no sprout inhibitors are allowed on the seed to prevent later problems with germination.

Tagging and shipping point inspection

When seed lots are sold, they are prepared for delivery to the buyer. No seed potatoes can be moved into the marketplace unless they have had an official shipping point inspection and are tagged as certified seed. Shipping point inspections are conducted to confirm each seed lot meets the grade standards for diseases and size, has contaminants (such as adhering soil) within tolerances, and has appropriate physiological condition (no excessive sprouting, bruising, or other issues indicative of tuber problems). It is during these inspections that samples are taken from the lot as it is being loaded into the shipping conveyance. After examining each sample, the inspector can make a determination about the seeds' status.

Common tuber diseases examined during inspection include wet rot, dry rot, and late blight, and skin conditions such as scab, silver scurf, *Rhizoctonia* damage, and bruising. Each of these must meet the appropriate tolerances for the class being shipped. It is also during this time that phytosanitary aspects are considered.

Depending on where the seed is destined, either in-country or as an international export, the phytosanitary restrictions for each location receiving seed must be met and officially documented by a government agent. Meeting region-specific phytosanitary conditions assures the buyers they will not be receiving unwanted or non-endemic diseases or pests in the seed potatoes. Seed potatoes can be an effective mechanism for transporting diseases and pests between regions. Therefore, it is critical that only seed meeting the specific phytosanitary restrictions of each region be moved.

The final step of the certification process is tagging the seed lot with an official certification document. Documentation includes a substantial amount of information describing the seed lot's identity, and assists in tracing the seed lot for re-certification eligibility.

Information on the tags (for individual sacks or smaller containers) or bulk certificates includes: the term "certified seed potatoes"; the country or state of origin; the seed grower's name and address; the cultivar; the unique lot identification; year of production; date of certification; grade of the potatoes (often designated by the color of the tag); the class (often designated by the color of the tag); net weight; and other pertinent information such as size or phytosanitary specifics. Usually, a seal with a unique identifying number for each load is attached to the conveyance (truck, bin, rail car, container, etc.) after shipping point inspection to prevent unwarranted additional inspections or tampering of the seed (NIVAP, 2005; SASA, 2013; Sather *et al.*, 2013). The seal is to be removed by the buyer or recipient of the seed, and together with the official paperwork, constitutes the final act in the certification process. Buyers of certified seed recognize these steps and expect each certification program will honor their commitment to this process.

8.5 Sanitation

Sanitation is one of the single, most critical components of any seed potato production operation (Secor *et al.*, 1985). Sanitation at every stage of production is not only essential but also keeps pathogens and pests from moving into the operation. If the field soil is contaminated with a pest, sanitation will not help the potatoes within that field, but can help keep the pest from moving from one field to the next. Each phase of a seed operation should have sanitation steps, but they are particularly important in phases where the tubers come into contact with other tubers and/or machinery, such as in sorting, cutting, planting, harvesting, and loading.

In general, pathogens and pests move from one tuber to the other when an infected tuber comes into contact with either equipment or other tubers. A prime example is the spreading of soft rot through active bacteria on a rotting potato, and subsequent tuber-to-tuber contact during movement operations, which transfers the bacteria between tubers. If conditions are right, then the previously non-infested tuber can become infected and start to decompose (Bohl and Johnson, 2010).

Proper sanitation starts with the cleaning of equipment, storage units, floors, and building surfaces with steam to remove the adhering soil, and is followed with the proper application of an effective disinfectant. Disinfectants must be applied to all surfaces at the appropriate concentration and left for the recommended exposure time (usually 5–10 min). There are many disinfectants available, so it behooves the producer to understand the effectiveness of each disinfectant (e.g. which pathogens it controls), whether or not it is deactivated by organic matter, hard water, or is corrosive to metal, and the safety information. When used properly on equipment and storage surfaces, effective sanitation can go a long way toward stopping the spread of several serious pathogens on seed potatoes, including bacteria, viruses, fungi, and certain pests.

8.6 Early-Generation Stocks: Types and Production Systems

The starting point for global certification programs is the use of tissue culture *in vitro* stocks. The stocks are tested for pathogens or line selections that have been isolated and tested for pathogens themselves. In developed regions, most certification programs use tissue culture stocks and either the plantlets or minitubers produced from these cuttings.

The recent advances in manipulating *in vitro* potato stocks through a variety of methods has made it possible to propagate large numbers of fully tested stocks to use as the origin for field production (Dodds, 1988). *In vitro* stocks are a standard mechanism for introducing new germplasm into the certification system. Sources of new germplasm are important because they bring improved varietal characteristics and disease resistance to the potato producers, whether seed or commercial (de Bokx and van der Want, 1987).

Sources of starting material for pre-basic seed

Pre-basic, non-field-raised seed stocks can come from a variety of methods. *In vitro* cuttings and plants are important for minituber production in protected greenhouses or screenhouses (production in field soil protected by insect-proof screening which lets in air but keeps out insects). Microtubers can be produced *in vitro* and used either in the greenhouse or field. Cuttings from disease-free microprogated plantlets grown *in vivo* can be used to produce more plants cheaply for minituber production. Sprouts can be broken off of mother stock/minitubers to be rooted and planted in the field. Finally, tissue culture plantlets can be sown directly in the field for the first field generation.

Basic micropropagation techniques

Micropropagation of tissue culture plantlets has been used extensively in crop production since the early 1970s, but was adopted by the potato industry in the 1980s (Roca *et al.*, 1978; Dodds, 1988). Tissue culture multiplication begins with plantlets obtained from mother stocks. The mother stocks are derived from either breeding material or field-grown tubers that have been selected for appearance and yield and verified as the true-to-type cultivar. Tubers are cleaned, disinfected, and sprouted. The sprouts are then placed into artificial media under aseptic conditions (Murashige and Skoog, 1962; Xie *et al.*, 2004) (Fig. 8.2).

After growth, each mother plant is tested for pathogens (mainly viruses and bacteria). Typically, up to ten individual plantlets representing the clone bank for the cultivar will be kept for the long term. These will be tested on an

Fig. 8.2. Micropropagated tissue culture plantlets in a growth room.

annual basis and have periodic grow-outs in the greenhouse to verify they have not mutated and are true to type (Stark and Love, 2003; Sather *et al.*, 2013). These mother stocks then become the source for increasing the cultivar via micropropagation.

The multiplication takes place by cutting between the individual nodes of each plant and inserting these nodes into artificial media under aseptic conditions. New plantlets grow from each axillary bud within a short 2- to 3-week period and the plantlets are available for cutting and propagating. The multiple cycles of propagation make it quite easy to increase plant numbers rapidly without the spread of pathogens. Using these techniques, it is possible to increase one plantlet to over 2500 within a period of 3–4 months. After the desired quantity is achieved, the producer can move the plantlets into the greenhouse or screenhouse to produce minitubers. Alternatively, they may be moved directly into the field as plantlets after hardening, or for other uses as described above.

It is also possible to increase the number of plantlets by cutting the plantlets while they are in a seedbed substrate. This is a reliable method for multiplying seed tubers if the *in vitro* plantlets are of high quality. In Dalat, Vietnam, growers cut the plantlets from the seedbed over 20 times through a 6-month growing period (Uyen and van der Zaag, 1985). Producing plant cuttings in the seedbed can also reduce space while reducing the overall costs of the production of minitubers.

The methods used in tissue culture have not changed greatly over the decades, and information is readily available online at industry websites such as the Food and Agriculture Organization (FAO) and the International Potato Center (CIP). There have been many innovations and improvements specific to potato tissue culture, such as the use of containers with new shapes (square or round boxes, tubes, etc.), the use of improved media (liquid and solid) for growing specific cultivars, media formulations (with or without sugar), automatic filling of the vessels for large-scale production, use of energy-saving lamps or natural lighting, and light-reflecting shelves. With innovations and improvements, the efficiency of producing *in vitro* plantlets has increased greatly, while simultaneously decreasing the costs of multiplication. There is much potential for a greater use of the technology around the world, even in relatively poor regions.

Minitubers

The most common form of tissue culture material used is the minituber (Fig. 8.3). Minitubers are produced in greenhouses from clean plantlets using sterile soils, grown in systems using a nutrient film technique (Wheeler *et al.*, 1990) or a hydroponic system (Rolot and Seutin, 1999; Chang *et al.*, 2012). All methods allow the producer to use fewer tissue culture plantlets for the process and still produce high numbers of minitubers (multiple tubers per plant) grown in a soil-less environment, thus removing the threat of soilborne diseases. Additionally, the tubers tend to be very uniform in size.

However, the disadvantage to either system is the high investment cost to start. The proliferation of various techniques to produce minitubers has been beneficial in reducing the costs of the initial starting material. Research has been conducted to increase the yields of minitubers. Milinkovic *et al.* (2012) discuss the increase in the yield of tubers when an induction period has

Fig. 8.3. Minituber production from a greenhouse facility in Colorado, USA.

been used between the *in vitro* stages of plantlet growth and minituber production. The growing period was extended and a hardening period was introduced with low temperatures during the *in vitro* growth, resulting in up to 97% more minitubers than when no induction period was used. Van der Veeken and Lommen (2009) demonstrated that planting density affected minituber production; at low planting densities, higher numbers of tubers were initiated, but they were somewhat smaller.

When produced in a greenhouse, the facility must be cleaned and disinfected, as well as items used for planting, including benches, pots, trays, etc. The facility must be sanitary and insect free, and must be maintained with strict sanitation for all personnel entering the greenhouse. The production cycle can take place at any time during the year, but often the costs associated with maintaining the greenhouse become prohibitive during the cooler winter months due to heating, and plant growth may not be as successful. Supplemental lighting may help with growth during the off-season times. Potting mixes are varied, but must originate from sterile sources with no soilborne pathogens or weeds. Insect control is essential to prevent the movement of viruses into the stocks. Additionally, certification officials must inspect the plants (usually twice during their growth) to verify cultivar, growing conditions, greenhouse suitability, and tuber identification, among other considerations (Sather *et al.*, 2013).

At harvest, representative tuber samples are taken for pathogen testing. Harvested material is stored in open mesh bags, typically in large coolers with appropriate humidity and airflow, similar to typical commercial potato storage. It is understood by the producers that the minitubers may need longer storage times to enhance dormancy break and/or to fit the timing of greenhouse production with subsequent field production. A good discussion of these factors is presented by Stark and Love (2003).

Minitubers should be sized prior to planting, for ease in planting and to keep emergence uniform. Producers should be aware of minitubers' slower emergence than tubers from field-grown crops, due to their longer dormancy. Also, strict sanitation and isolation are critical to keep diseases and pests from the initial field crop. Sanitation also includes using land suitable for production of the earliest-generation material, soil that has been tested and found free of major soilborne pests, rotating land formerly used for field-grown potatoes, and incorporating proper insect control techniques.

Microtubers

Improving the efficiency of minituber production by microtubers is a reliable, year-round laboratory solution, and negates the use of plantlets. The method was first studied in the late 1970s and early 1980s, and was implemented shortly thereafter in China and the Republic of Korea (Donnelly *et al.*, 2003). Microtuber production is a method of producing small tubers *in vitro* from tissue culture plantlets. They have many uses, and today most countries and institutions use microtubers in commercial seed production.

Microtubers can be produced year-round, have high productivity per square meter (200,000 microtubers m^{-2}), are a convenient size for storage and transportation, and when replacing plantlets, maintain uniformity during emergence in a greenhouse setting. However, the microtubers should be past dormancy break and ready to grow. One method described by Ewing *et al.* (1987) was simply to cut the tubers in half to assist in dormancy break.

Microtubers have also been used in the field rather than minitubers. While research has been conducted on the use of microtubers in this way (Kawakami *et al.*, 2003; Pruski *et al.*, 2003), they tended to be later in growth and had reduced yields. Other disadvantages of using microtubers for field growth are the need for a longer growing period due to increased dormancy and the requirement for specialized techniques for planting. Also, there tends to be a lower success rate in adapting microtubers to field conditions because of their size (Liu *et al.*, 2006; Qui, 2009).

Plantlets for field production

Plantlets may be used for field production, and although risky due to adverse weather and other factors, can be quite successful. Growers often use plantlets when there is a need for a quick turnaround time frame for producing early-generation material. Plantlets are easy to bulk up, but must be hardened properly prior to transplanting them in the field. Producers must take care in transplanting the stocks, to avoid root damage, and to place the plantlet properly in the field soil. Plants must be hydrated and a transplant fertilizer used during the planting to help the roots grow. Transplants must send roots into the fertilizer zone within a short period of time to be successful. Normally, plants in the 12–15 cm range with good roots are appropriate. Plants should be planted with only the top two to three leaves showing above the soil line. Irrigation should be immediate and proper transplant techniques used until the plants have begun to grow in the field and reached the 25–35 cm range in height.

Since plantlets are extremely susceptible to extreme weather such as hail, row covers may be used during their early growth. After plants are actively growing, producers can make rows, hand weed, etc. Once the plants are fully established, producers should avoid mechanical operations to prevent disease spread and root pruning (Stark and Love, 2003). One major concern regarding the use of plantlets in the field is the fact that they tend to remain juvenile longer into the season and usually have only one stem. This makes them ideal candidates for vectors and in-season spread of viruses like PVY.

Single leaf cuttings to produce minitubers

Single leaf cuttings from tissue culture, disease-tested plantlets can also be used to produce minitubers. Haapala (2005) and Haapala *et al.* (2008) discuss the cost-effectiveness and minituber increase potential by using *in vivo* production of single-node leaf buds or other small plant material. The cuttings are placed into growing rooms in flats containing non-sterile soil in a low-light, protected environment and allowed to root and grow, and then cut again within about 3 weeks. Because of their size, space requirements are minimal (up to 500 cuttings m^{-2}), with up to 100,000 progeny produced within 6 months from a single cutting. After obtaining the needed numbers, the plants can be planted into a greenhouse or similar clean production environment to produce minitubers. It is essential that starting material originate from disease-free, *in vitro* plantlets produced under normal aseptic conditions.

Sprouts derived from tubers

After clean, disease-free tubers (such as minitubers) have sprouted, the young sprouts can be removed from the mother tubers and transplanted in the greenhouse to produce plants or in the field to produce seed potatoes. In specialized greenhouse or screenhouse facilities, the tubers can be sprouted and the sprouts collected for planting. When enough seed tubers are not available, this method can increase the multiplication rate four to five times, allowing up to 525 kg of seed tubers planted for each hectare (Yan *et al.*, 1999).

8.7 Field Production

Field production of seed potatoes is more complicated than raising commercial table-stock potatoes. While many of the principles are similar, the seed requires additional steps to ensure disease issues do not become problematic. Also, seed potato growers must produce a crop suitable for seed the following year while meeting the requirements of the certification process.

The use of early-generation, fully tested tissue culture stocks has greatly improved the chances of a grower starting properly, with large numbers of pre-basic or nuclear seed stocks (first year in the field). However, the high cost of production, which includes the laboratory phase (whichever system is used), the growth phase (greenhouse or screenhouse), and the small size of the minitubers, which limits yield, results in pre-basic seed that is useful only for the seed producer. Additionally, having seed stocks of the appropriate physiological age with uniform growth characteristics and quick germination is critical for commercial operations (Iritani and Thornton, 1984; Bryan, 1989).

Field selection is the first criteria a seed grower must meet. In many countries, the soil must first be tested for the presence of specific pests such as potato cyst nematode or root-knot nematode (NAK, 2013; Sather *et al.*, 2013). If the results are negative, then a grower may be able to use the field for seed production. Additionally, most regions have requirements for isolation of early-generation stocks, 2- to 4-year crop rotation requirements between potato crops, and regulations allowing only certain fields to be planted to seed.

Production practices for seed potatoes in the field are similar to those used to produce commercial potato crops, with some notable exceptions. Often, seed lots are separated by one or two blank rows, to maintain lot identity and reduce cultivar mixing at harvest. Lots are planted by generation or class, with the earliest-generation lots grouped together to provide both isolation and identity, and the later-generation lots planted separately to prevent potential disease spread to the pre-basic material. The spread of disease in later-generation lots is common, since they tend to require larger areas of land than the early-generation material and more mechanization for field operations (Stark and Love, 2003).

A second requirement is the need to control adequately insects that act as vectors to problematic diseases like *Potato leaf roll virus* and the mosaic viruses (Radcliffe and Ragsdale, 2002). The use of an effective preventive insecticide early season with periodic applications of other insecticides or crop oils during the season is the cornerstone for any effective insect management scheme (Doring *et al.*, 2006). Other pests such as Colorado potato beetle, leafhoppers, and psyllids also must be controlled during the season.

Seed producers can use strategies to help earlier-generation material stay clean by isolating spatially (away from diseased fields) or intragenerationally (within the field surrounded by plants either resistant to virus spread or that are low spread risks). Isolation can be especially effective to help reduce the spread of vector-borne viruses such as PVY into the material. Additionally, growers can manipulate the planting and vine killing dates of seed to avoid the times in the season when aphid vectors are more prevalent (Davidson *et al.*, 2013). Another strategy is to use a green border of a non-potato crop, such as winter wheat or sudan grass, to allow aphid vectors to feed and remove the virus particles from their stylet (sucking mouth parts), as in the case of PVY (Radcliffe and Ragsdale, 2002).

A third difference between commercial and seed production is the use of roguing, to remove infected plants or varietal mixtures from a lot to help reduce the number of diseased or incorrect plants within a seed lot. Roguing takes place typically prior to the first inspection and up to the second inspection. Certain diseases cannot be rogued from a lot, such as bacterial ring rot, blackleg, brown rot and soilborne problems like nematodes. Roguing is normally used to remove virus-infected plants, but is not useful if the number of infected plants is too high (usually above 1–2% of the crop).

Finally, it is critical that producers control disease spread within the crop and from sources outside of the crop. There are numerous sites within an operation that need attention. As described earlier, all surfaces that may come into contact with the tubers must be cleaned and disinfected on a regular basis and between operations or when switching seed lots.

Other steps include using sterile footwear and leggings when walking in the fields planted with seed lots, and using dip pans with an effective disinfectant at the entrance to all storages, laboratories, and fields. Also, use of timely fungicide applications can reduce fungal problems such as late or early blight, and can help to maintain the foliage for proper growth and yield, as well as allowing certification personnel the opportunity to conduct useable crop inspections for the certification process.

Harvest and storage are also similar to commercial production, with the exception that each seed lot must be harvested separately and stored in a manner to maintain its identity. Following harvest, producers may sort the seed for size to remove and market the larger commercial-type potatoes. Heavy grading to remove diseased or bruised tubers and field soil/debris is essential to putting a good seed crop into the bin and setting up the storage for the long haul (NAK, 2013).

Problems that may impact the storability of the seed must be identified and a strategy developed to minimize potential storage issues and/or reduce the likelihood for tuber breakdown during the storage season. It is also essential to monitor the seed storage facilities routinely and to check for any tuber issues during the storage season. At any stage of the harvest and storage cycle, seed tubers must be handled to keep bruises and wounds to a minimum, since these are pathways to pathogens entering the tubers and causing infections. Finally, sprout inhibitors may not be used at any time, because they can have a major impact on the ability of the seed to sprout and grow normally the following season.

8.8 Summary

Seed potato production is a complicated business. To be successful, there are basic guidelines one must follow—whether using an informal grower-saved seed source or a more formal certification program. Producers should raise seed in regions low in disease pressure and conducive to growth of the desired tuber size, quality, and variety. Producers need ample land for rotating crops, as crop rotation is an essential tool in reducing disease and pest build-up. An infrastructure must exist or be established so that it provides good equipment and storage units and timely shipping of seed to buyers. Regions or countries need the mechanisms to conduct proper inspections of the material while it grows and is stored. Also, pathogen-testing programs are necessary to verify the overall quality of seed stocks.

If conducting seed certification, countries must have the infrastructure for a high-quality certification program, with rules and regulations in place. Field techniques minimizing pests and diseases must be enforced and followed. Growers must have the ability to introduce new cultivars into their systems via a formal certification framework that uses clean tissue culture starting material. Finally, seed producers must make sound production and marketing decisions to maintain financial stability.

Production of disease-free, high-quality seed potatoes is the first and most important component in the production of high-quality, high-yielding table-stock potatoes. Poor seed quality can financially ruin commercial producers, reduce the food supply, and damage an industry. Global potato production relies heavily on the availability of high-quality seed to feed a growing world sustainably.

References

Almekinders, C.J.M., Chujoy, E. and Thiele, G. (2009) The use of true potato seed as pro-poor technology: the efforts of an international agricultural research institute to innovating potato production. *Potato Research* 52, 275–293.

Bohl, W.H. and Johnson, S.B. (2010) Commercial potato production in North America (www.potatoassociation.org, accessed 18 June 2014).

Bryan, J.E. (1989) Breaking dormancy of potato tubers. *CIP Research Guide*, No 16. International Potato Center, Lima, pp. 1–12 (www.cipotato.org/publications-research-guide, accessed 18 June 2014).

Chang, C.C., Park, C.S., Kim, S.Y. and Lee, Y.B. (2012) Growth and tuberization of hydroponically grown potatoes. *Potato Research* 55, 69–81.

CIP (International Potato Center) (2013) World Potato Atlas (https://research.cip.cgiar.org/confluence/display/wpa/Home, accessed 30 April 2014).

Clark, M.F. and Adams, A.N. (1977) Characteristics of a microplate method of enzyme-linked immunosorbent assay for the detection of plant viruses. *Journal of General Virology* 34, 475–483.

Crosslin, J.M. (2013) PVY: an old enemy and a continuing challenge. *American Journal of Potato Research* 90, 2–6.

Davidson, R.D., Houser, A.J., Sather, K. and Haslar, R. (2013) Controlling PVY in seed: what works and what does not. *American Journal of Potato Research* 90, 28–32.

de Bokx, J.A. and van der Want, J.P.H. (1987) *Viruses of Potatoes and Seed-Potato Production*. Center for Agricultural Publishing and Documentation (Pudoc), Wageningen, the Netherlands.

Dodds, J.H. (1988) Tissue culture technology: practical application of sophisticated methods. *American Potato Journal* 65, 167–180.

Donnelly, D.J., Coleman, W.K. and Coleman, S.E. (2003) Potato microtuber production and performance: a review. *American Journal of Potato Research* 80, 103–115.

Doring, T.R., Schrader, J. and Schuler, C. (2006) Representation of *Potato virus Y* control strategies in current and past extension literature. *Potato Research* 49, 225–239.

Ewing, L.L., McMurry, S.E. and Ewing, E.E. (1987) Cutting as a method of breaking dormancy in microtubers produced *in vitro*. *American Potato Journal* 64, 329–332.

Gildemacher, P.R., Demo, P., Barker, I., Kaguongo, W., Woldegiorgis, G., Wagoire, W.W., *et al.* (2009) A description of seed potato systems in Kenya, Uganda, and Ethiopia. *American Journal of Potato Research* 86, 373–382.

Gildemacher, P.R., Schulte-Geldermann, E., Borus, D., Demo, P., Kinyae, P., Mundia, P. *et al.* (2011) Seed potato quality improvement through positive selection by smallholder farmers in Kenya. *Potato Research* 54, 253–266.

Grousset, F. and Smith, I.M. (1998) EPPO certification scheme for seed potatoes. *Bulletin OEPP/EPPO Bulletin* 28, 561–567.

Haapala, T. (2005) Use of single-leaf cuttings of potato for efficient mass propagation. *Potato Research* 48, 201–214.

Haapala, T., Cortbaoui, R. and Chujoy, E. (2008) Production of disease-free seed tubers (www.fao.org/potato-2008/en/potato/seedtubers.html, accessed 18 June 2014).

Hane, D.C. and Hamm, P.B. (1999) Effects of seedborne *Potato virus Y* infection on two potato cuitivars expressing mild disease symptoms. *Plant Disease* 83, 43–45.

Harris, P. (1992) *The Potato Crop: The Scientific Basis for Improvement*, 2nd edn. Chapman and Hall, New York.

Iritani, W.M. and Thornton, R.E. (1984) *Potatoes: Influencing Seed Tuber Behavior*. Pacific Northwest Extension Publication (PNW #248). Pacific Northwest Extension Publishing, Oregon State University, Corvallis, Oregon, pp. 1–15.

Kawakami, J., Iwama, K., Hasegawa, T. and Jitsuyama, Y. (2003) Growth and yield of potato plants grown from microtubers in fields. *American Journal of Potato Research* 80, 371–378.

Knutson, K.W. (1998) Implications of new technologies for seed potato certification programs and seed growers. *American Potato Journal* 65, 229–235.

Lambert, D.H., Currier, A.I. and Olanya, M.O. (1998) Transmission of *Phytophthora infestans* in cut potato seed. *American Journal of Potato Research* 75, 257–263.

Liu, J., Nie, B.H., Cai, X.K., Chen, L. and Xie, C.H. (2006) Establishment of two-year seed potato system and improvement of key techniques. *Chinese Potato Journal* 20(6), 321–325.

Malagamba, P., White, J., Wiersema, S., Accatino, P., Sadik, S. and Monares, A. (1983) True potato seed: an alternative method for potato production. International Potato Center CIP slide training series III-1-E-83. International Potato Center, Lima.

Milinkovic, M., Horstra, C.B., Rodoni, B.C. and Nicolas, M.E. (2012) Effects of age and pretreatment of tissue-cultured potato plants on subsequent minituber production. *Potato Research* 55, 15–25.

Murashige, T. and Skoog, F. (1962) A revised medium for rapid growth and bioasays with tobacco tissue cultures. *Plant Physiology* 15, 473–497.

NAK (Dutch General Inspection Service for Seed of Agricultural Crops) (2013) The production of high-quality seed potatoes in the Netherlands (www.nak.nl, accessed 9 June 2013).

NIVAP (Netherlands Potato Consultative Foundation) (2005) The inspection of Dutch seed potatoes; the importance of approval (www.aardappelpagina.nl/uk, accessed 18 June 2014).

Pruski, K., Astatkie, T., Duplessis, P., Stewart, L., Nowak, J. and Struick, P.C. (2003) Manipulation of microtubers for direct field utilization in seed production. *American Journal of Potato Research* 80, 173–181.

Qui, C.L. (2009) The improvement of virus-free potato microtuber production technology. *Heilongjiang Agricultural Sciences* 2, 52–53.

Radcliffe, E.B. and Ragsdale, D.W. (2002) Aphid-transmitted potato viruses: the importance of understanding vector biology. *American Journal of Potato Research* 79(5), 353–386.

Roca, W.M., Espinoza, N.O., Roca, M.R. and Bryan, J.E. (1978) A tissue culture method for the rapid propagation of potatoes. *American Potato Journal* 55, 691–701.

Rolot, J.L. and Seutin, H. (1999) Soilless production of potato minitubers using hydroponic technique. *Potato Research* 42, 457–469.

SASA (Science and Advice for Scottish Agriculture) (2013) Scottish seed potato classification scheme explanatory leaflet (www.sasa.gov.uk/, accessed 9 June 2013).

Sather, K., Haslar, R. and Davidson, R.D. (2013) Colorado potato certification rules and regulations for the certification of seed potatoes. Colorado Potato Certification Service document published annually, 39 pp.

Schulte-Geldermann, P., Gildemacher, P.R. and Struik, P.C. (2012) Improving seed health and seed performance by positive selection in three Kenyan potato varieties. *American Journal of Potato Research* 89, 429–437.

Secor, G.A., Gudmestad, N.E. and Preston, D.A. (1985) *Disease Control Guidelines for Seed Potato Selection, Handling and Planting*. North Dakota State University Cooperative Extension Publication, Fargo, North Dakota, 877 pp.

Singh, M., Singh, R.P., Fageria, M.S., Nie, X., Coffin, R. and Hawkins, G. (2013) Optimization of a real-time RT-PCR assay and its comparison with ELISA, conventional RT-PCR and the grow-out test for large scale diagnosis of *Potato virus Y* in dormant potato tubers. *American Journal of Potato Research* 90, 43–50.

Stark, J.C. and Love, S.L. (2003) *Potato Production Systems*. University of Idaho Agricultural Communications, Moscow, Idaho.

Struik, P.C. and Wiersema, S.G. (1999) *Seed Potato Technology*. Wageningen Press, Wageningen, the Netherlands.

UNECE Specialized Section on Standardization of Seed Potatoes (2010) Trilingual list of diseases and pests (http://www.unece.org/trade/agr/standard/potatoes/pot_e.html, accessed 18 June 2014).

UNECE Specialized Section on Standardization of Seed Potatoes (2011) Reproductive cycles (http://www.unece.org/trade/C/wp.7/GE.6/2011/INF.4, accessed 9 June 2013).

UNECE Specialized Section on Standardization of Seed Potatoes (2013a) International Standard for Certification and Marketing of Seed Potatoes (http://www.unece.org/trade/agr, accessed 9 June 2013).

UNECE (2013b) Standard for Seed Pototoes S-1; Working Party on Agricultural Quality Standards (http://www.unece.org/trade/agr/meetings/ge.06/2013-in-session.html, accessed 9 June 2013).

USDA/State National Harmonization Plan (2006) MOU Template for Seed Potatoes (www.nationalplantboard.org, accessed 9 June 2013).

Uyen, N.V. and van der Zaag, P. (1985) Potato production using tissue-culture in Vietnam: the status after four years. *American Potato Journal* 62, 237–241.

van der Veeken, A.J.H. and Lommen, W.J.M. (2009) How planting density affects number and yield of potato minitubers in a commercial greenhouse production system. *Potato Research* 52, 105–119.

Wheeler, R.M., Mackowiak, C.L., Sager, J.C., Knott, W.M. and Hinkle, C.R. (1990) Potato growth and yield using nutrient film technique (NFT). *American Potato Journal* 67, 177–187.

White, G.D., Potts, M.J. and Hunt G.L.T. (1988) A computer aid to potato seed production, storage, and distribution in developing countries. *American Potato Journal* 65, 551–560.

Whitworth, J.L., Hamm, P.B. and McIntosh, C.S. (2010) Effect of *Potato virus Y* on yield of a clonal selection of Russet Norkotah. *American Journal of Potato Research* 87(3), 310–314.

Xie, K.Y., Jine, L.P., Pang, W.F., Bian, C.S., Duan, S.G. and Qu, D.Y. (2004) Distribution of infected tubers from virus infected potato plants and selection of materials for virus elimination. *Acta Agriculturae Boreali-Sinica* 19(5), 168–170.

Yan, Z.G., Wang, G.M., Zhen, Y., Zhao, G.Q. and Guo, Q.L. (1999) Accelerate the multiplication of basic seed by breaking sprouts off from seed tubers repeatedly. *Inner Mongolia Agricultural Science and Technology* 1, 22–23.

9 Insect Pests of Potato

Timothy D. Waters[1]* and Andrew S. Jensen[2]
[1]*Washington State University Extension, Pasco, Washington, USA;*
[2]*Potato Commissions of Idaho, Oregon, and Washington, Eagle, Idaho, USA*

Potato is affected by a wide range of insects and other arthropods throughout the world. Some of these species specialize on Solanaceae and have colonized new areas as the crop has been moved from place to place by humans. Examples of this colonization are found in the Colorado potato beetle (*Leptinotarsa decemlineata* (Say)) and potato tuberworm (*Phthorimaea operculella* (Zeller)). Other important pests are generalists and nearly cosmopolitan in distribution, such as the green peach aphid (*Myzus persicae* (Sulzer)) and the two-spotted spider mite (*Tetranychus urticae* (Koch)). Most regions in the world are also affected by regionally important pests such as the beet leafhopper (*Circulifer tenellus* (Baker)) in the north-west USA and the twenty-eight-spotted ladybird (*Henosepilachna vigintioctopunctata* (Fabr.)) in China.

Insects and other arthropods can be divided into several categories based on how they impact the crop: (i) chewing insects affecting the foliage and thereby causing yield loss; (ii) chewing insects in the soil that cause yield loss by trimming roots; (iii) sucking insects that cause yield loss and in many cases transmit severely damaging plant pathogens; and (iv) chewing insects that cause direct damage to tubers. In addition, in the food processing industry such as French fry production, insects and insect parts can be considered contaminants in finished products.

The types of insects and insect-related problems of greatest concern for any given production region or situation are related to the end-use markets (e.g. subsistence agriculture, table stock, processing, long-term storage, etc.) and the level of industrialization and access to insecticides and other crop protection technology. For example, potato tuberworm can be a pest of life-threatening proportions for subsistence farmers using rustic storage for food preservation, whereas it is a pest controlled easily by one or two well-timed insecticide treatments for technologically advanced farms with refrigerated storage. Conversely, an aphid-related disease such as *Potato virus Y* (PVY) may be a minor yield-reducing concern for subsistence farming, but represents a nearly intractable problem and the difference between profit and loss for technologically intensive farms.

Insect management can be a major driver of the cost of production and/or the entire crop production scheme. In low-technology production systems affected by major chewing insects such as the Colorado potato beetle or the tuberworm, extensive manual labor may be required to eliminate beetles or to sort infested tubers, respectively. Crop rotation and efforts to avoid overlapping potato crops may be important management factors that constrain total food production. High-technology farms often face

* E-mail: twaters@wsu.edu

extremely low damage tolerances in their harvested product and/or have small profit margins. This often drives aggressive pest management regimes aimed at avoiding small percentages of crop damage and maximizing yield per hectare.

Potato is an important food crop around the world, resulting in extensive research and general publications on potato entomology over the past 100 years. Some valuable publications include a recent book covering potato entomology (Giordenengo *et al.*, 2013), the insect-related portions of the recently revised *Potato Health Management* (Johnson, 2008), and the basic biology of several insect-transmitted diseases as discussed in Stevenson *et al.* (2001). Nowadays, of course, much valuable information is available online. Good starting points for finding online information on potato entomology include the websites of the Potato Association of America (http://potatoassociation.org/), the European Association for Potato Research (http://www.eapr.net/), and the International Potato Center (http://cipotato.org/).

Our aims for this chapter are to introduce pests and categories of pests of nearly global significance, to discuss important management considerations, and to provide a set of useful references as starting points in the potato entomology literature.

9.1 Chewing Pests

Colorado potato beetle (Coleoptera: Chrysomelidae)

Recognition

The Colorado potato beetle, *Leptinotarsa decemlineata*, is the most widely recognized pest of potato. Colorado potato beetles develop through four larval instars before pupating. The larvae are orange to red in color, with black head and legs, and two rows of black dots along the sides (Fig. 9.1a–b). Adults are 10 mm long and 7 mm wide, pale yellow in color, with five black stripes on each wing, and covered with black markings on the head and pronotum (Fig. 9.1c). Eggs are laid in clusters of 20–40 on the underside of leaves, with each egg being 1.5 mm long and yellow when newly laid, but turning to orange prior to hatching. Female Colorado potato beetles can lay 300–800 eggs in their lifetime.

Fig. 9.1. Colorado potato beetle early (a) and late (b) larval instars and adult (c). (Drawings by Dylan Vermeul, Richland, Washington, USA.)

Distribution

The Colorado potato beetle is believed to be native to the central highlands of Mexico, where the primary host plant is believed to be buffalobur, *Solanum rostratum* (Casagrande, 1987). It was first reported as a major pest to cultivated potato in 1859 near Omaha, Nebraska, USA (Jacques, 1988). Though the species is capable of significant dispersal, most of the spread of Colorado potato beetle has been attributed to, or inadvertently facilitated by, humans. Currently, the Colorado potato beetle is an established pest of potato in North America, Mexico, Europe, Central Asia, Asia Minor, western China, and Iran (Weber, 2003). The wide distribution and colonization of colder climates by the Colorado potato beetle can be attributed largely to facultative overwintering diapause.

This diapause, or resting stage, is triggered by short days, cold temperatures, and reduced host food quality (de Kort, 1990). When diapause is triggered, adults burrow 10–25 cm deep into the soil. This behavior allows survivorship past harsh winter temperatures. Diapause lasts at least 3 months, and exit from diapause and resumption of normal life-stage activity is triggered by soil temperatures above 10°C. Males and females exit diapause synchronously and begin mating prior to colonizing host plants. Not all individuals exit diapause at the same time, and some individuals have been noted to remain in diapause for more than 2 years.

Damage

Both larvae and adult Colorado potato beetle are voracious feeders of potato foliage, and in extreme infestations they will feed on stems and exposed tubers. Ferro *et al.* (1985) have demonstrated that a single larva can remove 10 cm^2 of foliage during maturation and adults can remove 10 cm^2 foliage day^{-1}. The scientific literature notes that defoliation by Colorado potato beetle can be significant before yield is impacted negatively, depending on the growth stage of the potato crop at the time of damage. A study by Zehnder *et al.* (1995) concludes that controls need not be implemented until 20% defoliation has occurred from potato emergence to early bloom. The same study indicates that 30% and 60% defoliation can be incurred at the early to late bloom and late bloom to harvest stages, respectively, before control needs to be implemented. Scientific studies on yield loss in potato tend to be highly variable, and as such, commercial potato farmers cannot rely strictly on the abovementioned thresholds before implementing control strategies.

Sampling and monitoring

Colorado potato beetle monitoring should begin just after plants emerge. Scouting for adults exiting diapause and locating newly laid eggs on the underside of leaves near field edges will work best for early detection. Potato fields near where infestations occurred during the previous season should be monitored closely. Look for egg clusters mostly on the underside of leaves in the upper half of the plant canopy. Adults and larvae can be detected by visual inspection, sweep nets, or beating sheets.

Wireworm (Elateridae)

Recognition

Adult wireworm, or click beetles, are generally dark in color, range from 12 to 25 mm long, and live aboveground on the surface of soil or on plants. They are narrow, rounded on both ends, with sharp backward-facing corners on the posterior of the pronotum. Eggs, larvae, and pupae are subterranean. Eggs are usually laid beneath the soil surface and are light in color and spherical. As larvae, wireworms are slender and shiny, with a segmented appearance and relatively hard exoskeleton (Fig. 9.2a). In addition to wireworm larva, Fig. 9.2b–d shows other subterranean larvae commonly found in potato fields, including a carabid beetle larva (b) (Carabidae; a beneficial beetle), a white grub larva (c) (Scarabaeidae), and a crane fly larva (d) (Tipulidae).

Newly hatched wireworm larvae (~6 mm long) are usually light to white in color and turn yellowish brown in later instars (up to 38 mm long). Wireworm larvae have functional legs, well-developed mandibles, and the terminal abdominal segment sometimes resembles a pair of prolegs. Pupae are initially light in color and will darken as they mature to the adult stage.

The long life cycle and difficulty in sampling make wireworm a troublesome pest to predict in potato production systems. The larva is the life stage that causes economic damage to potato, but larvae go through stages where they feed less actively. Furlan (2004) indicates that larvae spend as little as 24% of the time in each larval stage feeding. Larval development for pestiferous wireworms ranges from 30 days to 6 years.

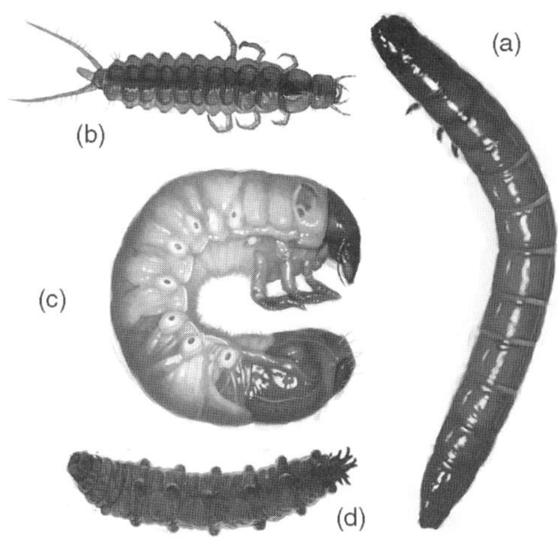

Fig. 9.2. Subterranean larvae commonly found in potato fields. (a) Larva of wireworm (Elateridae); (b) larva of beneficial carabid beetle (Carabidae); (c) larva of white grub (Scarabaeidae); (d) larva of crane fly (Tipulidae). (Drawings by Dylan Vermeul, Richland, Washington, USA.)

Numerous species of wireworm attack potato in different growing regions, and Jansson and Seal (1994) noted at least 39 species of wireworm attack potatoes. In most regions, two or more species are generally considered to be the primary pest wireworms of potato, but due to complexity in sampling and the difficulty in morphological identification, there is not always a strong consensus as to the species involved. Further complicating the matter, oftentimes more than one pest species of wireworm can be found in individual fields.

Damage

Wireworms are unique pests because they are not transient in nature; rather, they are long-term residents of fields into which potatoes are planted. Wireworm larvae damage potatoes by feeding on roots, but more important economic damage comes from feeding on tubers. Larval feeding early in tuber development results in holes that grow in size and suberize as the tubers grow, leading to misshapen tubers. Larval feeding later in tuber development results in small round holes. Single holes caused by wireworm render fresh market tubers unmarketable, while tubers that will be processed or used as seed can incur a higher number of feeding holes per tuber without significant economic loss. Under some conditions, wireworm damage may occur immediately prior to harvest, with the insects still inside the tubers.

Crop history plays a critical role in predicting the risk of damage by wireworm. Most wireworms have a broad host range, but prefer undisturbed soils with grass hosts. As such, production fields coming out of pasture or rangeland can be particularly active sites for wireworm. Miles and Cohen (1938) noted that waiting to plow pastures just prior to planting potato could reduce damage as compared to plowing well in advance of planting. It has been assumed that in the above scenario, wireworm larvae feed on the decomposing plant material rather than the potato crop, but once the decomposing sod is reduced, feeding on the potato crop will ensue.

Additionally, for many years following World War II, wireworms were not a significant issue due to the environmentally persistent insecticides, such as organochlorines, organophosphates, and carbamates, used to control them. Because they were effective, little research was conducted on wireworm biology and control. In recent years, however, many insecticides of those classes have been banned from use, removed from the marketplace, or are significantly restricted. Wireworms have resurged as a significant pest problem, resulting in more current studies on their biology and control with newer insecticides.

Sampling and monitoring

With most of the life stages of the wireworm being subterranean and the distribution of wireworms in fields uneven, the development of economic thresholds and accurate sampling for risk assessment have been challenging. Wireworms can be found throughout the soil profile, from the surface down to depths greater than 1.5 m, and can move easily and quickly within that profile, depending on soil temperature and moisture, and life stage. There are two general methods employed for sampling of wireworm larvae, the absolute method and the relative method. Both methods have strengths and weaknesses.

The absolute sampling method aims to estimate the size of the wireworm population and its distribution within the field by removing and inspecting different soil strata by excavation or soil cores. With the absolute method, a large number of samples or cores per field must be evaluated in order to assess the risk of wireworm damage. The samples are simply extracted and screened or sorted by hand to detect the number of larvae per unit area of soil.

The relative sampling method employs the use of an attractant or bait in order to measure the abundance of wireworm, and therefore the risk of damage. Many different types of baits have been used, but the primary cue that attracts wireworm to the different baits is the emission of CO_2. In the field, decomposing plant material, germinating seeds, and plants respiring all emit CO_2, and therefore attract pest insects, including wireworm. The relative method requires fewer samples per field, but does risk overestimating populations, since the baits attract wireworm from some distance. Additionally, several factors may affect attractiveness to bait and overall trap count numbers, including the amount of CO_2 generated, feeding status or stage of the wireworm, duration of the baiting period, soil temperature, soil moisture, soil texture, and the source of the CO_2 attractant. Due to all the variables associated with wireworm sampling, Horton (2006) reported that it remained challenging to predict damage, and hence to make management recommendations based on wireworm baiting.

Adult wireworm can be sampled using pheromone traps or pitfall traps, but in most situations, adult sampling will just alert one to the presence of wireworm in a general area and not specifically address the risk of damage to the potato crop by wireworm larvae.

Potato tuber moths (Gelechiidae)

Recognition

Three species of Gelechiidae make up the potato tuber moth complex that attacks potato crops. The three species are the potato tuber moth, *Phthorimaea operculella* (Zeller), the Guatemalan potato tuber moth, *Tecia solanivora* (Povolny), and the Andean potato tuber moth, *Symmetrischema tangolias* (Gyen).

The potato tuber moth is cosmopolitan in distribution and will therefore be the main focus of this discussion. Adult potato tuber moths are light brown in color, 8 mm in length, and when fully extended, the wings span 15 mm. The forewings have two to three dark bands along the rear edge, and the rears of the forewing, as well as the entire hindwing, are fringed. The long antennae resemble strands of thick hair, and there are three longitudinal dark stripes on the thorax. Eggs are white when newly laid, but turn yellow as they approach maturing to neonate larvae. The eggs are oblong (0.5×0.35 mm) and deposited singly, or in strings, on plant tissue. Each female can produce 100–300 eggs over her lifetime, depending on food resources and temperature (Broodryk, 1971). Larvae when newly hatched are 1 mm long and grow to nearly 13 mm in the fourth instar, just prior to pupation. Larvae feeding on leaf tissue will be purple to green in color, while those feeding on tubers will be white to purple. Larvae have a dark head capsule, obvious segmentation, six legs, and five pairs of prolegs on the rear abdominal segments. Pupae are 8 mm long and begin brown in color but darken to nearly black prior to emergence in the adult stage. Depending on location, the potato tuber moth can produce anywhere from four to ten generations per year, with the range due largely to temperatures in the different regions. Development from egg to adult also ranges with temperature, but at 26°C, it takes approximately 23 days, and adult longevity ranges from 6 to 15 days (Broodryk, 1970).

Distribution

The potato tuber moth is believed to have originated in the tropical mountains of South America, near the origin of the potato (Graf, 1917), and is currently found in over 90 countries across the globe. Adults are weak fliers, and the cosmopolitan distribution and range expansion of potato tuber moth is probably due largely to movement of larvae-infested seed. The potato tuber moth is a pest of potato from temperate to tropical regions, but tends to be less damaging in regions with cold winters that reduce survivorship and the number of generations per growing season. The lower developmental threshold for the potato tuber moth has been estimated as low as 4.25°C, possibly accounting for expansion and establishment in cool temperate regions.

The Guatemalan potato tuber moth is a pest of potato in Central and South America, and Guatemala is believed to be the place of origin (Povolny, 1973). The Andean potato tuber moth is believed to have originated in Peru and Bolivia, and can now be found in South America, Australia, and Indonesia.

Damage

The potato tuber moth, like many other Gelechiidae larvae, typically feeds as leafminers, which in potato rarely causes economically important damage. Potato tuber moths tend to reach higher levels of infestation during warm and dry weather, in crops that are not irrigated with sprinklers and in places where potato is grown throughout the year. Conversely, they tend to be lower in numbers during wet weather and in crops grown with sprinkler irrigation. When foliage senesces or is killed with desiccants prior to harvest, larvae will drop to the soil and feed on potato tubers, which is where significant economic damage occurs. When infested tubers are harvested, the development of larvae and further damage to tubers can occur in storage, depending on storage temperature. Oftentimes, tubers near the surface of a pile in storage are most drastically affected. Damage to tubers can be reduced when the time between vine kill and harvest is reduced, or when larvae are controlled with pesticides prior to vine kill and harvest. Harvest in warmer weather conditions has more significant risk of tuber infestation by potato tuber moth than harvest during cooler weather. Additionally, regular hilling and moist soil tend to reduce tuber damage by creating a physical barrier to infestation. Potato tuber moths can be transported inadvertently in infested seed, harvest bags, shipping containers, or tare dirt.

Sampling and monitoring

To minimize economic damage to the tubers, it is essential to monitor the potato tuber moth population, starting just prior to vine kill and continuing until the crop has been harvested. Adult populations of potato tuber moth can be monitored using pheromone traps, which attract male moths with an aroma similar to that coming from the female moth. This type of sampling can provide a general idea of the relative population, and therefore risk to the potato crop. Traps are most effective when placed upwind of the potato crop or storage. The threshold for the number of moths caught per trap per day will vary based on the potato crop stage. In the north-west USA, the authors have observed that commercially available pheromone lures capture several other unidentified gelechiid moths, presumably species native to the shrub-steppe habitat where potatoes are grown. The resulting mix of species in pheromone traps substantially complicates monitoring efforts in that region. Larval numbers in foliage can be assessed by looking for leaf mines on potato leaves. To assess larval infestation of tubers, a random sample of harvested tubers can be obtained and examined visually. More precise detection may be possible in the field prior to harvest by examining tubers that are naturally protruding from the soil surface. These surface-exposed tubers are most likely to have larval infestation by potato tuber moths.

9.2 Sucking Pests

Aphids (Aphididae)

Recognition

Aphids are small (1–3 mm), soft-bodied, phloem-feeding insects characterized by a pair of unique abdominal appendages called siphunculi, phenotypic plasticity involving specialized adult morphs with and without wings, and parthenogenetic viviparous reproduction (all summer forms are female and give birth to live young without male fertilization) during most of the annual life cycle (Fig. 9.3a–b). Winged forms almost always hold the wings roof-like over the body. Aphids are pests of crops worldwide, greenhouses, and indoor ornamental plants.

Distribution and species

Blackman and Eastop (2000) list 15 species of aphids known to feed on potatoes worldwide. None of these is host specific to the crop, and many have wide host ranges and large or cosmopolitan distributions. Many other species that do not feed on potatoes can be important to crop production in their role as vectors of non-persistent viruses (see Chapter 12 of this volume).

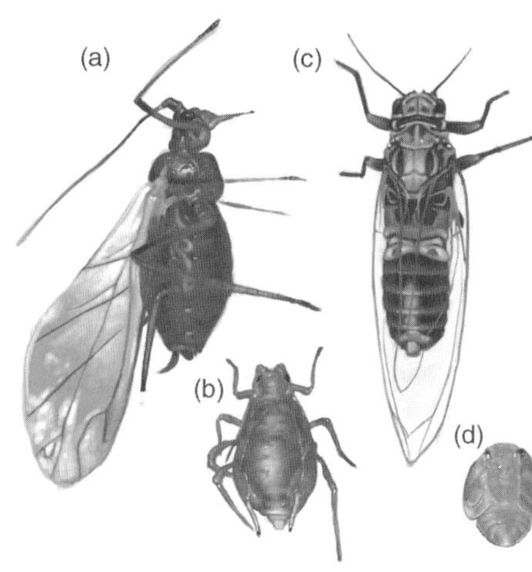

Fig. 9.3. Foliar pests. (a) Adult winged potato aphid; (b) adult wingless green peach aphid; (c) adult potato psyllid; (d) juvenile potato psyllid. (Drawings by Dylan Vermeul, Richland, Washington, USA.)

On a global basis, two species are by far the most important, *Myzus persicae* (Sulzer) (green peach aphid) and *Macrosiphum euphorbiae* (Thomas) (potato aphid). For more detail on the identification and basic biology of aphid species on potato and other crops, see Blackman and Eastop (2000).

The green peach aphid has one of the widest host ranges of any aphid, is probably originally from East Asia but is now found worldwide. In colder climates, it is heteroecious, using species of *Prunus* (Rosaceae) as overwintering hosts, where eggs survive the winter, and migrating to a wide variety of summer hosts in dozens of plant families. In warmer climates, green peach aphids are anholocyclic, and therefore reproduce parthenogenetically throughout the year. The body length is 1.2–2.3 mm, color is extremely variable, including white, yellow, green of various shades, pale pink to darker red, with a tendency toward darker colors in cooler conditions.

The potato aphid is of North American origin but has spread almost worldwide, and has a host range almost as wide as green peach aphid. Its life cycle in cooler parts of North America is heteroecious, usually with eggs overwintering on roses (*Rosa*: Rosaceae), and migrating to a wide range of summer hosts, including many crops. In most other parts of the world, and in all warm/tropical regions, the potato aphid is entirely anholocyclic. Larger than the green peach aphid, the potato aphid is usually 2–4 mm in length, and can be a wide variety of colors similar to green peach aphid. In western North America, the potato aphid is more common on the potato crop in cooler regions and seasons than the green peach aphid, which does well on potato in midsummer.

Another cosmopolitan aphid that will commonly be found on potato is *Aulacorthum solani* (Kaltenbach) (glasshouse-potato aphid, or foxglove aphid). It generally does not reach damaging levels in field-grown potatoes. In eastern North America, *Aphis nasturtii* (Kaltenbach) can be locally very important. Throughout the world, various polyphagous aphids will appear on potatoes, but usually not in damaging numbers.

Damage

Aphids are phloem feeders, and in large numbers can cause substantial yield loss or even plant death, due to energy diversion from plant growth to aphid growth and reproduction. Aphid populations can expand rapidly due to their viviparous parthenogenetic reproduction. In practice, population explosions of aphids in potatoes are usually related to the application of inappropriate or excessive insecticides for the control of aphids or other pests. For example, some potato entomologists intentionally foment aphid populations in research plots by applying pyrethroid insecticides. The presumed mechanism behind such aphid population growth is that broad-spectrum insecticides eliminate aphid natural enemies and result in unchecked reproduction. On a related note, Lagnaoui and Radcliffe (1998) showed that treatment of potatoes with fungicides aimed at late blight resulted in higher aphid numbers, and presented evidence this was caused by inadvertent control of entomopathogenic fungi.

By far the most important aphid-related damage in potatoes is caused by the plant viruses they transmit, including PVY and *Potato leaf roll virus*. These and other viruses transmitted by aphids are discussed in Chapter 12 of this volume. In intensive production regions such as western North America, the control of aphid-transmitted viruses via aphid management drives a large fraction of the cost of production. Producers of seed potatoes often focus even more diligently on aphid control because the value of a seed crop is highly correlated to its virus content.

Sampling and monitoring

Aphids in the potato crop are most commonly found on the middle to lower leaves, especially in the case of green peach aphid. Methods for monitoring aphids in potatoes include inspection of leaves, beating sheet/tray techniques, yellow pan traps, green tile traps, and suction traps. Each method has its devoted adherents, with methods such as pan and suction traps being employed successfully in certain regions, while in other regions those methods are unhelpful due to plant and aphid phenology. A widely accepted and effective method for aphid detection and monitoring is systematic sampling and visual inspection of the middle and lower leaves. Beating sheet sampling can be effective in cultivars with upright growth habits, if implemented consistently. There are no broadly applicable treatment thresholds for aphids in potatoes.

Leafhoppers and relatives: Cicadellidae, Delphacidae, Cixiidae

Recognition

Leafhoppers are small (2–5 mm) sucking insects that feed on cell contents or phloem. They are usually broadest at the front, and taper to a narrow caudal end, holding the wings roof-like over the body. Eggs are deposited in plant tissue by the female's specialized ovipositor. Nymphs are active runners and jumpers. Around the world there are many local and regional species of leafhoppers (Cicadellidae) and planthoppers (Delphacidae, Cixiidae) that can be pests in potatoes. These are divided into those that prefer potato as a host and cause direct damage to the plant, primarily species of *Empoasca*, and those that transmit pathogens, especially phytoplasmas, to potato during temporary residency in the crop. A few relatively well-studied examples are discussed briefly below. Brief mentions of many other regionally and locally important species are made by Munyaneza and Henne (2013).

Distribution and species

The name "potato leafhopper" is applied to the cicadellid species, *Empoasca fabae* (Harris) in mid-western and eastern North America, where it is an important pest. It is thought to be native to North America, where it overwinters in southern regions of the USA and migrates northward each spring to infest potato crops, and then returns south in the fall (as summarized by Munyaneza and Henne, 2013).

The potato leafhopper has a wide host range, but reproduces well in potato. As it feeds, the leafhopper weakens plants and causes a syndrome known as hopper burn that results in stunting, yellowing, and eventually necrosis of affected leaves. Severity of hopper burn is related directly to population density in the crop. Biological and cultural control options for this leafhopper are few, and management typically consists of well-timed insecticides. Other species of *Empoasca* can be locally abundant, such as *Empoasca filamenta* (DeLong) in the Pacific Northwest of the USA, and also cause hopper burn.

Beet leafhopper, *Circulifer tenellus* (Baker), is the first well-studied example of a leafhopper that affects potato in its role as vector of a phytoplasma. This leafhopper is thought to be native to the Mediterranean region, but has become established throughout the Americas, where it locally damages crops and transmits pathogens such as the phytoplasma called the beet leafhopper-transmitted virescence agent (BLTVA), which causes a potato purple top disease. Purple top in potato consists of a suite of symptoms including leaf yellowing, then purpling, curling, elongation of axial buds, aerial tubers, and early plant senescence.

Beet leafhopper is common in the naturalized vegetation in the arid and semi-arid potato producing regions of the USA. It does not reproduce in the potato crop, but transmits BLTVA to potato during its movements through the landscape. Aster leafhopper, *Macrosteles fascifrons* (Stål), transmits aster yellows phytoplasma to potato in the eastern USA. It migrates from more southern overwintering grounds to spend the summer in potato producing regions. Much like beet leafhopper, aster leafhopper does not reproduce on potato, but transmits phytoplasma during casual visits to the crop.

Sampling and monitoring

Sweep nets are commonly and effectively used for leafhopper sampling, especially for species resident in the crop. For *Empoasca* and related leafhoppers causing direct damage, sampling for field colonization by adults is useful. When large influxes are detected, treatments are commonly started when the first nymphs appear (Ragsdale *et al.*, 2008). One reason for the delay until nymphal emergence is because leafhopper eggs are inserted into plant tissues, affording protection from many insecticides.

In the case of non-resident leafhoppers that transmit pathogens, yellow sticky traps are often used to detect the flights of each adult generation. Placement of traps is specific to the target leafhopper. For example, yellow sticky traps deployed to capture beet leafhopper must be mounted close to the ground in sparse vegetation. There are not well-established treatment thresholds for most leafhoppers in potatoes, with general guidelines often in practice regionally and locally.

Psyllids (Psylloidea)

Recognition

Psyllids are small (2–5 mm) sucking insects similar to winged aphids in size and appearance as adults (Fig. 9.3c–d). Unlike aphids, they reproduce sexually in every generation, lay eggs, are active jumpers, and the nymphs are usually flattened and relatively sedentary. Nymphs are also similar to whitefly nymphs, with the chief distinguishing characteristics being distinct wing pads and fully functional legs in psyllids, as opposed to no wing pads and entirely immobile nymphs (beyond the first instar) in whiteflies.

The predominant psyllid affecting potato in the Americas and New Zealand, *Bactericera cockerelli* (Šulc), is known as the potato psyllid. Munyaneza and Henne (2013) mention other species, including *Bactericera nigricornis* (Förster) in Iran and *Russeliana solanicola* (Tuthill) in South America. The potato psyllid has been studied thoroughly since early in the 20th century, due to its ability to cause a plant syndrome known as psyllid yellows and much more recently the advent of a new potato disease known as "zebra chip", which is associated with the bacterium *Candidatus* Liberibacter solanacearum and is vectored by *B. cockerelli* (Munyaneza, 2012).

Distribution and species

The potato psyllid is about 2–3 mm long, the body is mostly dark with prominent white markings on the abdomen, and the wings are clear without dark markings. Nymphs are flat, light green, and camouflaged on the leaf and stem surfaces where they feed. Eggs are yellow to orange and mounted on a short stalk.

Native to North America, especially along the Rio Grande in the southern USA and northern Mexico, the psyllid now occurs throughout much of Central America and western and north-western USA. Host range for reproduction and development is largely limited to species of the Solanaceae and Convolulaceae, but adults have been found on reserve hosts in many plant families (Wallis, 1955).

In central USA, the potato psyllid is known to overwinter in the south near the US/Mexico border and to migrate north each year to neighboring regions and states. Since an outbreak of zebra chip in north-western USA in 2011, potato psyllid has been documented as overwintering in potato producing parts of that region in association with the perennial vine, *Solanum dulcamara* L. (Murphy *et al.*, 2013). Further, Swisher *et al.* (2013) presented DNA-based evidence that the psyllid occurring in the north-western states was distinct from those in regions such as California and Texas, and has been a resident of the north-west since at least 1998.

Potato psyllid can have several generations per year, developing from egg to adult in as little as 3 weeks under ideal conditions. It performs best at moderate temperatures, and fails to develop and/or dies at high summer temperatures in the 32–35°C range.

Damage

Potato psyllid in a crop can lead to three main types of damage: general plant weakening from direct feeding, psyllid yellows, and zebra chip. Direct feeding damage is generally slight, since it would only be expected under low psyllid pressure. Higher psyllid pressure results in psyllid yellows disease, characterized by a suite of symptoms including leaf yellowing, upward rolling, purpling, short and thick terminal internodes, axillary branches, and aerial tubers. These symptoms are similar to other diseases that affect phloem transport, such as purple top caused by phytoplasmas, zebra chip associated with *Candidatus* Liberibacter solanacearum, and even stem-girdling caused by the fungus, *Rhizoctonia solani*.

An interesting feature of the psyllid yellows disease is that the affected plants will recover if psyllids are removed. The largest, most damaging psyllid populations in potatoes have often been associated with heavy, but inappropriate, insecticide use aimed at psyllids or other insects. For example, insecticides in the pyrethroid class have very poor field efficacy against psyllids, and also eliminate beneficial arthropods. Their use when psyllids are present can lead to crop failure from psyllid yellows.

Zebra chip disease was first detected in Mexico in the mid-1990s and has since spread throughout most of Mexico, Central America, and the USA west of the Mississippi River. The putative zebra chip pathogen, *Candidatus* Liberibacter solanacearum, is closely related to the Liberibacter associated with citrus greening disease, which is also vectored by psyllids. The name "zebra chip" derives from the symptoms in tubers, consisting of dark stripes in fried tissue, and flecks and stripes in fresh tissue. Symptoms can be absent in fresh tubers, but appear very strongly when fried. Zebra chip also causes foliar symptoms in plants very similar to psyllid yellows and purple top, as mentioned above. Crop damage from zebra chip can be intense despite low psyllid population density. At times, zebra chip is found in crops despite psyllids having never been detected in the field. This difficult circumstance has resulted in extensive insecticide use for psyllid control, and associated risk of insecticide resistance development. A thorough review of zebra chip research and information has been provided by Munyaneza (2012).

Sampling and monitoring

Monitoring psyllid colonization and development in potato fields has proven to be exceptionally challenging since the advent of zebra chip. This is due to the low populations required to transmit damaging levels of zebra chip. In practice, a combination of techniques is used, including yellow sticky traps deployed inside the potato fields near edges, leaf sampling concentrating on field margins, and in some cases, vacuum sampling machines of various designs. Yellow sticky cards are most effective at detecting early adult colonists, but very poor at assessing resident populations in the field. Leaf samples are used to detect and monitor levels of reproduction and nymphal development. A majority of psyllid nymphs and eggs will be found in the middle of the plant canopy.

9.3 Miscellaneous Pests

The potato can be colonized or can serve as a temporary host for numerous insect species that can impact the crop negatively. The insects detailed above are the most common pests of potato that have broad distribution, but numerous pests are regionally important in various areas where potatoes are grown. Table 9.1 lists some of these regionally important pests, with some helpful references that can serve as a starting point for further research on those species. Additionally, some of these less common pests or beneficial insects that can be confused with common pests can be seen in Fig. 9.2 (b–d). It is important to remember that not all insects found in potato are detrimental to crop development and that the identification of insects is crucial before taking curative measures.

9.4 Integrated Crop Management

The potato is a high-value crop with many arthropod pests that can damage foliage, reduce crop yield, and even feed on the tubers, thereby rendering the crop unmarketable. As such, pests that impact plant growth and damage tubers should be managed to avoid economic loss. In order to avoid crop damage, potato producers should be good practitioners of integrated pest management (IPM).

IPM programs aim to coordinate production and pest management in a fashion that reduces pest damage to the crop, with low hazards to humans and the environment. There are four key principles to IPM programs: pest identification, monitoring, control action guidelines or economic injury levels, and management of the pest. In order to implement IPM, practitioners must be able to characterize and assess the pest population accurately and make sound management decisions.

Correct identification of the pest is the first step in an IPM program for any crop; the other three IPM principles are set in motion on the pest's identification. Proper identification is also important, because many insects benefit the crop by preying on pest species. Insects that prey on pest species are often referred to as "beneficial insects". Monitoring for pests can be species specific, as is the case with pheromone traps that typically attract only one species. Economic thresholds are almost always species specific. The number of pests per sample unit that inflict economic damage is dependent on the pest species, since they consume different amounts of foliage or affect the crop in different ways. Management tactics, particularly pesticide use, rely on proper identification, as many insecticides control only specific types of insects.

Once the pest has been identified, monitoring should be conducted to determine the density and life stages of the pest present in the crop. Monitoring tactics will vary based on the species monitored and can include: sweep net sampling,

Table 9.1. A selection of pests of regions or special crop circumstances. Below are references for more information on the pest. If one of these pests is of interest, we suggest consulting the Internet for images of the taxon and using the reference listed to find more information.

Taxa	Type of damage	Useful references
Acari (mites)		
Tetranychidae *Tetranychus urticae*	Leaf stippling, yellowing, bronzing, wilting	Fathi and Nouri-Ganbalani, 2010
Tarsonemidae *Polyphagotarsonemus latus*	Leaf stippling, yellowing, bronzing, wilting	Peña and Bullock, 1994
Symphyla (symphylans)		
Scutigerellidae *Scutigerella immaculata*	Root trimming, shallow pits in tuber skin	Umble and Fisher, 2003
Insecta		
Coleoptera: Chrysomelidae *Epitrix* flea beetles	Shot-holes in foliage, shallow pits and tunnels in tubers	Eyre and Giltrap, 2012
Coleoptera: Scarabaeidae White grubs	Root trimming, broad shallow holes and pits in tubers	Chandel *et al.*, 2013
Diptera: Anthomyiidae *Delia* e.g. seedcorn maggot	Seed tuber damage that often leads to rot	
Diptera: Tipulidae Craneflies	Broad shallow holes and pits in tubers	Blackshaw and Kerry, 2008
Hemiptera: Aleyrodidae Whiteflies	General plant weakening, rarely reaching damaging levels	
Hemiptera: Miridae *Lygus* plant bugs	Leaf or leaflet flagging, often little visible damage	Wheeler, 2001
Hemiptera: Pentatomidae Stink bugs	Flagging of leaves or entire stems, but sometimes without visible damage	
Hymenoptera: Formicidae Ants	Burrowing in stems and tubers	Chandel *et al.*, 2013
Isoptera: Termitidae Termites	Burrowing in stems and tubers	Chandel *et al.*, 2013
Lepidoptera: Crambidae *Ostrynia* European corn borer	Tunneling in aboveground stems	Ziems *et al.*, 2008
Lepidoptera: Noctuidae Loopers, cutworms	General leaf defoliation	
Thysanoptera: Thripidae Thrips	Leaf stippling and pitting	Abad *et al.*, 2005 Fathi and Nouri-Ganbalani, 2010

yellow sticky traps, beating sheets, pheromone traps, vacuum devices, water pan traps, pitfall traps, soil core sampling, and baited devices. The type of sampling method employed varies by production region and the comfort level of each sampling type by the practitioner. Many insects are small and cryptic, making active sampling difficult for most consultants and producers. The sampling method of choice is often one that attracts pests by color or odor to a site where they can be trapped and assessed more easily.

Monitoring a pest species is important, but without the concept of the economic injury level, monitoring does little for an IPM program. Stern *et al.* (1959) developed the concept of the economic injury level, the components of which are economic damage, economic injury level, and the economic threshold. In most simple terms, economic damage is defined as the amount of injury to the crop that justifies the costs associated with implementing control measures. The economic injury level is the lowest pest population density with the potential to cause economic crop damage. The economic threshold is defined as the population density at which control of the pest should be initiated in order to prevent the pest from reaching the economic injury level.

The term "action threshold" is at times used synonymously with the term "economic threshold". Unfortunately, the scientific process to develop pest densities associated with the economic injury level concept takes a great deal of effort and resources. As a result, many potato pests are without a widely accepted and adopted economic threshold. Often, practitioners must rely on past experience to develop economic thresholds for their farm or region in relation to the end use and value of their crop.

Management of arthropod pests of potato varies significantly based on the pest, growing region, end use of the crop, and potential crop value. Below, we discuss general concepts relative to good cultural and chemical pest management as a component of IPM.

Cultural management practices

Use of high-quality, certified potato seed, free from insect pests or insect-vectored pathogens, is essential in reducing pest build-up. Insects like potato tuberworm can be spread across the world quickly and easily via shipment of larvae-infested seed. To minimize seed-related insect spread, the commercial producer should visit the seed grower's farm during the growing and storage season prior to a seed purchase. In conjunction with regional certification, personal inspection of the seed grower's fields and storage units can contribute towards the movement of clean seed. Following planting and harvest, all growers should dispose of waste potatoes properly. Deep burial or rapid destruction and removal of waste potatoes will reduce the chances that a tuber-borne pest such as the potato tuberworm will survive and infest potatoes in the future.

Crop rotation is also important in reducing potential pest damage. Because insects like the Colorado potato beetle overwinter as pupae in the soil, potatoes planted back to the same field or in close proximity to that field during consecutive years may experience unnecessary pressure from the pest. By planting different crops each year in a particular field, high pest populations may be avoided. One must also consider, however, which rotational crop to plant prior to producing potatoes. The previous crop may impact the next potato crop negatively by providing food and safe haven for pests like the wireworm. When potatoes are planted following grass or long-term pasture, wireworms are more likely to be a pest than if the potatoes followed maize, sugarbeets, and many other crops.

Proper planting depth and post-planting tillage will minimize surface-exposed tubers, leaving them less vulnerable to infestation from insects like the potato tuberworm. Potato tuberworm damage can also be reduced by decreasing the time between vine kill and tuber harvest. The shortened pre-harvest interval does not allow tuberworm larvae adequate time to move from the foliage to the soil, where they can infest tubers near the surface. Although cultivar choice has a limited impact on reducing pest-related issues, some cultivars may fare better than others when pressured by certain pests. Potato cultivars with large canopies may be able to sustain more damage from chewing-type insects than those with small canopies. Cultivars that mature early may be harvested before insects like the potato tuber moth become an issue during a given season. Planting and harvest timing with all cultivars should be used when possible to reduce pest abundance and infestation issues. Utilizing cultural management of pest reduction is important, but it must often be combined with pesticide use under typical, modern farming systems.

9.5 Chemical Management

Due to the combination of risk-averse producers and the high crop value of potato, insecticides are oftentimes used to manage insect pests. Insecticides should be applied according to label recommendations, adhering to proper timing, application rates, use of surfactants when appropriate, and safety recommendations for handlers and bystanders. Judicious use of pesticides should be practiced to avoid the development of insect populations that are resistant or tolerant to specific pesticide compounds. Resistance or tolerance to a pesticide simply means the pesticide compound intended to result in insect mortality no longer causes mortality at the concentration or rate it did originally. Resistance typically occurs by making multiple applications of the same compound to successive insect generations. Such repeated applications select for naturally occurring genetic mutations in the insect population, which confer tolerance or resistance to the insecticide's mode of action. The surviving individuals possess a genetic mutation and then reproduce. The proportion of individuals with the mutation is greater in the subsequent generation.

Typically, resistance is noticed as a rate response, where it takes a higher and higher concentration of the pesticide to cause the same percentage mortality. Nearly complete resistance can result. Species that are multivoltine, or have multiple generations per crop season, are more likely to develop resistance. Colorado potato beetle has developed resistance to several different pesticides in most regions. Insecticides with a single-site mode of action are most likely to develop resistance. Resistance has been shown to pyrethroid-type insecticides in many different insect taxa. For multivoltine pests, it is very important to rotate or alternate pesticides with different modes of action in order to reduce the risk of insect populations developing resistance to commonly used pesticides. In some regions, insecticide cost and availability have a large impact on the phenomenon of pesticide resistance.

Newer insecticides tend to have shorter environmental persistence and are narrower in spectrum of control; that is, they are usually toxic to one or a few types of insects. Such products tend to be safer to the environment and non-target insects, but are not immune to insects developing resistance to them. Additionally, the use of broad-spectrum insecticides has detrimental effects on insect natural enemies and can cause flares of secondary pests, or pests typically not associated with economic crop damage. An example of this situation is the use of pyrethroid insecticides to control midseason pests in the western USA. Repeated use of pyrethroids can cause late-season flares of two-spotted spider mites, *Tetranychus urticae*. This likely occurs because the broad-spectrum pyrethroids are toxic to predatory mites and insects that typically keep two-spotted spider mite populations in check, but those pyrethroids are not particularly toxic to two-spotted spider mites.

9.6 Summary

Throughout most of the world, potato crops are attacked by several serious pest insects. Regardless of a practitioner's location, it is critical to become familiar with the pest insects in his or her area. We have given starting points on important insects and sampling and management approaches, but nothing substitutes for experience in the field watching crop growth and insect activity.

As mentioned above, pest identification is crucial to IPM, especially since many of the insects found in potato fields may be natural enemies to pest insects. Conservation of beneficial insects will reduce the likelihood of having to apply insecticides for pests. Potato fields minimally treated with pesticides can harbor a complex food web of insects, spiders, and other arthropods. These include a diverse set of decomposers on and in the soil, predators and parasitoids in and on the soil and on the foliage, and major and minor pests such as those discussed above.

A major goal of some potato entomology research has been to understand how best to manage the crop to maximize the use of this biodiversity and limit pest damage. During the 1990s, transgenic potatoes, containing the gene from *Bacillus thuringiensis* that kills Colorado potato beetle, were developed and briefly commercialized. This innovation allowed detailed study of the arthropod community in a potato agroecosystem with little to no insecticides (Reed *et al.*, 2001), and showed that eliminating the primary defoliating pest without insecticides preserved a wide range of beneficial generalist predators to control secondary pests. Koss *et al.* (2005) compared the arthropod community in commercial fields under three management regimes: broad-spectrum insecticides, selective insecticides, and certified organic management. They showed the use of selective insecticides resulted in similar levels of predators as in organic management, but with fewer pests.

Beneficial insects will be present in potato fields in almost all production regions. Growers and field staff should learn to recognize the main predators and parasites of the pests in their regions. This allows a fine-tuning of decisions to use insecticides. For example, knowledge of a large predator population in a field infested with pests may encourage a grower to wait before applying an insecticide, and thereby potentially realize all needed control without the insecticide.

A combination of the general principles of potato insect management and direct observations of local agroecosystems will give crop managers most of the knowledge necessary to minimize crop damage successfully and maintain profitability.

References

Abad, J.A., Holmes, G.A., Cubeta, M.A., Moyer, J.W. and Kennedy, G.G. (2005) *Tomato spotted wilt virus* on potato in eastern North Carolina. *American Journal of Potato Research* 82, 255–261.

Blackman, R.L. and Eastop, V.F. (2000) *Aphids on the World's Crops: An Identification and Information Guide*, 2nd edn. Wiley, Chichester, UK.

Blackshaw, R.P. and Kerry, B.R. (2008) Root herbivory in agricultural ecosystems. In: Johnson, S.N. and Murray, P.J. (eds) *Root Feeders: An Ecosystem Perspective*. Cromwell, Trowbridge, UK, pp. 35–53.

Broodryk, S.W. (1970) Dimension and development values for potato tuber moth *Phthorimaea operculella* in South Africa. *Phytophylactica* 2, 213–214.

Broodryk, S.W. (1971) Ecological investigations on the potato tuber moth, *Phthorimaea operculella* (Zeller) (Leptidoptera: Gelechiidae). *Phytophylactica* 3, 73–84.

Casagrande, R.A. (1987) The Colorado potato beetle: 125 years of mismanagement. *Bulletin of the Entomological Society of America* 33, 142–150.

Chandel, R.S., Chandla, V.K., Verma, K.S. and Pathania, M. (2013) Insect pests of potato in India: biology and management. In: Giordanengo, P., Vincent, C. and Alyokhin, A. (eds) *Insect Pests of Potato: Global Perspectives on Biology and Management*. Academic Press, Waltham, Massachusetts, pp. 227–268.

de Kort, C.A.D. (1990) Thirty-five years of diapauses research with the Colorado potato beetle. *Entomologia Experimentalis et Applicata* 56, 1–13.

Eyre, D. and Giltrap, N. (2012) *Epitrix* flea beetles: new threats to potato production in Europe. *Pest Management Science* 69, 3–6.

Fathi, S.A.A. and Nouri-Ganbalani, G. (2010) Assessing the potential for biological control of potato field pests in Ardabil, Iran: functional responses of *Orius niger* (Wolf.) and *O. minutus* (L.) (Hemiptera: Anthocoridae). *Journal of Pest Science* 83, 47–52.

Ferro, D.N., Logan, J.A., Voss, R.H. and Elkington, J.S. (1985) Colorado potato beetle (Coleoptera: Chrysomelidae) temperature-dependent growth and feeding rates. *Environmental Entomology* 14, 343–348.

Furlan, L. (2004) The biology of *Agriotes sordidus* Illinger (Coleoptera: Elateridae). *Journal of Applied Entomology* 128, 696–706.

Giordanengo, P., Vincent, C. and Alyokhin, A. (2013) *Insect Pests of Potato: Global Perspectives on Biology and Management*. Academic Press, Waltham, Massachusetts.

Graf, J.E. (1917) *The Potato Tuber Worm*. Bulletin 4217, 56. US Department of Agriculture, Washington, DC.

Horton, D.R. (2006) Quantitative relationship between potato tuber damage and counts of Pacific coast wireworm (Coleoptera: Elateridae) in baits: seasonal effects. *Journal of the Entomological Society of British Columbia* 103, 37–48.

Jacques, R.L. (1988) *The Potato Beetles*. E.J. Brill, Leiden, the Netherlands.

Jansson, R.K. and Seal, D.R. (1994) Biology and management of wireworm on potato. In: Zender, G.W., Powelson, M.L., Jansson, R.K. and Raman, K.V. (eds) *Advances in Potato Pest Biology and Management*. American Phytopathological Society Press, St Paul, Minnesota, pp. 31–53.

Johnson, D.A. (2008) *Potato Health Management*, 2nd edn. The American Phytopathological Society, St Paul, Minnesota.

Koss, A.M., Jensen, A.S., Schreiber, A., Pike, K.S. and Snyder, W.E. (2005) Comparison of predator and pest communities in Washington potato fields treated with broad-spectrum, selective, or organic insecticides. *Environmental Entomology* 34, 87–95.

Lagnaoui, A. and Radcliffe, E.B. (1998) Potato fungicides interfere with entomopathogenic fungi impacting population dynamics of green peach aphid. *American Journal of Potato Research* 75, 19–25.

Miles, H.W. and Cohen, M. (1938) Investigations on wireworms and their control. Annual Report 1937, Entomology Field Station, Warburton, UK.

Munyaneza, J.E. (2012) Zebra chip disease of potato: biology, epidemiology, and management. *American Journal of Potato Research* 89, 329–350.

Munyaneza, J.E. and Henne, D.C. (2013) Leafhopper and psyllid pests of potato. In: Giordanengo, P., Vincent, C. and Alyokhin, A. (eds) *Insect Pests of Potato: Global Perspectives on Biology and Management*. Academic Press, Waltham, Massachusetts, pp. 65–102.

Murphy, A.F., Rondon, S.I. and Jensen, A.S. (2013) First report of potato psyllids, *Bactericera cockerelli*, overwintering in the Pacific Northwest. *American Journal of Potato Research* 90(3), 294–296.

Peña, J.E. and Bullock, R.C. (1994) Effects of feeding of broad mite (Acari: Tarsonemidae) on vegetative plant growth. *Florida Entomologist* 77, 180–184.

Povolny, D. (1973) *Scrobipalpopsis solanivora* sp. n.—a new pest of potato (*Solanum tuberosum*) from Central America. *Acta Universitatis Agriculturae, Facultas Agronomica, Brno* 21, 133–146.

Ragsdale, D.W., Radcliffe, E.B. and Flanders, K.L. (2008) Managing aphids and leafhoppers. In: Johnson, D.A. (ed.) *Potato Health Management*, 2nd edn. The American Phytopathological Society, St Paul, Minnesota, pp. 149–159.

Reed, G.L., Jensen, A.S., Riebe, J., Head, G. and Duan, J.J. (2001) Transgenic Bt potato and conventional insecticides for Colorado potato beetle management: comparative efficacy and non-target impacts. *Entomologia Experimentalis et Applicata* 100, 89–100.

Stern, V.M., Smith, R.F., van den Bosch, R. and Hagen, K.S. (1959) The integrated control concept. *Hilgardia* 29, 81–101.

Stevenson, W.R., Loria, R., Franc, G.D. and Weingartner, D.P. (2001) *Compendium of Potato Diseases*, 2nd edn. The American Phytopathological Society, St Paul, Minnesota.

Swisher, K.D., Munyaneza, J.E. and Crosslin, J.M. (2013) Temporal and spatial analysis of potato psyllid haplotypes in the United States. *Environmental Entomology* 42, 381–393.

Umble, J.R. and Fisher, J.R. (2003) Influence of below-ground feeding by garden symphylans (Cephalostigmata: Scutigerellidae) on plant health. *Environmental Entomology* 32, 1251–1261.

Wallis, R.L. (1955) *Ecological Studies on the Potato Psyllid as a Pest of Potatoes*. Technical Bulletin 1107. US Department of Agriculture, Washington, DC.

Weber, D. (2003) Colorado beetle: pest on the move. *Pesticide Outlook* 14, 256–259.

Wheeler, A.G. (2001) *Biology of the Plant Bugs (Hemiptera: Miridae) Pests, Predators, Opportunists.* Comstock Publishing Associates, Ithaca, New York.

Zehnder, G., Vencill, A.M. and Speese, J. III (1995) Action thresholds based on plant defoliation for management of Colorado potato beetle (Coleoptera: Chrysomelidae) in potato. *Journal of Economic Entomology* 88, 155–161.

Ziems, J.R., Hoback, W.W., Higley, L.G., Hunt, T.E., Fernandes, O.A., Bastos, C., *et al.* (2008) Second generation European corn borer injury and Irish potato physiology, yield, and quality. *Agronomy Journal* 100, 720–725.

10 Plant Parasitic Nematodes of Potato

Juan E. Palomares-Rius,[1,2]* Claudio M.G. Oliveira[3] and Vivian C. Blok[2]
[1]Institute for Sustainable Agriculture (IAS), Spanish National Research Council (CSIC), Córdoba, Spain; [2]Cell and Molecular Sciences, James Hutton Institute, Invergowrie, UK; [3]Instituto Biológico, Campinas, Brazil

The Phylum Nematoda is highly diverse in terms of species richness and is one of the most abundant metazoan groups on Earth (Hugot *et al.*, 2001). Nematodes are essentially aquatic organisms, the majority of which are microscopic in size (0.3–3.0 mm), found in habitats ranging from oceans to the microscopic film of water surrounding soil particles (De Ley, 2000). Based on their different feeding habits, terrestrial nematodes can be divided into functional (trophic) groups (Yeates *et al.*, 1993). The plant parasitic nematode (PPN) group is regarded as one of the most important due to the direct damage they cause to crops, the indirect damage resulting from subsequent fungal and bacterial infection, and for some species the transmission of plant viruses. Other non-parasitic, free-living nematode species are also of great economic importance in agricultural systems due to their involvement in nutrient cycling and maintaining soil health.

Global annual crop losses from PPNs due to reduced yield and quality, and additional management costs, have been estimated at approximately 12% (Chitwood, 2003), corresponding to monetary losses to world agriculture of approximately US$100 billion. The lack of specific diagnostic symptoms resulting from PPN infestations, difficulties in effective sampling (PPNs typically have patchy distributions within the field and can be highly dispersed within the soil profile), the occurrence of PPNs as mixtures of species, and their microscopic size with limited morphological features for species identification makes determining economic damage from specific PPNs difficult. Lack of nematological expertise and resources can also be a significant constraint in assessing the impact of nematode problems. Eradication of PPNs, once established, is costly and hard to achieve due to their distribution within the soil matrix, the persistence of some species in dormant stages, and the incomplete effectiveness of control methods such as nematicides. Effective measures to prevent new introductions are therefore paramount.

Aboveground symptoms are generally nonspecific and may involve stunted plant growth, yellowing, wilting, and occasionally plant death. Infestations can also occur with no aboveground symptoms, but yields are reduced and/or tuber quality is affected. The PPNs that are regarded globally as most important in potato production are *Globodera* spp., *Meloidogyne* spp., *Naccobus aberrans* (Thorne, 1935) Thorne & Allen, *Pratylenchus* spp. and *Trichodorus* spp. (Nicol *et al.*, 2011). Regionally, particular species predominate, reflecting agronomic practices, climatic conditions, and historical distributions/introductions. For example, *Globodera pallida* (Stone, 1973) Behrens, 1975, was recognized in 2006 for the first time in Idaho, USA (Skantar *et al.*, 2007), and *Scutellonema*

* E-mail: palomaresje@ias.csic.es

bradys (Steiner and Lehew, 1933) Andrássy, 1958, was reported as a new potential threat to potato production in West Africa (Coyne and Claudius-Cole, 2009; Coyne *et al.*, 2011). In the tropics, several species are found associated with potato, but the root-knot nematodes (RKNs) (*Meloidogyne incognita* (Kofoid and White, 1919) Chitwood and *Meloidogyne javanica* (Treub, 1885) Chitwood, 1949) and lesion nematodes (*Pratylenchus brachyurus* (Godfrey, 1929) Filipjev & Schuurmans Stekhoven, 1941, and *Pratylenchus coffeae* Goodey, 1951) are particularly important. The shifting of the main regions for potato production to Asia, Africa, and Latin America, in line with the growing importance of potato as a staple crop in these regions, movement of commodities through trade, which could lead to the introduction of new pests and diseases, and climate change are all factors that will determine nematode threats in the future. This chapter describes the current main nematode parasites of potato and discusses options for their management. There are many sources of additional information in the literature, and distribution maps and host range information can be accessed at http://www.cabi.org/dmpd.

10.1 Plant Parasitic Nematodes of Potato

Potato cyst nematodes, *Globodera rostochiensis*, *Globodera pallida* and *Globodera ellingtonae*

The yellow and white potato cyst nematodes (PCNs), *Globodera rostochiensis* (Wollenweber, 1923) Behrens, 1975, and *G. pallida* have coevolved with Solanaceae plants in the Andes in South America, where they are indigenous in this mountainous region with a temperate climate (Stone, 1979, 1985). They were probably introduced into Europe around 1850 with potato breeding material (Jones and Jones 1974), from where they were further dispersed around the world. A pathotype scheme was devised by Kort *et al.* (1977), and further pathotypes were described in South America by Canto Saenz and De Scurrah (1977), which was consistent with the latter being the center of diversity for the species and only a subset of this diversity being present in Europe. Molecular characterization of *G. pallida* populations in the Andean region has localized the origin of European populations to the region around Lake Titicaca in southern Peru (Plantard *et al.*, 2008). Recently another species that infects potato, *Globodera ellingtonae* n. sp. (Handoo *et al.*, 2012), was found in Oregon and Idaho, USA and similar isolates are reported from Chile and Argentina. The pathogenicity of this species is currently under investigation. Given the diversity of these cyst nematodes in S. America it is likely that further species will be described in the future.

PCNs have a host range restricted to the Solanaceae that includes potato, tomato, and aubergine, and weeds such as deadly nightshade, and have highly specialized interactions with their hosts, requiring the formation of a feeding site (syncytium). The cyst, which is formed from the dead body wall of the female, forms a protective container for the 200–500 eggs. The second-stage juveniles within the eggs can remain dormant for many years, until stimulated to hatch by root diffusate from the host. The juveniles enter the root near the tip, puncturing the epidermal cell walls with their stylet, and then continue through internal cells before becoming sedentary and establishing a feeding site from a cortical parenchyma or endodermis cell (Sobczak and Golinowski, 2011). The syncytium develops following the partial degradation of cell walls adjacent to the initial cell, creating a multinucleate, metabolically active sink. They feed from the syncytium, developing into males or females after the fourth molt, depending on environmental conditions (Ellenby, 1954; den Ouden, 1960). The vermiform males leave the roots and fertilize the rotund females that protrude from the roots, fertilized eggs develop within the female, the female dies, and then her tanned cuticle forms the cyst (Fig. 10.1). The cysts facilitate persistence and dispersal, adhering to tubers, soil, farm machinery and equipment, and within which the eggs can remain viable in the soil for many years.

PCNs are estimated to cause yield losses of up to 10% worldwide (Oerke *et al.*, 1994). Despite efforts to prevent their spread, they are now found in many potato growing regions worldwide, though major production areas including China and Brazil have not reported having PCNs. They are designated as A2 quarantine organisms in Europe and EPPO-region countries, and measures are taken to prevent their import and regulate their spread. Heavily infested fields can show serious plant damage; however, poor growth in patches is more typical, with yellowing, wilting, or death of the leaves. Damage can also be

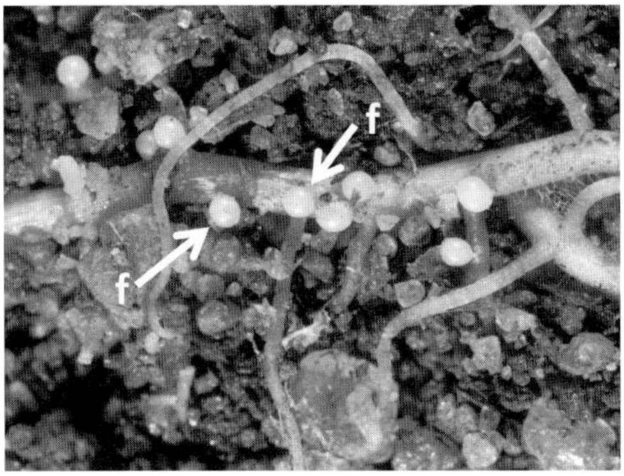

Fig. 10.1. Females of *Globodera pallida* (f) on roots of Desiree potato. (Photo courtesy of J.E. Palomares-Rius.)

associated with bacteria or fungi such as the wilt fungus *Verticillium*. The presence of PCNs on roots can be seen with the appearance of pinhead-sized, white-to-yellow colored females around the time of flowering. The life cycle takes ~5–7 weeks, depending on conditions such as soil temperature and host genotype, and in some circumstances two generations can occur within the growing season (Vovlas, 1996). Both species are difficult to differentiate morphologically; characters used are the shape of the stylet knobs, which are smaller and rounded in *G. rostochiensis* and more robust and hooked in *G. pallida*, and *G. pallida* has a longer stylet. Cyst characters are also used to distinguish between PCN species and to distinguish PCNs from other cyst nematodes (Fleming and Powers, 1998). Molecular tools provide valuable additional species diagnostics, and various methods have been developed for species identification and quantification (Reid *et al.*, 2010) and within-species differentiation (Plantard *et al.*, 2008; Hoolahan *et al.*, 2012). For further information on PCNs, see Subbotin *et al.* (2010). Recently *G. ellingtonae* has been described from fields in Oregon and Idaho, USA which parasitizes potato (Handoo *et al.*, 2012) and the ITS sequence has high similarity to a sample from Chile.

Root-knot nematodes, *Meloidogyne* spp.

The RKNs, *Meloidogyne* spp., are globally the most economically important PPNs, particularly in warm regions. Wide host ranges and the capability of producing multiple generations within one crop season can lead to severe damage from some species. At least six *Meloidogyne* species are considered to be of global economic importance as potato pests: in the tropics, *M. incognita* (southern root-knot nematode) and *M. javanica* are widely distributed and cause substantial damage (Scurrah *et al.*, 2005), whereas, *Meloidogyne hapla* Chitwood, 1949 (northern root-knot nematode), *Meloidogyne chitwoodi* (Golden, O'Bannon, Santo and Finley, 1980) (Columbia root- knot nematode) and *Meloidogyne fallax* Karssen, 1996, are more important in temperate regions (Karssen, 1996). *M. chitwoodi* and *M. fallax* are quarantine organisms in Europe, and recently another species, *Meloidogyne minor* Karssen, Bolk, van Aelst, van den Beld, Kox, Korthals, Molendijk, Zjilstra, Van Hoof & Cook, 2004, has been described in northwestern Europe, which can cause severe quality damage in potato (De Weerdt *et al.*, 2011). *Meloidogyne* spp can be disseminated by infested soil adhering to farm machinery and in irrigation water, but seed potato tubers are the most efficient way to spread these nematodes to new areas. Thus, RKNs are a permanent concern in the movement of seed potato tubers to countries that have quarantine restrictions for these species.

RKNs are sedentary endoparasites with four juvenile stages. The females produce ~500 eggs; the first-stage juvenile develops in the egg and undergoes the first molt, giving rise to the second-stage juvenile. This parasitic stage invades near the root tips, induces the formation of giant cells, and then begins feeding, undergoing three molts until the adult stage. In both tubers and roots, feeding sites are characterized by giant cells containing granular cytoplasm and many hypertrophied nuclei. Males can be present even in parthenogenetic species, but they do not feed.

For *Meloidogyne* species, the aboveground symptoms are not specific, and symptoms can occur in patches of reduced growth with plants that are stunted with chlorotic leaves and wilting, particularly with moisture stress. Complete plant loss can occur with high infestations. However, infestations may occur without causing any aboveground symptoms, mainly in potato crops that receive high levels of fertilizers. Roots infected by *M. incognita* or *M. javanica* can exhibit large characteristic galls. However, roots infected by *M. chitwoodi* and *M. fallax* have very small galls, while those infected with *M. hapla* may show more evident galls and a proliferation of lateral roots around the galls. *M. minor* also causes similar symptoms on tubers to *M. chitwoodi* and *M. fallax* (De Weerdt *et al.*, 2011). Mature female nematodes can also be found in the galls formed in the tubers, below the potato skin. Necrotic spots can be observed in the region between the tuber surface and the vascular ring. In all cases, the tuber surfaces could be deformed with swellings, which may have a warty appearance, resulting in a reduction of tuber quality and making tubers unmarketable (Fig. 10.2). Usually, more than one generation can occur during the season, with the first generation occurring in the roots, while the subsequent generations are able to attack potato tubers (Charchar *et al.*, 2009).

Diagnostics and identification for these species is difficult. However, several references for the genus are very useful for species identification based on morphology (Hirschmann, 1985; Hunt and Handoo, 2010). Additionally, female isoenzyme characterization and molecular methods based on polymerase chain reaction (PCR) have proven to be very useful in species identification (Blok and Powers, 2010).

False root-knot nematodes, *Nacobbus aberrans*

The false RKN, *N. aberrans* is a species complex that has a wide host range including potatoes, sugarbeets, beans, peppers, crucifers, and tomato, and in which host specialization is found. They occur in temperate and tropical regions of North and South America. Different pathotypes have been described, and the potato pathotype is specialized on Solanaceous hosts and occurs in Mexico and the Andean regions of Argentina, Bolivia, Chile, and Peru. This potato pathotype can cause severe damage to potato, with yield losses reported averaging 65% in the Andean regions of Latin America and 36% for bean in Mexico (Otazú *et al.*, 1985; Manzanilla-López *et al.*, 2002). Aboveground symptoms are not specific, but plants are weak and stunted. Infested roots show strings of galls that may look similar to *Meloidogyne*-infested roots. False RKNs can survive under the skin of the tuber, acting as a source of new infestations. These nematodes are listed as quarantine organisms by many countries because of the severe damage they can cause to various crops, and especially potato.

The false RKNs are endoparasitic and migratory, invading roots and tubers where they feed, migrate, and develop, causing lesions, necrosis, and cavities. Females reinvade healthy roots, where they become sedentary and establish permanent feeding sites, producing root galls, and following fertilization, producing egg masses. The third- and fourth-stage juveniles can become dormant in roots and tubers, surviving in adverse conditions and facilitating their dispersal (Manzanilla-López *et al.*, 2002).

Root-lesion nematodes – *Pratylenchus* spp.

Ten root-lesion nematode species, *Pratylenchus* spp., are known to parasitize potato: *P. brachyurus*, *P. coffea*, *Pratylenchus crenatus* Loof, 1960; *Pratylenchus neglectus* (Rensch, 1924) Filipjev and Schuurmans-Stekhoven, 1941; *Pratylenchus thornei* Sher and Allen, 1951; *Pratylenchus scribneri*

Fig. 10.2. Potato tubers infected by *Meloidogyne javanica* exhibiting the characteristic galls and deformed surface. (Photo courtesy of C.M.G. Oliveira.)

Steiner in Sherbakoff and Stanlet, 1943; *Pratylenchus andinus* Lordello, Zamith and Boock, 1961; *Pratylenchus penetrans* (Cobb, 1917) Filipjev and Schuurmans-Stekhoven, 1941; *Pratylenchus flakkensis* Seinhorst, 1968; and *Pratylenchus vulnus* Allen and Jensen, 1951 (Scurrah *et al.*, 2005). In North America, *P. penetrans* is a severe pest of potatoes, whereas in tropical regions only *P. brachyurus* and *P. coffeae* have a wide distribution, and both are known to damage potato. Root-lesion nematodes cause damage by feeding directly on roots or tubers and moving through cortical tissues. They are often found associated with bacterial and fungal infections, causing potato early dying. Generally, plants parasitized by root-lesion nematodes are stunted and exhibit pronounced leaf chlorosis and root shedding. *Pratylenchus* spp. cause extensive root cortical lesions and produce cortex destruction. Initially, reddish-brown lesions are caused on the roots, but these later turn black. On tubers, lesion nematodes can penetrate through lenticels, parasitizing the adjacent tissue and producing lesions of variable size, depending on the population level. Small lesions may be confused with lenticels, but large ones, although superficial, can affect tuber quality. Infected tubers may have pimples, pustules, or wart-like protuberances affecting their marketability. Marketed tubers have a shorter shelf life when infected by *Pratylenchus* spp.

All root-lesion nematodes are endoparasitic and migratory. Both juveniles and adults enter and reproduce in the host roots and tubers. In general, the *Pratylenchus* life cycle is similar to that of other PPNs, comprising eggs, four juvenile stages (J1 through J4), and adults (Fig. 10.3). Males are rare in those species that reproduce by mitotic parthenogenesis, such as *P. brachyurus*, or abundant in amphimitic ones, such as *P. coffeae*. Eggs are laid singly in the roots or in the soil.

Kubo *et al.* (2001) reported that *P. brachyurus*, *Pratylenchus jaehni* (*P. coffeae* population K5), and *P. coffeae* caused brown and round lesions surrounding the lenticels of tubers of cultivar Monalisa and cultivar Bintje (Fig. 10.4), and reported several degrees of susceptibility in potato cultivars. In Norway, Holgado *et al.* (2009) reported that the yield of potato cultivar Saturna was reduced by 50% in the area of the field affected by a large population of *P. penetrans*. Transect sampling showed plant growth to be correlated negatively with densities of *P. penetrans*, and suggested a damage threshold of potato to the nematode of 100 specimens per

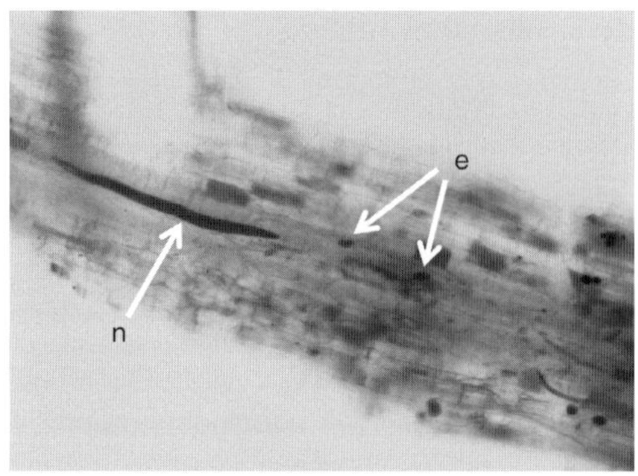

Fig. 10.3. Stained *Pratylenchus coffeae* (n) and eggs (e) inside the cortical tissues of potato cultivar Atlantic. (Photo courtesy of C.M.G. Oliveira.)

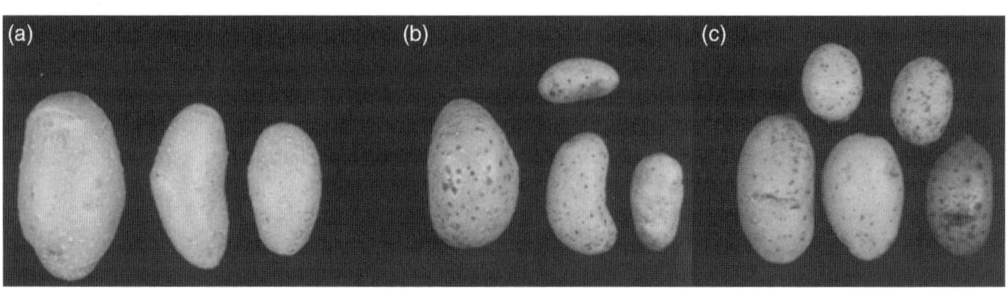

Fig. 10.4. Tubers of potato cultivar Monalisa. (a) Healthy, without nematodes; (b) lesions caused by *Pratylenchus coffeae*; (c) lesions caused by *Pratylenchus brachyurus*. (Photo courtesy of R.K. Kubo, Instituto Biológico, Campinas, Brazil.)

250 g of soil. However, when a combined infection occurs with other pathogens, the symptoms and yield losses are more severe. For example, the combined infection of *Verticillium* spp. and *Pratylenchus* spp., known as potato early dying, has long been recognized as a problem of potato production areas in the USA, often resulting in yield reductions of 30–50% (Powelson and Rowe, 1993). In a greenhouse experiment, Holgado *et al.* (2009) found that *P. penetrans* alone induced tuber cross lesions similar to those of common scab (*Streptomyces scabies*), but the combined inoculation of the bacterium and the nematode enhanced symptom expression.

The presence and survival of root-lesion nematodes in tubers is a source for further spread of these nematodes via seed potatoes into new areas. Timing of harvesting to limit damage, cold storage of tubers to reduce damage, and ensuring seed is not infected are very important. Traditionally, experience is required to identify these nematodes using morphological characters. Several identification keys are available that take into account biological (presence/absence of males) and morphological/morphometric characteristics for identification (Castillo and Vovlas, 2007). Molecular techniques can be very valuable in identification, and several methods based on PCR have been developed (Al-Banna *et al.*, 2004; Oliveira *et al.*, 2009).

Stubby-root nematodes, *Paratrichodorus* spp. and *Trichodorus* spp.

The stubby-root nematodes (*Paratrichodorus* spp. and *Trichodorus* spp.) have wide host ranges and feed on root surfaces, damaging the growing tips, resulting in the formation of numerous stubby roots. They are obligate ectoparasites feeding primarily on the meristem cells of root tips. As well as the direct damage caused to roots, which results in yield and quality reduction, some trichodorid nematodes can transmit plant viruses. Symptoms of stunting, wilting, and lodging are usually patchy within the field, and tend to be more severe in sandy soils. At least 13 species of trichodorid nematodes can transmit *Tobacco rattle virus* (TRV), which causes spraing or corky ringspot in tubers (Fig. 10.5).

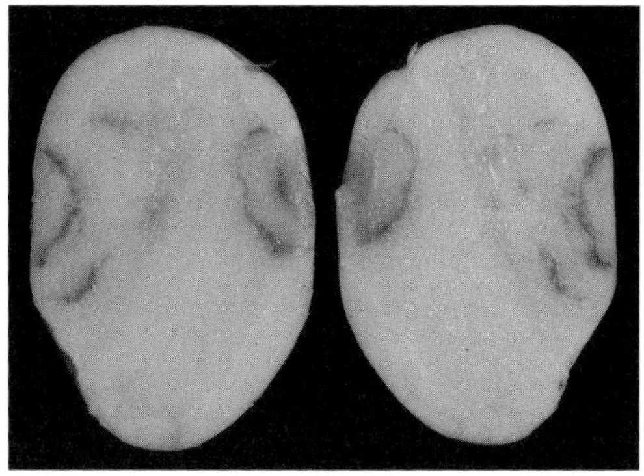

Fig. 10.5. Potatoes infected by *Tobacco rattle virus*. (Photo courtesy of R. Neilson, The James Hutton Institute, Invergowrie, UK.)

Spraing appears as noticeable brown rings on the surface and/or brown arcs or flecking within the tuber and mottle in the stems and leaves of susceptible potato cultivars (Lamberti and Roca, 1987). Spraing symptoms in tubers, even at relatively low levels, can render entire crops unsaleable, for both the fresh and the processing industries (Brodie *et al.*, 1993; Scurrah *et al.*, 2005).

Trichodorid nematodes comprise different taxonomic groups that are difficult to distinguish visually, but which vary significantly in terms of their distribution, pathogenicity, and virus transmission frequencies and specificity with the serotypes of virus they transmit (Ploeg *et al.*, 1992). The problem of their identification is complicated by the coexistence of different species in the same field. *Trichodorus* spp. occur predominantly in the more temperate regions, whereas *Paratrichodorus* spp. are more prevalent in tropical and subtropical regions (Luc *et al.*, 1990). The removal of approved nematicides and the expected population increases with more conducive conditions resulting from climate change are expected to exacerbate difficulties with their management in temperate regions. Soil samples should be managed carefully because of the nematode's delicate cuticle. Diagnosis of these nematodes is extremely difficult and is based on morphological/morphometrical characters (Decraemer, 1995). *Paratrichodorus* spp. are more difficult to identify than *Trichodorus* spp. (Duarte *et al.*, 2011). New diagnostic methods based on DNA (specific primers, qPCR, and RFLP PCR) have been developed for some species (Boutsika *et al.*, 2004; Holeva *et al.*, 2006; Duarte *et al.*, 2011).

Potato rot nematode or potato tuber nematode, *Ditylenchus destructor*

Potato rot nematode or potato tuber nematode, *Ditylenchus destructor* Thorne, 1945, is widely distributed in temperate climates; it is found in localized areas of North America and Europe, the Mediterranean region, and Asia. It has a wide host range including crops, ornamentals, and weeds, making control by crop rotation difficult. Potato tuber nematodes are migratory endoparasites of roots and tubers. The nematodes enter the tuber through lenticels, feed, and multiply, rapidly completing their life cycle in three weeks at 20–27°C. Dispersal is primarily via infected tubers, infested soil, and irrigation water, though survival overwintering as adults, larvae, or eggs is also reported. At temperatures from 15 to 20°C, combined with high relative humidity, it can be an important pest of potato and is particularly important for damaging tubers during storage if temperatures are not kept low. Potato rot nematode is a regulated quarantine pest in many countries.

There are no obvious symptoms on the aerial parts of potato infected with potato rot nematodes, although plants may be weak and die. Infected tubers initially have small, off-white spots below the peel, which enlarge and darken and become surrounded by rotten tissue that becomes infected with secondary organisms, leading to general rot during storage. The surface of the tuber develops slightly sunken areas; the skin may be wrinkled, cracked, and detached in places. The decaying tissue may have a discolored and dry, mealy appearance (Fig. 10.6).

Several morphological characters are used to distinguish *D. destructor* from related species. Wendt *et al.* (1993) originally developed a molecular diagnostic based on PCR-ITS (internal transcriber spacer)-RFLP to distinguish *D. destructor* from other *Ditylenchus* spp., and species-specific primers have also been designed in this ITS region. Subbotin *et al.* (2011) subsequently examined many populations of *D. destructor* from around the world and found that there was significant length variation in the ITS region, suggesting this species was unusually diverse.

Fig. 10.6. Potato tubers infected with *Ditylenchus destructor*. (Photo courtesy of L. Kuzmina, Laboratory of Plant Health and Microbiology, Saku, Estonia.)

The yam nematode, *Scutellonema bradys*

Coyne and Claudius-Cole (2009) reported that potato could be affected by *Scutellonema bradys* in West Africa, and further studies were undertaken by Coyne *et al.* (2011) to assess its potential pathogenicity on potato under greenhouse and field conditions, and on marketed tubers. This study provided strong evidence of the pest potential of *S. bradys* on potato, supporting the initial report of its occurrence on the crop. At low densities, the nematode was recovered from the majority of tubers sampled from northern Nigeria, indicating that the nematode could reproduce in the potato field conditions prevailing there and that it could be spread with infected tubers. Potato plants inoculated with *S. bradys* produced tubers with substantial cracking and evident tuber rot. Symptoms of nematode infection on tubers included a scaly appearance, surface cracking, as well as deeper tissue cracks, distortions, and darkened surface patches. In most cases, these patches were related to subsurface rot.

10.2 Management of Plant Parasitic Nematodes in Potato

Management includes all of the measures (agronomical, legal, and genetic) employed to reduce or maintain PPN numbers under the economic threshold level. Losses induced by nematodes depend on the nematode species, the type of damage induced, and the final destination of the potato crop (seed, fresh market, or processing). Plant parasitic nematodes also may contribute to disease complexes that result in synergistic interactions with other pests or pathogens, including the loss of resistance to other diseases, as found with *Globodera–Verticillium dahliae* and *Pratylenchus–V. dahliae* (Back *et al.*, 2002), *Meloidogyne* spp. and fungus (*Verticillium*, *Fusarium*, and *Rhizoctonia* spp.), or bacteria (*Ralstonia solanacearum* and *Erwinia* spp.) in warm regions (Brodie *et al.*, 1993). Potato viruses transmitted by stubby-root nematodes are also a major concern in some potato growing areas. The impact of these interactions with other pathogens is related to nematode numbers and the infective capacity of the nematode during the crop cycle; thus, a primary goal is controlling nematode population levels in the field. In the UK, £9 million is spent annually on nematicides to control PCNs (DEFRA, 2004), and in the USA, US$20 million is spent in Washington State to control *M. chitwoodi* and *M. hapla* (Santo, 1994). Cropping history, nematode soil analysis (identification and quantification), experience with local soils, and crop benefit are all important factors to consider in making nematode management decisions. Reliable identification based on an integrative view of classical and molecular techniques is decisive in planning strategies, as are crop rotation or the use of resistant cultivars.

Management strategies

Prevention of spread

Prevention of spread is one the most important strategies for PPN management, because once a field is infested, complete nematode eradication is difficult to achieve and may incur significant short- and long-term costs. Nematodes move short distances in the soil, so spread from infested to non-infested areas is associated mainly with human activities such as the movement of contaminated soil attached to farm equipment, boots, and the hooves of animals, infected plant material including farm-saved seed, and in reused irrigation water (Santo, 1994). Root-knot, potato cyst, stem and root-lesion nematodes and other endoparasites or semi-endoparasites can be introduced into the field by seed potatoes; therefore, the use of certified potato seed is essential to avoid new introductions. Special attention is required to identify symptomless seed potatoes that are infected with *M. javanica* and/or *M. chitwoodi* or *M. hapla* in the cooler areas of production because of the shorter growing seasons (Santo, 1994). Also, viruses such as TRV can be spread by symptomless potato clones, and different virus–nematode vector combinations show specific interactions with cultivars (Xenophontos *et al.*, 1998). For this reason, sensitive techniques like RT-PCR are the most appropriate test in seed certification against TRV.

Agronomical measurements

Crop rotation

Crop rotation can be useful, depending on the nematode species concerned. For the major PPNs of potato (*Globodera* spp. and *Meloidogyne* spp.), this strategy is less effective, and long rotation periods must be used due to the long survival of dormant eggs within the cysts of *G. pallida* and *G. rostochiensis*, or because of the wide host ranges for *Meloidogyne* spp. The average decline rates for PCNs are approximately 20–30% per year in UK conditions (Clayton *et al.*, 2008), and a 7-year rotation between potato crops is a typical recommendation (EPPO, http://www.eppo.int/QUARANTINE/nematodes/Globodera_pallida/HETDSP_ds.pdf). However, the rate of decline is different for each of the PCN species, with *G. pallida* decline rates being particularly prolonged. Decline is also affected by the susceptibility/tolerance of the variety, and by environmental factors such as soil type (Hockland, 2010). Crop rotation can be combined with nematicide application and/or the use of resistant potato varieties in order to maintain the population at manageable levels over a shorter rotation period. Amendments such as poultry litter compost (in the case of *P. penetrans*) are also reported to increase the efficacy of these measures (Everts *et al.*, 2006). However, commercially attractive cultivars with full resistance are not always available, as in the case of *G. pallida*, and both crop rotation and granular nematicides are less effective at controlling this species than *G. rostochiensis* (Minnis *et al.*, 2004).

Rotations that include a non-host crop for RKNs, such as maize or wheat, will minimize damage from some *Meloidogyne* spp. (Krishna Prasad, 2008). However, some RKN species such as *M. chitwoodi* can attack cereals and can survive in potato tubers stored for more than 2 years at 1°C, and in tubers left in the field over winter (Finley, 1981). Cereal rotations (2–3 years) have proven useful in the management of *D. destructor* in potato fields, despite their wide host range (Kornobis and Stefan, 1991). Each strategy must consider the specific attributes of the target species, as well as rotation options.

Potato volunteers in the crop rotation or tillage period must be taken into account for successful nematode reduction, as well as the possible weed hosts for particular species. TRV can be maintained by weeds in rotation of tillage; for example, hairy nightshade (*Solanum sarrachoides* Sendtn.) in rotation crops of lucerne or Scotch spearmint (*Mentha* × *Gracillis*) (Boydston *et al.*, 2008).

Early harvesting

Growing early-maturing potato varieties or harvesting tubers as soon as possible after allowing for proper maturation may help in lowering the severity of tuber damage by *Meloidogyne* spp. However, *M. chitwoodi* females continue to develop within infected tubers in storage at lower temperature, causing problems with the marketability of the crop (Santo, 1994). This strategy must be based on nematode development rates, which are determined from the accumulated degrees above a basal development temperature for each nematode species. However, this approach can lead to specific species selection if this measurement is done routinely in fields with mixed species, as has been shown with *G. pallida* (more favored) and *G. rostochiensis* (Webley and Jones, 1981). Knowledge of the degree days required by the different PCN species to develop through their life-cycle stages in each potato cultivar could be used to plan the harvest time to precede cyst development, and thereby minimize population reproduction (Alonso *et al.*, 2011). Additionally, deep plowing and exposing soil to summer heat could also diminish egg viability inside the cysts (Krishna Prasad, 2008), though this is mainly applicable in warm climates.

Green manure

Green manure amendments for the management of PPNs and improving soil physical characteristics, especially water infiltration and penetration resistance, are currently being investigated. Additionally, some green manures could be applied with other strategies, as nematicides, in order to improve their efficacy. Sorghum sudangrass (*Sorghum bicolor* (L.) × *Sorghum arundinaceum* (Desv.) Stapf var. sudanense (Stapf) Hitchc.), grown annually as a green manure in combination with poultry litter/tillage/fallow, has been effective in the control of RKNs and lesion nematodes (Kratochvil *et al.*, 2004). MacGuidwin *et al.* (2012) have demonstrated that green manure amendments can be used to reduce damage from *V. dahliae* and *P. penetrans*. Crop rotation for 3 years, with poor or non-hosts of *M. chitwoodi* interspersed with green manure amendments without nematicides, provided nematode control comparable to soil fumigation (Riga *et al.*, 2003). Some more local manure amendments such as wastes from the logging and wood industries (freshly-crushed conifer bark and sodium lignosulfonates) possess nematicidal properties, and significant reduction of PCN multiplication has been observed (Matveeva *et al.*, 2010).

Trap cropping

Solanum sisymbriifolium Lamarck can be an effective trap crop with strong hatch stimulation and total resistance to both *G. rostochiensis* and *G. pallida* (Scholte, 2000); nematodes do not develop further than J3 (Sasaki-Crawley *et al.*, 2010). It can be plowed in as green manure. However, this plant needs to accumulate a certain level of dry matter to achieve sufficient control, which restricts its use in cold climates (Hockland, 2010). Trap cropping using potato plants could also be useful if the plants were incorporated into the soil before nematodes reproduced (about 300 day degrees at a threshold of 4°C in *G. pallida*) and using vigorous rooting potato varieties such as Cara (Lane and Trudgill, 1999). However, the timing, costs, and risks of this strategy must be considered carefully. Other crops in more restricted conditions (Andean region), such as *Lupinus mutabilis* Sweet, oca (*oxalis tuberosa* Molina), isano (*Tropaeolum tuberosum* Ruiz & Pav.), ollucus (*Ullucus tuberosus* Caldas), quinuas (*Chenopodium quinua* Willd.), and barley (*Hordeum vulgare* L.), have proved successful for PCN control (Franco *et al.*, 1999). However, the high genetic variability in these crops needs to be considered.

The possibility of reproduction (or not) in the plant by the target nematodes or other PPNs must be established in order to use trap cropping as a strategy. Additionally, with PCNs, the use of trap cropping during the same year of the potato crop will not be effective, because of the diapause these nematodes undergo. Trap cropping is an option where PCN infestations are very large and need to be reduced quickly, because the technique is costly and requires very careful management (Lane and Trudgill, 1999).

Seed pretreatment and disinfestation

The possibility of potato disinfestation has been studied for *P. brachyurus* by treating the tubers at 50°C for 40–60 min (Koen, 1969). Excess soil adhering to the tubers must be left in the field, and removal of soil by high-pressure washing or brushing until practically free of soil is not wholly effective but should minimize the risk of spread (Hockland, 2010). Low-temperature pretreatment of potato tubers for short periods has been reported to increase plant tolerance of subsequent low field temperatures, decrease PCN multiplication, and increase crop productivity (Matveeva *et al.*, 2010).

Nematicides

Nematicides, if used, must be included in an integrated pest management program (IPM) and not used as the only solution to PPNs. Also, due to toxicological and environmental issues, their use is being restricted in many countries and their availability reduced. National requirements in terms of nematicides can also differ; for example, in Spain the use of fumigants is not permitted, only ethoprophos and oxamyl (MAAMA, 2012).

Two main kinds of nematicides are available in potato: fumigant and non-fumigant. They have a restricted range of application conditions (i.e. soil type, water conditions, temperature), and these conditions are generally more restrictive for fumigant than for non-fumigant nematicides. Fumigant nematicides release toxic gases that kill dormant juveniles inside cysts, egg masses, or active nematodes, and they must be used prior to planting, while granular nematicides can be applied at planting. These compounds prevent the hatched juvenile nematodes from finding and invading the growing potato roots. Fumigants are better suited for reducing high nematode population levels (Lane and Trudgill, 1999). However, nematicides (particularly fumigants) are not specific and can have broad effects on the soil communities, which in turn can have detrimental effects on plant health and development. Non-fumigant nematicides (granular) are more influenced by how well the granules are incorporated, and the best way to do that is by using a powered rotator cultivator (Lane and Trudgill, 1999). Global positioning systems have been investigated for use with targeted application of nematicides in "hot-spots" in the field; however, this has been of limited success because the nematode infestation is likely to be present elsewhere in the field below detectable levels. Non-fumigant nematicides are usually more effective than fumigants in the control of stubby-root nematodes because they are applied deeper in the soil. The best results are obtained at planting, when most stubby-root nematodes are in the upper soil layers, provided that irrigation or rain does not wash them into deeper levels (Strand, 1986).

Control of *G. pallida*, which has an extended hatching period compared with *G. rostochiensis*, influences the efficacy of nematicide control (Hockland, 2010). For this reason, it is important to identify correctly the PCN species found in the field. In other cases, the vertical migration of species can be an important factor in the effective performance of pesticides applied in soil. Populations of *M. chitwoodi* have been detected as deep as 180 cm (Santo *et al.*, 1985).

Plant resistance

Host-plant resistance is an environmentally friendly method of nematode control; however, cultivars are often not available with the appropriate agronomic characters, and repeated use of resistant cultivars in the same field can lead to the selection of other PPN species, or virulent pathotypes, negating their utility. PCNs and

RKNs are the two major nematological problems in potato, and for both, breeders have been developing resistant varieties. Resistant cultivars carrying the gene *H1*, derived from *Solanum tuberosum* L. ssp. *andigena* (CPC 1673), have been extremely effective at reducing populations of *G. rostochiensis* throughout Europe, providing resistance to pathotypes Ro 1 and Ro 4 (Dale *et al.*, 2010). Additional resistance genes against *G. rostochiensis* have been detected in other genetic sources (Gebhardt and Valkonen, 2001). However, it has become apparent that these cultivars are primarily useful when *G. pallida* is absent or present at very low levels, though in the latter situation gradual increases in *G. pallida* populations are probable. Fewer cultivars are available with high levels of resistance for *G. pallida*, though in the Netherlands, and to a lesser extent in Germany, there are some specialized cultivars for starch production with resistance (Hockland *et al.*, 2012). For fresh consumption, a few cultivars have also been registered, i.e. cultivar Amanda in Germany in 2006 and cultivar Iledher in 2009 in France (Hockland *et al.*, 2012). No cultivars with very high levels of resistance to all *G. pallida* Pa1, 2/3 pathotypes are currently listed in the European Potato Database (www.europotato.org). Producent and Sante are reported to have very high levels of resistance to Pa2; Cyprian, Innovator, Pallidia, Seresta, Skawa, Zagloba, and Zeus with high levels of resistance to Pa3; and Arela, Benol, Morag and Vales Everest with high to very high levels of resistance to Pa1. On the UK Potato Database (http://www.varieties.potato.org.uk/), Ambassador, Crisps4all, and Vales Everest are classified as resistant to *G. pallida* pathotypes Pa2/3, 1. In the future, combining resistances to both species and to different pathotypes should provide an improved spectrum of cultivars. Cultivars with partial resistance, such as Nadine, Valor, Rocket, and Sante, allow only a proportion of juveniles to reach maturity, and they could be used in PCN management programs, but their use requires careful monitoring for the selection of new nematode genotypes that are more virulent (Hockland *et al.*, 2012).

Several resistance loci have been mapped that confer resistance to *G. pallida* pathotypes Pa2 and Pa3. These loci are derived from the diploid relatives of cultivated potato, including *Solanum vernei* Bitter & Wittm., *Solanum sparsipilum* (Bitter) Juz. & Bukasov, *Solanum spegazinii* Bitter, *Solanum tarijense* Hawkes and the tetraploid *S. tuberosum* ssp. *andigena* CPC1673 (only pathotype Pa2) and CPC 2802 (pathotypes Pa2 and Pa3) (Moloney *et al.*, 2009), and no single resistance gene has yet been identified that gives complete resistance to *G. pallida*. The higher diversity of the *G. pallida* populations compared to *G. rostochiensis* populations also makes it more difficult to incorporate effective broad-spectrum resistance (Schnick *et al.*, 1990; Folkertsma *et al.*, 1994). However, the diversity of both species found in Europe represents only a small amount of the diversity in comparison to their center of origin in South America (Andes) (Hockland *et al.*, 2012). The resistance sources used in European breeding programs are directed to the particular populations introduced into Europe, and for this reason it is important to avoid the introduction of more virulent pathotypes from their center of origin (Hockland *et al.*, 2012).

Efficient potato breeding strategies against PCNs have been developed (Bradshaw *et al.*, 2003), with the emphasis placed on a progeny test to assess and select or discard whole progenies before starting conventional within-progeny selection at the unreplicated small-plot stage (Dale *et al.*, 2010). However, the use of molecular markers will assist in the selection for important disease resistance and quality traits, allowing the development of lines with multiple resistance sources (Dale *et al.*, 2010). In some cases, the combination of different resistance loci can result in apparent additive effects on resistance levels (for example in pathotypes Pa5 and Pa6 in *G. pallida*) (Rouppe van der Voort *et al.*, 2000). The use of parental material which is triplex or quadruplex for a single resistance gene will also help to ensure that the progeny have the resistance gene, eliminating the need to test for resistance. New methods using pyrosequencing or single-nucleotide polymorphism markers have been developed in order to determine the dosage of the resistance gene (in this case *H1*) (Dale *et al.*, 2010), and recently *Avr* genes have been used as probes to screen for resistance in progeny.

Several *Solanum* wild species have resistance against *Meloidogyne* spp. The introgression of several genes against RKNs has been developed from some wild parentals. Resistance introgression into cultivated potato for *M. chitwoodi* has been achieved using *Solanum bulbocastanum*

Donal in somatic hybridization (Brown *et al.*, 1996); *Solanum hougasii* Correll against race 1 and race 2 of *M. chitwoodi* and *M. fallax* (Brown *et al.*, 1999); *Solanum stoloniferum* Schltdl. & Bouché and *Solanum fendleri* Schltdl. & Bouché, both resistant to *M. fallax* and *M. chitwoodi* (Janssen *et al.*, 1997); *Solanum chacoense* Bitter against *M. fallax* and *M. hapla*; *S. tarijense* against *M. hapla* (Draaistra, 2006); and *S. sparsipilum* against *M. fallax* (Abou Bakari *et al.*, 2006). However, the combining of resistance genes from *S. tarijense* and *S. chacoense* does not result in an increase of resistance against *M. hapla* (Tan *et al.*, 2009). Some clones of *S. chacoense* and *S. tuberosum* have been found to be resistant to Italian populations of *Meloidogyne arenaria*, *M. hapla*, *M. incognita*, and *M. javanica* (Di Vito *et al.*, 2003), opening the possibility of breeding for resistance to more tropical species.

Resistances for other nematode species have been studied less intensively. Several native *S. tuberosum* ssp. *andigena* with resistance have been identified (ex. Andean landraces Azul, Huaca Lajra, and Tunti Imilla), as well as resistant wild tuber-bearing *Solanum* species (*Solanum megistracrolobum* Bitter, *Solanum acaule* Bitter, and *S. sparsipilum*) against *N. aberrans* (Cahuana *et al.*, 1975; Inserra *et al.*, 1985; Suarez *et al.*, 2009). However, resistance seems to be dependent on the race and/or population. Other nematodes such as *Pratylenchus* spp. or *D. destructor* have presented fewer possibilities with resistance or tolerant cultivars.

New management strategies

Biocontrol agents

Not many biocontrol agents have been developed for potato. However, as more nematicides will be withdrawn from the market, new possibilities for these agents to be incorporated are being considered. Several agents look promising against some PPNs. *Paecilomyces lilacinus* (Thom) Samson and *Pochonia chlamydosporia* (Goddard) Zare & W. Gams have offered good results against PCNs (Jacobs *et al.*, 2003; Tobin *et al.*, 2008). For these organisms, more research with different environmental conditions will help to determine their potential for use as part of an IPM strategy. Other fungi have been isolated from PCN eggs (Clovis and Nolan, 1983), and up to 93% infection of eggs has been achieved in laboratory experiments with *Verticillium suchlasporium* Gams & Dackman (Dackman, 1990). *Beauveria brogniartii* (Sacc.) Petch showed good control in laboratory and greenhouse experiments against *N. aberrans* on potatoes (Balderrama, 1992).

Transgenic potatoes

Several transgenic clones have been produced, and they represent a new prospect for nematode management. One has used the expression of a proteinase inhibitor against a broad spectrum of nematodes. Transgenic plants derived from commercial cultivars have been tested against *G. pallida* and *M. incognita*, including localizing expression using root-specific promoters (Lilley *et al.*, 2004). The combination of natural R genes and cystatin genes in the commercial cultivars Sante and Maria Haunca produced high levels of resistance to virulent and avirulent populations of *G. pallida* in Sante (Urwin *et al.*, 2003). Cystatin-expressing potato plants do not harm non-target insect aerial feeders, their natural parasitoid enemies, or the soil microbial community (Fuller *et al.*, 2008). In the future, as more information about natural plant proteinase inhibitors and their possible nematode targets increases with better designs of synthetic proteinase inhibitors, the efficiency of these techniques may improve even further.

Another transgenic plant strategy is the expression *in planta* of double-stranded RNA (dsRNA), which can interfere with the gene expression of the targeted nematode gene following its ingestion. Transgenic potatoes using this technology are not yet well developed, but several potential genes for silencing have been studied in some of the major PPNs of potato, as well as with model plants (Lilley *et al.*, 2007). Recently, other approaches using nematode-repellent peptides have been developed in potato (Lilley *et al.*, 2011), and this interesting technique has also been implemented using specific expression in the cells of the root cap with the promoter AtMDK4-20. Resistance levels of 95% were achieved against *G. pallida* using these transgenic plants. The combination of high efficiency in PPN control combined with specific expression using promoters makes these plants interesting for future field applications.

These new approaches offer new alternatives for controlling PPN, which may be particularly useful with nematode populations that exhibit high intraspecific diversity (e.g. *G. pallida*), or where the target is a broader spectrum of nematode species. However, food web soil analysis and extensive food toxicological experiments are needed before these genetically modified (GM) approaches receive approval. The possibility of gene flow to wild relatives in the main center of biodiversity for potato is a barrier for their use in certain situations. The use of the male sterile cultivar Revolucion, engineered to provide nematode resistance, could eliminate the possibility of this route of gene flow (Celis *et al.*, 2004). The appearance of nematode resistance and the effects on other non-target organisms must also be examined.

Models for population dynamics and yield losses

The ability to predict the losses expected from a given nematode population is essential in making management decisions (Barker *et al.*, 1985). Practical models have been included in programs to help growers and crop managers to decide on the use of chemicals, crop rotation, and how other factors such as the type of soil, the initial soil nematode inoculum, and potato cultivar will affect yield and population dynamics. The software-based nematode advisory system, NemaDecidePlus™ (Been *et al.*, 2005), is a program that can compute scenarios for the control of PCNs, the Columbia root-knot nematode, the root-lesion nematode, and for mixed infestations of all these nematodes. This program takes into account whether the crop is intended for seed, ware, or starch potatoes. Decision support with mixed populations is very interesting to predict the long-term effects of management decisions and to anticipate future management problems. Another program of interest was developed by the Scottish Crop Research Institute and financed by the British Potato Council (UK), called "Integrated control of potato cyst nematode (PCN) *Globodera pallida*, the white potato cyst nematode" (British Potato Council, 2011). This program takes into account the level of resistance or tolerance of the cultivar used in the rotation, the estimated maximum yield, soil type, population levels at planting, and the possibility of some nematicide treatment in order to obtain an acceptable potato yield and satisfactory nematode population control. Both programs are available under license. These programs were developed for specific scenarios and geographic locations, so they require validation for their application in different agricultural and climatic conditions.

IPM in plant parasitic nematodes

The definition of this strategy is encompassed in the definition by the University of California:

> A pest management strategy that focuses on long-term prevention or suppression of pest problems through a combination of techniques such as encouraging biological control, use of resistant varieties, and adoption of alternate cultural practices such as modification of irrigation or pruning to make the habitat less conducive to pest development. Pesticides are used only when careful monitoring indicates they are needed according to pre-established guidelines, treatment thresholds, or to prevent pests from significantly interfering with the purposes for which plants are being grown.
>
> (http://www.ipm.ucdavis.edu/PMG/glossary.html#I)

Specifically for nematodes, each region with different ecological conditions and pests must have its own IPM system in order to obtain the best profit of all available actions.

The development of nematode integrated management programs requires analysis of the impact of each individual tool on a nematode population, as well as the determination of cost–benefit ratios for grower acceptance (Sikora *et al.*, 2005). From this perspective, measures to control nematodes are not to eradicate but to control the population under threshold damage levels, and these measures must be economically viable. Some tools used in IPM can be applied at different times in a cropping cycle or in a rotation sequence, and they have mainly been listed in the points discussed above in this chapter. However, it is vital for sustainable management in IPM that correct identification and quantification is performed for the PPNs that could be a risk for crop management.

Several books and articles could be consulted in order to implement IPM (Strand, 1986; Barker *et al.*, 1998; Whitehead, 1998; Sikora *et al.*, 2005; Krishna Prasad, 2008), and various IPM management strategies are included in the technical rules of Integrated Production Certification in certain countries, or other private certifications (for example, GLOBALG.A.P., http://www.globalgap.org).

References

Abou Bakari, K., Kerlan, M.C., Caromel, B., Dantec, J.P., Fouville, D., Manzanares-Dauleux, M., *et al.* (2006) A major gene mapped on chromosome XII is the main factor of a quantitatively inherited resistance to *Meloidogyne fallax* in *Solanum sparsipilum*. *Theoretical and Applied Genetics* 112, 699–707.

Al-Banna, L., Ploeg, A.T., Williamson, V.M. and Kaloshian, I. (2004) Discrimination of six *Pratylenchus* species using PCR and species-specific primers. *Journal of Nematology* 36, 142–146.

Alonso, R., Alemany, A. and Andres, M.F. (2011) Population dynamics of *Globodera pallida* (Nematoda: Heteroderidae) on two potato cultivars in natural field conditions in Balearic Islands, Spain. *Spanish Journal of Agricultural Research* 9, 589–596.

Back, M.A., Haydock, P.P.J. and Jenkinson, P. (2002) Disease complexes involving plant parasitic nematodes and soilborne pathogens. *Plant Pathology* 51, 683–697.

Balderrama, F. (1992) *Beauvaria brogniartii* como biocontrolador del nematodo del rosario de la papa, *Nacobbus averrans* (Thorne, 1935) Thorne y Allen, 1944. PhD thesis, Universidad Mayor de San Simon, Cochabamba, Bolivia.

Barker, K.R., Schmitt, D.P. and Imbriani, J.L. (1985) Nematode population dynamics with emphasis on determining damage potential to crops. In: Barker, K.R., Carter, C.C. and Sasser, J.N. (eds) *An Advanced Treatise on Meloidogyne. Volume II Methodology*. North Carolina State University Graphics, Raleigh, North Carolina, pp. 135–148.

Barker, K.R., Pederson, G.A. and Windham, G.L. (eds) (1998) *Plant and Nematode Interactions*. American Society of Agronomy, Inc, Madison, Wisconsin.

Been, T.H., Schomaker, C.H. and Molendijk, L.P.G. (2005) NemaDecide: a decision support system for the management of potato cyst nematodes. In: Haverkort, A.J. and Struik, P.C. (eds) *Potato in Progress*. Academic Publishers, Wageningen, the Netherlands, pp. 154–167.

Blok, V.C. and Powers, T.O. (2010) Biochemical and molecular identification. In: Perry, R.N., Moens, M. and Starr, J.L. (eds) *Root-Knot Nematodes*. CABI, Wallingford, UK, pp. 98–118.

Boutsika, K., Phillips, M.S., MacFarlane, S.A., Brown, D.J.F., Holeva, R.C. and Blok, V.C. (2004) Molecular diagnostics of some trichodorid nematodes and associated *Tobacco rattle virus*. *Plant Pathology* 53, 110–116.

Boydston, R.A., Mojtahedi, H., Crosslin, J.M., Brown, C.R. and Anderson, T. (2008) Effect of hairy nightshade (*Solanum sarrachoides*) presence on potato nematodes, diseases, and insect pests. *Weed Science* 56, 151–154.

Bradshaw, J.E., Dale, M.F.B. and Mackay, G.R. (2003) Use of mid-parent values and progeny tests to increase the efficiency of potato breeding for combined processing quality and disease and pest resistance. *Theoretical and Applied Genetics* 107, 36–42.

British Potato Council (2011) Integrated control of potato cyst nematode (PCN) *Globodera pallida*, the white potato cyst nematode PCN MODEL USER GUIDE NOTES (http://www.potato.org.uk/publications/integrated-control-potato-cyst-nematode-pcn-globodera-pallida-white-potato-cyst, accessed 6 May 2014.

Brodie, B.B., Evans, K. and Franco, J. (1993) Nematode parasites of potatoes. In: Evans, K., Trudgill, D.L. and Webster, J.M. (eds) *Plant Parasitic Nematodes in Temperate Agriculture*. CAB International, Wallingford, UK, pp. 87–132.

Brown, C.R., Yang, C.-P., Mojtahedi, H. and Santo, G.S. (1996) RFLP analysis of resistance to Columbia root-knot nematode derived from *Solanum bulbocastanum* in a BC2 population. *Theoretical and Applied Genetics* 92, 572–576.

Brown, C.R., Mojtahedi, H. and Santo, G.S. (1999) Genetic analysis of resistance to *Meloidogyne chitwoodi* from *Solanum hougasii* into cultivated potato. *Journal of Nematology* 31, 264–271.

Cahuana, A., Cornejo, O.W. and Huanco, S.V. (1975) Sources of resistance to the false root-knot nematode (*Nacobbus* spp.) among 50 potato clones. *Fitopatologia* 10, 71.

Canto Saenz, M. and De Scurrah, M.M. (1977) Races of the potato cyst nematode in the Andean region and a new system of classification. *Nematologica* 23, 340–349.
Castillo, P. and Vovlas, N. (2007) Pratylenchus *(Nematoda: Pratylenchidae): Diagnosis, Biology, Pathogenicity and Management* (Hunt, D.J. and Perry, R.N. (eds) Series). Nematology Monographs and Prespectives, Vol 6. Brill Academic Publishers, the Netherlands.
Celis, C., Scurrah, M., Cowgill, S., Chumbiauca, S., Green, J., Franco, J., *et al.* (2004) Environmental biosafety and transgenic potato in a centre of diversity for this crop. *Nature* 432, 222–225.
Charchar, J.M., Oliveira, V.R. and Moita, A.W. (2009) Penetration, development, and reproduction of *Meloidogyne incognita* race I and *M. javanica* on potato cultivars in the field. *Nematologia Brasileira* 33, 147–153.
Chitwood, D.J. (2003) Research on plant-parasitic nematode biology conducted by the United States Department of Agriculture-Agricultural Research Service. *Pest Management Science* 59, 748–753.
Clayton, R., Storey, M., Parker, B., Ballingal, M. and Davies, K. (2008) *Impact of Reduced Pesticide Availability on Control of Potato Cyst Nematodes and Weeds in Potato Crops.* Agriculture and Horticulture Development Board, Potato Council, Oxford, UK.
Clovis, C.J. and Nolan, R. (1983) Fungi associated with cysts, eggs and juveniles of the golden cyst nematode (*Globodera rostochiensis*) in Newfoundland. *Nematologica* 29, 346–356.
Coyne, D.L. and Claudius-Cole, A. (2009) *Scutellonema bradys* newly reported affecting Irish potato (*Solanum tuberosum*) in Nigeria. *Plant Pathology* 58, 805.
Coyne, D.L., Akpheokhai, L.I. and Adeniran, A.F. (2011) The yam nematode (*Scutellonema bradys*), a potential threat to potato (*Solanum tuberosum*) production in West Africa. *Plant Pathology* 60, 992–997.
Dackman, C. (1990) Fungal parasites of the potato cyst nematode *Globodera rostochiensis*: isolation and reinfection. *Journal of Nematology* 22, 594–597.
Dale, M.F.B., McLean, K. and Bryan, G. (2010) The use of molecular markers linked to nematode resistance within potato breeding programmes. In: Haydock, P.P. (ed.) *Proceedings of the 3rd Symposium on Potato Cyst Nematodes*. The Association of Applied Biology, Harper Adams University College, Newport, UK, pp. 63–66.
DEFRA (2004) Development of a robust assay to estimate the viability of potato cyst nematodes Globodera spp. – HP0143 (http://randd.defra.gov.uk/Default.aspx?Menu=Menu&Module=More&Location=None&Completed=0&ProjectID=10913, accessed 6 May 2014).
De Ley, P. (2000) Lost in worm space: phylogeny and morphology as road maps to nematode diversity. *Nematology* 2, 9–16.
De Weerdt, M., Kox, L., Waeyenberge, L., Viaene, N. and Zijlstra, C. (2011) A real-time PCR assay to identify *Meloidogyne minor*. *Journal of Phytopathology* 159, 80–84.
den Ouden, H. (1960) Periodicity in spontaneous hatching of *Heterodera rostochiensis* in the soil. *Nematologica suppl.* 11, 101–105.
Decraemer, W. (1995) *The Family Trichodoridae: Stubby Root and Virus Vector Nematodes. Developments in Plant Pathology, 6*. Kluwer Academic Publishers, Dordrecht, the Netherlands.
Di Vito, M., Greco, N., Carputo, D. and Frusciante, L. (2003) Response of wild and cultivated potato clones to Italian populations of root knot nematodes *Meloidogyne* spp. *Nematropica* 33, 65–72.
Draaistra, J. (2006) Genetic analyses of root-knot nematode resistance in potato. PhD thesis, Wageningen University, the Netherlands.
Duarte I.M., Almeida, M.T.M., Duarte, M.M., Brown, D.J.F. and Neilson, R. (2011) Molecular diagnosis of trichodorid species from Portugal. *Plant Pathology* 60, 586–594.
Ellenby, C. (1954) Environmental determination of the sex ratio of a plant parasitic nematode. *Nature* 174, 1016–1017.
Everts, K.L., Sardanelli, S., Kratochvil, R.J., Armentrout, D.K. and Gallagher, L.E. (2006) Root-knot and root-lesion nematode suppression by cover crops, poultry litter, and poultry litter compost. *Plant Disease* 90, 487–492.
Finley, A.M. (1981) Histopathology of *Meloidogyne chitwoodi* (Golden *et al.*) on Russet Burbank potato. *Journal of Nematology* 13, 486–491.
Fleming, C.C. and Powers, T.O. (1998) Potato cyst nematode diagnostics: morphology, differential hosts and biochemical techniques. In: Marks, R.J. and Brodie, B.B. (eds) *Potato Cyst Nematodes*. CAB International, Wallingford, UK, pp. 91–114.
Folkertsma, R.T., Rouppe van der Voort, J.N., Gent-Pelzer, M.P., de Groot, K.E., Rieneke, W.J., Schots, A., *et al.* (1994) Inter-and intraspecific random variation between populations of *Globodera rostochiensis* and *G. pallida* revealed by amplified polymorphic DNA. *Phytopathology* 84, 807–811.

Franco, J., Main, G. and Oros, R. (1999) Trap crops as a component for the integrated management of *Globodera* spp. (potato cyst nematodes) in Bolivia. *Nematropica* 29, 51–60.

Fuller, V.L., Lilley, C.J. and Urwin, P.E. (2008) Nematode resistance. *New Phytologist* 180, 27–44.

Gebhardt, C. and Valkonen, J.P.T. (2001) Organization of genes controlling disease resistance in the potato genome. *Annual Review of Phytopathology* 39, 79–102.

Handoo, Z.A., Skantar, A.M., Carta, L.K. and Chitwood, D.J. (2012) Description of *Globodera ellingtonae* n. sp. (Nematoda: Heteroderidae) from Oregon. *Journal of Nematology* 44, 40–57.

Hirschmann, H. (1985) The genus *Meloidogyne* and morphological characters differentiating its species. In: Sasser, J.N. and Carter, C.C. (eds) *An Advanced Treatise on Meloidogyne, Biology and Control* (Vol 1). International *Meloidogyne* Project. North Carolina State University, Raleigh, North Carolina, pp. 79–93.

Hockland, S. (2010) Implementation of the new PCN directive in England and Wales. In: Haydock, P.P. (ed.) *Proceedings of the 3rd Symposium on Potato Cyst Nematodes*. The Association of Applied Biology, Harper Adams University College, Newport, UK, pp. 17–22.

Hockland, S., Niere, B., Grenier, E., Blok, V., Phillips, M., Den Nijs, L., *et al*. (2012) An evaluation of the implications of virulence in non-European populations of *Globodera pallida* and *G. rostochiensis* for potato cultivation in Europe. *Nematology* 14, 1–13.

Holeva, R., Phillips, M.S., Neilson, R., Brown, D.J.F., Boutsika, K. and Blok, V.C. (2006) Real-time PCR detection and quantificiaton of vector trichodorid nematodes and Tobacco rattle virus. *Molecular and Cellular Probes* 20, 203–211.

Holgado, R., Oppen Skau, K.A. and Magnusson, C. (2009) Field damage in potato by lesion nematode *Pratylenchus penetrans*, its association with tuber symptoms and its survival in storage. *Nematologia mediterranea* 37, 25–29.

Hoolahan, A.H., Blok, V.C., Gibson, T. and Dowton, M. (2012) A comparison of three molecular markers for the identification of populations of *Globodera pallida*. *Journal of Nematology* 44, 7–17.

Hugot, J.P., Baujard, P. and Morand, S. (2001) Biodiversity in helminths and nematodes as a field of study: an overview. *Nematology* 3, 199–208.

Hunt, D.J. and Handoo, Z.A. (2010) Taxonomy, identification and principal species. In: Perry, R.N., Moens, M. and Starr, J.L. (eds) *Root-Knot Nematodes*. CAB International, Wallingford, UK, pp. 55–97.

Inserra, R.N., Griffin, G.D. and Anderson, J.L. (1985) The false root-knot nematode *Nacobbus aberrans*. *Utah State University Research Bulletin* 510.

Jacobs, H., Gray, S.N. and Crump, D.H. (2003) Interactions between nematophagous fungi and consequences for their potential as biological agents for the potato cyst nematodes. *Mycological Research* 107, 47–56.

Janssen, G.J.W., van Norel, A., Verkerk-Bakker, B., Janssen, R. and Hoogendoorn, J. (1997) Introgression of resistance to root-knot nematodes from wild Central American *Solanum* species. *Theorical and Applied Genetics* 95, 490–496.

Jones, F.G.W. and Jones, M.G. (1974) *Pests of Field Crops*. Arnold, London, 448 pp.

Karssen, G. (1996) Description of *Meloidogyne fallax* n.sp. (Nematoda: Heteroderidae), a root-knot nematode from the Netherlands. *Fundamental and Applied Nematology* 19, 593–599.

Koen, H. (1969) Thermotherapeutic control of *Pratylenchus brachyurus* in potato tubers. *Phytophylactica* 1, 67–70.

Kornobis, S. and Stefan, K. (1991) Plant parasitic nematodes as pest of potatoes in Poland. *EPPO Bulletin* 21, 333–339.

Kort, J., Ross, H., Rumpenhorst, H.J. and Stone, A.R. (1977) An international scheme for identifying and classifying pathotypes of poato cyst-nematodes *Globodera rostochiensis* and *G. pallida*. *Nematologica* 23, 333–339.

Kratochvil, R.J., Sardanelli, S., Everts, K. and Gallagher, E. (2004) Evaluation of crop rotation and other cultural practices for management of root-knot and lesion nematodes. *Agronomy Journal* 96, 1419–1428.

Krishna Prasad, K.S. (2008) Management of potato nematodes: an overview. *Journal of Horticultural Sciences* 3, 89–106.

Kubo, R.K., Inomoto, M.M., Silva, R.A. and Antedomênico, S.R. (2001) Symptoms caused by *Meloidogyne incognita*, *Pratylenchus brachyurus* and *P. coffeae* on tubers of potato cvs. Monalisa and Bintje. *Summa Phytopathologica* 27, 331–333.

Lamberti, F. and Roca, F. (1987) Present status of nematodes as vectors of plant viruses. In: Veech, J.A. and Dickson, D.W. (eds) *Vistas on Nematology*. Society of Nematologists Inc, Hyattsville, Maryland, pp. 321–328.

Lane, A. and Trudgill, D. (1999) *Potato Cyst Nematode: A Management Guide*. Barker, B. (ed.). MAFF Publications, London.

Lilley, C.J., Urwin, P.E., Johnston, K.A. and Atkinson, H.J. (2004) Preferential expression of a plant cystatin at nematode feeding sites confers resistance to *Meloidogyne incognita* and *Globodera pallida*. *Plant Biotechnology Journal* 2, 3–12.

Lilley, C.J., Bakhetia, M., Charlton, W.L. and Urwin, P.E. (2007) Recent progress in the development of RNA interference for plant parasitic nematodes. *Molecular Plant Pathology* 8, 701–711.

Lilley, C.J., Wang, D., Atkinson, H.J. and Urwin, P.E. (2011) Effective delivery of a nematode-repellent peptide using a root-cap-specific promoter. *Plant Biotechnology Journal* 9, 151–161.

Luc, M., Bridge, J. and Sikora, R.A. (1990) Reflections on nematology in subtropical and tropical agriculture. In: Luc, M., Bridge, J. and Sikora, R.A. (eds) *Plant Parasitic Nematodes in Subtropical and Tropical Agriculture*. CAB International, Wallingford, UK, pp. 11–17.

MAAMA (Ministerio de Agricultura, Alimentación y Medio Ambiente) (2012) http://www.magrama.gob.es/ (accessed 13 May 2014).

MacGuidwin, A.E., Knuteson, D.L., Connell, T., Bland, W.L. and Bartelt, K.D. (2012) Manipulating inoculum densisties of *Verticillium dahliae* and *Pratylenchus penetrans* with green manure amendments and solariztion influence potato yield. *Phytopathology* 102, 519–527.

Manzanilla-López, R., Costilla, M.A., Doucet, M., Franco, J., Inserra, R.N., Lehman, P.S., *et al.* (2002) *Nacobbus* species: systematics, distribution, biology and management. *Nematropica* 32, 149–227.

Matveeva, E.M., Gruzdeva, L.I., Sysoeva, M.I. and Sherudilo, E.G. (2010) Management practices for potato cyst nematode in the north-west of Russia. In: Haydock, P.P. (ed.) *Proceedings of the 3rd Symposium on Potato Cyst Nematodes*. The Association of Applied Biology, Harper Adams University College, Newport, UK, pp. 67–72.

Minnis, S.T., Haydock, P.P.J. and Evans, K. (2004) Control of potato cyst nematodes and economic benefits of application of 1,3-dichloropropene and granular nematicides. *Annals of Applied Biology* 145, 145–156.

Moloney, C., Griffin, D., Jones, P.W., Bryan, G.J., McLean, K., Bradshaw, J.E., *et al.* (2009) Development of diagnostic markers for use in breeding potatoes resistant to *Globodera pallida* pathotype Pa2/3 using germplasm derived from *Solanum tuberosum* ssp. *andigena* CPC 2802. *Theorical and Applied Genetics* 120, 679–689.

Nicol, J.M., Turner, S.J., Coyne, D.L., den Nijs, L., Hockland, S. and Tahna Maafi, Z. (2011) Current nematode threats to world agriculture. In: Jones, J.T., Gheysen, G. and Fenoll, C. (eds) *Genomics and Molecular Genetics of Plant–Nematode Interactions*. Springer, Dordrecht, Germany, pp. 21–44.

Oerke, E.C., Dehne, H.W., Schönbeck, F. and Weber, A. (1994) *Crop Production and Crop Protection: Estimated Losses in Major Food and Cash Crops*. Elsevier Science BV, Amsterdam.

Oliveira, C.M.G., Machado, A.C.Z., Kubo, R.K. and Harakava, R. (2009) Diagnose de *Aphelenchoides fragariae* e *Pratylenchus* spp. pela aplicação da tecnologia do código de barras do DNA. *Nematologia Brasileira* 33, 218–225.

Otazú, V.R., Hoopes, G., Caero, C. and Huayta, I. (1985) El rosario de la papa causado po *Nacobbus aberrans* (Thorne, 1935) Thorne & Allen 1944. Su efecto en el rendimiento y algunos aspectos que inciden en su propagación y prevalencia en Bolivia. *Fitopatología* 20, 65–70.

Plantard, D., Picard, D., Valette, S., Scurrah, M., Grenier, E. and Mugniéry, D. (2008) Origin and genetic diversity of Western European populations of the potato cyst nematode (*Globodera pallida*) inferred from mitochondrial sequences and microsatellite loci. *Molecular Ecology* 17, 2208–2018.

Ploeg, A.T., Brown, D.J.F. and Robinson, D.J. (1992) The association between species of *Trichodorus* and *Paratrichodorus* vector nematodes and serotypes of tobacco rattle tobravirus. *Annals of Applied Biology* 121, 619–630.

Powelson, M.L. and Rowe, R.C. (1993) Biology and management of early dying of potatoes. *Annual Review of Phytopathology* 31, 111–126.

Reid, A., Kenyon, D.M., Evans, F.F., Mulholland, V., Pickup, J., Blok, V.C., *et al.* (2010) Development of a high-throughput method for the detection and species determination of PCN. In: Haydock, P.P. (ed.) *Proceedings of the 3rd Symposium on Potato Cyst Nematodes*. The Association of Applied Biology, Harper Adams University College, Newport, UK, pp. 11–14.

Riga, E., Mojtahedi, H., Ingham, R.E. and McGuire, A.M. (2003) Green manure amendments and management of root knot nematodes on potato in the Pacific Northwest of USA. *Nematology Monographs and Perspectives* 2, 151–158.

Rouppe van der Voort, J., Van der Vossen, E., Bakker, E., Overmars, H., Van Zandvoort, P., Hutten, R., *et al.* (2000) Two additive QTLs conferring broad-spectrum resistance in potato to *Globodera pallida* are localized on resistance gene clusters. *Theoretical and Applied Genetics* 101, 1122–1130.

Santo, G.S. (1994) Biology and management of root-knot nematodes on potato in the Pacific Northwest. In: Zehnder, G.W., Powelson, M.L., Jansson, R.K. and Raman, K.V. (eds) *Advances in Potatoes: Pest, Biology and Management*. American Phytopathological Society, St Paul, Minnesota, pp. 193–201.

Santo, G.S., Ponti, R.P. and Wilson, J.H. (1985) Control of *Meloidogyne chitwoodi* on potato with DD. Fungicide and nematicide tests: results. *American Phytopathological Society (USA)* 40, 106–107.

Sasaki-Crawley, A., Curtis, R., Birkett, M., Powers, S., Papadopoulos, A., Pickett, J., *et al*. (2010) Signalling and behaviour of potato cyst nematode in the rhizosphere of the trap crop, *Solanum sisymbriifolium*. In: Haydock, P.P. (ed.) *Proceedings of the 3rd Symposium on Potato Cyst Nematodes*. The Association of Applied Biology, Harper Adams University College, Newport, UK, pp. 45–51.

Sasser, J.N. and Freckman, D.W. (1987) A world perspective on nematology, the role of the society. In: Veech, J.A. and Dickson, D.W. (eds) *Vistas on Nematology*. Society of Nematologists, Hyattsville, Maryland, pp. 7–14.

Schnick, D., Rumpenhorst, H.J. and Burgermeister, W. (1990) Differentiation of closely related *Globodera pallida* (Stone) populations by means of DNA restriction fragment length polymorphisms (RFLPs). *Journal of Phytopathology* 130, 127–136.

Scholte, K. (2000) Screening of non-tuber bearing Solanaceae for resistance to and induction of juvenile hatch of potato cyst nematodes and their potential for trap cropping. *Annals of Applied Biology* 136, 239–246.

Scurrah, M., Niere, B. and Bridge, J. (2005) Nematode parasites of potato and sweet potato. In: Luc, M., Sikora, R. and Bridge, J. (eds) *Plant Parasitic Nematodes in Subtropical and Tropical Agriculture*, 2nd edn. CAB International, Wallingford, UK, pp. 193–221.

Sikora, R.A., Bridge, J. and Starr, J.L. (2005) Management practices: an overview of integrated nematode management technologies. In: Luc, M., Sikora, R.A. and Bridge, J. (eds) *Plant Parasitic Nematodes in Subtropical and Tropical Agriculture,* 2nd edn. CAB International, Wallingford, UK, pp. 793–826.

Skantar, A.M., Handoo, Z.A., Carta, L.K. and Chitwood, D.J. (2007) Morphological and molecular identification of *Globodera pallida* associated with potato in Idaho. *Journal of Nematology* 39, 133–144.

Sobczak, M. and Golinowski, W. (2011) Cyst nematodes and syncytia. In: Jones, J.T., Gheysen, G. and Fenoll, C. (eds) *Genomics and Molecular Genetics of Plant–Nematode Interactions*. Springer, Dordrecht, Germany, pp. 61–82.

Stone, A.R. (1979) Co-evolution of nematodes and plants. *Symbolae Botanicae Upsalienses* 22, 46–61.

Stone, A.R. (1985) Co-evolution of potato cyst nematodes and their hosts: implications for pathotypes and resistance. *EPPO Bulletin* 15, 131–137.

Strand, L. (1986) *IPM for Potatoes in the Western United States.* 1st edn. UC Division of Agriculture and Natural Resources, Oakland, California, USA.

Suarez, S.I., Chaves, E.J., Clausen, A.M. and Franco, J. (2009) Solanum tuber-bearing species resistance behaviour against *Nacobbus aberrans*. *Journal of Nematology* 41, 5–10.

Subbotin, S.A., Mundo-Ocampo, M. and Baldwin, J.G. (2010) *Systematics of Cyst Nematodes (Nematoda: Heteroderinae) Parts A and B*. Nematology Monographs and Perspectives Vol 8. Brill, Leiden, the Netherlands.

Subbotin, S.A., Mohammad Deimi, A., Zheng, J. and Chizhov, V.N. (2011) Length variation and repetitive sequences of Internal Transcribed Spacer of ribosomal RNA gene, diagnostics and relationships of populations of potato rot nematode, *Ditylenchus destructor* Thorne, 1945 (Tylenchida: Anguinidae). *Nematology* 12, 773–785.

Tan, M.Y.A., Alles, R., Hutten, R.C.B., Visser, R.G.F. and van Eck, H.J. (2009) Pyramiding of *Meloidogyne hapla* resistance genes in potato does not result in an increase of resistance. *Potato Research* 52, 331–340.

Tobin, J.D., Haydock, P.P.J., Hare, M.C., Woods, S.R. and Crump, D.H. (2008) Effect of the fungus *Pochonia chlamydosporia* and fosthiazate on the multiplication rate of potato cyst nematodes (*Globodera pallida* and *G. rostochiensis*) in potato crops grown under UK field conditions. *Biological Control* 46, 194–201.

Urwin, P.E., Green, J. and Atkinson, H.J. (2003) Expression of plant cystatin confers partial resistance to *Globodera*, full resistance is achieved by pyramiding a cystatin with natural resistance. *Molecular Breeding* 12, 263–269.

Vovlas, N. (1996) Infestation by the cyst forming nematode *Globodera pallida* of potato tubers in Southern Italy. *European Journal of Plant Pathology* 102, 743–746.

Webley, D.P. and Jones, F.B.W. (1981) Observations on *Globodera pallida* and *G. rostochiensis* on early potatoes. *Plant Pathology* 30, 217–224.

Wendt, K.R., Vrain, T.C. and Webster, J.M. (1993) Separation of three species of *Ditylenchus* and some host races of *D. dipsaci* by restriction fragment length polymorphism. *Journal of Nematology* 25, 555–563.

Whitehead, A.G. (1998) *Plant Nematode Control*. CAB International, Wallingford, UK.

Xenophontos, S., Robinson, D.J., Dale, M.F.B. and Brown, D.J.F. (1998) Evidence for persistent, symptomless infection of some potato cultivars with *tobacco rattle virus*. *Potato Research* 41, 255–265.

Yeates, G.W., Bongers, T., Goede, R.G.M., De Freckman, D.W. and Georgieva, S.S. (1993) Feeding habits in soil nematode families and genera – an outline for soil ecologists. *Journal of Nematology* 25, 315–331.

11 Fungal and Bacterial Disease Aspects in Potato Production

William W. Kirk[1]* and Phillip S. Wharton[2]
[1]Department of Plant, Soil and Microbial Science, Michigan State University, East Lansing, USA; [2]University of Idaho, Aberdeen Research and Extension Center, Aberdeen, USA

Potatoes (*Solanum tuberosum* L.) are seriously impacted by disease. Crop losses due to late blight alone have been estimated as high as US$210.7 million in the USA, with control costs totaling US$77.1 million for fungicides. This amounts to US$507 ha^{-1} on average, not including non-fungicide control practices (Guenthner *et al.*, 2001). This chapter presents the disease cycles for various common and economically significant pathogens of the potato crop and emphasizes where in the cycle interventions are thought to be most effective.

In developing economies, apart from late blight as the top-ranked problem, other destructive pathogens have high incidence. For example, in a survey, in East Africa, *Ralstonia solanacearum*, the cause of bacterial wilt of potatoes, was found on 74% of potato farms (Gildemacher *et al.*, 2009). The International Potato Center (CIP) identified other factors that could be influenced by the global term "disease". These included improved cultivars for high and stable yield potential that were suitable for consumption; improved supply of quality potato seed (seed systems management and seed production); control of viruses and their vectors (through varietal resistance, seed production, and crop management); and control of bacterial wilt (through varietal resistance and crop management). These five needs were ranked high in all or most of the potato growing regions of Asia, Africa, and Latin America for which the respondents reported (Fuglie, 2007).

Potato is ranked fourth in the world's food crop production after wheat, maize, and rice (Bradshaw and Ramsay, 2009). Potato production has increased worldwide since the early 1990s, due largely to increases in the developing economies of Asia and South America (Guenthner, 2010). Most potatoes were grown and consumed in Europe, North America, and countries of the former Soviet Union until the early 1990s. Recently, the increase in demand for potatoes in Asia, Africa, and Latin America has resulted in increased production, from less than 30 to more than 165 million tonnes (Mt) from the early 1960s to 2007 (http://www.fao.org/potato-2008/en/world/index.html). Almost 33% of all potatoes are grown and harvested in China and India. However, disease incidence in both the field and storage remains the major limiting factor in profitable potato production (Secor and Gudmestad, 1999).

Potato production throughout the year is not feasible in North America and other northern climates (Sonnewald, 2001); hence, long-term storage is essential to maintain year-round delivery of fresh market and processing potatoes.

* E-mail: kirkw@msu.edu

Similarly, seed potato tubers are stored for at least 6 months before they are shipped for planting (Olsen, 2010; Frost *et al.*, 2013). Therefore, good storage conditions are essential to maintain the quality and health of potatoes, which can be compromised by disease development (Secor and Gudmestad, 1999).

Sustainable agriculture practices have been developed over millennia of food and fiber production. Practices for managing pests and diseases include changing plant and crop architecture, burning, adjusting crop density, depth or time of planting, planting at increased elevation, planting diverse crops, fallowing, flooding, mulching, multiple cropping, planting without tillage, using organic amendments, planting in raised beds, rotation, sanitation, and manipulating shade (Thurston, 2004). Potatoes, like many other crops, have been grown for food for at least 8000 years, and several of the practices mentioned were used to avoid outbreaks of disease. Modern potato production in North America and Europe demands uniformity in quality and economic return for the grower and the processor. Recent trends suggest the use of potatoes has shifted from consumption of table-stock potatoes to processed potato products (Pavlista and Feuz, 2005; Loy *et al.*, 2011). However, in developing economies the emphasis is often on increasing or maintaining yield where there are limitations to nutritional inputs and rotational options and certified seed (Allemann *et al.*, 2004; Scott, 2011). In both developed and developing economies, the basic objectives for potato production include system stability, resilience, predictable productivity, and efficiency (Johnson and Rowe, 2007). However, several diseases annually threaten these goals. For example, potato late blight caused by the oomycete, *Phytophthora infestans* (Mont. De Bary), initiated devastating losses across Europe and New England, USA, in the mid-19th century (Turner, 2005), and continues to threaten production even into the 21st century (Fry, 2008; Hu *et al.*, 2012).

Historically, most potato pathogens have likely been transported within and between countries during the course of trade over time. More recently, the impact of climate change has had a number of observed, anticipated, or possible consequences on crop health worldwide, as discussed in several reviews (Hijmans *et al.*, 2000; Sparks *et al.*, 2008, 2010; Haverkort *et al.*, 2009; Garrett *et al.*, 2011; Savary *et al.*, 2011; Sturrock *et al.*, 2011). Global population increase and changes in sources of supply and demand for natural and processed food products from local to global (Padgham, 2009; Von Braun, 2009) enforce further pressures on the distribution of diseases of all crops, including the potato crop. The interaction between global population dynamics and climate change may have far-reaching effects on plant and potato health in the future (Savary *et al.*, 2011). These factors could have interacted already, as manifested by the recent appearance of the disease, zebra chip, caused by *Candidatus* Liberibacter solanacearum, which is vectored by the potato psyllid (*Bactericera cockerelli* (Hemiptera: Triozidae)), and has impacted potato production internationally (Crosslin *et al.*, 2011, 2012; Munyaneza *et al.*, 2011; Pitman *et al.*, 2011; Buchman *et al.*, 2012).

According to the American Phytopathological Society publication, *Compendium of Potato Diseases*, 2nd edition, there are 35 economically important bacterial, fungal, and oomycete pathogens of potato worldwide (Stevenson *et al.*, 2001). The most common and serious bacterial, oomycete and fungal pathogens of potatoes worldwide are profiled in Table 11.1. The objectives of this chapter are to describe how some of the main potato diseases affect production and markets, how the diseases perpetuate and disseminate, and to discuss control strategies and prospects for the future.

11.1 Survival and Spread of Potato Diseases

Seed health management

As in all crops, the occurrence of any disease in the potato crop is a function of the interaction of host, pathogen, and the environment. Shifts in production techniques, such as importing seed from new international sources, may lead to the appearance of non-indigenous diseases. Shortening of rotation intervals between crops may favor the build up of soilborne inoculum of tuber pathogens, leading to a crop health status that requires the development of new management options. In extreme cases, if a new pathogen is

Table 11.1. Important bacterial, oomycete, and fungal diseases of the potato.

Causal agent	Common name; synonyms (previous names), other species and races	Distribution	Symptoms or signs: Foliage	Symptoms or signs: Tubers
BACTERIA				
Candidatus Liberibacter solanacearum	Zebra chip	Latin America, New Zealand, south-western and Pacific north-western USA	Yellowing	Discoloration, especially of processed tubers
Clavibacter michiganensis subsp. *sepedonicus*	Ring rot	Europe, North America	Wilt, rolling of leaflet margins, mottling, necrosis	Discoloration of vascular ring and adjacent tissue with bacterial ooze
Pectobacterium carotovora	Blackleg, soft rot	Worldwide	Wilt	Soft rot
Pseudomonas fluorescens	Pink eye	Worldwide		Pink discoloration, especially around lenticels and eyes
Ralstonia solanacearum (Smith) Yabuuchi *et al.* *Bacterium solanacearum* (Smith) Chester *Burkholderia solanacearum* (Smith) Yabuuchi *et al.* (1992) *Pseudomonas solanacearum* (Smith) Smith	Brown rot; Race 1 and 2 ("warm biotypes"); Race 3 Biovar2 "temperate biotype"	R1: Africa, southern USA, southern Asia, Middle East R2: Central America, North Africa, South America, southern Asia R3B2: Africa, USA, Central America, South America, Russia, India, China, Australasia	Wilt	Discoloration of vascular ring and adjacent tissue with bacterial ooze
Streptomyces scabies, *Streptomyces acidiscabies*, *Streptomyces turgidiscabies*, *Streptomyces europaeiscabiei*	Common scab	Worldwide	None	Superficial blemishes to deep pitted or raised lesions
OOMYCETE				
Phytophthora infestans (Mont.) de Bary.	Potato late blight	Worldwide	Progressive necrosis and ± defoliation	Tuber rot

Continued

Table 11.1. Continued.

Causal agent	Common name; synonyms (previous names), other species and races	Distribution	Symptoms or signs	
			Foliage	Tubers
Phytophthora erythroseptica Pethybr.	Pink rot; *Phytophthora cryptogea*, *Phytophthora drechsleri*, *Phytophthora megasperma*, *Phytophthora nicotianae* var. *parasitica*	Worldwide	Wilt, premature senescence	Characteristic pink tuber rot or can be brownish if *P. nicotianae*
Pythium ultimum Trow.	Leak or watery wound rot; *Pythium aphanidermatum*, *Pythium deliense*	Worldwide	None	Wet rot often accompanied by bacterial soft rot
FUNGI				
Alternaria alternata (Fries.) Keissler = *Alternaria tenuis* Ness.	Brown leaf spot and tuber black pit	Worldwide	Progressive necrosis and eventual defoliation	Skin lesions and some pitting
Alternaria solani (E&M) Jones & Grout	Early blight	Worldwide	Progressive necrosis and eventual defoliation	Tuber pitting
Botrytis cinerea Pers.:Fr. *Botryotinia fuckeliana* (de Bary) Whetzel [teleomorph]	Gray mold or tan spot	Worldwide	Leaf necrosis and stem decay	Skin lesions
Colletotrichum coccodes (Wallr.) S.J. Hughes = *Colletotrichum atramentarium* (Berk. & Broome) Taubenhaus	Black dot	Worldwide	Stem lesions may be associated with Potato Early Die Syndrome	Superficial skin lesions
Erysiphe cichoracearum	Powdery mildew	Arid regions	Late-season, white foliar pustules on leaflets	
Fusarium spp.	Dry rot; *Fusarium solani*, *Fusarium sambucinum*, *Fusarium avenaceum*, *Fusarium oxysporum*, *Fusarium culmorum*, *Fusarium coeruleum*, *Fusarium acuminatum*, *Fusarium graminerum*, *Fusarium cerealis*, *Fusarium equiseti*, *Fusarium crookwellense*	Worldwide	Non-emergence and wilt	Seed rot and general tuber rot

Helminthosporium solani Dur. & Mont.	Silver scurf	Cosmopolitan	None	Superficial skin lesions
Macrophomina phaseolina (Tassi)	Charcoal rot; *Macrophomina phaseoli* (Maubl.) *Sclerotium bataticola* Taubenhaus	India, subtropical regions	Stem rot	Tuber rot
Phoma solanicola f. *foveata* (Foister) Malcolmson Boerema	Gangrene; *Phoma exigua* Desm. f. sp. *foveata* (Foister) Malcolmson & Gray *P. exigua* Desm. var. *exigua*	Europe, North America, Australasia	Non-emergence	Tuber rot
Polyscytalum pustulans (Owen & Wakef.) Ellis	Skin spot	Europe, North America, Australasia	Non-emergence (coalescing lesions kill eyes)	Tuber rot
Rhizoctonia solani Kühn, *Thanatephorus cucumeris* (A.B. Frank) Donk [teleomorph]	Rhizoctonia stem and stolon canker and black scurf	Worldwide	Delayed emergence, upper leaf rolling, stem and stolon canker	Tuber black scurf (sclerotia)
Sclerotinia sclerotiorum (Lib.) de Bary	White mold	Worldwide	Stem rot	None
Sclerotium rolfsii Sacc. *Athelia rolfsii* (Curzi) Tu & Kimbrough [teleomorph]	Stem rot (southern blight)	Southern USA	Stem rot	None
Spongospora subterranea (Wallr.) Lagerh. f. sp. *subterranea* Tomlinson	Powdery scab	Worldwide	Root galls	Superficial pustules that may coalesce
Synchytrium endobioticum (Schilb.) Perc	Wart	Europe, North and South America, South Africa, Asia	Petiole and leaf deformation	Deformation
Angiosorus solani (Barrus) Thirum. & O'Brien = *Thecaphora solani* Barrus	Thecaphora smut	Central and South America		Tuber malformation and internal necrosis
Verticillium albo-atrum Reinke & Berthier and *Verticillium dahliae* Kleb.	Verticillium wilt	Worldwide	Chlorosis and premature senescence	Vascular browning

introduced, the range of management options may decline because of quarantines that could be imposed by government agencies. Examples of this include potato wart caused by *Synchytrium endobioticum* (Franc, 2001), bacterial wilt and tuber brown rot caused by *R. solanacearum* race 3 biovar 2 (Norman *et al.*, 2009), which can be introduced on imported cuttings of Geranium (Swanson *et al.*, 2007), and more recently, *Dickeya* spp. in Scotland (Toth *et al.*, 2011). As many diseases have tuber-borne phases, the increase in trade between countries enhances the risk of introducing new pathogens or genotypes of existing pathogens and the chance of pathogen mutation. Seed certification and importation regulations are therefore the first line of defense against the introduction of new diseases. However, the increase in trade, coupled with the decrease in investment in inspection services, reduces the likelihood of intercepting infected potatoes at ports of entry. Between 2010 and 2011, funding for inspection in the USA was cut by US$73 million, and in 2012 it was cut by US$39 million (http://www.obpa.usda.gov/budsum/FY12budsum.pdf).

Seed certification

Pathogens may infect foliage, roots, stolons, and tubers; therefore, disease management is an important factor throughout the crop cycle. Although true seed is used increasingly in some developing economies (Ierna and Tenorio, 2011), vegetative seed tubers produced through meristematic stem culturing systems and further propagated by aeroponic or hydroponic systems (Chang *et al.*, 2012) are the main initial source of propagation material for commercial potato production. Within the tissue culture system, plants are produced under aseptic conditions, tested for various bacterial and viral diseases, and transplanted into greenhouses to produce the nuclear class of seed, or what is commonly known as minitubers. This involves meristem-derived tissue culture seed stock development, after which minitubers or tissue culture plantlets are planted into a field(s) as the initial source of certified seed potato lots. The seed lots are then multiplied and increased to reach a sufficient quantity for commercial use for up to six generations (Field Year 1–6 Class Seed), after which the seed lot is no longer eligible for recertification. As an example, in Michigan, USA, a limited seed tuber generation program was adopted in 2002 and used tissue culture material (Wang and Hu, 1985). The seed lots entered for certification in Michigan are tested in the winter in Hawaii, USA. On passing this test, the seed lots are evaluated by Michigan Department of Agriculture and Rural Development (MDARD) seed inspectors prior to shipping seed tubers from the Michigan seed farms. The seed lots that pass this stage achieve "blue tag" certification (US Standards for Seed Potato Grades Regulation No 628) from the Michigan Seed Potato Association (MPSA, PO Box 1865, Gaylord, MI 49734, USA) and the MDARD. Seed and commercial potato growers receive seed as whole tubers to grow for recertification as seed or for commercial production. Similar systems of seed for the production of tubers for processing, starch, ware, or table-stock are used throughout the world (Janse and Wenneker, 2002; Simakov *et al.*, 2008; Hirpa *et al.*, 2010) and have been reviewed extensivley by Frost *et al.* (2013). However, much of the less-developed world, including China, a very large producer, still does not have countrywide certification programs, and the cost of using certified seed is too high for most growers. There are concomitant yield losses due to increased pathogen (especially viral) load in repeatedly saved seed.

As seed lots are grown in fields then stored between growing seasons, the tubers are exposed to soilborne diseases that can remain latent on the seed. Seed tubers and daughter tubers alike are exposed to pathogens that are both seed- and soilborne, as well as propagules of pathogens that are washed from the foliage into the soil. Seed tubers are particularly vulnerable to diseases when they are cut into seed pieces. Several studies have shown that infection can occur during seed cutting in the case of potato late blight (Lambert *et al.*, 1998; Kirk *et al.*, 1999; Glass *et al.*, 2001). The pathogen can also be transmitted from the seed piece to cause disease on the stems, stolons, and daughter tubers in the case of Rhizoctonia stem canker (Secor and Gudmestad, 1999; Tsror, 2010) and prevent emergence through the sprout rotting in the case of Fusarium dry rot (Wharton *et al.*, 2007b) and Rhizoctonia diseases (Woodhall *et al.*, 2007).

11.2 The Role of the Tuber in Potato Disease

Infection and transmission of many potato diseases, including late blight of potato caused by the oomycete *P. infestans* (Mont. De Bary), and the soilborne diseases Fusarium dry rot and Rhizoctonia stem and stolon canker, are dependent on survival in or on potato tubers. An example of a common disease cycle, that of Rhizoctonia stem and stolon canker, is shown in Fig. 11.1 to illustrate the role of the tuber in the disease cycle. Most diseases have at least a transitory tuber phase, and for many diseases, such as pink rot (*Phytophthora erythroseptica*) (Venkataramana *et al.*, 2010; Taylor *et al.*, 2011), the major effect is on the tuber. Tuber late blight (*P. infestans*) results in tuber rotting, both in the field and later in storage, in tubers either intended for seed or for consumption (Bonde and Schultz, 1943; Johnson and Cummings, 2009; Johnson, 2010). Seed tubers infected with *P. infestans* will either rot in storage after planting in the field or survive and initiate new epidemics of potato late blight (Doster *et al.*, 1989; Stevenson *et al.*, 2007; Olanya *et al.*, 2009). The epidemiology of the foliar phase of the disease is correlated to infection in the tuber phase and vice versa (Fairclough *et al.*, 1997; Cooke *et al.*, 2011). Tubers are usually infected by inoculum produced on the plant foliage that is subsequently washed down to the soil by water movement resulting from rainfall and irrigation (Andrivon, 1995; Porter *et al.*, 2005). Tubers can become blighted shortly after the disease is established on the foliage. *P. infestans* survives in tubers, where it acts as the primary source of inoculum for infection in the following growing season (Bonde and Schultz, 1943). The transmission of potato late blight from infected seed tubers into developing sprouts and consequently to the foliage was first suggested by Berkeley (1846), and was demonstrated by Appel *et al.* (2001) using molecular techniques.

The pathway by which *P. infestans* progresses from infected tubers to the newly formed plant is not fully understood, but a current understanding

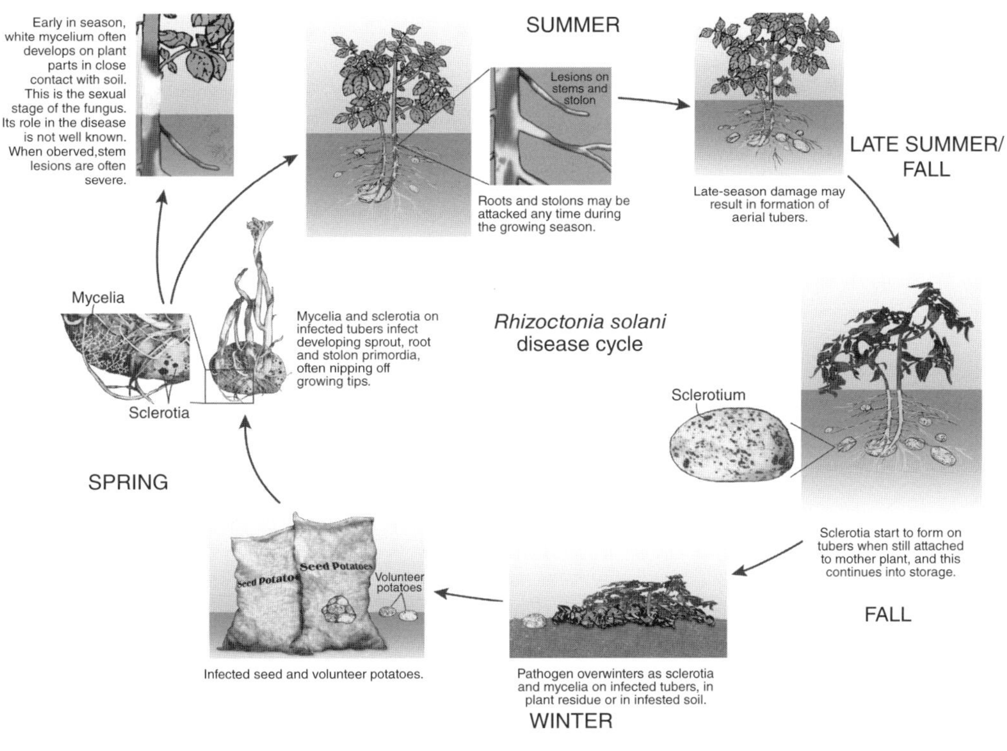

Fig. 11.1. The disease cycle of *Rhizoctonia solani*, causal agent of Rhizoctonia stem canker and black scurf.

is described schematically in Fig. 11.2. The relationship between foliar late blight severity and tuber blight severity has been investigated in several studies (Hirst and Stedman, 1962; Lacey, 1965; Boyd, 1972), all of which were most likely done using the US-1 genotype of *P. infestans*. In spite of the frequent strong correlation between foliar blight severity and tuber blight, some of these studies demonstrated cases of high incidence of tuber blight in the absence of foliar symptoms, and in contrast, Hirst *et al.* (1965) demonstrated low incidence of tuber blight after severe foliar blight epidemics. Potato late blight in tubers is therefore of significant importance in commercial production. Stored and freshly harvested table-stock (ware) and processing crops are directly impacted. Seed tubers may initiate epidemics by transmission from harboring *P. infestans* infected tubers to sprouts, stems, and foliage, and may also directly reduce plant stand and delay plant emergence, resulting in further problems that affect yield and tuber size distribution (Bussan *et al.*, 2007). The dynamics of potato blight development in tubers is influenced largely by temperature (Kirk *et al.*, 2001) and can result in the non-emergence of plants due to seed and sprout rot (Lambert and Currier, 1997; Kirk *et al.*, 2009; Johnson, 2010). Transmission studies need to continue because new genotypes are now predominant and there is a gap in our understanding of these more aggressive genotypes (Haas *et al.*, 2009; Moller *et al.*, 2009; Vargas *et al.*, 2009; Young *et al.*, 2009; Runno-Paurson *et al.*, 2010; Hu *et al.*, 2012; Rojas *et al.*, 2014).

Fusarium dry rot of potato (*Solanum tuberosum* L.) is a devastating postharvest disease worldwide and is caused by several species of *Fusarium* (Boyd, 1972; Secor and Salas, 2001). Dry rot affects both tubers in storage and seed tuber pieces in the field (Choiseul *et al.*, 2001; Wharton *et al.*, 2007b). Losses associated with dry rot have been estimated to range from 6 to 25%, and occasional losses as great as 60% have been reported during long-term storage (Seppanen, 1981; Estrada *et al.*, 2010).

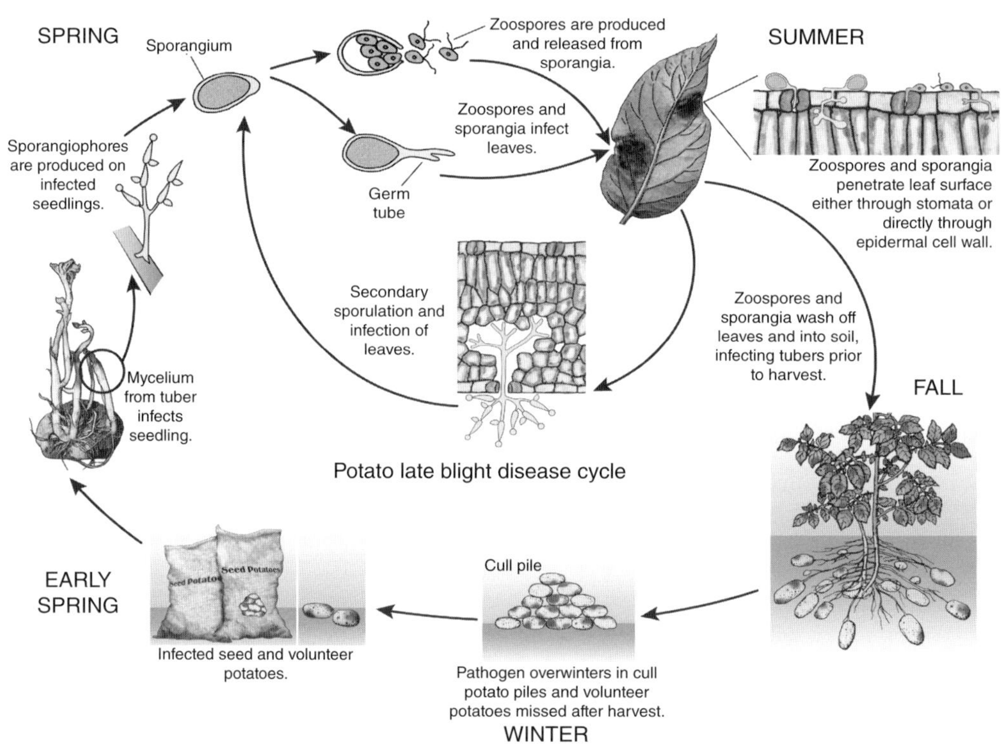

Fig. 11.2. The disease cycle of *Phytophthora infestans*, causal agent of late blight.

Infection of potato tubers by pathogens that cause dry rot occurs through wounds inflicted during harvesting, grading, cutting, and handling of tubers (Secor and Salas, 2001). Thirteen species of *Fusarium* have been implicated in fungal dry rots of potatoes worldwide (Hanson *et al.*, 1996; Cullen *et al.*, 2005; Peters *et al.*, 2008a), of which eight species have been reported in the northern USA (Hanson *et al.*, 1996). The most important species infecting potatoes are *Fusarium sambucinum* Fuckel; ("syn" *Fusarium sulphureum* Schlechtend: "tel": *Gibberella pulicaris*), *Fusarium solani* (Mart.) Sacc. var. *coeruleum* (Lib. ex Sacc.) C. Booth ("syn" *Fusarium coeruleum*; "tel": *Nectria haematocca*), and *Fusarium oxysporum* Schlechtend. Fr. (Hanson *et al.*, 1996; Ocamb *et al.*, 2007). Other species reported in the northern USA include *Fusarium avenaceum* (Fr.) Sacc., *Fusarium culmorum* (W.G. Smith), *Fusarium acuminatum* Ellis & Everh., *Fusarium equiseti* (Corda), and *Fusarium crockwellense* L.W. Bugess, P.E. Nelson & Ravenel (*Fusarium cerealis*) (Hanson *et al.*, 1996). Most of these species were also recovered in a study in the Pacific region of the USA, with *F. sambucinum* being the most prevalent (Ocamb *et al.*, 2007). Recently, *Fusarium graminearum* was reported to be the prevalent *Fusarium* species causing potato dry rot in North Dakota (Estrada *et al.*, 2010).

Fusarium dry rot on seed potatoes can be controlled at two phases during the potato growth cycle (Fig. 11.3): (i) control of seed piece decay prior to planting, which is achieved primarily by seed treatment (Wharton *et al.*, 2007b); and (ii) postharvest control of dry rot on both stored commercial tubers and seed tubers (Nolte *et al.*, 2003). Fludioxonil (Maxim® Seed Potato Protectant; Syngenta Crop Protection, Greensboro, North Carolina, USA), a phenylpyrrole, is among the few fungicides registered for potato seed treatment against Fusarium dry rot in the USA and Canada. Recently, fludioxonil-resistant strains of *Fusarium* spp. were reported in Canada and the USA, including strains of *F. sambucinum*, *F. oxysporum*, and *F. coeruleum* (Peters *et al.*, 2008b; Gachango *et al.*, 2011). Sensitivity to fludioxonil in other *Fusarium* spp. that cause potato seed piece decay is not known.

Postharvest management of potato dry rot has been achieved primarily by reducing tuber bruising, providing conditions for rapid wound healing (Secor and Johnson, 2008), and applying thiabendazole (TBZ), a benzimidazole fungicide (Mertect 340F®, Syngenta Crop Protection), as tubers are placed into storage (Hide *et al.*, 1992; Ocamb *et al.*, 2007). Isolates of *F. sambucinum* resistant to TBZ were first found in Europe (Hide *et al.*, 1992), and subsequently in the USA (Desjardins *et al.*, 1993; Desjardins, 1995). Fungicide resistance has reduced the efficacy of TBZ in controlling *F. sambucinum*, but may be useful where TBZ-sensitive isolates still occur. To counteract the reduced effectiveness of TBZ and, potentially, of fludioxonil associated with resistant strains of pathogens, new registrations of efficacious postharvest fungicides with differing mechanisms of action (Fungicide Resistance Action Committee (FRAC) groups) are needed. One fungicide that has been tested with promising results is difenoconazole (FRAC Group 3; Inspire®; Seed Potato Protectant; Syngenta Crop Protection), a sterol biosynthesis inhibitor (Adaskaveg and Förster, 2010; Gachango *et al.*, 2012a).

Role of soil inoculum in potato disease

The occurrence and development of potato disease depends on the mutual concurrence of several diverse factors affecting either the pathogen or the plant, but is fully dependent on the survival and spread of inoculum. The role of the tuber in some potato diseases has already been discussed. Long-term survival of pathogens is also dependent on the survival of inoculum in soil and on other types of media, such as in storage structural material. Fungal spores such as conidia of *Helminthosporium solani* (silver scurf) may survive in dust in storages, but may colonize wood and other materials such as foam insulating sealant (FastFoam™) and plywood (Frazier *et al.*, 1998). Although storage structural materials and mechanical devices used in potato production are important locations and vehicles for the dissemination of inoculum, the most common medium for inoculum survival is the soil (Carnegie *et al.*, 2003; Al-Mughrabi *et al.*, 2007; Gudmestad *et al.*, 2007; Johnson and Dung, 2010; Hannukkala, 2011; Fiers *et al.*, 2012). Inoculum of many pathogens survives in the soil independently of tubers (Fiers *et al.*, 2012). Favorable conditions for survival, germination, and infection of many potato pathogens correspond to

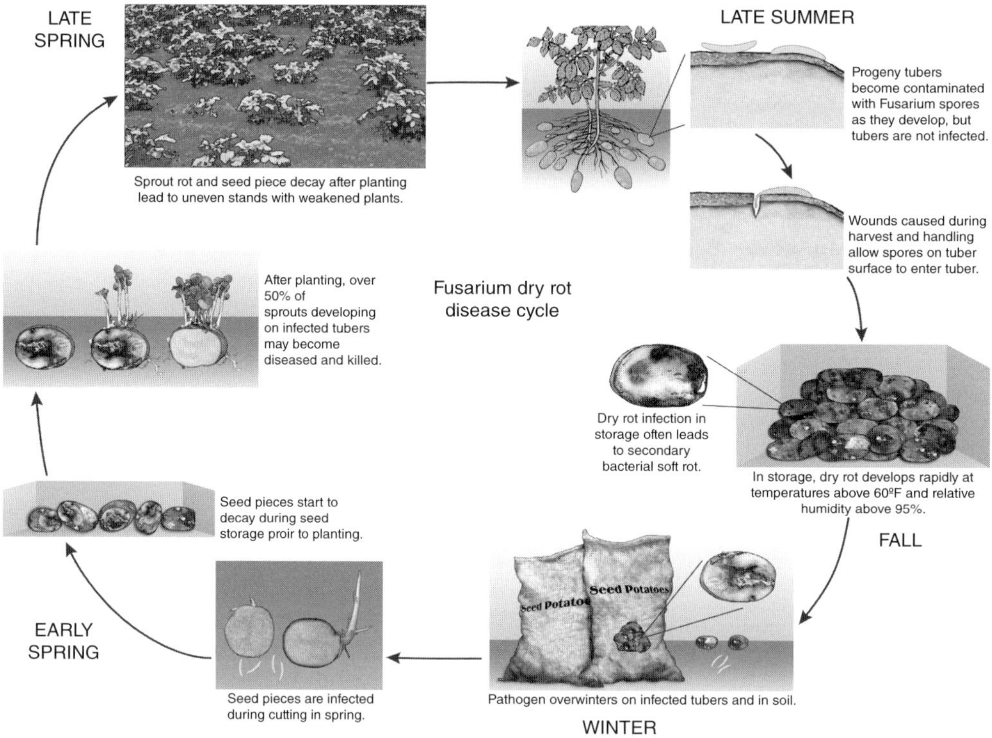

Fig. 11.3. The disease cycle of *Fusarium sambucinum*, causal agent of Fusarium dry rot.

the conditions favorable for potato growth, such as soil temperature between 10 and 25°C, 6–7 pH, and soil O_2 content of about 15–20%. Cultural practices that favor inoculum survival include rotations shorter than 4 years, inadequate water management, rotation with alternative susceptible hosts, shortening of intervals between haulm/vine killing and harvest, and harvesting when soil and tuber pulp temperatures are outside the optimal range (variety dependent but generally from 10 to 25°C; Fiers *et al*., 2012).

Several pathogens have resting spores that can survive in the soil for extensive periods in the absence of known hosts. Examples of such pathogens that cause potato diseases include *Colletotrichum coccodes* (black dot), *Fusarium* spp. (dry rot), *Rhizoctonia solani* (black scurf), *H. solani* (silver scurf), *Spongospora subterranea* (powdery scab), *Streptomyces* spp. (common scab), *P. erythroseptica* and *Phytophthora nicotianae* (pink rot), *Pythium ultimum* (Pythium leak), *Sclerotinia sclerotiorum* (white mold and stem rot) and *S. endobioticum* (potato wart). In the following paragraphs, some major diseases with soil survival phases will be addressed.

C. coccodes (Wallr.) S. Hughes is the causal agent of black dot. The disease is named for the minute, pinprick-sized sclerotia that develop and appear as small, black dots on host tissues, including roots, tubers, and stems. Up to 30% yield reductions have been reported due to infection of roots by the pathogen (Tsror *et al*., 1999). Infected tubers exhibit silver to brown lesions with sclerotia. These blemishes reduce tuber quality, particularly in cultivars destined for the fresh market (Read and Hide, 1995). *C. coccodes* is an effective saprophyte and is common in soils of all potato growing regions (Read and Hide, 1995). Both seed- and soilborne inoculum of *C. coccodes* can cause disease (Read and Hide, 1988; Lees *et al*., 2010; Pasche *et al*., 2010). In several studies, soilborne inoculum resulted in a greater host plant colonization than seedborne inoculum (Dashwood *et al*., 1992; Denner *et al*., 1998). Furthermore, soilborne inoculum has been correlated highly to black dot severity on daughter tubers (Wale *et al.*, 2008). Although soilborne inoculum appears to be more important than seedborne inoculum in black dot disease dissemination, infected seed is the principal way the pathogen is introduced into uncontaminated soil in the first place. As a result, in some countries, table-stock (ware) producers have imposed tight seed tolerances on the level of black dot infection in order to limit soil contamination.

Rhizoctonia diseases of potato are caused by the fungus, *R. solani* Kühn (teleomorph *Thanatephorus cucumeris* (A.B. Frank) Donk), and can be found on all underground parts of the plant at different times during the growing season. *R. solani* has many synonyms and is divided into subgroups called anastomosis groups (AGs), in which isolates are categorized according to their ability to anastomose with one another (Johnk *et al.*, 1993; Carling *et al.*, 2002; Woodhall *et al.*, 2007). Rhizoctonia isolates can be assigned to one of thirteen AGs; AG-3PT is the group of Rhizoctonia isolates that is most commonly associated with disease in potato. AG-3PT is relatively specific to potato, and sclerotia on tubers belong almost exclusively to this AG. However, AG2-2 IIIB and AG4 HGII, which are more commonly associated with sugarbeet crown and root rot, have recently been shown to cause stem canker on potato stems in Idaho and Michigan, USA (Woodhall *et al.*, 2012). AG2-1, 4, 5, and 8 have also been isolated from potato in the UK (Woodhall *et al.*, 2008; Ritchie *et al.*, 2009).

R. solani causes black scurf on tubers (Fig. 11.1) and stem and stolon canker in potatoes, and occurs wherever potatoes are grown. However, *R. solani* causes economically significant damage primarily in cool, wet soils (Banville, 1989). Losses from Rhizoctonia are sporadic and only occur when the weather is cold and wet in the weeks following planting. In northern areas, where growers must often plant in cold soils, Rhizoctonia is a more consistent problem (Bandy *et al.*, 1988). Poor stands, stunted plants, reduced tuber number and size, and misshapen tubers are characteristic of diseases caused by *R. solani*. The symptoms of the disease are found on both above- and belowground portions of the plant. Black scurf (Fig. 11.1) is the most conspicuous sign of Rhizoctonia disease. In this phase of the disease, the fungus forms dark brown to black hard masses (sclerotia) on the surface of the tuber. These structures are the resting bodies of the fungus. Sclerotia are superficial and irregularly shaped, ranging from small, flat, barely visible blotches to large, raised lumps. Although these structures adhere tightly to the tuber skin, they do not penetrate or damage the tuber, even in storage. However, they will perpetuate the disease and inhibit the establishment of potato plants if infected tubers are used as seed. Currently, it is not possible to control Rhizoctonia diseases completely, but severity may be limited by following a combination of cultural and crop protection strategies. Effective management of this disease requires the implementation of an integrated disease management approach and knowledge of each stage of the disease. Although the most important measures are cultural, chemical controls should also be used.

The potato early die complex (PED) has been recognized as a key limiting factor in potato production since the early 1970s (Davis *et al.*, 1972) and remains a challenge in current production (Davis *et al.*, 2010a), especially in the sandy soils of north central USA (MacGuidwin and Rouse, 1990; MacGuidwin *et al.*, 2012). This infectious disease is caused by an interaction between the root-lesion nematode (*Pratylenchus penetrans*) and the Verticillium wilt fungus (*Verticillium dahliae*) (Davis *et al.*, 1972), and is often complicated by interaction with other pathogens such as the black dot pathogen (*C. coccodes*) (Davis and Howard, 1976; Azad *et al.*, 1985; Barkdoll and Davis, 1992) and fertility regimes (Davis *et al.*, 1990, 1994). In various studies, the impact of green manures and maize has resulted in suppression of PED in Idaho (Davis *et al.*, 2010a,b), and may be useful in other regions. Yield losses range from 10 to 50% and about half of Michigan's potato acreage is known to have the problem. Throughout the years, PED has been managed with soil fumigation (1,3-D or metam) and non-fumigant nematicides (aldicarb, oxamyl, ethoprop, or phenamiphos), while chloropicrin is used in Wisconsin. Currently, metam sodium applied at a nematicidal rate is used extensively for PED control (MacGuidwin *et al.*, 2012). Crop rotation has not been successful for PED management, and multi-year searches for PED-resistant germplasm have also not been successful. Currently, there is an urgent need to identify new PED management practices for use in potato production because of the cost of soil fumigants, their potential negative impacts on soil health, and new regulations (MacGuidwin *et al.*, 2012).

A number of recurrent and persistent bacterial diseases are soilborne and have long plagued potato production systems. These include diseases such as bacterial soft rot and common scab, and are discussed in detail below.

11.3 Epidemiology and Control of Important Diseases

Efforts to prevent disease include the use of host resistance (Douches *et al.*, 1996; Whitworth and Novy, 2007), disease prediction models (Wallin and Schuster, 1960; Hijmans *et al.*, 2000; Baker and Kirk, 2007; Wharton *et al.*, 2008; Skelsey *et al.*, 2009a,b, 2010; Baker *et al.*, 2010), crop protection measures with fungicides and biofungicides (Olanya *et al.*, 2009; Al-Mughrabi, 2010; Deliopoulos *et al.*, 2010; Frenkel *et al.*, 2010; Horsfield *et al.*, 2010; Khandaker *et al.*, 2010; Liljeroth *et al.*, 2010; Nyankanga *et al.*, 2011), and cultural practices such as management of inoculum sources (Zwankhuizen *et al.*, 1998) and soil amendments (Goicoechea, 2009). Methods of delivery of crop protection materials have expanded since the introduction and common use of fungicides in the latter part of the 20th century. In the potato crop, they include applications to seed tubers at storage and planting, in-furrow applications of fungicides (Al-Mughrabi *et al.*, 2007), aerial and chemigation application systems (Geary *et al.*, 1999; Hamm and Clough, 1999), and the development of machinery for potato handling and storage (Finckh, 2008). Spending on disease control varies among potato production regions and crop uses, and is dependent on crop value and the number of diseases that need to be controlled. Import restrictions on seed crops require further control measures. More recent restrictions on allowable pesticide residues and genetic make-up of the seed and non-seed tubers for export affect how effectively diseases are controlled.

The demand for the improved appearance and quality of potato tubers has imposed a strict set of constraints that growers must achieve to command a market and sustain their businesses successfully. Diseases that cause skin blemishes, such as silver scurf (*H. solani*), black dot (*C. coccodes*), common scab (*Streptomyces* spp.), and black scurf (*R. solani*), impact appearance and can have serious consequences for the marketability of the seed and commercial crops.

Potato disease cycles mostly follow a common pattern in that most pathogens survive on or in potato tubers between growing seasons. The tubers may be intended for seed, or they may survive as volunteers or "groundkeepers" in the soil. Some other pathogens such as the sexual stage of oomycetes, *Streptomycetes* spp., *S. endobioticum* (Wart), *Fusarium* spp., and *R. solani* (stem canker and black scurf) survive for significant periods in the soil or on plant debris. Some pathogens are spread by association with fungi and cause disease with both foliar and tuber phases; for example, *Potato mop-top virus* (a virus vectored by the fungus *S. subterranea*, the cause of potato powdery scab); corky ring spot (*Tobacco rattle virus* vectored by *Trichodorus* and *Paratrichodorus* spp. of stubby-root nematodes), zebra chip (a bacterium *Candidatus* Liberibacter solanacearum vectored by the potato psyllid, *B. cockerelli*).

Epidemiology and management of bacterial diseases

Soft rot bacteria

The soft rot pathogens of potato (*Pectobacterium* spp. and *Dickeya* spp.) are generally thought to be disseminated on seed (Perombelon, 2002), but also may be soilborne, mechanically disseminated, and/or waterborne (Toth *et al.*, 2003; Laurila *et al.*, 2010; Fig. 11.4). *Pectobacterium* spp. are important bacterial potato pathogens and can be aggressive on tubers and stems, causing wilting and eventual plant death. *Dickeya* species (formerly *Erwinia chrysanthemi*) have emerged as a new threat to potato production in Europe (Toth *et al.*, 2011). *Dickeya* spp. were first reported on potato in the Netherlands in the 1970s. However, since 2004–2005, a new pathogen with the proposed name, *Dickeya solani*, has been spreading across Europe on seed tubers. Toth *et al.* (2011) described the symptoms as "indistinguishable from those of the more established blackleg pathogen *Pectobacterium* spp.". However, *Dickeya* spp. are different from *Pectobacterium* spp. epidemiologically, in that they can start disease in potatoes from lower inoculum levels, spread more readily

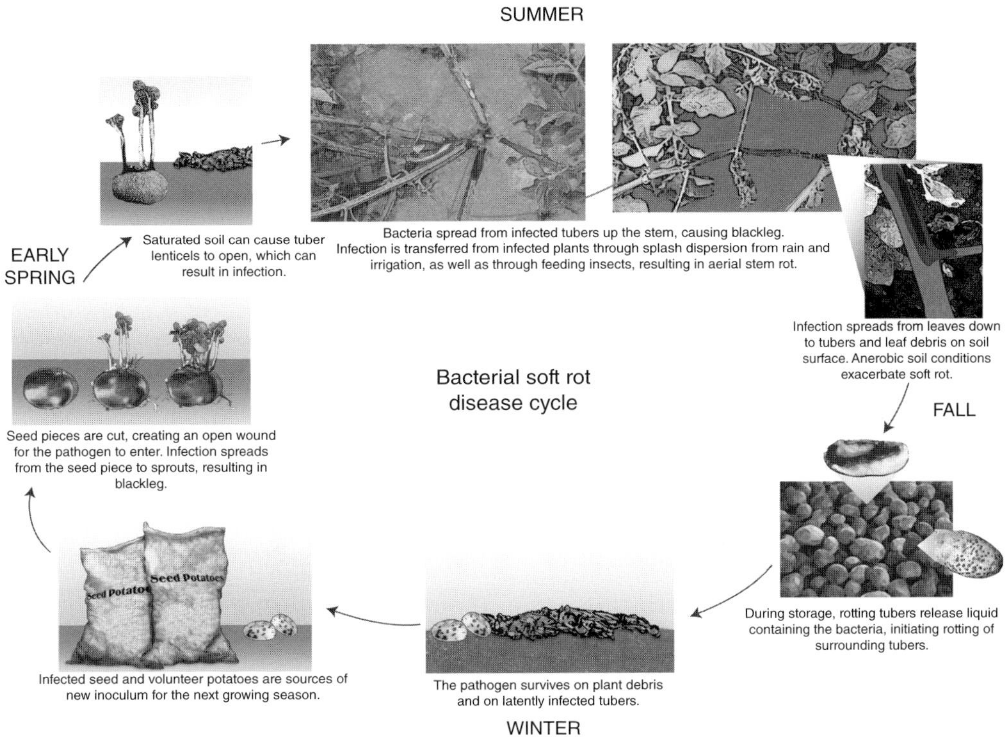

Fig. 11.4. The disease cycle of soft rot bacteria (*Pectobacterium* spp.).

through the plant's vascular tissue, are more aggressive, and have higher optimal temperatures for disease development. *Dickeya* spp. also appear to be less robust than *Pectobacterium* spp. in soil and other environments. Like *Pectobacterium* spp., the principal control for *Dickeya* spp. involves seed tuber certification and consequent rejection of infected seed lots. If, however, seed lots are contaminated uniformly, as reported for *Pectobacterium* spp. (Perombelon, 1992), improved diagnostic techniques would have to be interpreted carefully to determine critical inoculum loads.

The pathogens that cause soft rot of tubers, wilting, and stem and foliar necrosis may be introduced as secondary infecting pathogens after the plant has been compromised. For example, bacterial soft rots of tubers can be introduced after fungal infection, or through wounds caused by mechanical damage (Fig. 11.4). Practices that reduce exposure to damage during harvest, storage, and post-storage are important in the control of soft rots. These were reviewed by Knowles and Plissey (2007) and included a checklist of preharvest factors such as washing digging equipment, timing of crop desiccation (dependent on canopy and tuber maturity), storage preparation including inspection, repair, and cleaning of insulation, ducts, fans and humidifiers, doors, and sensors and control panels. Modifications to harvest equipment are important factors in managing soft rot bacteria and other pathogens that enter through damaged tuber periderm. Knowles and Plissey (2007) identified the harvester as being the major source of mechanical damage to tubers and made recommendations that could reduce damage. These included adjustments to the digging blade, reducing drop heights, use of oblique flow angles, and use of rubberized soft surfaces to decrease bruising. The potential for damage continues from harvest to the loading of tubers on to bulk trucks and to the storage-bin piler, and similar recommendations based on the use of improved padding and flow speeds have been made. Preferably, crops should be harvested when tuber pulp temperature is in the range of 7–18°C, to eliminate temperature gradients that can promote the development of condensation, which in turn enhance the conditions that are conducive for the development of soft rot. Some other important factors that reduce the risk of soft rot developing during the early storage period were described by Knowles and Plissey (2007) and included:

- Limiting the pile size to a height of 5–5.5 m.
- Quickly cooling the tubers to the final storage temperature (3, 5, 7, and 10°C for seed, table-stock, French fry and chip (crisp) processing, respectively).
- No humidification.
- Fans run to dry the tubers as much as possible.
- Daily monitoring for high-risk areas with elevated temperature and/or moisture.

The effective management of soft rot ensures that tubers destined for consumption are high quality and profitable. Healthy potato seed tubers are vital for sustaining the continued supply of potatoes between seasons (Toth *et al.*, 2003).

Potato common scab

The potato crop is produced mainly for fresh, chipping, frozen, or starch markets. The fresh and chipping markets require a high quality in finished appearance. Recurrent and persistent bacterial soilborne diseases have long plagued potato production systems, such as common scab (PCS, *Streptomyces* spp.; Stevenson *et al.*, 2001). Typically, PCS impacts tuber quality and marketable yield, but in seed can cause poor plant vigor, or even death (Stevenson *et al.*, 2001; Loria *et al.*, 2006; Wanner, 2009). PCS is an annual production concern for commercial potato growers (Loria *et al.*, 1997; Slack, 1992) and has been identified as the highest priority for research, and as an intractable and endemic problem by potato research and commodity groups.

Of the more than ~800 identified species in the genus *Streptomyces*, only a fraction are considered plant pathogenic. PCS may be caused by several soil-dwelling plant pathogenic bacterial species in this genus (Wanner, 2007; Hao *et al.*, 2009). These include *Streptomyces scabies*, *Streptomyces europeaiscabiei* (Wanner, 2009), *Streptomyces acidiscabies*, and *Streptomyces turgidiscabies*. In particular, *S. scabies* has been well documented in causing scab lesions (Loria *et al.*, 1997, 2006; Wanner, 2005, 2006, 2009; Bignell *et al.*, 2010). *S. scabies* infects a number of root-grown crops (Loria *et al.*, 1997), including radish (*Raphanus sativus*), parsnip (*Pastinaca sativa*), beet (*Beta vulgaris*), carrot (*Daucus carota*), as well as potato (*S. tuberosum* L.). The disease may be found wherever potatoes are grown worldwide. Since the marketplace for potatoes is quality driven, the presence of scab lesions, especially those that are pitted, significantly lessens the marketability for both table-stock and processing varieties. The genus *Streptomyces* is also noted for its ability to produce medicinally important antibiotics. More than two-thirds of naturally derived antibiotics currently in use today are produced as secondary metabolites by *Streptomyces* species.

The symptoms of common potato scab are present on the surface of the potato tuber. The disease forms several types of cork-like lesions including surface, raised, and pitted lesions (Fig. 11.5). Sometimes, surface lesions are also referred to as russeting, particularly on round white tubers, as the general appearance resembles that of a russet tuber skin. Pitted lesions vary in their range of depth, although on average they extend 3–4 mm deep. Infection begins with the onset of tuberization. Once the pathogen penetrates through several layers of the cells of a young tuber, it is able to derive its nutrient needs from the dead cell material it has infected. This becomes the source of a scab lesion. Under extreme infection, a noticeable reduction in vine growth and yield may be observed.

Potato common scab is a saprophyte that may overwinter either in soil or on the surface of tubers, thereby serving as next year's inoculum source (Fig. 11.6). The inoculum is spread by water, wind, on seed tubers, and on farm equipment with leftover soil residue. Once present in the soil, *S. scabies* may survive indefinitely, since it finds refuge in organic matter. The disease is known to persist in soils with a pH between 5.2 and 7.0. There are other species that may survive and function at even lower pH. *S. acidiscabies*, as its name implies, has the ability to cause symptoms identical to those of *S. scabies* at a pH as low as 4.5.

S. scabies has branched mycelium, and sporogenous hyphae develop into corkscrew-like spiral

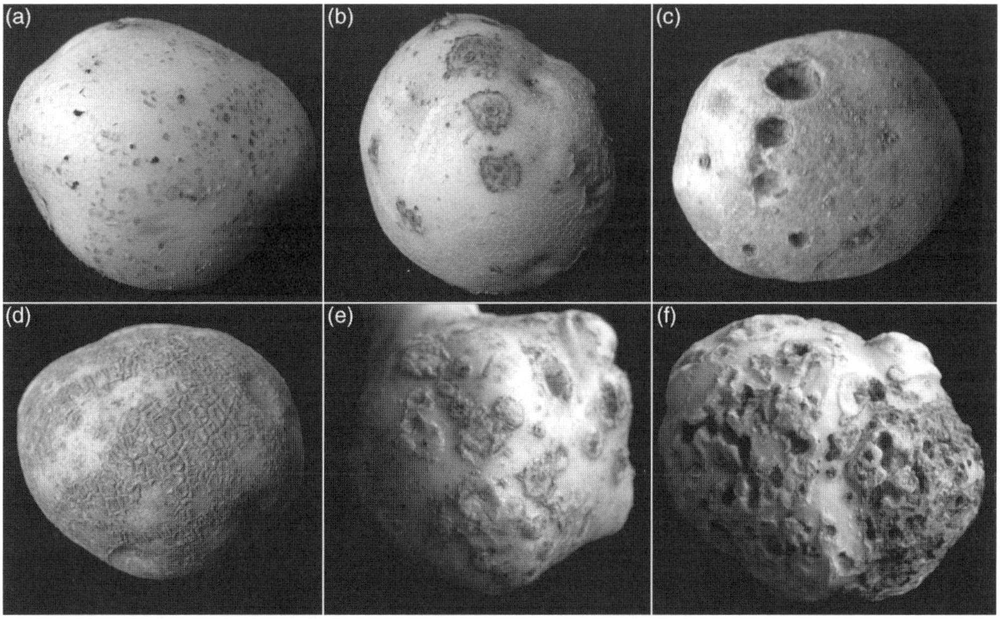

Fig. 11.5. The different lesion types of common potato scab caused by *Streptomyces scabies*. (a) Superficial discrete lesions formed on the surface of the tuber. (b) Raised discrete lesions. (c) Pitted discrete lesions varying in depth from 3 to 5 mm may develop from raised discrete lesions on the tuber surface. (d) Coalescing superficial lesions, often referred to as russeting, as when they cover the entire surface of the tuber, the appearance resembles that of the skin of a russet tuber. (e) Raised coalescing lesions. (f) Pitted coalescing lesions.

chains with cross walls that eventually constrict and break off into individual spores. When the spores mature, they develop gray or melanin pigmentation. When the spore comes into contact with a suitable host, it will germinate and the infection process begins (Fig. 11.6). The pathogen can infect the tuber directly, but may also invade lenticels or any open wound on the surface of the potato tuber. On recognizing infection, the potato produces several layers of cork cells as a defense. The pathogen may then re-infect the tuber through its protective cork layer. Lesion size will also vary, depending on when infection occurs. The earlier a tuber becomes infected, generally the larger the lesion. The phytotoxin, thaxtomin, is strongly believed to be linked to the infection process; however, its mode of action is not known (Fry and Loria, 2002; Loria *et al.*, 2008). Preliminary studies have indicated a necrotic response on tuber tissue when thaxtomin is applied directly.

Common scab infection is characterized by the formation of raised suberized lesions containing cross-linked hydroxycinnamic acid amides and feruloyltyramine trimers in the tuber periderm (King and Calhoun, 2005, 2010). These compounds contribute to the formation of a polyphenolic barrier to further attack. Not all polyphenolics were effective as anthocyanins, a class of pigmented polyphenolics, and were not found enriched in lesions, and tubers high in anthocyanins were also not resistant to common scab (Tai *et al.*, 2013). Polyphenolic biosynthetic enzyme gene expression is regulated by myeloblastosis (MYB) and basic helix-loop-helix (b-HLH) transcription factors and expression of two *MYB* and three *b-HLH* genes was correlated highly with common scab resistance (Tai *et al.*, 2013).

The use of irrigation as a way to manage this disease has been observed since the early 1920s, although it is difficult to maintain adequate soil moisture (Lapwood *et al.*, 1973). Soil moisture may also enhance infection risk by other pathogens. In semi-arid growing conditions, potato plants must be watered for 6–8 weeks after the onset of tuberization to achieve a lasting degree

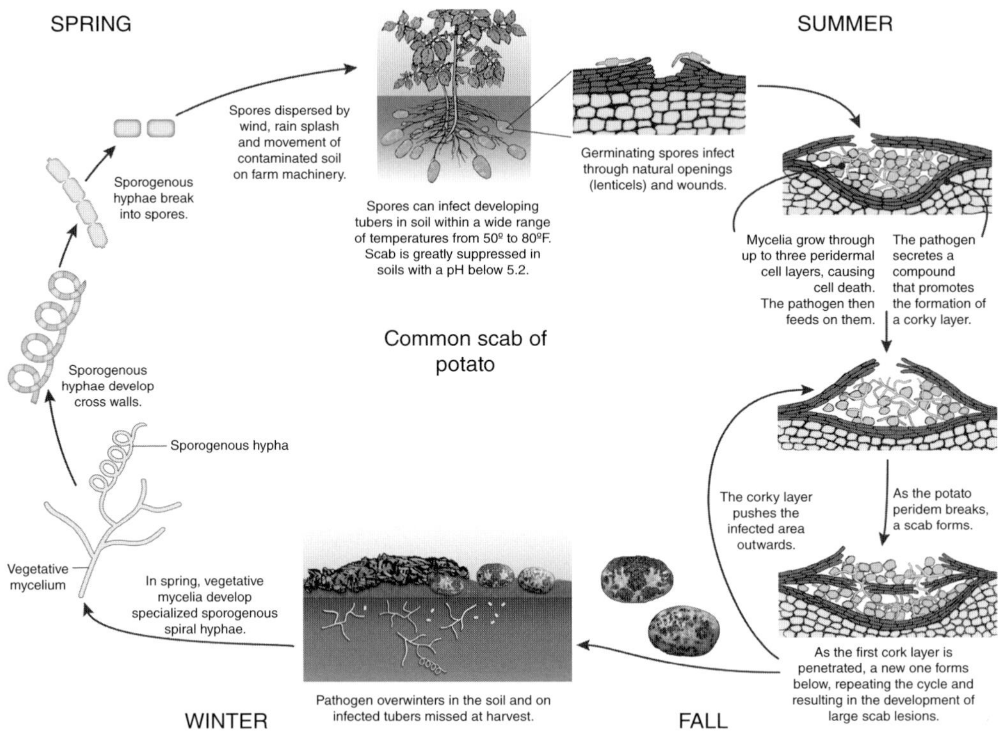

Fig. 11.6. The disease cycle of the common scab pathogen, *Streptomyces scabies*.

of control. Although irrigation may provide an effective means for managing scab on a small scale, it may not always be the most practical, especially for soils with low water-holding capacity. Soil amendments including biological agents, manure, lime, and cover crops have produced inconsistent results for the control of scab (Lazarovits, 2001, 2004; Abbasi *et al.*, 2007; Conn and Lazarovits, 2007; Lazarovits *et al.*, 2008). It has been noted that the use of swine, chicken, and cattle manures has produced an initial reduction in the incidence of common potato scab. However, in subsequent years, the incidence has increased (Conn and Lazarovits, 2007). Aside from amendments, soil type and microbial populations play an integral part of disease presence.

Chemical and other antimicrobial compounds have been used and vary in their degree of success. Chemical treatments such as 3,5-D or benzoic and picolinic acids (McIntosh *et al.*, 1988) tend to cause plant injury. Antimicrobial compounds have also been tested for their efficacy on common potato scab. A few have been effective at lower rates, but none is registered for use in the USA. The chemical, pentachloronitrobenzene (PCNB), has been used with some degree of success (Webster *et al.*, 2011). The use of antagonistic *Streptomyces* spp. and other biocontrol approaches have been shown to decrease the amount of pathogenic *S. scabies* present in the soil and to reduce common scab on tubers (Wanner *et al.*, 2014). These antagonistic species have an antibiotic effect on the pathogenic species (Meng *et al.*, 2012).

Planting cultivars resistant to common potato scab is probably the best and easiest way to combat the disease. Genetic improvement from diploid species may be done for a variety of agronomically important traits, including disease resistance. It is known that certain accessions of wild potato species have resistance to common potato scab, although this is not universal within diploid species. These species include *Solanum commersonii*, *Solanum chacoense*, *Solanum caldasii* var. *glabrescens*, and *Solanum jamesii*. Although there is no variety that is immune to common scab, resistant varieties can be grown to minimize the rate of infection (Haynes *et al.*, 2010).

Zebra chip

(Association between *Candidatus* Liberibacter solanacearum and potato psyllid, *B. cockerelli*.)

A new defect of potato, "zebra chip", was first reported in 2007 in the USA, Mexico, and in Central America (Munyaneza *et al.*, 2007). The disease was named for the characteristic symptoms that developed in fried chips (crisps) from infected potato tubers. This defect was economically significant, causing several millions of dollars in losses to potato producers and processors. The symptoms of zebra chip in the plant resemble those caused by potato purple top and psyllid yellows diseases. Initially, it was thought there might be an association between phytoplasmas or viruses and the potato psyllid (Siller *et al.*, 2008), but further research indicated that the association was between *Candidatus* Liberibacter solanacearum and the potato psyllid (Li *et al.*, 2008; Abad *et al.*, 2009; Henne *et al.*, 2009; Munyaneza *et al.*, 2009; Secor *et al.*, 2009). Munyaneza *et al.* (2007) conducted experiments to elucidate the association between the psyllid *B. cockerelli* (Sulc) (Homoptera: *Psyllidae*) and zebra chip by exposing clean potato plants to this insect under greenhouse and field conditions. Potato plants and tubers exhibiting zebra chip symptoms were tested for phytoplasmas by polymerase chain reaction. Potato psyllids collected from infected potato fields were also tested. Results indicated that there was an association between the potato psyllid and zebra chip. Plants exposed to psyllids in the greenhouse and field developed zebra chip. In the greenhouse, 25.8 and 59.2% of tubers exhibited zebra chip symptoms in the raw tubers and fried chips, respectively. In the field, 15 and 57% of tubers showed symptoms in raw tubers and chips, respectively. No zebra chip was observed in tubers from plants that had not been exposed to psyllids, either in the greenhouse or the field.

Other bacterial diseases

Several other diseases and disorders may also be associated with bacterial pathogens such as stem rotting, initially caused by *Botrytis cinerea* and *Sclerotinia sclerotiorum*, and the disorder toxic seed piece syndrome (Johnson, 1991). Although not covered in this section, bacterial diseases such as bacterial ring rot (*Clavibacter michiganensis* subsp. *sepedonicus*) and brown rot (*Ralstonia solanacearum* race 3, biovar 2), which is active in temperate conditions, may develop into major problems in potato production in the future as trade in seed and commercial potatoes expands.

Epidemiology and management of seed- and soilborne fungal diseases

A number of important potato diseases are transmitted by seed- and soilborne inoculum. The epidemiology of these diseases has been discussed above. Chemical control of pathogens including *C. coccodes* and *R. solani* can be inconsistent (Wale, 2004). However, several products have been developed specifically for the control of seedborne fungal diseases of potato (Atkinson *et al.*, 2010) and offer broad-spectrum control for Rhizoctonia, silver scurf, and Fusarium dry rot, and to some extent black dot. These seed treatments include fludioxonil and flutolonil formulations with or without mancozeb. The general impact of these seed treatments is noted in improved plant stand and crop vigor, but occasionally application of seed treatments in combination with cold and wet soils can result in delayed emergence. The delay is generally transient and the crop normally compensates. The additional benefit of the inclusion of mancozeb is for the prevention of seedborne late blight (Wharton *et al.*, 2012b). Applications of fungicides in-furrow at planting have resulted in significant improvement in Rhizoctonia disease control in potatoes. Fungicides such as flutolonil and azoxystrobin applied in-furrow at planting have given consistent and excellent control of Rhizoctonia diseases of potatoes (Atkinson *et al.*, 2010, 2011).

Epidemiology and management of foliar diseases

Late blight

Potato late blight (*P. infestans*) is considered the most important disease affecting potato worldwide, and is also a major disease of tomatoes (Fry and Goodwin, 1997b; Guenthner *et al.*, 2001). The resurgence of late blight has occurred despite the introduction of improved fungicides and application technologies, improved diagnostics and seed certification systems, and improved host resistance. This is attributable largely to changes in *P. infestans* populations (Goodwin *et al.*, 1996; Smart and Fry, 2001; Fry, 2008; Kirk *et al.*, 2009; Hu *et al.*, 2012). The most important of these changes worldwide, and especially in North America, have been the occurrence of A_2 mating type (Deahl *et al.*, 1991), the resistance to phenylamide fungicides in the field (Shattock, 1988), and the appearance of increasingly aggressive genotypes of *P. infestans* (Fry and Goodwin, 1997a; Hu *et al.*, 2012).

Before 1993, only the US-1 genotype had been found in the USA. The US-1 genotype is mating type A_1. *P. infestans* requires two mating types, A_1 and A_2, to come into contact to produce a sexual spore, known as an oospore. Oospores are resistant to freezing and other environmental extremes, and can survive in diseased leaves and stems or free in soil. With the arrival of the A_2 (e.g. US-8) and new A_1 (e.g. US-23) mating types in North America, the potential exists for the production of oospores that can survive between seasons, though this has not yet been proven to occur (Hu *et al.*, 2012). If the pathogen does start to produce overwintering oospores, it may be able to overwinter more easily and may become a factor in virtually every growing season when conditions favor the development of late blight.

P. infestans can survive only in living potato tissue or, as oospores, survives as mycelia from year to year in infected tubers placed in storage, in piles of cull potatoes, or in infected tubers missed during harvest that remain unfrozen over the winter (volunteer potatoes). If infected tubers freeze and die over winter, the disease cycle is broken, and very often the disease does not appear even when the weather conditions are favorable. If the pathogen survives within tubers, in the spring it can be transmitted from infected tubers to potato foliage by airborne spores (Fig. 11.2). Some infected tubers may rot in the soil before emergence, and not every plant that emerges from an infected tuber will contract late blight. The fact that many reproductive cycles are possible within a season accounts for the rapid increase in disease once it becomes established in a field. This is responsible for the spread to other fields, and the spread within a field can take place by wind and rain within a single round of sporangia production. *P. infestans* favors wet weather with moderate temperatures (15–26°C), high humidity, and frequent rainfall. Environmental conditions during the growing season in the many potato growing regions of North America are sporadically but frequently conducive to the development of epidemics of potato late blight, and significant financial costs, in terms of crop protection (up to US\$700 ha^{-1}) and crop losses (up to US\$5000 ha^{-1}), are incurred when intervention measures to control potato late blight are unsuccessful (Guenther *et al.*, 2001).

Many interacting variables, including meteorological factors such as climatic change and increasing tolerance of *P. infestans* to colder temperatures (Kirk, 2003), represent a serious situation for the potato industry in the USA (Hijmans *et al.*, 2000; Baker *et al.*, 2004). It is predicted that global climate change will result in a significant increased risk of late blight epidemics on all continents where potatoes are grown (Hijmans *et al.*, 2000). For example, from 1950 to 2001, climatic conditions in Michigan, USA, became steadily more conducive for the initiation and development of potato late blight epidemics (Baker *et al.*, 2005), although this may not be the case for all regions. Traditional thresholds for spray initiation as predicted in BLITECAST (Krause *et al.*, 1975) were reached earlier in the season, and conditions during midsummer were also cooler and wetter and resulted in more days conducive for late blight development in Michigan (Baker *et al.*, 2005).

Potato late blight prediction models have been used to estimate environmental conditions that are favorable for epidemic risk and fungicide recommendations appropriate to that risk for more than 50 years (Wallin and Schuster, 1960; Wallin, 1962; Mackenzie, 1981). Such models attempt to limit grower expenditures and reduce the amount of chemical used while achieving optimal control of potato late blight. One of the downfalls of these models is that they use weather data that are at best real time. The incorporation of extended range forecast data into a disease risk system would render these systems even more valuable by providing a prediction of risk conditions up to several days in advance of their occurrence. Extended range forecast model output statistics (MOS) including 192-h maximum and minimum daily temperatures have been produced by the US National Weather Service (NWS) since 1994 with the Global Forecast System (GFS) numerical model (Carrol and Erickson, 2004). Baker and Kirk (2007) developed a method to derive hourly microclimate variables associated with potato late blight risk from the available MOS produced by the NWS. Data were then fed into a neural network computer algorithm that generated accurate predictions of late blight disease risk up to 5 days in the future. Beginning in 2010, the new Baker and Kirk (2007) model was incorporated into the Michigan State University Late Blight Risk Management website (http://www.lateblight.org/forecasting.php).

Early blight and brown leaf spot

Early blight caused by the pathogen, *Alternaria solani* Sorauer, and brown leaf spot caused by *Alternaria alternata* (Fries.) Keissl are economically important to potato crops around the world, as they cause defoliation late in the growing season and reduce potato yield (Rotem, 1994). Both are found annually in most potato growing areas. The timing of their appearance and rate of disease progress help determine their impact on the potato crop. The diseases occur over a wide range of climatic conditions and depend in large part on the frequency of foliage wetting from rainfall, fog, dew, or irrigation, on the nutritional status of foliage, and on cultivar susceptibility. While losses rarely exceed 20%, if left uncontrolled these diseases can be very destructive. Intensive, effective fungicide treatment helps contain losses to less than 5%. Early blight rarely develops early in the season, and usually appears on mature foliage. Yield losses can be reduced with protective, timely fungicide applications (Pasche *et al.*, 2004; Pasche and Gudmestad, 2008; Rosenzweig *et al.*, 2008a).

Foliar symptoms of early blight first appear as small, irregular to circular dark brown spots on the lower (older) leaves and progress into characteristic dark, alternating concentric angular rings that are generally restricted by large veins (Stevenson *et al.*, 2007). As Alternaria brown leaf spot progresses, the lesions coalesce across large veins and the whole leaf may become chlorotic, with infected areas turning brown and disintegrating, causing the edges of the leaf to roll up. The differences between the disease symptoms are shown in Fig. 11.7. Early blight lesions may range in size from a pinpoint to 0.3 cm diameter. As the spots enlarge, they become restricted by leaf veins and take on an angular shape. Early in the growing season, early blight lesions on young, fully expanded succulent leaves may be larger, up to 1.3 cm in diameter, and may, due to their size, be confused with late blight lesions. A narrow band of chlorotic tissue often surrounds each lesion of both diseases, and extensive chlorosis of infected foliage develops over time. Elongated superficial brown or black early blight lesions may also form on stems and petioles. Severe infection of foliage by the early to mid-bulking period can result in

Fig. 11.7. Foliar symptoms of early blight (a) caused by *Alternaria solani*, and brown leaf spot (b) caused by *Alternaria alternata*.

smaller tubers, yield loss, and lower tuber dry matter content (Stevenson *et al.*, 2007).

Tuber symptoms of early blight include circular to irregular lesions that are slightly sunken and often surrounded by a raised purple to dark brown border (Fig. 11.8a). The underlying tissues are leathery to corky in texture, dry and usually dark brown and surrounded by a yellow halo (Fig. 11.8b). These lesions reduce the quality and marketability of fresh market tubers. Tuber infection also presents a challenge to processors, as tuber lesions often require additional peeling to remove the darkened lesions and underlying tissues (Stevenson *et al.*, 2007). Early blight tuber lesions (Fig. 11.8a,b) can be distinguished easily from late blight tuber lesions (Fig. 11.8c,d), which are characterized by discoloration of the potato skin (Fig. 11.8c) and a tan to reddish-brown, dry, granular rot underneath the discolored skin (Fig. 11.8d).

Overwintering spores and mycelia of *A. solani* and *A. alternata* are melanized (darkly pigmented) and can withstand a wide range of environmental conditions, including exposure to sunlight and repeated cycles of drying, freezing, and thawing. In spring, spores (conidia) serve as primary inocula to initiate disease (Kirk and Wharton, 2012). Plants grown in fields or adjacent to fields where potatoes have been infected with brown leaf spot during the previous season are most prone to infection, since large quantities of overwintering inoculum are likely to be present from the previous crop. Initial inoculum is moved readily within and between fields, as air currents, windblown soil particles, splashing rain, and irrigation water carry the spores easily.

Alternating wet and dry periods with temperatures between 5 and 30°C (the optimum is 20°C) in this range favor *Alternaria* spore production. Few spores are produced on plant tissue that is continuously wet or dry (Stevenson *et al.*, 2007). Many cycles of early blight and brown leaf spot spore production and lesion formation occur within a single growing season once primary infections are initiated (Fig. 11.9). Secondary spread of these pathogens begins when spores are produced on foliar lesions and carried to

Fig. 11.8. Tuber symptoms of early blight, caused by *Alternaria solani* (a,b) and late blight (c,d). Early blight lesions on tubers (a) are usually circular to irregular and slightly sunken, often surrounded by a raised, purple to dark brown border. (b) Underlying tissues are leathery to corky in texture and usually surrounded by a yellow halo. Late blight tuber lesions are characterized by discoloration of the potato skin (c) and a tan to reddish-brown, dry, granular rot underneath the discolored skin (d).

neighboring leaves and plants. Early blight and brown leaf spot are largely diseases of older plant tissues and are more prevalent on senescing tissues on plants that have been subjected to stresses induced by injury, poor nutrition, insect damage, or other types of stress. The rate of disease spread increases after flowering and can be quite rapid later in the season during the bulking period and during periods of plant stress.

On potato tubers, germinated spores penetrate the tuber epidermis through lenticels and mechanical injuries to the skin. Tubers often become contaminated with *A. solani* and *A. alternata* spores during harvest. These spores may have accumulated on the soil surface, or may have been dislodged from desiccated vines during harvest. Infection is most common on immature tubers and those of white- and red-skinned cultivars, since they are highly susceptible to abrasion and skinning during harvest. Course-textured soil and wet harvest conditions also favor infection. In storage, individual lesions may continue to develop but secondary spread does not occur. Infected tubers may shrivel through excessive water loss, depending on storage conditions and disease severity (Stevenson *et al.*, 2007).

Effective management of these diseases requires implementation of an integrated disease management approach (MacDonald *et al.*, 2007; Horsfield *et al.*, 2010). These diseases are controlled primarily through the use of cultural practices, resistant cultivars, and foliar fungicides (Pasche *et al.*, 2004; Pasche and Gudmestad, 2008). Cultural practices, such as crop rotation, removal and burning of infected plant debris, and eradication of weed hosts, helps reduce the inoculum level for subsequent plantings (Rotem, 1994). Since *A. solani* and *A. alternata* persist in plant debris in the field from one growing season to the next, rotation with non-host crops (e.g. small grains, maize or soybean) may reduce the amount of initial inoculum available for disease initiation. Other cultural control measures may include the following (Stevenson *et al.*, 2007; Wharton and Kirk, 2007):

- Avoid irrigation in cool cloudy weather and time irrigation to allow plants time to dry before nightfall.
- Use certified, disease-free seed.
- Use tillage practices, such as fall plowing, that bury plant refuse.

To minimize tuber infection after harvest, tubers should be stored under conditions that promote rapid suberization, as *A. solani* is unable to infect through intact periderm.

Cultivars with moderate to good levels of field resistance to both diseases are available. However, no immunity to early blight or brown leaf spot has been found in commercial potato cultivars or in their wild parents. Highly susceptible cultivars such as Red Norland, Norchip, and Superior should be avoided in locations where early blight or brown leaf spot is prevalent and disease pressure is high. Field resistance to foliage infection is associated with plant maturity. Thus, late-maturing cultivars are usually more resistant than early-maturing cultivars, and therefore one should avoid planting early and late cultivars in the same or adjacent fields.

The most common and effective method for the control of early blight is through the application of foliar fungicides (MacDonald *et al.*, 2007; Horsfield *et al.*, 2010). Protectant fungicides recommended for late blight control (e.g. maneb, mancozeb, chlorothalonil, and triphenyl tin hydroxide) are also effective against early blight when applied at approximately 7- to 10-day intervals. Currently, there are a wide variety of fungicides to choose from for the control of early blight. However, the strobilurin or quinone outside inhibitor (QoI) fungicides are often favored because they offer broad-spectrum protection against a wide range of fungal and oomycete diseases, have reduced environmental impact and reduced toxicity to mammals and bees compared with conventional protectant fungicides (e.g. chlorothalonil, mancozeb, and mefenoxam) used to control early blight (Rosenzweig *et al.*, 2008a,b). Other products that have shown efficacy against early blight include pyrimethamil (FRAC group 9), boscalid (FRAC group 7), and fluopyram (FRAC group 7). Spraying should commence at the first sign of disease or immediately after bloom, or when spray initiation thresholds have been reached (e.g. the threshold of 200 P-days (disease severity values for scheduling initial fungicide applications for early blight), as described by Sands *et al.*, 1979). The frequency of subsequent sprays should be determined according to the genotype and age-related resistance of the cultivar.

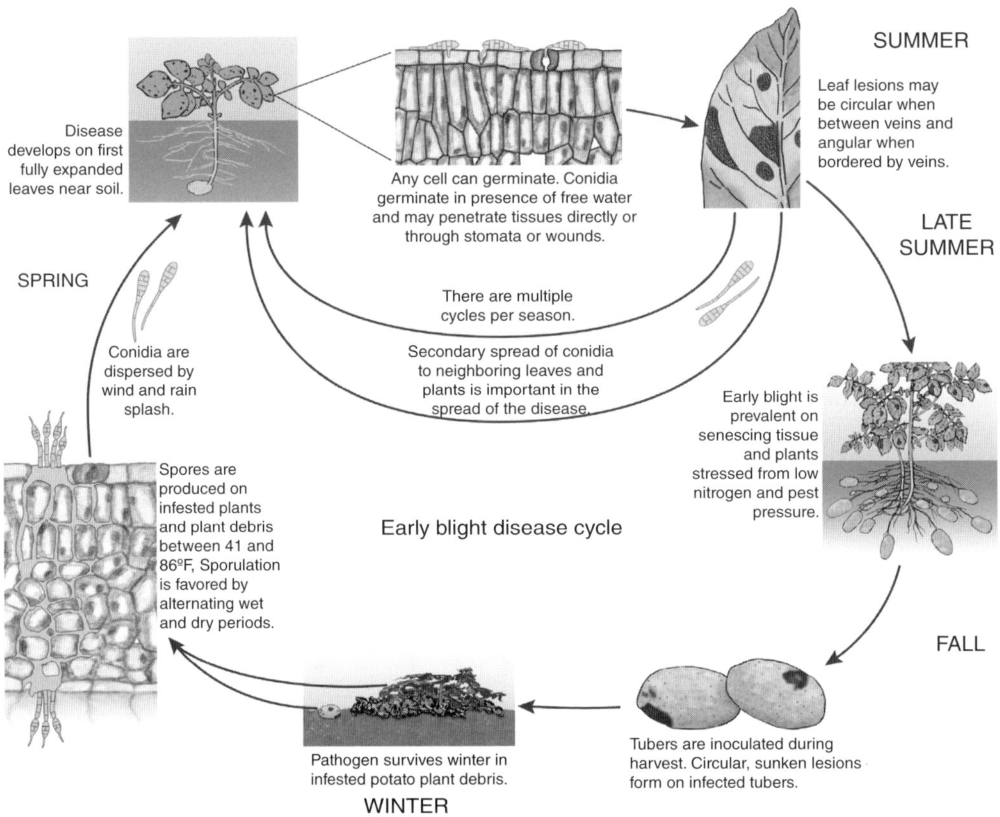

Fig. 11.9. The disease cycle of the early blight pathogen, *Alternaria solani*.

Protectant fungicides should be applied initially at relatively long intervals, and subsequently at shorter intervals as the crop ages.

Epidemiology and management of postharvest diseases

Potato growers aim to produce and harvest a healthy and high-quality crop. However, there are several diseases that infect potatoes in storage, thus compromising their quality and creating a potential economic loss (Secor and Gudmestad, 1999). The major storage diseases include late blight caused by *P. infestans*, dry rot caused by *F. sambucinum* and *Fusarium* spp. (Secor and Salas, 2001; Peters *et al.*, 2008a,b; Gachango *et al.*, 2012a), Pythium leak caused by *P. ultimum* (Gudmestad *et al.*, 2007), pink rot caused by *P. erythroseptica* (Taylor *et al.*, 2004; Fitzpatrick-Peabody and Lambert, 2011), and silver scurf caused by *H. solani* (Rodriguez *et al.*, 1996; Errampalli *et al.*, 2001; Miller *et al.*, 2004, 2011). In addition, bacteria causing soft rot can impact postharvest, especially when harvest conditions are wet (Elphinstone and Perombelon, 1986; Perombelon, 1992; Toth *et al.*, 2003, 2011; De Boer *et al.*, 2007; Fiers *et al.*, 2012). In the field, tubers may be wounded during harvest, transport to storage, and during storage activities. The duration of this process, depending on the distance of the field to the storage (up to 150 km in Michigan, USA) and the handling capacity of the system, can take up to 24 h during September–October in temperatures ranging from 15 to 25°C, especially when equipment breaks down (Gachango *et al.*, 2012a).

Currently, the primary control for these diseases in storage facilities includes the elimination of infected tubers prior to storage and storage management using forced air ventilation, and controlled temperature and humidity feedback systems (Knowles and Plissey, 2007). There is a shortage of postharvest fungicides or effective disinfectant products to control these pathogens completely (Olsen *et al.*, 2003; Slininger *et al.*, 2007; Olsen, 2010). The few compounds available for potato tuber treatment in storage include chlorine-based disinfectants such as sodium hypochlorite, calcium hypochlorite, and chlorine dioxide, and mixtures of hydrogen peroxide and peroxyacetic acid (Wharton *et al.*, 2007a). Hydrogen peroxide (H_2O_2; Storox™, hydrogen dioxide 27%; BioSafe Systems, Glastonbury, Connecticut, USA) is a broad-spectrum disinfectant that is able to provide immediate control of storage pathogens (Olsen and Miller, 2005). Hydrogen peroxide is environmentally friendly, with its activities based on the oxidation of fungi and bacteria, and has been reported to control silver scurf (Afek *et al.*, 2001), pink rot (Al-Mughrabi, 2006), and early blight (Al-Mughrabi, 2005) successfully. However, use of disinfectants does not control storage pathogens completely (Olsen *et al.*, 2003; Miller *et al.*, 2006), hence other postharvest products in combination with proper storage management are recommended (Powelson and Rowe, 2008).

In recent years, several new biocontrol fungicides based on bacteria, *Bacillus subtilis* (QST 713; Serenade; AgraQuest, Inc, California) and *Bacillus pumilis* (QST 2808; Sonata; AgraQuest, Inc, California, USA) have been registered for control of potato pathogens. These products are used for organic potato production to complement copper products and have successfully shown a reduction of foliar late blight disease development (Stephan *et al.*, 2005). *B. subtilis* produces three groups of lipopeptides that work together to stop spores of plant pathogens from germinating, disrupt germ tube and mycelial growth, and inhibit attachment of the plant pathogen to the leaf surface (Marrone, 2002). Another product registered for postharvest use is Bio-Save 10LP (*Pseudomonas syringae*: ESC-10), which has been reported to reduce Fusarium dry rot and silver scurf when applied to tubers prior to storage (Hopkins and Hirnyck, 2008). *B. subtilis* effectively controlled postharvest diseases of potatoes in combination with field-applied biofungicides and fungicides (Gachango *et al.*, 2012c).

Phosphorous acid (Phostrol mono- and dibasic sodium, potassium and ammonium salts of phosphorous acid; Nufarm Americas, Inc, AGT-Division, Burr Ridge, Illinois, USA) was recently registered for postharvest use in potato production (Powelson and Rowe, 2008). The US Environmental Protection Agency (US-EPA) considers it to be a systemic fungicide but not a biochemical (http://www.epa.gov/opp00001/chem_search/reg_actions/registration/fs_PC-076416_1-Oct-98.pdf). The direct effects of phosphorous acid include inhibition of mycelial growth and inhibition of particular metabolic processes, and the indirect effects include stimulation of the natural defense responses of the plant (Guest and Bompeix, 1990). Phosphorous acid has been reported to control effectively tuber late blight and pink rot in storage (Inglis *et al.*, 2004; Johnson *et al.*, 2004; Gachango *et al.*, 2012b).

Control of dry rot in storage has been achieved primarily through reducing tuber bruising, providing conditions for rapid wound healing (Secor and Johnson, 2008), and applying TBZ (Mertect 340-F, Syngenta, Greensboro, North Carolina, USA), a benzimidazole fungicide, as tubers are loaded into storage (Hide *et al.*, 1992). However, *F. sambucinum* resistant to TBZ and other benzimidazole fungicides was discovered in Europe in 1973 (Hide *et al.*, 1992) and in the USA in 1992 (Desjardins, 1995), leading to reduced effectiveness in controlling dry rot (Staub, 1991). To counteract this loss, registration of other chemistries is imperative, and recently Stadium®, a three-way mixture of fungicides from three different FRAC groups (azoxystrobin (11), fludioxonil (12), and difenoconazole (3)) was registered in the USA by Syngenta Crop Protection for control of dry rot and silver scurf (Kirk *et al.*, 2013).

11.4 Future Prospects

Climate change

The viability of potato production is affected by spatial and temporal variability in soils and weather and the availability of water resources where supplementary irrigation is required (Savary *et al.*, 2011; Daccache *et al.*, 2012). Soil characteristics, markets, and agroclimatic conditions influence greatly the choice of cultivar, agronomic practices, and crop protection decisions. Daccache *et al.* (2012) reported that by the 2050s, the area of land that was currently well or moderately suited for rain-fed production would decrease by 88 and 74%, respectively, due to the increased likelihood of dry conditions.

As mentioned above, it is predicted that global climate change will result in a significant increased risk of late blight epidemics on all continents where potatoes are grown (Hijmans *et al.*, 2000). In some areas of the USA, the patterns of rainfall are changing and winters are getting milder, leading to conditions that are more conducive for the initiation and development of potato late blight epidemics; for example, in Michigan (Baker *et al.*, 2004). One way to counter this problem will be through the use of disease risk forecasting models (Wharton *et al.*, 2008).

Fungicide resistance

Fungicide resistance in potato pathogens has become a major issue in recent years, as the industry has moved to mostly specialized single-site mode of action and translaminar fungicides (MacDonald *et al.*, 2007; Olanya *et al.*, 2009; Horsfield *et al.*, 2010; Venkataramana *et al.*, 2010; Cooke *et al.*, 2011; Taylor *et al.*, 2011). This has been typified by the rapid development of resistance to the QoI fungicides (FRAC group 11) in the *Alternaria* pathogens on potato. The QoI fungicides work by disrupting respiration in fungal cells by binding to the so-called Qo site (the outer quinol-oxidation site) of the respiratory enzyme complex, cytochrome bc_1. As a result of this narrow mode of action, there is an inherent risk for the development of pathogen resistance. Since 1999, the persistent and intensive use of azoxystrobin and other QoI fungicides alternated with chlorothalonil on commercially grown potato crops in the Midwest and throughout the USA has increased the selection pressure for the development of QoI fungicide resistance in *A. solani* (Pasche *et al.*, 2004). As a result, the efficacy of these fungicides has started to decline, and isolates of *A. solani* with reduced sensitivity or complete resistance to QoI fungicides have now been identified (Pasche and Gudmestad, 2008). Furthermore, field surveys have shown that these resistant isolates are now widespread throughout the USA (Fairchild *et al.*, 2013) and Canada.

In 2009, *A. solani* isolates from potato were discovered with resistance to boscalid (Wharton *et al.*, 2012a). The succinate dehydrogenase inhibitor (SDHI) fungicide, boscalid (FRAC group 7), was released for use on potato in 2003 (Stammler *et al.*, 2007) and was readily adopted by many growers in the USA as an alternative to the QoI fungicides for the control of early blight. The emergence of resistance to boscalid is worrying as, recently, the US-EPA has registered several new fungicides containing active ingredients with similar chemistries to boscalid. Penthiopyrad (Fontelis, Vertisan, DuPont Crop Protection) and flyopyram (Luna, Luna Tranquility, Bayer CropScience) are two such compounds that are also in FRAC group 7. These studies show the need for increased education programs to promote the use of resistance management strategies in the application of these products.

Molecular disease diagnostics

Detection and identification of potato pathogens is dependent on a diverse range of techniques and skills, from traditional culturing and taxonomic skills to modern molecular-based methods. The wide range of methods employed reflects the great diversity of potato pathogens and the different parts of the plant that they may infect. A typical disease diagnostic laboratory usually has the capacity to perform morphological identification using microscopical methods, methods for detecting proteins from pathogens such as enzyme-linked immunosorbent assay (ELISA) or electrophoresis, methods for detecting fatty acids, molecular methods identifying the nucleic acids such as polymerase chain reaction (PCR), and last, traditional bioassays such as inoculation of test plants, or isolation on selective media followed by morphological identification. A decline in funding and the availability of trained staff to perform traditional techniques is a significant issue for maintaining the critical mass required to deliver this type of service.

Generic methods that can be used to identify a number of different pathogens can help with the capacity maintenance and sustainability of small diagnostics laboratories. Using a smaller number of techniques means less expertise is needed and training new staff is more readily achievable. With robust generic techniques, laboratory staff do not need to be highly qualified. This helps to keep running costs low, allows delivery of a cost-effective service, and makes sustainability more achievable. Real-time polymerase chain reaction (qPCR) is emerging as one such generic platform. It is also particularly well suited to high-throughput detection of a limited number of known target pathogens. Real-time PCR is exploited as a front-line diagnostic screening tool in human health, animal health, homeland security, and biosecurity, as well as plant health.

Although molecular techniques such as qPCR are now routinely used in many diagnostic laboratories to identify known plant pathogens, progress in developing generic molecular techniques for the identification of unknown plant pathogens has been less rapid (Boonham *et al.*, 2008). Recently, *de novo* sequencing methods have been used to good effect in identifying potential disease-causing agents. Recent reports (Cox-Foster *et al.*, 2007; Ledford, 2007) have shown that deep sequencing of cDNA and DNA generated using generic primer sets (e.g. 16S primers for bacteria) can be used to identify the presence of a pathogen sequence. Deep sequencing techniques can generate millions of sequences, yet require no *a priori* knowledge of the pathogens present in the host, and offer for the first time a completely generic diagnostic tool regardless of host or pathogen. Currently, however, the sequencing is expensive and the bioinformatics required to deal with the large amount of data is in its infancy; however, it is likely that it will be a widely used method in the future. With the emergence of novel potato pests and diseases such as zebra chip, systems such as these will be essential in keeping potato production cost-effective.

References

Abad, J.A., Bandla, M., French-Monar, R.D., Liefting, L.W. and Clover, G.R.G. (2009) First report of the detection of '*Candidatus liberibacter*' species in zebra chip disease-infected potato plants in the United States. *Plant Disease* 93, 108–109.

Abbasi, P.A., Conn, K.L. and Lazarovits, G. (2007) Managing soilborne diseases of vegetable crops with a pre-plant soil or substrate amendment of a corn distillation product. *Biocontrol Science and Technology* 17, 331–344.

Adaskaveg, J.E. and Förster, H. (2010) New developments in postharvest fungicide registrations for edible horticultural crops and use strategies in the United States. In: Prusky, D. and Gullino, M.L. (eds) *Postharvest Pathology*. Springer, New York, pp. 107–117.

Afek, U., Orenstein, J. and Jin Kim, J. (2001) Control of silver scurf disease in stored potato by using hydrogen peroxide plus (HPP). *Crop Protection* 20, 69–71.

Allemann, J., Laurie, S.M., Thiart, S. and Vorster, H.J. (2004) Sustainable production of root and tuber crops (potato, sweet potato, indigenous potato, cassava) in southern Africa. *South African Journal of Botany* 70, 60–66.

Al-Mughrabi, K. (2005) Efficacy of OxiDate™ for control of early blight (*Alternaria solani*) in potato storages. *Plant Pathology Journal* 4, 1–4.

Al-Mughrabi, K.I. (2006) Sensitivity to hydrogen peroxide *in vitro* of North American isolates of *Phytophthora erythroseptica*, the cause of pink rot of potatoes. *Plant Pathology Journal* 5, 7–10.

Al-Mughrabi, K.I. (2010) Biological control of Fusarium dry rot and other potato tuber diseases using *Pseudomonas fluorescens* and *Enterobacter cloacae*. *Biological Control* 53, 280–284.

Al-Mughrabi, K.I., Peters, R.D., Platt, H.W., Moreau, G., Vikram, A., Poirier, R., *et al.* (2007) In-furrow applications of metalaxyl and phosphite for control of pink rot (*Phytophthora erythroseptica*) of potato in New Brunswick, Canada. *Plant Disease* 91, 1305–1309.

Andrivon, D. (1995) Biology, ecology, and epidemiology of the potato late blight pathogen *Phytophthora infestans* in soil. *Phytopathology* 85, 1053–1056.

Appel, R., Adler, N. and Habermeyer, J. (2001) A method for the artificial inoculation of potato tubers with *Phytophthora infestans* and polymerase chain reaction assay of latently infected sprouts and stems. *Journal of Phytopathology* 149, 287–292.

Atkinson, D., Thornton, M.K. and Miller, J.S. (2010) Development of *Rhizoctonia solani* on stems, stolons and tubers of potatoes I. Effect of inoculum source. *American Journal of Potato Research* 87, 374–381.

Atkinson, D., Thornton, M.K. and Miller, J.S. (2011) Development of *Rhizoctonia solani* on stems, stolons and tubers of potato II. Efficacy of chemical applications. *American Journal of Potato Research* 88, 96–103.

Azad, H.R., Davis, J.R., Schnathorst, W.C. and Kado, C.I. (1985) Differential-effects of verticillium wilt resistant and susceptible potato genotypes on populations of rhizosphere bacteria. *Phytopathology* 75, 1301.

Baker, K.M. and Kirk, W.W. (2007) Comparative analysis of models integrating synoptic forecast data into potato late blight risk estimate systems. *Computers and Electronics in Agriculture* 57, 23–32.

Baker, K.M., Kirk, W.W., Andresen, J. and Stein, J.M. (2004) A problem case study: influence of climatic trends on late blight epidemiology in potatoes. *Acta Horticulturae (ISHS)* 638, 37–42.

Baker, K.M., Kirk, W.W., Stein, J.M. and Andresen, J.A. (2005) Climatic trends and potato late blight risk in the Upper Great Lakes region. *HortTechnology* 15, 510–518.

Baker, K., Wharton, P., Stein, J., Paz, J. and Kirk, W. (2010) Building true forecasts into crop disease management systems. In: Bundgaard, K. and Isaksenpp, L. (eds) *Agriculture Research and Technology*. Nova Science Publishers, Hauppauge, New York, pp. 281–301.

Bandy, B.P., Leach, S.S. and Tavantzis, S.M. (1988) Anastomosis Group-3 is the major cause of rhizoctonia disease of potato in Maine. *Plant Disease* 72, 596–598.

Banville, G.J. (1989) Yield losses and damage to potato plants caused by *Rhizoctonia solani* Kuhn. *American Potato Journal* 66, 821–834.

Barkdoll, A.W. and Davis, J.R. (1992) Distribution of *Colletotrichum coccodes* in Idaho and variation in pathogenicity on potato. *Plant Disease* 76, 131–135.

Berkeley, M.J. (1846) *Observations, Botanical and Physiological, on the Potato Murrain*. American Phytopathological Society, East Lansing, Michigan.

Bignell, D.R.D., Huguet-Tapia, J.C., Joshi, M.V., Pettis, G.S. and Loria, R. (2010) What does it take to be a plant pathogen: genomic insights from Streptomyces species. *Antonie Van Leeuwenhoek International Journal of General and Molecular Microbiology* 98, 179–194.

Bonde, R. and Schultz, E.S. (1943) Potato cull piles as a source of late blight infection. *American Potato Journal* 20, 112–118.

Boonham, N., Glover, R., Tomlinson, J. and Mumford, R. (2008) Exploiting generic platform technologies for the detection and identification of plant pathogens. *European Journal of Plant Pathology* 121, 355–363.

Boyd, A.E.W. (1972) Potato storage diseases. *Review of Plant Pathology* 51, 297–321.

Bradshaw, J.E. and Ramsay, G. (2009) Potato origin and production. In: Jaspreet, S. and Lovedeep, K. (eds) *Advances in Potato Chemistry and Technology*. Academic Press, San Diego, California, pp. 1–26.

Buchman, J.L., Fisher, T.W., Sengoda, V.G. and Munyaneza, J.E. (2012) Zebra chip progression: from inoculation of potato plants with liberibacter to development of disease symptoms in tubers. *American Journal of Potato Research* 89, 159–168.

Bussan, A.J., Mitchell, P.D., Copas, M.E. and Drillas, M.J. (2007) Evaluation of the effect of density on potato yield and tuber size distribution. *American Journal of Potato Research* 81, 443–448.

Carling, D.E., Baird, R.E., Gitaitis, R.D., Brainard, K.A. and Kuninaga, S. (2002) Characterization of AG-13, a newly reported anastomosis group of *Rhizoctonia solani*. *Phytopathology* 92, 893–899.

Carnegie, S.F., Choiseul, J.W. and Roberts, A.M.I. (2003) Detection of *Colletotrichum coccodes* and *Helminthosporium solani* in soils by bioassay. *Plant Pathology* 52, 13–21.

Carrol, K.L. and Erickson, M.C. (2004) Improvements in extended-range temperature and probability of precipitation guidance. Symposium on the 50th Anniversary of Operational Numerical Weather Prediction. American Meteorological Society, College Park, Maryland.

Chang, D.C., Park, C.S., Kim, S.Y. and Lee, Y.B. (2012) Growth and tuberization of hydroponically grown potatoes. *Potato Research* 55, 69–81.

Choiseul, J.W., Allen, L. and Carnegie, S.F. (2001) The role of stem inoculum in the transmission of *Fusarium sulphureum* to potato tubers. *Potato Research* 44, 165–172.

Conn, K.L. and Lazarovits, G. (2007) Reduction of potato scab with acidified liquid swine manure soil amendment. *Canadian Journal of Plant Pathology* 29, 440.

Cooke, L.R., Schepers, H.T.A.M., Hermansen, A., Bain, R.A., Bradshaw, N.J., Ritchie, F., *et al.* (2011) Epidemiology and integrated control of potato late blight in Europe. *Potato Research* 54, 183–222.

Cox-Foster, D.L., Conlan, S., Holmes, E.C., Palacios, G., Evans, J.D., Moran, N.A., *et al.* (2007) A metagenomic survey of microbes in honey bee colony collapse disorder. *Science* 318, 283–287.

Crosslin, J.M., Lin, H. and Munyaneza, J.E. (2011) Detection of '*Candidatus Liberibacter solanacearum*' in the potato psyllid, *Bactericera cockerelli* (Sulc), by conventional and real-time PCR. *Southwestern Entomologist* 36, 125–135.

Crosslin, J.M., Hamm, P.B., Eggers, J.E., Rondon, S.I., Sengoda, V.G. and Munyaneza, J.E. (2012) First report of Zebra Chip disease and '*Candidatus Liberibacter solanacearum*' on potatoes in Oregon and Washington State. *Plant Disease* 96, 452–453.

Cullen, D.W., Toth, I.K., Pitkin, Y., Boonham, N., Walsh, K., Barker, I., *et al.* (2005) Use of quantitative molecular diagnostic assays to investigate Fusarium dry rot in potato stocks and soil. *Phytopathology* 95, 1462–1471.

Daccache, A., Keay, C., Jones, R.J.A., Weatherhead, E.K., Stalham, M.A. and Knox, J.W. (2012) Climate change and land suitability for potato production in England and Wales: impacts and adaptation. *Journal of Agricultural Science* 150, 161–177.

Dashwood, E.P., Fox, R.A. and Perry, D.A. (1992) Effect of inoculum source on root and tuber infection by potato blemish disease fungi. *Plant Pathology* 41, 215–223.

Davis, J.R. and Howard, M.N. (1976) Presence of *Colletotrichum atramentarium* in Idaho and relation to Verticillium Wilt (*Verticillium dahliae*). *American Potato Journal* 53, 397–398.

Davis, J.R., Callihan, R.H. and Groskopp, M.D. (1972) Efficacy of potassium azide for control of *Rhizoctonia solani* and Verticillium Wilt of potato. *Plant Disease Reporter* 56, 400–406.

Davis, J.R., Sorensen, L.H., Stark, J.C. and Westermann, D.T. (1990) Fertility and management-practices to control verticillium wilt of the Russet Burbank potato. *American Potato Journal* 67, 55–65.

Davis, J.R., Stark, J.C., Sorensen, L.H. and Schneider, A.T. (1994) Interactive effects of nitrogen and phosphorus on verticillium wilt of Russet Burbank potato. *American Potato Journal* 71, 467–481.

Davis, J.R., Huisman, O.C., Everson, D.O., Nolte, P., Sorensen, L.H. and Schneider, A.T. (2010a) The suppression of verticillium wilt of potato using corn as a green manure crop. *American Journal of Potato Research* 87, 195–208.

Davis, J.R., Huisman, O.C., Everson, D.O., Nolte, P., Sorensen, L.H. and Schneider, A.T. (2010b) Ecological relationships of verticillium wilt suppression of potato by green manures. *American Journal of Potato Research* 87, 315–326.

Deahl, K.L., Goth, R.W., Young, R., Sinden, S.L. and Gallegly, M.E. (1991) Occurrence of the A2 mating type of *Phytophthora infestans* in potato fields in the United States and Canada. *American Potato Journal* 68, 717–725.

De Boer, S.H., Elphinstone, J.G. and Saddler, G.S. (2007) Molecular detection strategies for phytopathogenic bacteria. In: Punja, Z.K.,De Boer, S.H. and Sanfacon, H. (eds) *Biotechnology and Plant Disease Management*. CAB International, Cambridge, Massachusetts, pp. 165–194.

Deliopoulos, T., Kettlewell, P.S. and Hare, M.C. (2010) Fungal disease suppression by inorganic salts: a review. *Crop Protection* 29, 1059–1075.

Denner, F.D.N., Millard, C.P. and Wehner, F.C. (1998) The effect of seed- and soilborne inoculum of *Colletotrichum coccodes* on the incidence of black dot on potatoes. *Potato Research* 41, 51–56.

Desjardins, A.E. (1995) Population structure of *Gibberella pulicaris* (Anamorph *Fusarium sambucinum*) from potato tuber dry rot in North America and Europe. *American Potato Journal* 72, 145–156.

Desjardins, A.E., Christharned, E.A., Mccormick, S.P. and Secor, G.A. (1993) Population structure and genetic-analysis of field-resistance to thiabendazole in *Gibberella pulicaris* from potato tubers. *Phytopathology* 83, 164–170.

Doster, M.A., Sweigard, J.A. and Fry, W.E. (1989) The influence of host resistance and climate on the initial appearance of foliar late blight of potato from infected seed tubers. *American Potato Journal* 66, 226–233.

Douches, D.S., Maas, D., Jastrzebski, K. and Chase, R.W. (1996) Assessment of potato breeding progress in the USA over the last century. *Crop Science* 36, 1544–1552.

Elphinstone, J.G. and Perombelon, M.C.M. (1986) Contamination of progeny tubers of potato plants by seed-borne and leaf-borne *Erwinia carotovora*. *Potato Research* 29, 77–93.

Errampalli, D., Saunders, J.M. and Holley, J.D. (2001) Emergence of silver scurf (*Helminthosporium solani*) as an economically important disease of potato. *Plant Pathology* 50, 141–153.

Estrada, R., Gudmestad, N.C., Rivera, V.V. and Secor, G.A. (2010) *Fusarium graminearum* as a dry rot pathogen of potato in the USA: prevalence, comparison of host isolate aggressiveness and factors affecting aetiology. *Plant Pathology* 59, 1114–1120.

Fairchild, K.L., Miles, T.D. and Wharton, P.S. (2013) Assessing fungicide resistance in populations of *Alternaria* in Idaho potato fields. *Crop Protection* 49, *31–39.*

Fairclough, R.W., Bain, R.A. and Holmes, S.J. (1997) The infection of potatoes by *Phytophthora infestans* during simulated washing and its control using a hot water treatment. *Potato Research*, 40, 91–99.
Fiers, M., Edel-Hermann, V., Chatot, C., Le Hingrat, Y., Alabouvette, C. and Steinberg, C. (2012) Potato soil-borne diseases. A review. *Agronomy for Sustainable Development* 32, 93–132.
Finckh, M.R. (2008) Integration of breeding and technology into diversification strategies for disease control in modern agriculture. *European Journal of Plant Pathology* 121, 399–409.
Fitzpatrick-Peabody, E.R. and Lambert, D.H. (2011) Methodology and assessment of the susceptibility of potato genotypes to *Phytophthora erythroseptica*, causal organism of Pink Rot. *American Journal of Potato Research* 88, 105–113.
Franc, G.D. (2001) Wart. In: Stevenson, W.R., Loria, R., Franc, G.D. and Weingartner, D.P. (eds) *Compendium of Potato Diseases*. American Phytopathological Society, St Paul, Minnesota, pp. 46–47.
Frazier, M.J., Shetty, K.K., Kleinkopf, G.E. and Nolte, P. (1998) Management of silver scurf (*Helminthosporium solani*) with fungicide seed treatments and storage practices. *American Journal of Potato Research* 75, 129–135.
Frenkel, O., Yermiyahu, U., Forbes, G.A., Fry, W.E. and Shtienberg, D. (2010) Restriction of potato and tomato late blight development by sub-phytotoxic concentrations of boron. *Plant Pathology* 59, 626–633.
Frost, K.E., Groves R.L. and Charkowski, A.O. (2013) Integrated control of potato pathogens through seed potato certification and provision of clean seed potatoes. *Plant Disease* 97, 1268–1280.
Fry, B.A. and Loria, R. (2002) Thaxtomin A: evidence for a plant cell wall target. *Physiological and Molecular Plant Pathology* 60, 1–8.
Fry, W. (2008) *Phytophthora infestans*: the plant (and R gene) destroyer. *Molecular Plant Pathology* 9, 385–402.
Fry, W. and Goodwin, S. (1997a) Re-emergence of potato and tomato late blight in the United States. *Plant Disease* 81, 1349–1357.
Fry, W. and Goodwin, S. (1997b) Resurgence of the Irish potato famine fungus. *Bioscience* 47, 363–371.
Fuglie, K.O. (2007) Priorities for potato research in developing countries: results of a survey. *American Journal of Potato Research* 84, 353–365.
Gachango, E., Kirk, W., Hanson, L., Rojas, A., Tumbalam, P. and Shetty, K. (2011) First report of *in vitro* fludioxonil-resistant isolates of *Fusarium* spp. causing potato dry rot in Michigan. *Plant Disease* 95, 228.
Gachango, E., Hanson, L.E., Rojas, A., Hao, J.J. and Kirk, W.W. (2012a) *Fusarium* spp. causing dry rot of seed potato tubers in Michigan and their sensitivity to fungicides. *Plant Disease* 96, 1767–1774.
Gachango, E., Kirk, W. and Schafer, R. (2012b) Effects of in-season crop-protection combined with postharvest applied fungicide on suppression of potato storage diseases caused by oomycete pathogens. *Crop Protection* 41, 42–48.
Gachango, E., Kirk, W., Wharton, P.S. and Schafer, R. (2012c) Evaluation and comparison of biocontrol and conventional fungicides for control of postharvest potato tuber diseases. *Biological Control* 63, 115–120.
Garrett, K.A., Forbes, G.A., Savary, S., Skelsey, P., Sparks, A.H., Valdivia, C., *et al.* (2011) Complexity in climate-change impacts: an analytical framework for effects mediated by plant disease. *Plant Pathology* 60, 15–30.
Geary, B., Johnson, D.A., Hamm, P.B. and Cummings, T.F. (1999) Fungicide application for late blight management: a boom attached to a center pivot irrigation system. *Plant Disease* 83, 512–515.
Gildemacher, P.R., Demo, P., Barker, I., Kaguongo, W., Woldegiorgis, G., Wagoire, W.W., *et al.* (2009) A description of seed potato systems in Kenya, Uganda and Ethiopia. *American Journal of Potato Research* 86, 373–382.
Glass, J.R., Johnson, K.B. and Powelson, M.L. (2001) Assessment of barriers to prevent the development of potato tuber blight caused by *Phytophthora infestans*. *Plant Disease* 85, 521–528.
Goicoechea, N. (2009) To what extent are soil amendments useful to control Verticillium wilt? *Pest Management Science* 65, 831–839.
Goodwin, S., Sujkowski, L. and Fry, W. (1996) Widespread distribution and probable origin of resistance to metalaxyl in clonal genotypes of *Phytophthora infestans* in the United States and western Canada. *Phytopathology* 86, 793–800.
Gudmestad, N.C., Taylor, R.J. and Pasche, J.S. (2007) Management of soilborne diseases of potato. *Australasian Plant Pathology* 36, 109–115.
Guenthner, J., Michael, K. and Nolte, P. (2001) The economic impact of potato late blight on US growers. *Potato Research* 44, 121–125.
Guenthner, J.F. (2010) Utilization. In: Bohl, W.H. and Johnson, S.B. (eds) *Commercial Potato Production in North America (Potato Association of America Handbook)*. Potato Association of America, Orono, Maine, pp. 15–17.

Guest, D. and Bompeix, G. (1990) The complex mode of action of phosphonates. *Australasian Plant Pathology* 19, 113–115.

Haas, B.J., Kamoun, S., Zody, M.C., Jiang, R.H.Y., Handsaker, R.E., Cano, L.M., *et al.* (2009) Genome sequence and analysis of the Irish potato famine pathogen *Phytophthora infestans*. *Nature* 461, 393–398.

Hamm, P.B. and Clough, H. (1999) Comparison of application methods on deposition and redistribution of chlorothalonil in a potato canopy and potential for control of late blight. *Plant Disease* 83, 441–444.

Hannukkala, A.O. (2011) Examples of alien pathogens in Finnish potato production – their introduction, establishment and consequences. *Agricultural and Food Science* 20, 42–61.

Hanson, L.E., Schwager, S.J. and Loria, R. (1996) Sensitivity to thiabendazole in *Fusarium* species associated with dry rot of potato. *Phytopathology* 86, 378–384.

Hao, J.J., Meng, Q.X., Yin, J.F. and Kirk, W.W. (2009) Characterization of a new Streptomyces Strain, DS3024, that causes potato common scab. *Plant Disease* 93, 1329–1334.

Haverkort, A.J., Struik, P.C., Visser, R.G. and Jacobsen, F.E. (2009) Applied biotechnology to combat late blight in potato caused by *Phytophthora infestans*. *Potato Research* 52, 249–264.

Haynes, K.G., Wanner, L.A., Thill, C.A., Bradeen, J.M., Miller, J., Novy, R.G., *et al.* (2010) Common scab trials of potato varieties and advanced selections at three US locations. *American Journal of Potato Research* 87, 261–276.

Henne, D., Workneh, F., Gudmestad, N. and Rush, C. (2009) Role of '*Candidatus Liberibacter*'-infected seed tubers in epidemiology of potato zebra chip. *Phytopathology* 99, S52.

Hide, G.A., Read, P.J. and Hall, S.M. (1992) Resistance to thiabendazole in *Fusarium* species isolated from potato-tubers affected by dry rot. *Plant Pathology* 41, 745–748.

Hijmans, R.J., Forbes, G.A. and Walker, T.S. (2000) Estimating the global severity of potato late blight with GIS-linked disease forecast models. *Plant Pathology* 49, 697–705.

Hirpa, A., Meuwissen, M.P.M., Tesfaye, A., Lommen, W.J.M., Lansink, A.O., Tsegaye, A., *et al.* (2010) Analysis of seed potato systems in Ethiopia. *American Journal of Potato Research* 87, 537–552.

Hirst, J.M. and Stedman, O.J. (1962) The infection of potato foliage and tubers by *Phytophthora infestans*. *Rothamsted Experimental Station Report for 1961*. Rothamsted Experimental Station, Harpenden, UK, pp. 113–114.

Hirst, J.M., Stedman, O.J., Lacey, J. and Hide, G.A. (1965) The epidemiology of *Phytophthora infestans*. IV. Spraying trials, 1959–1963, and the infection of tubers. *Annals of Applied Biology* 55, 373–395.

Hopkins, B. and Hirnyck, R. (2008) Organic potato production. In: Johnson, D.A. (ed.) *Potato Health Management*. APS Press, St Paul, Minnesota, pp. 101–108.

Horsfield, A., Wicks, T., Davies, K., Wilson, D. and Paton, S. (2010) Effect of fungicide use strategies on the control of early blight (*Alternaria solani*) and potato yield. *Australasian Plant Pathology* 39, 368–375.

Hu, C.H., Perez, F., Donohoo, R., McLeod, A.I., Myers, K., Ivors, K., *et al.* (2012) Recent genotypes of *Phytophthora infestans* in eastern USA reveal clonal populations and reappearance of mefenoxam sensitivity. *Plant Disease* 96, 1323–1330.

Ierna, A. and Tenorio, J. (2011) Effects of pre-sowing treatment on plant emergence and seedling vigour in true potato seed. *Journal of Horticultural Science and Biotechnology* 86, 467–472.

Inglis, D.A., Miller, J.S. and Johnson, D.A. (2004) Efficacy of phosphorous acid against pink rot and Pythium leak on potato. *Phytopathology* 94, S44.

Janse, J.D. and Wenneker, M. (2002) Possibilities of avoidance and control of bacterial plant diseases when using pathogen-tested (certified) or -treated planting material. *Plant Pathology* 51, 523–536.

Johnk, J.S., Jones, R.K., Shew, H.D. and Carling, D.E. (1993) Characterization of populations of *Rhizoctonia solani* AG-3 from potato and tobacco. *Phytopathology* 83, 854–858.

Johnson, D.A. (2010) Transmission of *Phytophthora infestans* from infected potato seed tubers to emerged shoots. *Plant Disease* 94, 18–23.

Johnson, D.A. and Cummings, T.F. (2009) Latent infection of potato seed tubers by *Phytophthora infestans* during long-term cold storage. *Plant Disease* 93, 940–946.

Johnson, D.A. and Dung, J.K.S. (2010) Verticillium wilt of potato – the pathogen, disease and management. *Canadian Journal of Plant Pathology* 32, 58–67.

Johnson, D.A. and Rowe, R. (2007) Preface. In: Johnson, D.A. (ed.) *Potato Health Management*. American Phytopathological Society Press, St Paul, Minnesota, pp. v–vi.

Johnson, D.A., Inglis, D.A. and Miller, J.S. (2004) Control of potato tuber rots caused by oomycetes with foliar applications of phosphorous acid. *Plant Disease* 88, 1153–1159.

Johnson, S.B. (1991) Toxic seed piece syndrome. *American Potato Journal* 68, 131.

Khandaker, M.M., Bhuiyan, M.K.A. and Khair, A. (2010) *In vitro* evaluation of fungicidal responses on the growth of pathogenic *Rhizoctonia solani* Kuhn, antagonistic binucleate *Rhizoctonia* and *Trichoderma harzianum* Rifai. *Bangladesh Journal of Botany* 39, 107–110.
King, R.R. and Calhoun, L.A. (2005) Characterization of cross-linked hydroxycinnamic acid amides isolated from potato common scab lesions. *Phytochemistry* 66, 2468–2473.
King, R.R. and Calhoun, L.A. (2010) A feruloyl trimer isolated from potato common scab lesions. *Phytochemistry* 71, 17–18.
Kirk, W.W. (2003) Tolerance of mycelium of different genotypes of *Phytophthora infestans* to freezing temperatures for extended periods. *Phytopathology* 93, 1400–1406.
Kirk, W. and Wharton, P. (2012) *Brown Leaf Spot*. Michigan State University Extension Bulletin E-3182 (http://www.potatodiseases.org/pdf/brown-leaf-spot-July-2012-E3182.pdf, accessed 7 May 2014).
Kirk, W.W., Niemira, B.A., Stein, J.M. and Hammerschmidt, R. (1999) Late blight (*Phytophthora infestans* (Mont) De Bary) development from potato seed-pieces treated with fungicides. *Pesticide Science* 55, 1151–1158.
Kirk, W.W., Niemira, B.A. and Stein, J.M. (2001) Influence of storage temperature on rate of potato tuber tissue infection caused by different biotypes of *Phytophthora infestans* (Mont.) de Bary estimated by digital image analysis. *Potato Research* 44, 86–96.
Kirk, W., Abu-El Samen, F., Tumbalam, P., Wharton, P., Douches, D., Thill, C., *et al.* (2009) Impact of different US genotypes of *Phytophthora infestans* on potato seed tuber rot and plant emergence in a range of cultivars and advanced breeding lines. *Potato Research* 52, 121–140.
Kirk, W., Gachango, E., Schafer, R. and Wharton, S. (2013) Effects of in-season crop-protection combined with postharvest applied fungicide on suppression of potato storage diseases caused by Fusarium pathogens. *Crop Protection* 51, 77–84.
Knowles, N.R. and Plissey, E.S. (2007) Maintaining tuber health during harvest, storage, and post-storage handling. In: Johnson, D.A. (ed.) *Potato Health Management*. American Phytopathological Society Press, St Paul, Minnesota, pp. 79–99.
Krause, R.A., Massie, L.B. and Hyre, R.A. (1975) BLITECAST, a computerized forecast of potato late blight. *Plant Disease Reporter* 59, 95–98.
Lacey, J. (1965) The infectivity of soils containing *Phytophthora infestans*. *Annals of Applied Biology* 59, 363–380.
Lambert, D.H. and Currier, A.I. (1997) Differences in tuber rot development for North American clones of *Phytophthora infestans*. *American Potato Journal* 74, 39–43.
Lambert, D.H., Currier, A.I. and Olanya, M.O. (1998) Transmission of *Phytophthora infestans* in cut potato seed. *American Journal of Potato Research* 75, 257–263.
Lapwood, D.H., Wellings, L.W. and Hawkins, J.H. (1973) Irrigation as a practical means to control potato common scab (*Streptomyces scabies*) – final experiment and conclusions. *Plant Pathology* 22, 35–41.
Laurila, J., Hannukkala, A., Nykyri, J., Pasanen, M., Helias, V., Garlant, L., *et al.* (2010) Symptoms and yield reduction caused by *Dickeya* spp. strains isolated from potato and river water in Finland. *European Journal of Plant Pathology* 126, 249–262.
Lazarovits, G. (2001) Management of soil-borne plant pathogens with organic soil amendments: a disease control strategy salvaged from the past. *Canadian Journal of Plant Pathology* 23, 1–7.
Lazarovits, G. (2004) Managing soilborne plant diseases through selective soil disinfestation by a knowledge-based application of soil amendments – Guest Editorial. *Phytoparasitica* 32, 427–432.
Lazarovits, G., Conn, K.L., Abbasi, P.A., Soltani, N., Kelly, W., McMillan, E., *et al.* (2008) Reduction of potato tuber diseases with organic soil amendments in two Prince Edward Island fields. *Canadian Journal of Plant Pathology* 30, 37–45.
Ledford, H. (2007). Rapid sequencer puts virus in the frame for deaths. *Nature* 447, 12–13.
Lees, A.K., Brierley, J.L., Stewart, J.A., Hilton, A.J., Wale, S.J., Gladders, P., *et al.* (2010) Relative importance of seed-tuber and soilborne inoculum in causing black dot disease of potato. *Plant Pathology* 59, 693–702.
Li, X., De Boer, S.H., Secor, G., Gourley, J., Ross, P., Rivera, V., *et al.* (2008) Are bacteria the causal agent of the zebra chip disease of potato? *Canadian Journal of Plant Pathology* 30, 377.
Liljeroth, E., Bengtsson, T., Wiik, L. and Andreasson, E. (2010) Induced resistance in potato to *Phytophthora infestans*-effects of BABA in greenhouse and field tests with different potato varieties. *European Journal of Plant Pathology* 127, 171–183.
Loria, R., Bukhalid, R., Fry, B. and King, R. (1997) Plant pathogenicity in the genus Streptomyces. *Plant Disease* 81, 836–846.

Loria, R., Kers, J. and Joshi, M. (2006) Evolution of plant pathogenicity in Streptomyces. *Annual Review of Phytopathology* 44, 469–487.

Loria, R., Bignell, D.R.D., Moll, S., Huguet-Tapia, J.C., Joshi, M.V., Johnson, E.G., *et al.* (2008) Thaxtomin biosynthesis: the path to plant pathogenicity in the genus Streptomyces. *Antonie Van Leeuwenhoek International Journal of General and Molecular Microbiology* 94, 3–10.

Loy, J.P., Riekert, S. and Steinhagen, C. (2011) Potato prices as affected by demand and yearly production: a German perspective. *American Journal of Potato Research* 88, 195–198.

MacDonald, W., Peters, R.D., Coffin, R.H. and Lacroix, C. (2007) Effect of strobilurin fungicides on control of early blight (*Alternaria solani*) and yield of potatoes grown under two N fertility regimes. *Phytoprotection* 88, 9–15.

MacGuidwin, A.E. and Rouse, D.I. (1990) Role of *Pratylenchus penetrans* in the potato early dying disease of Russet Burbank potato. *Phytopathology* 80, 1077–1082.

MacGuidwin, A.E., Knuteson, D.L., Connell, T., Bland, W.L. and Bartelt, K.D. (2012) Manipulating inoculum densities of *Verticillium dahliae* and *Pratylenchus penetrans* with green manure amendments and solarization influence potato yield. *Phytopathology* 102, 519–527.

McIntosh, A.H., Bateman, G.L. and Chamberlain, K. (1988) Substituted benzoic and picolinic acids as foliar sprays against potato common scab. *Annals of Applied Biology* 112, 397–401.

Mackenzie, D. (1981) Scheduling fungicide applications for potato late blight. *Plant Disease* 65, 394–399.

Marrone, P.G. (2002) An effective biofungicide with novel modes of action. *Pesticide Outlook* 13, 193–194.

Meng, Q.X., Yin, J.F., Rosenzweig, N., Douches, D. and Hao, J. (2012) Culture-based assessment of microbial communities in soil suppressive to potato common scab. *Plant Disease* 96, 712–717.

Miller, J.S., Hamm, P.B., Olsen, N., Geary, B.D. and Johnson, D.A. (2004) Managing silver scurf on potatoes in storage with post-harvest fungicides and disinfestants. *Phytopathology* 94, S71.

Miller, J.S., Olsen, N., Woodell, L., Porter, L.D. and Clayson, S. (2006) Postharvest applications of zoxamide and phosphite for control of potato tuber rots caused by oomycetes at harvest. *American Journal of Potato Research* 83, 269–278.

Miller, J.S., Hamm, P.B., Olsen, N., Geary, B.D and Johnson, D.A. (2011) Effect of post-harvest fungicides and disinfestants on the suppression of silver scurf on potatoes in storage. *American Journal of Potato Research* 88, 413–423.

Moller, K., Dilger, M., Habermeyer, J., Zinkernagel, V., Flier, W.G. and Hausladen, H. (2009) Population studies on *Phytophthora infestans* on potatoes and tomatoes in southern Germany. *European Journal of Plant Pathology* 124, 659–672.

Munyaneza, J.E., Crosslin, J.M. and Upton, J.E. (2007) Association of *Bactericera cockerelli* (Homoptera : Psyllidae) with 'zebra chip', a new potato disease in southwestern United States and Mexico. *Journal of Economic Entomology* 100, 656–663.

Munyaneza, J.E., Sengoda, V.G., Crosslin, J.M., De la Rosa-Lozano, G. and Sanchez, A. (2009) First Report of '*Candidatus Liberibacter psyllaurous*' in potato tubers with zebra chip disease in Mexico. *Plant Disease* 93, 552.

Munyaneza, J.E., Buchman, J.L., Sengoda, V.G., Fisher, T.W. and Pearson, C.C. (2011) Susceptibility of selected potato varieties to zebra chip potato disease. *American Journal of Potato Research* 88, 435–440.

Nolte, P., Bertram, M., Bateman, M. and McIntosh, C.S. (2003) Comparative effects of cut and treated seed tubers vs untreated whole seed tubers on seed decay, Rhizoctonia stem canker, growth, and yield of Russet Burbank potatoes. *American Journal of Potato Research* 80, 1–8.

Norman, D.J., Zapata, M., Gabriel, D.W., Duan, Y.P., Yuen, J.M.F., Mangravita-Novo, A., *et al.* (2009) Genetic diversity and host range variation of *Ralstonia solanacearum* strains entering North America. *Phytopathology* 99, 1070–1077.

Nyankanga, R.O., Olanya, O.M., Ojiambo, P.S., Wien, H.C., Honeycutt, C.W. and Kirk, W.W. (2011) Validation of tuber blight (*Phytophthora infestans*) prediction model. *Crop Protection* 30, 547–553.

Ocamb, C.M., Hamm, P.B. and Johnson, D.A. (2007) Benzimidazole resistance of *Fusarium* species recovered from potatoes with dry rot from storages located in the Columbia basin of Oregon and Washington. *American Journal of Potato Research* 84, 169–177.

Olanya, O.M., Ojiambo, P.S., Nyankanga, R.O., Honeycutt, C.W. and Kirk, W.W. (2009) Recent developments in managing tuber blight of potato (*Solanum tuberosum*) caused by *Phytophthora infestans*. *Canadian Journal of Plant Pathology* 31, 280–289.

Olsen, N. (2010) Storage. In: Bohl, W.H. and Johnson, S.B. (eds) *Commercial Potato Production in North America (Potato Association of America Handbook)*. Potato Association of America, Orono, Maine, p. 81.

Olsen, N. and Miller, J.S. (2005) Storage management option for disease control. Report. University of Idaho, Twin Falls and Aberdeen, Idaho.

Olsen, N., Kleinkopf, G. and Woodell, L. (2003) Efficacy of chlorine dioxide for disease control on stored potatoes. *American Journal of Potato Research* 80, 387–395.

Padgham, J. (2009) *Agricultural Development Under a Changing Climate: Opportunities and Challenges for Adaptation*. The World Bank, Washington, DC (http://www.preventionweb.net/files/11832_climatechangecombined1.pdf, accessed 7 May 2014).

Pasche, J.S. and Gudmestad, N.C. (2008) Prevalence, competitive fitness and impact of the F129L mutation in *Alternaria solani* from the United States. *Crop Protection* 27, 427–435.

Pasche, J.S., Wharam, C.M. and Gudmestad, N.C. (2004) Shift in sensitivity of *Alternaria solani* in response to Q(o)I fungicides. *Plant Disease* 88, 181–187.

Pasche, J.S., Taylor, R.J. and Gudmestad, N.C. (2010) Colonization of potato by *Colletotrichum coccodes*: effect of soil infestation and seed tuber and foliar inoculation. *Plant Disease* 94, 905–914.

Pavlista, A.D. and Feuz, D.M. (2005) Potato prices as affected by demand and yearly production. *American Journal of Potato Research* 82, 339–343.

Perombelon, M.C.M. (1992) Potato blackleg – epidemiology, host pathogen interaction and control. *Netherlands Journal of Plant Pathology* 98, 135–146.

Perombelon, M.C.M. (2002) Potato diseases caused by soft rot erwinias: an overview of pathogenesis. *Plant Pathology* 51, 1–12.

Peters, J.C., Lees, A.K., Cullen, D.W., Sullivan, L., Stroud, G.P. and Cunnington, A.C. (2008a) Characterization of *Fusarium* spp. responsible for causing dry rot of potato in Great Britain. *Plant Pathology* 57, 262–271.

Peters, R.D., Platt, H.W., Drake, K.A., Coffin, R.H., Moorehead, S., Clark, M.M., *et al.* (2008b) First report of fludioxonil-resistant isolates of *Fusarium* spp. causing potato seed-piece decay. *Plant Disease* 92, 172.

Pitman, A.R., Drayton, G.M., Kraberger, S.J., Genet, R.A. and Scott, I.A.W. (2011) Tuber transmission of '*Candidatus Liberibacter solanacearum*' and its association with zebra chip on potato in New Zealand. *European Journal of Plant Pathology* 129, 389–398.

Porter, L.D., Dasgupta, N. and Johnson, D.A. (2005) Effect of tuber depth and soil moisture on infection of potato tubers in soil by *Phytophthora infestans*. *Plant Disease* 89, 146–152.

Powelson, M.L. and Rowe, H.C. (2008) Managing diseases caused by seedborne and soilborne fungi and fungus-like pathogens. In: Johnson, D.A. (ed.) *Potato Health Management*. APS Press, St Paul, Minnesota, pp. 183–195.

Read, P.J. and Hide, G.A. (1988) Effects of inoculum source and irrigation on black dot disease of potatoes (*Colletotrichum coccodes* (Wallr.) Hughes) and its development during storage. *Potato Research* 31, 493–500.

Read, P.J. and Hide, G.A. (1995) Development of black dot disease (*Colletotrichum coccodes* (Wallr.) Hughes) and its effects on the growth and yield of potato plants. *Annals of Appled Biology* 127, 57–72.

Ritchie, F., Bain, R.A. and McQuilken, M.P. (2009) Effects of nutrient status, temperature and pH on mycelial growth, sclerotial production and germination of *Rhizoctonia solani* from potato. *Journal of Plant Pathology* 91, 589–596.

Rodriguez, D.A., Secor, G.A., Gudmestad, N.C. and Francl, L.J. (1996) Sporulation of *Helminthosporium solani* and infection of potato tubers in seed and commercial storages. *Plant Disease* 80, 1063–1070.

Rojas, A., Gachango, E., Douches, D., Hanson, L. and Kirk, W. (2014) Tuber blight development in potato cultivars in response to different genotypes of *Phytophthora infestans*. *Journal of Phytopathology* 162, 33–42, doi:10.1111/jph.12153.

Rosenzweig, N., Atallah, Z.K., Olaya, G. and Stevenson, W.R. (2008a) Evaluation of QoI fungicide application strategies for managing fungicide resistance and potato early blight epidemics in Wisconsin. *Plant Disease* 92, 561–568.

Rosenzweig, N., Olaya, G., Atallah, Z.K., Cleere, S., Stanger, C. and Stevenson, W.R. (2008b) Monitoring and tracking changes in sensitivity to azoxystrobin fungicide in *Alternaria solani* in Wisconsin. *Plant Disease* 92, 555–560.

Rotem, J. (1994) *The Genus Alternaria – Biology, Epidemiology, and Pathogenicity*. The American Phytopathological Society, St Paul, Minnesota.

Runno-Paurson, E., Fry, W.E., Remmel, T., Mand, M. and Myers, K.L. (2010) Phenotypic and genotypic characterisation of Estonian isolates of *Phytophthora infestans* in 2004–2007. *Journal of Plant Pathology* 92, 381–390.

Sands, P.J., Hackett, C. and Nix, H.A. (1979) A model of the development and bulking of potatoes (*Solanum tuberosum* L.). *Field Crops Research* 2, 309–331.

Savary, S., Nelson, A., Sparks, A.H., Willocquet, L., Hodson, D., Duveiller, E., *et al.* (2011) International agricultural research tackling the effects of global and climate changes on plant diseases in the developing world. *Plant Disease* 95, 1204–1216.

Scott, G.J. (2011) Growth rates for potatoes in Latin America in comparative perspective: 1961–07. *American Journal of Potato Research* 88, 143–152.

Secor, G.A. and Gudmestad, N.C. (1999) Managing fungal diseases of potato. *Canadian Journal of Plant Pathology* 21, 213–221.

Secor, G.A. and Johnson, S.B. (2008) Seed tuber health before and during planting. In: Johnson, D.A. (ed.) *Potato Health Management*. American Phytopathological Society, St Paul, Minnesota, pp. 43–54.

Secor, G.A. and Salas, B. (2001) Fusarium dry rot and Fusarium wilt. In: Stevenson, W., Loria, R., Franc, G.D. and Weingartner, D.P. (eds) *Potato Health Management*. American Phytopathological Society Press, St Paul, Minnesota, pp. 23–25.

Secor, G.A., Rivera, V.V., Abad, J.A., Lee, I.M., Clover, G.R.G., Liefting, L.W., *et al.* (2009) Association of '*Candidatus Liberibacter solanacearum*' with zebra chip disease of potato established by graft and psyllid transmission, electron microscopy, and PCR. *Plant Disease* 93, 574–583.

Seppanen, E. (1981) Fusariums of the potato in Finland, III. Varietal resistance of potato tubers to some Fusarium species. *Annales Agriculturae Fenniae* 20, 177–183.

Shattock, R. (1988) Studies on the inheritance of resistance to metalaxyl in *Phytophthora infestans*. *Plant Pathology* 37, 4–11.

Siller, L.M.V., Covarrubias-Ramirez, J., Garzon-Tiznado, J., Zuniga-Estrada, M. and Garcia-Garza, S. (2008) Relationship between potato zebra chips symptom incidence and detections of Potato Virus Y (PVY), Potato Virus Y strain N (PVY(N)) and unbalance nutritional index. *Phytopathology* 98, S162.

Simakov, E.A., Anisimov, B.V., Yashina, I.M., Uskov, A.I., Yurlova, S.M. and Oves, E.V. (2008) Potato breeding and seed production system development in Russia. *Potato Research* 51, 313–326.

Skelsey, P., Kessel, G.J.T., Holtslag, A.A.M., Moene, A.F. and van der Werf, W. (2009a) Regional spore dispersal as a factor in disease risk warnings for potato late blight: a proof of concept. *Agriculture and Forest Meteorolology* 149, 419–430.

Skelsey, P., Kessel, G.J.T., Rossing, W.A.H. and van der Werf, W. (2009b) Parameterization and evaluation of a spatiotemporal model of the potato late blight pathosystem. *Phytopathology* 99, 290–300.

Skelsey, P., Rossing, W.A.H., Kessel, G.J.T. and van der Werf, W. (2010) Invasion of *Phytophthora infestans* at the landscape level: How do spatial scale and weather modulate the consequences of spatial heterogeneity in host resistance? *Phytopathology* 100, 1146–1161.

Slack, S. (1992) A look at potato leafroll virus and PVY: past, present, and future. *Valley Potato Grower* 57, 35–39.

Slininger, P.J., Schisler, D.A., Eirjcsson, L.D., Brandt, T.L., Frazier, M.J., Woodell, L.K., *et al.* (2007) Biological control of post-harvest late blight of potatoes. *Biocontrol Science and Technology* 17, 647–663.

Smart, C. and Fry, W. (2001) Invasions by the late blight pathogen: renewed sex and enhanced fitness. *Biological Invasions* 3, 235–243.

Sonnewald, U. (2001) Control of potato tuber sprouting. *Trends in Plant Science* 6, 333–335.

Sparks, A., Forbes, G., Hijmans, R. and Garrett, K. (2010) Mapping the future: metamodels for scaling potato late blight risk analysis in climate change scenarios. *Phytopathology* 100, S121.

Sparks, A.H., Raymundo, R., Simon, R., Forbes, G. and Garrett, K. (2008) Regional predictions of potato late blight risk in a GIS incorporating disease resistance profiles, climate change, and risk neighborhoods. *Phytopathology* 98, S149.

Stammler, G., Brix, H.D., Glattli, A., Semar, M. and Schoefl, U. (2007) Biological properties of the carboxamide boscalid including recent studies on its mode of action. In: *Proceedings of the XVI International Plant Protection Congress*, Glasgow Scotland, UK. British Crop Protection Council Publications, Alton, UK, pp. 16–21.

Staub, T. (1991) Fungicide resistance. Practical experience with antiresistance strategies and the role of integrated use. *Annual Review of Phytopathology* 29, 421–442.

Stephan, D., Schmitt, A., Carvalho, S.M., Seddon, B. and Koch, E. (2005) Evaluation of biocontrol preparations and plant extracts for the control of *Phytophthora infestans* on potato leaves. *European Journal of Plant Pathology* 112, 235–246.

Stevenson, W., Loria, R., Franc, G. and Weingartner, D. (2001) *Compendium of Potato Diseases*. American Phytopathological Society, St Paul, Minnesota.

Stevenson, W.R., Kirk, W.W. and Atallah, Z.K. (2007) Management of foliar disease, early blight, late blight and white mold. In: Johnson, D.A. (ed.) *Potato Health Management*. American Phytopathological Society Press, St Paul, Minnesota, pp. 209–222.

Sturrock, R.N., Frankel, S.J., Brown, A.V., Hennon, P.E., Kliejunas, J.T., Lewis, K.J., *et al.* (2011) Climate change and forest diseases. *Plant Pathology* 60, 33–149.

Swanson, J.K., Montes, L., Mejia, L. and Allen, C. (2007) Detection of latent infections of *Ralstonia solanacearum* race 3 biovar 2 in geranium. *Plant Disease* 91, 828–834.

Tai, H.H., Goyer, C. and Murphy, A.M. (2013) Targeted expression profiling of anthocyanin biosynthetic enzyme, *MYB* and *bHLH* transcription factor genes in potato clones with variation in anthcyanins and common scab resistance. *Botany* 91(10), 722–730.

Taylor, R.J., Salas, B. and Gudmestad, N.C. (2004) Differences in etiology affect mefenoxam efficacy and the control of pink rot and leak tuber diseases of potato. *Plant Disease* 88, 301–307.

Taylor, R.J., Pasche, J.S. and Gudmestad, N.C. (2011) Effect of application method and rate on residual efficacy of mefenoxam and phosphorous acid fungicides in the control of pink rot of potato. *Plant Disease* 95, 997–1006.

Thurston, H.D. (2004) Ten thousand years of experience with sustainable plant disease control. *Plant Disease* 88, 550–551.

Toth, I.K., Sullivan, L., Brierley, J.L., Avrova, A.O., Hyman, L.J., Holeva, M., *et al.* (2003) Relationship between potato seed tuber contamination by *Erwinia carotovora* ssp. *atroseptica*, blackleg disease development and progeny tuber contamination. *Plant Pathology* 52, 119–126.

Toth, I.K., van der Wolf, J.M., Saddler, G., Lojkowska, E., Helias, V., Pirhonen, M., *et al.* (2011) *Dickeya* species: an emerging problem for potato production in Europe. *Plant Pathology* 60, 385–399.

Tsror, L. (2010) Biology, epidemiology and management of *Rhizoctonia solani* on potato. *Journal of Phytopathology* 158, 649–658.

Tsror, L., Erlich, O. and Hazanovsky, M. (1999) Effect of *Colletotrichum coccodes* on potato yield, tuber quality, and stem colonization during spring and autumn. *Plant Disease* 83, 561–565.

Turner, R.S. (2005) After the famine: plant pathology, *Phytophthora infestans*, and the late blight of potatoes, 1845–1960. *Historical Studies in the Physical and Biological Sciences* 35, 341–370.

Vargas, A.M., Ocampo, L.M.Q., Cespedes, M.C., Carreno, N., Gonzalez, A., Rojas, A., *et al.* (2009) Characterization of *Phytophthora infestans* populations in Colombia: first report of the A2 mating type. *Phytopathology* 99, 82–88.

Venkataramana, C., Taylor, R.J., Pasche, J.S. and Gudmestad, N.C. (2010) Prevalence of mefenoxam resistance among *Phytophthora erythroseptica* Pethybridge isolates in Minnesota and North Dakota. *American Journal of Potato Research* 87, 521–530.

Von Braun, J. (2009) Addressing the food crisis: governance, market functioning, and investment in public goods. *Food Security* 1, 9–15.

Wale, S.J. (2004) Integrating knowledge of soilborne pathogens to minimise disease risk in potato production. *Australasian Plant Pathology* 33, 167–172.

Wale, S., Hilton, A., Lees, A., Breirley, J., Peters, J., Bradshaw, N., *et al.* (2008) Developing effective control measures for the control of black dot. Project Report, UK Potato Council, 2008. Agriculture and Horticulture Development Board, Oxford, UK.

Wallin, J.R. (1962) Summary of recent progress in predicting late blight epidemics in the United States and Canada. *American Potato Journal* 39, 306–312.

Wallin, J.R. and Schuster, M.L. (1960) Forcasting potato late blight in western Nebraska. *Plant Disease Reporter* 44, 896–900.

Wang, P.J. and Hu, C. (1985) Potato tissue culture and its applications in agriculture. In: Li, P.H. (ed.) *Potato Physiology*. Academic Press, Inc, London, pp. 503–577.

Wanner, L. (2005) Diversity in Streptomycetes causing potato common scab. *Phytopathology* 95, S109.

Wanner, L. (2006) A survey of genetic variation in Streptomyces isolates causing potato common scab in the United States. *Phytopathology* 96, 1363–1371.

Wanner, L. (2007) A new strain of Streptomyces causing common scab in potato. *Plant Disease* 91, 352–359.

Wanner, L. (2009) A patchwork of Streptomyces species isolated from potato common scab lesions in North America. *American Journal of Potato Research* 86, 247–264.

Wanner, L., Kirk, W. and Xinshun, Q. (2014) Field efficacy of non-pathogenic Streptomyces species against potato common scab. *Journal of Applied Microbiology* 116, 123–133, doi:10.1111/jam.12336.

Webster, B., Crockford, A. and Gevens, A.J. (2011) Evaluating fumigation and at-plant treatments for control of potato common scab in Wisconsin, 2011. *Plant Disease Management Reports*, American Phytopathological Society, St Paul, Minnesota, p. V096.

Wharton, P. and Kirk, W.W. (2007) *Early Blight*. Michigan State University Extension Bulletin E-2991 (http://www.potatodiseases.org/pdf/early-blight-bulletin.pdf, accessed 7 May 2014).

Wharton, P., Fairchild, K., Belcher, A. and Wood, E. (2012a) First report of *in-vitro* boscalid-resistant isolates of *Alternaria solani* causing early blight of potato in Idaho. *Plant Disease* 96, 454–455.

Wharton, P., Kirk, W., Berry, D. and Snapp, S. (2007a) *Rhizoctonia Stem Canker and Black Scurf of Potato. Potato Diseases*. Michigan State University Extension Bulletin, E- 2994. (http://www.potatodiseases.org/pdf/rhizoctonia-bulletin.pdf, accessed 7 May 2014).

Wharton, P.S., Kirk, W.W., Berry, D. and Tumbalam, P. (2007b) Seed treatment application-timing options for control of Fusarium decay and sprout rot of cut seedpieces. *American Journal of Potato Research* 84, 237–244.

Wharton, P.S., Kirk, W.W., Baker, K.M. and Duynslager, L. (2008) A web-based interactive system for risk management of potato late blight in Michigan. *Computers and Electronics in Agriculture* 61, 136–148.

Wharton, P.S., Kirk, W.W., Schafer, R.L. and Tumbalam, P. (2012b) Evaluation of biological seed treatments in combination with management practices for the control of seed-borne late blight in potato. *Biological Control* 63, 326–332.

Whitworth, J.L. and Novy, R.G. (2007) Potato breeding and certification in North America. *Phytopathology* 97, S153.

Woodhall, J.W., Lees, A.K., Edwards, S.G. and Jenkinson, P. (2007) Characterization of *Rhizoctonia solani* from potato in Great Britain. *Plant Pathology* 56, 286–295.

Woodhall, J.W., Lees, A.K., Edwards, S.G. and Jenkinson, P. (2008) Infection of potato by *Rhizoctonia solani*: effect of anastomosis group. *Plant Pathology* 57, 897–905.

Woodhall, J.W., Belcher, A.R., Peters, J.C., Kirk, W.W. and Wharton, P.S. (2012) First report of *Rhizoctonia solani* AG2-2IIIb infecting potato stems and stolons in the United States. *Plant Disease* 96, 460.

Young, G.K., Cooke, L.R., Kirk, W.W., Tumbalam, R., Perez, F.M. and Deahl, K.L. (2009) Influence of competition and host plant resistance on selection in *Phytophthora infestans* populations in Michigan, USA and in Northern Ireland. *Plant Pathology* 58, 703–714.

Zwankhuizen, G., Govers, F. and Zadoks, J.C. (1998) Development of potato late blight epidemics: disease foci, disease gradients, and infection sources. *Phytopathology* 88, 754–763.

12 Virus Disease Problems Facing Potato Industries Worldwide: Viruses Found, Climate Change Implications, Rationalizing Virus Strain Nomenclature, and Addressing the *Potato Virus Y* Issue

Roger A.C. Jones[1,2*]

[1]*School of Plant Biology and Institute for Agriculture, University of Western Australia, Crawley, Western Australia;* [2]*Department of Agriculture and Food, South Perth, Western Australia*

Potato is the fourth most important food crop globally after maize, wheat, and rice, and the most important root or tuber crop. It is a staple crop in many parts of the world with temperate climates, and at high altitude in tropical regions. It is also grown for food on a large scale in winter plantings in Mediterranean-type environments and subtropical regions. It plays a major role in feeding people in food-insecure regions of the world. As discussed in Chapter 2, nine different potato species are cultivated, with *Solanum tuberosum* the most important of these. *S. tuberosum* sub. sp. *andigena* is the dominant type of potato grown at the center of domestication of the crop in the Andean region of South America, but elsewhere only *S. tuberosum* sub. sp. *tuberosum* is cultivated. Although potato virus diseases and how to control them have been the subject of much research since the early days of plant virology at the beginning of the 20th century, they still limit its productivity in many countries (e.g. Jones, 1981a; de Bokx and van der Want, 1987; Loebenstein *et al.*, 2001; Stevenson *et al.*, 2001; Loebenstein and Thottappilly, 2003).

Several textbooks, compendia, or image set texts summarized the information available about potato viruses at the time they were written (e.g. de Bokx and van der Want, 1987; Loebenstein *et al.*, 2001; Stevenson *et al.*, 2001; Jones *et al.*, 2009). Also, many books, general articles, or reviews have been published on different aspects of potato virus research. Some of these focused on the diversity of potato viruses in the center of crop domestication (Fribourg, 1980; Jones, 1981a; Salazar, 2006), specific potato viruses (e.g. Jones, 1988; Singh *et al.*, 2008; Gray *et al.*, 2010), natural resistance to potato viruses (e.g. Cockerham, 1970; Valkonen, 1994; Valkonen *et al.*, 1996; Gebhart and Valkonen, 2001; Solomon-Blackburn and Barker, 2001a,b), or economic losses and biotechnological potential (Valkonen, 2007). This chapter emphasizes recent information or aspects previously not addressed fully. It starts by discussing briefly the viruses known to infect the crop, their relative importance, and the likelihood that more than the 50 or so currently known viruses will

* E-mail: roger.jones@uwa.edu.au

emerge as potato is increasingly grown in warmer, food-insecure parts of the world. It then discusses what is currently known about the likely influences of climate change in exacerbating the future incidence of potato virus diseases and altering their distributions in different potato growing regions. An analysis is provided of the current disagreement between biological and phylogenetic nomenclature systems for different strains of common potato viruses and how to approach a resolution, especially regarding *Potato virus Y* (PVY) strain group classification. The chapter then focuses on the reasons for the recent upsurge of problems with PVY in potato crops in many countries and what can be done to address this. Finally, it emphasizes areas where research on potato viruses is needed in the future, the need for surveillance over the emergence of additional potato viruses, and the importance of improving the effectiveness of seed potato health, especially in developing countries.

12.1 Viruses Infecting Potato

Like most vegetatively propagated crops, potato becomes infected with many virus diseases. These are of two main types: (i) those that primarily infect potato, and (ii) those that predominantly infect other plants but also invade potato. The first group coevolved with wild potato species in the Andean region and then invaded cultivated potatoes during their early domestication. Most of the second group only spread to the potato when it was introduced following its dispersal to new locations after European colonization of South America (e.g. Jones, 1981a, 2009; Valkonen, 2007). Currently, about 50 viruses and one viroid, *Potato spindle tuber viroid*, are known to infect the crop globally (Table 12.1). As the technology available for virus identification becomes increasingly sophisticated (e.g. Boonham *et al.*, 2003, 2014; Donaire *et al.*, 2009; Kreuze *et al.*, 2009; Jones, 2014), the numbers found are expected to continue to increase. Compounding this effect, climate change is projected to increase new encounters between introduced crops and native flora, increasing the rate of emergence of novel viruses capable of invading potato crops (e.g. Jones, 2009; Navas-Castillo *et al.*, 2011; Jones and Barbetti, 2012).

Well-known potato viruses

Some of the well-known potato viruses found in most parts of the world where potatoes are grown have become less common over time, for example *Potato virus M* and *Potato virus A* (PVA), or rare, for example *Potato aucuba mosaic virus* and *Potato yellow dwarf virus*. However, PVY, *Potato leaf roll virus* (PLRV), *Potato virus X* (PVX), and *Potato virus S* (PVS) continue to occur commonly on a global basis (Table 12.1). PLRV and PVY both cause substantial yield and tuber quality losses in many countries (e.g. Valkonen, 2007). When present alone, PVX and PVS normally cause minor yield losses, but become more important when present in mixed infections with each other, or especially when PVX co-infects plants with PVY (rugose mosaic), PVA (crinkle) or PLRV (Wright, 1977; de Bokx and van der Want, 1987; Stevenson *et al.*, 2001; Valkonen, 2007; Nyalugwe *et al.*, 2012). In the past, PLRV was generally considered the most important potato virus, especially when potato crops were grown in warmer climates (e.g. de Bokx and van der Want, 1987; Jones *et al.*, 2009). This continues to be the case in many regions with warmer climates, such as Australia (e.g. Holmes and Teakle, 1980; Wilson and Jones, 1990; Holland and Jones, 2005). In many other regions, however, PVY has recently overtaken PLRV in importance (e.g. Blanchard *et al.*, 2008; Rolland *et al.*, 2008; Gray *et al.*, 2010).

Viruses causing tuber necrosis

Viruses causing potato tuber necrosis, including necrotic rings, continue to cause serious damage in some parts of the world. In regions with warmer climates, for example in southern Australia, South Africa, Brazil, and Argentina, *Tomato spotted wilt virus* (TSWV) is an important cause of tuber necrosis (Table 12.1) (e.g. Wilson, 2001; Jones *et al.*, 2009). In regions with cold climates,

Table 12.1. Viruses found infecting the potato in different parts of the world.

Virus	Genus	Transmission	Distribution	Notes
Viruses not restricted to South America				
Potato leaf roll virus (PLRV)	Polerovirus	Aphids	Worldwide	Most important virus in warmer countries
Potato virus Y (PVY)	Potyvirus	Aphids	Worldwide	Overtaking PLRV in importance in many countries
Potato virus X (PVX)	Potexvirus	Contact	Worldwide	Common
Potato virus S (PVS)	Carlavirus	Contact, aphids	Worldwide	Common
Potato virus M (PVM)	Carlavirus	Aphids	Worldwide	Now common only in central and Eastern Europe
Potato virus A (PVA)	Potyvirus	Aphids	Worldwide	Restricted to certain cultivars
Potato virus V (PVV)	Potyvirus	Aphids	Andes, Europe	Restricted to certain cultivars
Potato aucuba mosaic virus (PAMV)	Potexvirus	Contact, aphids	Worldwide	Has become very rare
Potato mop-top virus (PMTV)	Pomovirus	Fungus	Europe, North and South America, Asia	Favored by cool, moist conditions
Potato latent virus (PLV)	Carlavirus	Aphids	North America	Rarely found
Potato virus H (PVH)	Carlavirus	Contact?	East Asia	Recently found in China
"Potato stunt virus" (PStV)	?	?	Europe	Found in Scotland, little studied
Potato yellow dwarf virus (PYDV)	Nucleorhabdovirus	Leafhoppers	North America	Has become rare
"Potato yellow mosaic virus" (PYMV)	Begomovirus	Whiteflies	Caribbean, Venezuela	= *Tomato yellow mosaic virus*
Alfalfa mosaic virus (AlMV)	Alfamovirus	Aphids	Worldwide	Spreads sporadically to potato
Beet curly top virus (BCtV)	Curtovirus	Leafhoppers	North and South America, Europe, Asia	Spreads to potato in arid environments
Cherry leaf roll virus (CLRV)	Nepovirus	Pollen?, TPS, nematodes	Europe, North and South America	Reported recently from wild potato
Cherry rasp leaf virus (CRLV)	Cheriavirus	Nematodes	North America	Found only in a potato cDNA library
Cucumber mosaic virus	Cucumovirus	Aphids	Worldwide	Only infects potato occasionally
Eggplant mottled dwarf virus (EMDV)	Nucleorhabdovirus	Aphids	Europe, North Africa, Asia	Infects potato occasionally
Groundnut bud necrosis virus (GBNV)	Tospovirus	Thrips	India, South-east Asia	Favored by subtropical conditions
Groundnut ringspot virus (GRSV)	Tospovirus	Thrips	Africa, South and North America	Infects potato occasionally

Viruses not restricted to South America				
Impatiens necrotic spot virus (INSV)	Tospovirus	Thrips	Worldwide	Recently reported infecting potato
"SB29 virus"	Potyvirus	Aphids	North America	Found in Canada, little studied
Sowbane mosaic virus (SbMV)	Sobemovirus	Contact	Worldwide	Rarely infects potato
Tobacco mosaic virus (TMV)	Tobamovirus	Contact	Worldwide	Rarely infects potato
Tobacco necrosis virus (TNV)	Necrovirus	Fungus	Worldwide	Relatively uncommon in potato
Tomato black ring virus (TBRV)	Nepovirus	Nematode	Europe	Spreads sporadically to potato
Tomato chlorosis virus (ToCV)	Crinivirus	Whiteflies	Europe, North and South America, Africa, Asia	Recently reported infecting potato in Brazil and Spain
Tomato leaf curl New Delhi virus (TLCNDV)	Begomovirus	Whiteflies	Africa, Asia	Infects potato in India
Tomato mottle Taino virus (ToMoTV)	Begomovirus	Whiteflies	Caribbean	Infects potato in Cuba
Tomato mosaic virus (ToMV)	Tobamovirus	Contact	Europe, Andean region	Rarely found in potato
Tobacco rattle virus (TRV)	Tobravirus	Nematode	Worldwide	Common in cool climates
Tomato spotted wilt virus (TSWV)	Tospovirus	Thrips	Worldwide	Commonly affects potato in regions with warm conditions
Tobacco streak virus (TSV)	Ilarvirus	Pollen via thrips	Worldwide	Reported infecting potato in Brazil
Tomato yellow fruit ring virus (TYFRV)	Tospovirus	Thrips	Iran	Recently reported infecting potato in Iran
Viruses restricted to South America				
Andean potato latent virus (APLV)	Tymovirus	Contact, beetles, TPS[a]	Andean region	Common, causes symptoms despite its name
Andean potato mild mosaic virus (APMMV)	Tymovirus	Contact, beetles, TPS?	Andean region	Recently separated from APLV
Andean potato mottle virus (APMV)	Comovirus	Contact, beetles	Andean region, Brazil	Common in potato
Arracacha virus B – Oca strain (AVB-0)	Cheravirus	TPS, pollen	Peru, Bolivia	Rarely found in potato
Potato black ringspot virus (PBRSV)	Nepovirus	TPS, nematodes	Peru	= *Tobacco ringspot virus*, potato calico strain
Potato deforming mosaic virus (PDMV)	Begomovirus	Whiteflies	Argentina, Brazil	= *Tomato yellow vein streak virus*
Potato rough dwarf virus (PRDV)	Carlavirus	Contact?, Aphids	Brazil, Uruguay, Argentina	= *Potato virus P*
Potato virus T (PVT)	Trichovirus	TPS, pollen	Andean region	Reported occasionally
Potato virus U (PVU)	Nepovirus	TPS?, nematodes	Peru	Only reported once, little studied

Continued

Table 12.1. Continued.

Virus	Genus	Transmission	Distribution	Notes
Viruses restricted to South America				
Potato yellow vein virus (PYVV)	Crinivirus	Whiteflies	Northern and central Andes	Recently expanded its range south into Peru
Potato yellowing virus (PYV)	Ilarvirus	TPS, aphids	Andean region	Only reported occasionally
Tomato chlorotic spot virus (TCSV)	Tospovirus	Thrips	South America	Infects potato in Argentina and Brazil
Tomato severe rugose virus (ToSRV)	Begomovirus	Whiteflies	Brazil	Reported recently in potato
Solanum apical leaf curl virus (SALCV)	Begomovirus	Whiteflies?	Peru	Subtropical virus, little studied
"SB26 virus"	?	Psyllids	Peru	Only reported once
Wild potato mosaic virus (WPMV)	Potyvirus	Aphids	Peru	So far, found only in wild potatoes
Viroids infecting potato				
Potato spindle tuber viroid (PSTV)	Pospiviroid	TPS, contact	Worldwide	Widespread but mostly uncommon

Notes: For detection references and more detail, see: Jones and Fribourg, 1979; Fribourg, 1980; Jones, 1981a; de Bokx and van der Want, 1987; Jeffries, 1998; Loebenstein *et al*., 2001; Morales *et al*., 2001; Brattey *et al*., 2002; Cordero *et al*., 2003; Lambert *et al*., 2003; Thompson *et al*., 2004; Urbino *et al*., 2004; Usharani *et al*., 2004; Perry *et al*., 2005; Nisbet *et al*., 2006; Ribeiro *et al*., 2006; Salazar, 2006; Golnaraghi *et al*., 2007, 2008; Valkonen, 2007; Souza-Dias *et al*., 2008; Jones *et al*., 2009; Pappu *et al*., 2009; Crosslin *et al*., 2010; Silvestre *et al*., 2011; Fortes and Navas-Castillo, 2012; Freitas *et al*., 2012; Adams *et al*., 2013; Kreuze *et al*., 2013; Li *et al*., 2013; Souza-Richards *et al*., 2014. Jones *et al*. (2009) provides images of symptoms. [a]TPS = true potato seed.

northern Europe for example, *Potato mop-top virus* (PMTV) is an important cause, while in temperate regions more generally, *Tobacco rattle virus* (TRV) is often important. The disease syndrome that PMTV and TRV cause in tubers is called "corky ringspot" in North America and "spraing" in Europe (e.g. Harrison and Jones, 1971; Stevenson *et al.*, 2001; Xu *et al.*, 2004; Santala *et al.*, 2010). In the 1980s in Europe, a tuber quality defect called potato tuber necrotic ringspot disease (PTNRD) was recognized and associated with PVY infection (e.g. Beczner *et al.*, 1984; Steinbach and Hamann, 1989). PTNRD was subsequently found in North and South America, Asia, Africa, and New Zealand (e.g. McDonald and Singh, 1996; Oshima *et al.*, 2000; Boonham *et al.*, 2002; Fletcher and Lister, 2004; Sawazaki *et al.*, 2009; Djilani-Khouadja *et al.*, 2010; Gray *et al.*, 2010; Kogovsek *et al.*, 2010). PVY has now become an important cause of viral tuber necrosis disease in most parts of the world. In Australia, potato tuber necrosis caused by TSWV has always been important (e.g. Wilson, 2001), but PVY was traditionally considered unimportant (e.g. Holmes and Teakle, 1980; Wilson and Jones, 1990; Holland and Jones, 2005). This situation may change soon following the appearance of PTNRD caused by PVY in the south-east of the continent in 2003 (Rodoni, 2005; Kehoe and Jones, 2011).

Begomoviruses and tospoviruses

Begomoviruses evolve rapidly (e.g. Navas-Castillo *et al.*, 2011) and occur widely in subtropical regions where potato is increasingly being grown to help address food insecurity issues (Table 12.1). There are increasing numbers of examples where begomoviruses invaded potato after the crop was introduced to such regions. *Potato deforming mosaic virus* (PDMV) spread to potato when it was grown in Argentina and southern Brazil, where PDMV had been known to occur for 60 years (Calderoni, 1965; Faria *et al.*, 1997). PDMV is now considered to be the same begomovirus species as *Tomato yellow vein streak virus* (Ribeiro *et al.*, 2006). Thirty years ago, *Solanum apical leaf curl virus* invaded potato when it was introduced to the subtropical high jungle of Peru (Hooker and Salazar, 1983). Recently, two new tomato-infecting begomoviruses, *Tomato severe rugose virus* and *Tomato chlorosis virus*, invaded potato plantings in southern Brazil (Souza-Dias *et al.*, 2008; Freitas *et al.*, 2012). Three begomoviruses, *Tobacco mottle Taino virus* (ToMoTV), *Tomato leaf curl New Delhi virus* (TLCNDV), and *Potato yellow mosaic virus* (PYMV), provide examples where the potato crop has been invaded in subtropical zones outside South America. ToMoTV and PYMV infect potato in the Caribbean region, and TLCNDV in the Indian subcontinent (Cordero *et al.*, 2003; Urbino *et al.*, 2004; Usharani *et al.*, 2004; Valconen, 2007; Jones *et al.*, 2009).

Like begomoviruses, tospoviruses also invade new hosts readily. Initially, TSWV was the only tospovirus known to infect potato, but over the last two decades the number found infecting the crop increased (Table 12.1). In the 1990s in India, *Groundnut bud necrosis virus* (GBNV) was recognized as the cause of a stem necrosis disease resulting in considerable losses in potato crops (e.g. Pappu *et al.*, 2009). Also in the 1990s, *Groundnut ringspot virus* and *Tomato chlorotic spot virus* (TCSV) were found infecting potato in Argentina, and TCSV in Brazil (Nagata *et al.*, 1995; Granval de Millán *et al.*, 1998; Jeffries, 1998). More recently, *Impatiens necrotic spot virus* (INSV) and *Tomato yellow fruit ring virus* were both found infecting potato in Iran (Golnaraghi *et al.*, 2008; Pourrahim *et al.*, 2012), and INSV in greenhouse-grown potato in North America (Perry *et al.*, 2005; Crosslin and Hamlin, 2010).

Other viruses

As the potato crop becomes more widely grown in countries with warm climates, it will become increasingly exposed to new encounters with viruses spreading from native plants. Such new encounters provide new opportunities for rapid, adaptive virus evolution and host species jumps (Woolhouse *et al.*, 2005; Fargette *et al.*, 2006; Jones, 2009), so the number of viruses found infecting the crop is likely to increase further. Also, the increasing rapidity and volume of trade in plant produce and movement of potato cultivars around the world is likely to increase the distribution of potato viruses formerly of restricted distribution. For example, several viruses that infect cultivated and wild potato species in the Andean region have not yet become established in other continents (Table 12.1). Some of these occur commonly in the region and cause damaging potato diseases; for example, contact and beetle transmitted *Andean potato mottle virus* (APMoV) and *Andean potato latent virus* (APLV). Others are less widespread, for example *Potato virus T* (PVT), rarely found, for example *Potato virus U*, or occur only in wild potatoes, for example *Wild potato mosaic virus*. In addition, several are seedborne in true potato seed and so are likely to spread when true seed is exchanged between countries for breeding purposes (e.g. Jones, 1981a, 1982a; Valkonen *et al.*, 1992; Valkonen, 2007). Although quarantine regulations have apparently avoided their establishment outside South America so far, vigilance needs to be maintained. *Pepino mosaic virus* (PepMV) provides an example where an indigenous virus from a *Solanum* species (*Solanum muricatum*) growing in the Andean region remained localized for >25 years after its initial detection in 1973. In 1999, it suddenly appeared in Europe. PepMV then spread quickly to most other continents (e.g. Jones *et al.*, 1980; Mumford and Jones, 2005; Jones, 2009). Moreover, among the common potato viruses, although some important strains have already spread outside the Andes, others may still do so, including PVX resistance-breaking strain PVX^{HB} (Moreira *et al.*, 1980). The Y^{N} strain of PVY provides a classic example of this kind of spread. It was introduced into Europe inadvertently in potato germplasm from the Andes about 55 years ago (Todd, 1961; Brucher, 1969). It rapidly became widespread and was soon introduced into potato growing areas in other continents through seed potato exports.

12.2 Climate Change Predictions

The anticipated influences of climate change on diverse aspects of plant viral pathogens, their vectors, and the diseases they cause, have been reviewed recently (Canto *et al.*, 2009; Jones, 2009; Jones and Barbetti, 2012). Jones and Barbetti (2012) developed frameworks for (i) each important direct and indirect climate change parameter, and (ii) each significant biological (host, vector, and pathogen) pathosystem parameter. These frameworks were then cross-checked one against the other. This analysis revealed that alterations in climate were likely to modify diverse components of viral disease in many different ways, including viral pathogen geographic ranges and relative abundance, rates of spread, effectiveness of host resistances, physiology of host–virus interactions, rate of virus evolution and host adaptation, and effectiveness of control measures. The complication of needing to consider the effects of climate change parameters on different vector types (e.g. the aphid, thrips, whitefly, leafhopper, beetle, fungal and nematode vectors of potato viruses) added a significant additional variable. Another significant variable arose from the need to consider the likely effects of climate change parameters in increasing the frequency of new encounters between introduced crops and native plants, thereby accelerating the emergence of new viruses causing potentially damaging infections (e.g. Jones, 2009; Navas-Castillo *et al.*, 2011; Jones and Barbetti, 2012).

Overall effects at different latitudes and altitudes

Depending on the type of pathosystem and circumstances, in many instances climate change seems likely to enhance potato virus disease epidemics in higher latitude (temperate) regions. Temperate potato growing regions likely to be affected include ones in Canada, northern USA, northern Europe, north-east Asia and southern South America (Norse and Gommes, 2003; Jones and Barbetti, 2012). Where potatoes are unirrigated, climate change is likely to have the opposite effect in potato growing areas in drying mid-latitude regions, for example parts of southern Australia, Central America, the Middle East, southern Europe, and North Africa. Potato-infecting viruses best adapted to warmer regions (e.g. GBNV, *Tomato chlorosis virus* (ToCV), TLCNDV, PLRV, PYMV, and *Potato yellow vein virus* (PYVV)) are likely to expand their geographical ranges from the areas they currently occupy to areas of higher latitude previously too cool for them, and to formerly cooler higher elevations in mountainous regions within the tropics or subtropics. Conversely, the geographical distributions of viruses adapted to cooler regions (e.g. APLV, PVX, PVS, PVA, PMTV, and PVT) are projected to contract to areas of higher latitude or higher elevations in mountainous regions within the tropics and subtropics, including ones previously too cold for the potato crop. For vector-borne potato viruses, such alterations would be limited if the ranges of their vectors were to remain unchanged, but the opposite scenario is predicted for key temperate region potato virus insect vectors, such as aphids, or warmer climate vectors, such as whitefly or thrips (Jones and Barbetti, 2012).

Effects on vectors and potato viruses

Aphid vectors react strongly to small changes in mean temperatures due to their low developmental threshold temperatures, short generation times, and great capacity for reproduction (Harrington, 2002; Harrington *et al.*, 2007). An additional five generations of aphids per year is predicted in temperate zones from a warming of 2°C (Yamamura and Kiritani, 1998). The risk of serious epidemics of aphid-transmitted potato viruses therefore increases as their populations and activities increase. For example, PLRV is projected to become more widespread in temperate regions under increasing mean winter and summer temperature scenarios (Boland *et al.*, 2004; Jones, 2009). Whitefly vectors also react strongly to climatic changes due to their short generation times and great capacity for reproduction. With the important whitefly virus vector species, *Bemisia tabaci*, 25–28°C is optimal for development (Wagner, 1995), and much shorter adult-to-adult generation times occur at high (31–33°C) rather than low (17°C) temperatures (Muniz and Nombela, 2001). Thus, rising mean temperatures increase the risk of damaging epidemics of viruses transmitted by *B. tabaci* in formerly cooler regions. Interestingly, at lower altitudes in the Andean region, *B. tabaci* is displacing another whitefly species, *Trialurodes vaporariorum*, while the latter species is expanding its range at higher altitudes. In turn, this shift in vector distribution is being reflected in the distribution of PYVV, which is transmitted by *T. vaporariorum* but not *B. tabaci* (Salazar *et al.*, 2000; Barker *et al.*, 2007). Increased mean temperatures can also increase thrips vector populations by hastening their development rates, leading to more generations per year. However, as different thrips species have different thermal thresholds, a change in species composition is projected from global warming. Thus, the range of the tropical vector species, *Thrips palmi*, is expected to expand into areas formerly too cold for it, displacing vector thrips species adapted to cooler temperatures (Pappu *et al.*, 2009). This, in turn, is likely to cause expansion of damaging potato-infecting tospoviruses transmitted by *T. palmi* (e.g. GBNV) into regions formerly too cool for them.

Epidemics of potato viruses transmitted by fungi or nematodes in the soil are projected to alter in response to altered temperature and rainfall. For example, epidemics of the fungus-transmitted virus, PMTV, are likely to become prevalent over increasingly wide areas in temperate regions (Jones, 2009). This is because of increased activity and movement of vector zoospores resulting from increased soil moisture and temperature.

Some research has been undertaken on the effects of unusually high temperatures on symptoms or virus content of virus-infected potato plants or tubers. For example, PVX was detected when potato plants were held at 25°C after inoculation, but not at 30°C (Adams *et al.*, 1986). Also, PLRV foliar symptom severity was diminished in potato plants kept above 20°C (Jones, 1981b), and PLRV, *Alfalfa mosaic virus*, and *Tomato black ring virus* were eliminated from diseased tubers of several potato cultivars by hot-air treatment at 37°C, but similar hot-air treatments for up to 10 weeks did not eradicate PVY (Kassanis, 1949; Kaiser, 1980). Unusually high air temperatures can also reduce insect vector populations (Jones and Barbetti, 2012). Thus, there is a possibility that, in warm countries, prolonged heatwaves resulting from climate change might decrease the incidences of some viruses in surviving potato crops or in seed tubers stored without refrigeration. There was no evidence of temperature-sensitive resistance to PVX in potato plants (Adams *et al.*, 1986).

There is apparently no information on the responses of potato plants to virus infection when elevated greenhouse gas concentrations are present in the atmosphere, but elevated CO_2 increased resistance to PVY infection in tobacco plants (Matros *et al.*, 2006). Elevated CO_2 concentrations have the potential to influence insect vector numbers, but experiments with elevated ambient CO_2 had mixed effects on aphids, the results being highly species specific. Elevated ambient CO_2 concentrations had no effect on *B. tabaci*, but affected *T. vaporiarum* negatively. Elevated CO_2 had little effect on populations of the important thrips tospovirus vector species, *Frankliniella occidentalis*, but might increase population sizes of some other thrips vector species (Jones and Barbetti, 2012).

Implications for seed potato production

Rising temperatures in formerly cooler regions are likely to have serious implications for potato seed production. This is not only because of increased aphid vector activity but also because of factors like greater survival of volunteer potatoes arising from milder winters, or the introduction of other Solanaceous crops that act as infection reservoirs for aphids or potato viruses but require warmer summer growing conditions than were available formerly (e.g. tomato, pepper). Seed production areas in some parts of the world are likely to become unsuitable for high-quality seed tuber production, necessitating moves away from formerly cooler regions, such as ones at lower altitude or in windswept coastal areas, to higher-latitude regions formerly too cool for growing potatoes, or to cooler, higher altitudes in highland regions.

Epidemiological modeling approaches have been adopted to predict PVY spread in potato crops in northern Europe (Sigvald, 1986; Doring, 2011; Kirchner *et al.*, 2011). More such models are needed with other common potato viruses and in other regions (Jones *et al.*, 2010). Moreover, modeling studies are lacking on the possible effects of different climate change scenarios on the spread of common potato viruses and their worldwide geographical distributions (e.g. Jones and Barbetti, 2012).

12.3 Rationalizing Biological and Phylogenetic Potato Virus Strain Nomenclature

Before the advent of large-scale nucleic acid sequencing technologies, strains of common potato viruses were defined biologically or serologically. Distinct differences between the symptoms that different strains induced in potato plants or virus indicator hosts were used to distinguish strain groups (i.e. pathotypes) or strains. For example, the reactions of potato cultivars differing in single gene hypersensitive resistances (differentials) were used to distinguish the different strain groups of PVX, PVY, and PVA (Cockerham, 1955; Jones, 1982b, 1990; Valkonen *et al.*, 1995). With PVY, the ability to induce necrotic symptoms in tobacco was also used in addition to potato resistance genes. With PVS, systemic invasion of the indicator host, *Chenopodium quinoa* was used to distinguish Andean from ordinary strains (Hinostroza-Orihuela, 1973; Santillan, 1979; Santillan *et al.*, 1980). With PLRV, strains were defined by the severity of the symptoms they induced in potato plants or the diagnostic indicator host, *Physalis floridana*, mild strains sometimes causing latent infections (Wright *et al.*, 1967; Chiko and Guthrie, 1969). Serological differences or specific monoclonal antibodies (MAbs) were also used to distinguish strains of some potato viruses. For example, MAbs were used to distinguish strains of common potato viruses, such as PVY (Ellis *et al.*, 1977), or APLV and APMoV in the Andean region (Fribourg *et al.*, 1977a,b; Koenig *et al.*, 1979). When viral nucleic acid sequencing and phylogenetic analysis became widely available, grouping of nucleotide sequences provided another avenue to distinguish potato virus strains. A problem then arose due to attempts to use biologically derived potato virus strain nomenclature for phylogenetically distinguished strains without knowledge of whether the two systems actually coincided or which viral genes influenced symptom formation. The paragraphs that follow discuss examples and solutions to the nomenclature confusion that has resulted.

Potato virus S strains

With PVS, when the biologically defined Andean strain (PVS^A) was found outside the Andean region (e.g. in Europe, North America, and New Zealand) and the coat protein (CP) genes of three European isolates were sequenced, these grouped with ordinary strain PVS isolates (PVS^O). By contrast, when the CPs of three other PVS isolates defined as Andean biologically (from Peru, India, and an unspecified country) were sequenced, they grouped together in a separate clade also called "Andean" (PVS^A) (Cox and Jones, 2010a). The acronym PVS$^{O\text{-}CS}$ (CS = "*Chenopodium* systemic") was therefore suggested for isolates that invaded *Chenopodium* spp. systemically but behaved as PVS^O on phylogenetic analysis (Matoušek *et al.*, 2005). Should any isolates be found in the future that fit within PVS^O biologically but within PVS^A on phylogenetic analysis, the acronym PVS$^{A\text{-}CL}$ (CL = "*Chenopodium* localized") has been suggested (Cox and Jones, 2010a).

Potato virus X strains

With PVX, when the CP genes of isolates defined biologically by the symptoms elicited in the presence of hypersensitivity genes *Nx* and *Nb* were sequenced, the two clades found were named B and X. This was on the assumption that type B isolates, but not type X isolates, overcame *Nx* (Santa-Cruz and Baulcombe, 1995). Malcuit *et al.* (2000) found strain groups 1 and 3 within X, and strain groups 2 and 4 within B, which was still consistent with this suggestion. However, Cox and Jones (2010b) subsequently found that isolates belonging to biological strain group 4 (which overcomes *Nx* and *Nb*) could also occur in clade X, showing that the biologically and phylogenetically defined PVX strain groupings did not necessarily coincide. This finding supported the change in PVX clade nomenclature from X to I and from B to II made previously by others (e.g. Komatsu *et al.*, 2005; Yu *et al.*, 2008). Clade I isolates were from six different continents, but clade II subclades II-1 and II-2 were only from Europe (II-1) or North and South America (II-2) (Cox and Jones, 2010b). The PVX genes that elicit the different host PVX resistance genes are known: the PVX 25-kDa movement protein gene elicits hypersensitivity in plants with *Nb* but the CP gene elicits hypersensitivity in plants with *Nx*, and extreme resistance in plants with gene *Rx* (Kavanagh *et al.*, 1992; Goulden *et al.*, 1993; Bendahmane *et al.*, 1995; Malcuit *et al.*, 1999).

Potato virus Y strains

PVY and *Potato virus C* (PVC) were first recognized in 1931 and 1936, respectively, being distinguished because PVC caused a hypersensitive phenotype in some potato cultivars (Bawden, 1936). In 1944, PVC was included within PVY, becoming strain group Y^C (Bawden and Sheffield, 1944), and additional strain group PVY^N was recognized (Nobrega and Silberschmidt, 1944). Original or "ordinary" PVY then became strain group PVY^O (Singh *et al.*, 2008). In the 1980s–1990s, two further PVY strain groups were identified, PVY^Z (Jones, 1990), followed by PVY^{ZE} (Kerlan *et al.*, 1999), later renamed PVY^E (Singh *et al.*, 2008). These five strain groups are defined by phenotypes induced by inoculation to potato cultivar differentials with known hypersensitivity genes *Nc*, Ny_{tbr}, and (putative) *Nz*, and to tobacco plants. PVY^C elicits *Nc*, PVY^O elicits Ny_{tbr}, Y^Z elicits *Nz*, and PVY^N elicits necrosis in tobacco. In contrast, PVY^E overcomes *Nc*, Ny_{tbr}, and *Nz*, and does not induce necrosis in tobacco (Jones, 1990; Kerlan *et al.*, 1999, 2011; Singh *et al.*, 2008). The PVY gene that elicits host genes *Nc* and Ny_{tbr} is the helper component proteinase (HC-Pro) (Moury *et al.*, 2011). HC-Pro is also involved in eliciting the veinal necrosis phenotype in tobacco (Hu *et al.*, 2009b; Tian and Valkonen, 2013), but the nuclear inclusion (protease) gene (NIa) is involved in eliciting gene *Ry* (Mestre *et al.*, 2000, 2003). The PVY gene that elicits *Nz* remains unidentified. However, "biologically" defined, so-called "PVY^O" isolates, are often just PVY isolates that produce no necrosis in tobacco, because, with a few exceptions (e.g. Singh *et al.*, 2008; Dullemans *et al.*, 2011; Galvino-Costa *et al.*, 2011; Kehoe and Jones, 2011; Kerlan *et al.*, 2011; Karasev and Gray, 2013a,b; Tian and Valkonen, 2013), much of the recent PVY literature ignores strain groups PVY^C, PVY^Z, and PVY^E. In addition, isolates have often been defined as "PVY^O" by immunological tests with strain-specific antibodies without any recourse to inoculation to potato differentials (e.g. Nie *et al.*, 2004; Hu *et al.*, 2009a,b). Strain-specific MAbs do not distinguish between biologically defined strain groups PVY^O, PVY^Z, and PVY^E, and Karasev *et al.* (2010) documented PVY^N misidentification by immunological tests. Moreover, a group of isolates called PVY^{O5} (Ellis *et al.*, 1977, 1996; Baldauf *et al.*, 2006) was distinguished from others only by its reactions with MAbs. Because, as mentioned above, the proteinase (HC-Pro) rather than the CP gene is responsible for recognition of PVY by hypersensitive resistance genes or the veinal necrosis phenotype in tobacco (e.g. Tian and Valkonen, 2013), use of serological criteria alone is unlikely to be helpful in distinguishing PVY strain groups (e.g. Singh *et al.*, 2008).

When PTNRD was first recognized (see above), it was associated with PVY^{N} (e.g. Beczner *et al.*, 1984; Steinbach and Hamann, 1989; McDonald and Singh, 1996). PVY^{N} isolates causing PTNRD were therefore named PVY^{NTN} (Le Romancer *et al.*, 1994). When the nucleotide sequences of some PVY^{NTN} isolates were analyzed, they proved to be recombinants between "PVY^{O}" and PVY^{N} isolates in the CP encoding region, with additional recombination junctions in the HC-Pro and NIa and NIb encoding regions (e.g. Glais *et al.*, 1998, 2002; Boonham *et al.*, 1999, 2002). However, other PVY^{NTN} isolates had no recombination junctions within their CPs (Oshima *et al.*, 2000; Boonham *et al.*, 2002; Schubert *et al.*, 2007). The name PVY^{NTN} (or $PVY^{Eu\text{-}NTN}$) was used for isolates detected by a PVY^{NTN} primer set from Europe (Weidemann and Maiss, 1996). PVY^{NTN} isolates not detected by this primer set and that lacked recombination junctions within their CP were called North American PVY^{NTN} (Weilguny and Singh, 1998), or $PVY^{NA\text{-}N/NTN}$ (Nie and Singh, 2002, 2003a,b).

Based on their initial detection in Europe in cultivar Wilga, another group of PVY^{N} isolates that differed in virulence in potato was named $PVY^{N\text{-}Wi}$ (Chrzanowska, 1991). Later, $PVY^{N\text{-}Wi}$ isolates were found in North America. They resembled PVY^{NTN} in sometimes causing PTNRD and in being recombinants between "PVY^{O}" and PVY^{N}, but differed from it in reacting with PVY^{O} specific MAbs, despite causing veinal necrosis in tobacco (Singh *et al.*, 2008). $PVY^{N\text{-}Wi}$ isolates usually had one or two recombination junctions. Ones with a single recombination junction were renamed $PVY^{N:O}$ (Nie and Singh, 2003a,b). $PVY^{N\text{-}Wi}$ isolates then had two (Glais *et al.*, 2002) or, occasionally, four recombination junctions (Schubert *et al.*, 2007). However, the names $PVY^{N:O}$ and $PVY^{N\text{-}Wi}$ have sometimes been used interchangeably (e.g. Nie and Singh, 2002; Nie *et al.*, 2004; Piche *et al.*, 2004; Lorenzen *et al.*, 2006; Karasev *et al.*, 2011). In Europe, PVY^{NTN} has displaced "PVY^{O}" as the predominant PVY strain in the past 25 years (Rolland *et al.*, 2008). In North America, "PVY^{O}" is gradually being displaced by PVY^{NTN} and $PVY^{N\text{-}Wi}$ (Gray *et al.*, 2010; Karasev and Gray, 2013a,b). A further isolate grouping called $PVY^{NTN\text{-}NW}$ was recently found that falls between $PVY^{N\text{-}Wi}$ and PVY^{NTN} (Chikh Ali *et al.*, 2010). Other types of recombinants are also being found; for example, NE-11 is a recombinant between PVY^{N} and an unknown PVY variant (Lorenzen *et al.*, 2008), and isolate nnp is a PVY^{O}/PVY^{N}/PVY^{C} recombinant (Schubert *et al.*, 2007). Most recombinant isolates behave like PVY^{N} in inducing necrosis in tobacco (Singh *et al.*, 2008), but some do not (Hu *et al.*, 2009b; Robles-Hernandez *et al.*, 2010).

Although a primary characteristic of PVY^{NTN} isolates was considered to be PTNRD induction, infected indicator potato cultivars for this syndrome sometimes lacked PTNRD in some or all tubers (Beczner *et al.*, 1984; Xu *et al.*, 2005; Singh *et al.*, 2008). Moreover, not only can isolates of $PVY^{N\text{-}Wi}$ cause PTNRD, but also "PVY^{O}" and $PVY^{NTN\text{-}NW}$ isolates sometimes do this (Boonham *et al.*, 2002; Piche *et al.*, 2004; Baldauf *et al.*, 2006; Schubert *et al.*, 2007; Chikh-Ali *et al.*, 2010; Gray *et al.*, 2010). Thus, PVY^{NTN} is an unfortunate name for a phylogenetic grouping as the PTNRD syndrome is also caused by isolates belonging to several other PVY strain groupings, and PVY^{NTN} isolates may not cause it. Moreover, the environmental conditions responsible for reliable induction of necrotic tuber symptoms have yet to be defined, and information on which viral genes influence PTNRD formation is lacking (Singh *et al.*, 2008). CP and HC-Pro regions are unlikely to be involved (Glais *et al.*, 2002; Schubert *et al.*, 2007).

A misconception arose that PVY^{C} was not readily aphid-borne because an old attenuated stock culture, which had lost its ability to be aphid-borne during many years of serial transfers by sap inoculation, was used in well-publicized "helper" transmission studies with aphids and PVY^{O} (Kassanis and Govier, 1971a,b). In fact, PVY^{C} infects potato crops more often than currently appreciated, but is usually confused with PVY^{O} unless isolates are inoculated to potato differentials with gene *Nc* (e.g. Jones, 1987, 1990). Although relatively few PVY^{C} isolates from potato have been compared biologically and phylogenetically, so far there is apparent agreement between PVY^{C} isolates defined by inoculation to a potato differential with *Nc* and their phylogenetic placement within the PVY^{C} clade (e.g. Dullemans *et al.*, 2011; Kehoe and

Jones, 2011). Phylogenetic placement of biologically defined PVY^{Z} and PVY^{E} isolates was unknown until 2011, when whole genome sequences of PVY^{Z} isolate L26 (Kerlan *et al.*, 2011) and PVY^{E} isolate Mon (Galvino-Costa *et al.*, 2011) were placed within phylogenetically defined PVY^{NTN} and a separate clade between PVY$^{N/NA-N}$ and PVY^{NTN} lineages, respectively. PVY-L26 is a recombinant between "PVY^{O}" and PVY^{N} and PVY-Mon between PVY^{NTN} and PVY-NE11 (Hu *et al.*, 2009a,b; Galvino-Costa *et al.*, 2011; Kerlan *et al.*, 2011). The original PVY^{E} type isolate (Kerlan *et al.*, 1999; Singh *et al.*, 2008) remains unsequenced. However, when the CP genes of three original PVY^{Z} isolates were sequenced, they grouped with "PVY^{O}" or PVY^{N-Wi} (Kehoe and Jones, 2011). Thus, PVY^{Z} may fit into both recombinant and non-recombinant groupings.

Knowledge that (i) few phylogenetically defined "PVY^{O}" isolates have been shown to belong to biologically defined PVY^{O}, (ii) many isolates are recombinants between "PVY^{O}" and PVY^{N}, and (iii) that PVY^{Z} may fit within three distinct phylogenetic groupings which include "PVY^{O}" suggests that use of the name PVY^{O} for both resistance and phylogenetic PVY strain groupings should be reconsidered. Use of PVY^{O} as a resistance grouping has historical precedence. Kehoe and Jones (2011) suggested use of PVY^{O} be restricted to isolates defined biologically by their reactions with gene Ny_{tbr}, and that a letter not used previously be employed instead (Q), giving the designation PVY^{Q} for the phylogenetic grouping currently known as "PVY^{O}", and the designation PVY$^{N:Q}$ for the phylogenetic grouping PVY$^{N:O}$. Kerlan *et al.* (2011) combined resistance and phylogenetic groupings for PVY^{Z} isolate L26, designating it PVY$^{Z/NTN}$. This approach provides a useful means of avoiding confusion caused by the different nomenclature systems; for example, the historical PVY^{O} and PVY^{Z} isolates sequenced by Kehoe and Jones (2011) would then become PVY$^{O/Q}$, PVY$^{Z/Q}$, or PVY$^{Z/N-Wi}$. However, sequence characterization of whole PVY genomes is still needed with these historical PVY^{Z} isolates.

Singh *et al.* (2008) suggested that newly found PVY isolates should be described within the context of the original strain groupings based on host resistance, and not named using geographical, cultivar, or place-association designations. However, within a strain group, amendment of isolate names using additional codes to show that the isolate differed at the molecular, serological, or phenotypic level was acceptable. They recognized five strain groups based on host resistance (PVY^{C}, PVY^{O}, PVY^{E}, PVY^{Z}, PVY^{N}). They also recognized seven strains based on a combination of host resistance and whole genome sequencing (PVY^{C}, PVY^{O}, PVY^{E}, PVY^{Z}, PVY^{N}, PVY^{NTN}, PVY^{N-Wi}/PVY$^{N:O}$). The two recombinant strains (PVY^{NTN}, PVY^{N-Wi}/PVY$^{N:O}$) were placed within strain group PVY^{N}. They also suggested that a more systematic investigation and characterization of PVY at the biological and molecular levels should eventually result in a more meaningful strain concept. Karasev and Gray (2013a) suggested an increase in the number of PVY strains to nine (PVY^{C}, PVY^{O}, PVY^{E}, PVY^{Z}, PVY^{N}, PVY^{N-Wi}, PVY$^{N:O}$, PVY^{NA-N}, PVY-NE11). This change resulted from: (i) treating the ability to induce veinal necrosis in tobacco as a subordinate trait to reactions with potato cultivar differentials; (ii) including PVY^{NTN} within PVY^{Z}; (iii) separating the former PVY^{N-WI} strain into PVY^{N-WI} and PVY$^{N:O}$; and (iv) adding new strains PVY$^{NA:N}$ and PVY-NE11. Treating the ability to induce veinal necrosis in tobacco as a subordinate trait was because the viral genetic determinants involved in the induction of veinal necrosis in tobacco were shown to be distinct from the viral genetic determinants involved in the induction of hypersensitive resistance or PTNRD in potato (e.g. Hu *et al.*, 2009b; Galvino-Costa *et al.*, 2011; Kerlan *et al.*, 2011; Moury *et al.*, 2011; Faurez *et al.*, 2012; Tian and Valkonen, 2013). Combining PVY^{NTN} with PVY^{Z} was based on findings that potato differentials reacted similarly to both (Barker *et al.*, 2009; Galvino-Costa *et al.*, 2011; Kerlan *et al.*, 2011). Separation of the former PVY^{N-WI} strain into PVY^{N-WI} and PVY$^{N:O}$, and the separate classification of PVY^{NA-N} and PVY-NE11, was because these strains all had molecular characteristics clearly distinct from those of other PVY strains (Nie and Singh, 2003a; Lorenzen *et al.*, 2006, 2008). An improved multiplex IC-RT-PCR assay was developed that distinguished these nine PVY strains (Chikh-Ali *et al.*, 2013). However, PVY^{N-WI}, PVY$^{N:O}$, PVY^{NA-N}, and PVY−NE11 all still lacked biological typing data using potato cultivar differentials. As mentioned above, Kehoe and Jones

(2011) used the designation PVY^{Q} to distinguish phylogenetic "PVY^{O}" from biological PVY^{O}, and suggested the use of $PVY^{N:Q}$ for the phylogenetic grouping $PVY^{N:O}$. An alternative to using distinguishing letters for phylogenetic PVY groupings would be to employ Latinized numerals (e.g. I–V) and number subclades (e.g. I-1, I-2, etc.) instead, as currently done with PVX (see above). This approach would help eliminate the problems caused by the current phylogenetic PVY strain nomenclature, and provide a clearer picture of the biological characteristics of isolates based on potato differentials. Karasev and Gray (2013a) emphasize that further refinement of PVY classification systems may be expected in the future, once more information becomes available. Future research on PVY diversity should include PVY isolates from the Andean potato domestication center, as it is also the center of diversity of common potato viruses (e.g. Jones, 1981a; Spetz *et al.*, 2003). Sequencing of Andean PVY isolates and their strain group typing with potato differentials would help to establish the extent of this diversity and identify any additional PVY strains groups, for example the unknown PVY variant that recombined with Y^{N} within isolate NE-11 (Lorenzen *et al.*, 2006).

Providing a final resolution to the PVY strain nomenclature system issue is of critical importance to the PVY research community, and of considerable importance to post-entry quarantine testing and breeding to incorporate strain-specific PVY resistance genes (e.g. *Ny*) into new potato cultivars. However, it has less relevance to situations where large-scale, routine testing for PVY is undertaken, such as in standard seed potato certification, diagnostic laboratories, or breeding to incorporate extreme resistance gene *Ry*. This is because the priority for such activities is normally to establish virus occurrence regardless of strain, rather than determine what PVY strain is actually present. For this purpose, large-scale serological tests using generic PVY monoclonal or polyclonal antibodies are generally adequate to detect the virus reliably. In the future, as technological innovation advances, alternative approaches to large-scale, routine testing by serology are likely to evolve more efficient and sensitive assays that displace it (Boonham *et al.*, 2014; Jones, 2014).

12.4 Addressing the PVY Problem

The factors responsible for the evolution of recombinant PVY strains that cause tuber necrosis and the current upsurge in PVY incidence in potato crops in many countries have been discussed in recent reviews and research articles (e.g. Valkonen, 2007; Kerlan and Moury, 2008; Rolland *et al.*, 2008; Singh *et al.*, 2008; Gray *et al.*, 2010; Karasev and Gray, 2013a,b; Visser *et al.*, 2013; MacKenzie *et al.*, 2014). The main contributory factors in different parts of the world were: (i) lack of sufficient potato cultivars carrying comprehensive (non-strain-specific) PVY resistance (gene *Ry*); (ii) increased survival due to milder winters of volunteer potatoes that acted as PVY sources in seed potato fields; (iii) increased aphid vector numbers late in the growing season, resulting in unforeseen late PVY spread to seed tubers; (iv) complacency over the effectiveness of seed potato programs, resulting in less emphasis being placed on traditional virus control measures; (v) widespread release by potato breeding programs of new potato cultivars that were symptomless PVY carriers or developed very mild foliage symptoms; and (vi) an alteration in the genetic composition of PVY strains, leading to the development of genetic recombinants that caused mild or symptomless infection in potato foliage. Factors (iii–vi) meant that PVY infection was often missed on visual inspection of seed crops, resulting in virus incidences that built up undetected in each successive seed crop and spread to other cultivars, including sensitive ones in which substantial yield losses occurred (e.g. Jones, 2006; Valkonen, 2007; Gray *et al.*, 2010).

Controlling PVY effectively during seed potato production needs to focus on removing or minimizing the initial virus infection source and protecting the crop from migrant viruliferous aphids that can reintroduce the virus from external sources and spread it within seed crops. Relaxation of traditional virus control measures in response to commercial pressures to reduce seed production costs, especially isolation distances between generations or from non-certified potato crops, and lack of strict adherence to virus tolerance limits should be resisted at all costs (Jones, 2006). Wherever possible, seed production should be moved away from farms producing ware crops in more aphid-prone regions. Relying on visual inspections alone during seed certification will not be sufficient to identify all infected plants. This requires large-scale testing of samples from growing seed crops and tubers after harvest (e.g. Loebenstein *et al.*, 2001; Mortimer-Jones *et al.*, 2009; Cox and Jones, 2012). Commencing the flush-through system by starting the first generation with virus-tested minitubers should be adopted universally. Effective hygiene should be emphasized, involving timely roguing of symptomatic potato plants within crops, and prompt removal of volunteer potatoes, alternative virus or aphid weed hosts (especially Solanaceous weeds), and nearby finished crops of other cultivated plants that harbor aphids. Non-host border crops that act as virus-cleansing barriers and straw mulches that diminish aphid landing rates before foliage closes across rows can be deployed. Early foliage destruction to avoid PVY spread during late aphid flights can be practiced (e.g. de Bokx and van der Want, 1987; Saucke and Doring, 2004; Jones *et al.*, 2009; Gray *et al.*, 2010; MacKenzie *et al.*, 2014).

The development of predictive models that forecast the timing of aphid flights should be encouraged (e.g. Sigvald, 1986; Jones *et al.*, 2010), as they can help with decisions over deploying cultural control measures against PVY spread at planting time. They can also help with decisions over when to deploy foliar applications of mineral oils supplemented with insecticide to help suppress PVY spread within seed potato crops (MacKenzie *et al.*, 2014). Although insecticide applications used alone as seed tuber dressings or foliar sprays can be effective against PLRV, such treatments are generally ineffective at controlling non-persistently aphid-borne viruses like PVY. If timed to coincide with aphid flights from external sources, targeted insecticide application suppressed *Myzus persicae* numbers arriving at crop margins in large seed potato fields, thereby diminishing their movement deeper into the crop (Carroll *et al.*, 2009). This is likely to help diminish PLRV spread and might provide some benefits with PVY where other species are minor components of the migrant aphid population and large-sized seed fields with small perimeter:area ratios are planted with seed potatoes.

12.5 Future Research Needs

Although potato viruses have been the subject of much research, they still cause substantial losses in seed and ware potato production in many parts of the world. There are still many research areas where questions need to be answered and additional investigations undertaken. Some of the most important of these are:

1. Over the past 20 or so years, PVY has become an increasingly major concern for potato seed and ware production in many countries, yet the disagreement between its biological and phylogenetic strain group nomenclature systems remains unresolved. The relationship between these two classification systems needs to be unraveled. It would help to resolve this if (i) use of identical names for strain groupings within the two systems, and (ii) phylogenetic group names based on tuber symptoms or geographical regions were avoided. Moreover, further studies to reveal the full biological and phylogenetic diversity of PVY strains are required, as more are likely to be found, especially in the Andean domestication center where the greatest potato virus diversity exists, and little research of this kind has been done. Such research should include a search for biological strain groups that overcome more than one PVY resistance gene, as with PVX group 4 (Cockerham, 1955), and for additional strain-specific PVY resistance genes. Understanding of the key factors driving the rapid evolution of new PVY variants at the molecular level also needs to be improved (e.g. Visser *et al.*, 2013).

2. Breeding for virus resistance is likely to be highly beneficial, especially incorporation of resistance gene *Ry* against PVY in new cultivars destined for regions with temperate or Mediterranean climates, and for resistance to PVX, PVY, and PLRV in cultivars produced for food-insecure subtropical or tropical regions. There is therefore a need for potato breeding programs to refocus on releasing new cultivars with resistance to common potato viruses. Currently, breeding for virus resistance is rarely considered a priority for potato breeding programs, and knowledge of whether the parental lines they use in crosses possess virus resistance is often lacking (e.g. Nyalugwe *et al.*, 2012).

Moreover, due to lack of adequate screening beforehand, when new cultivars are released, information on whether any virus resistance genes are present is rarely provided, only less helpful descriptors, such as "moderate resistance" to PVX, PVY, PLRV, or PVA, or "low resistance" to one or other of these viruses.

3. Quantification of virus-induced yield losses needs to be revisited, as the data available for common potato viruses present singly, or in mixed infections, is mostly old, limited, and involves outmoded potato cultivars (e.g. Broadbent *et al.*, 1957). Furthermore, such yield data tend to be lacking for viruses found more recently. Comparisons between infected and healthy potted plants, spaced plants, or single-row plots are inadequate here. This is because, when virus spreads from infected plants within the crop, plants become infected at different growth stages, and the growth of neighboring healthy plants compensates for the growth of infected plants (e.g. Jones, 2006). Therefore, what is needed to quantify the consequences of infection is replicated field experiments with large plots, sensitive and tolerant cultivars, mild and severe virus strains, and introduced virus infection foci consisting of infected transplants or planted, secondarily infected tubers. Tuber quality should also be assessed so that marketable yield data can be provided.

4. The process of searching for and characterizing additional potato viruses in the Andean region recommenced on a small scale recently (Adams *et al.*, 2013; Kreuze *et al.*, 2013; Souza-Richards *et al.*, 2014). With the aid of innovations in molecular techniques, the time has arrived to revitalize the effort to find new potato viruses and to complete the characterization of ones identified earlier. Such research will help in identifying reasons why potato yields are not achieving their full potential in the Andean region. It will also help to avoid the introduction to other continents of potato viruses currently restricted to the Andes. It should include studies on virus transmission through true potato seed, as this is an important route via which Andean potato viruses might be introduced, when germplasm is being exchanged worldwide by plant breeders. National and international plant quarantine authorities need to remain vigilant to ensure that the risk of such introductions is minimized.

5. Since potato is being grown increasingly in warmer regions of the world, the possible emergence of new tospoviruses, begomoviruses, and other types of viruses via new encounter scenarios with native vegetation needs to be monitored carefully. This is necessary as part of the efforts to ensure food security and to avoid new virus problems building to serious levels without being noticed. The same applies to their possible spread to potato from other Solanaceous crops such as tomato, pepper, tobacco, or aubergine.

6. Improved diagnostic techniques suitable for mass virus detection in samples from seed potato fields or stored dormant tubers are increasingly likely to arise through medical research in virus diagnostics (e.g. Boonham *et al.*, 2014; Jones, 2014). These need to be exploited rapidly to benefit seed potato industries.

7. High incidences of plants with mixed infections with common potato viruses, especially PVY, PVX, PVS, and PLRV, are still common in potato fields in many countries without sophisticated seed potato production schemes. For example, in Kenya, mixed infections in seed potatoes ranged from 50 to 100% (Gildemacher *et al.*, 2009), and in Iran an extensive survey revealed an average 52% of plants with mixed infections (Pourrahim *et al.*, 2007). These mixed infections contributed to very low average potato yields in ware crops. It is important to continue to develop simple yet effective seed production systems that control potato viruses effectively for such countries.

8. Finally, looking to the future, forewarned is forearmed, so research is urgently needed to understand better the likely consequences of climate change on the impacts of virus disease on potato industries worldwide. This requires examining the effects of different climate change parameters alone or in combination on potato plants or stored tubers with single or mixed virus infections. It also requires modeling studies that establish where climate change is likely to result in damaging potato virus epidemics in regions where they were previously considered of minor importance. Similarly, more climate change scenario modeling is needed to establish when potato viruses or vectors are likely to invade regions where conditions were formerly unsuitable for them.

References

Adams, I.P., Glover, R., Souza-Richards, R., Bennet, S., Hany, U. and Boonham, N. (2013) Complete genome sequence of *Arracacha virus B*: a novel cheravirus. *Archives of Virology* 158, 909–913.

Adams, S.E., Jones, R.A.C. and Coutts, R.H.A. (1986) Effect of temperature on *Potato virus X* infection in potato cultivars carrying different combinations of hypersensitivity genes. *Plant Pathology* 35, 517–526.

Baldauf, P.M., Gray, S.M. and Perry, K.L. (2006) Biological and serological properties of *Potato virus Y* isolates in northeastern United States potato. *Plant Disease* 90, 559–566.

Barker, H., McGeachy, K.D., Toplak, N., Gruden, K., Zel., J. and Browning, I. (2009) Comparison of genome sequences of PVY isolates with biological properties. *American Journal of Potato Research* 86, 227–238.

Barker, I., Gamarra, H., Fuentes, S., Muller, G., Juarez, H. and Morales, F. (2007) Risk of spread and vector relations of *Potato yellow vein virus* in the Andes. In: *Proceedings of 10th International Plant Virus Epidemiology Symposium*. ICRISAT, Patancheru, India, p. 50 (Abstract).

Bawden, F.C. (1936) The viruses causing top necrosis (acronecrosis) of the potato. *Annals of Applied Biology* 23, 487–497.

Bawden, F.C. and Sheffield, F.M.L. (1944) The relationships of some viruses causing necrotic diseases of the potato. *Annals of Applied Biology* 31, 33–40.

Beczner, L., Horvath, J., Romhanyi, I. and Förster, H. (1984) Studies on the etiology of tuber necrotic ringspot disease in potato. *Potato Research* 27, 339–352.

Bendahmane, A., Köhm, B.A., Dedi, C. and Baulcombe, D.C. (1995) The coat protein of potato virus X is a strain-specific elicitor of Rx1-mediated virus resistance in potato. *Plant Journal* 8, 933–941.

Blanchard, A., Rolland, M., Lacroix, C., Kerlan, C. and Jacquot E. (2008) *Potato virus Y*: a century of evolution. *Current Topics in Virology* 7, 21–32.

Boland, G.J., Melzer, M.S., Hopkin, A., Higgins, V. and Nassuth, A. (2004) Climate change and plant diseases in Ontario. *Canadian Journal of Plant Pathology* 26, 335–350.

Boonham, N., Hims, M., Barker, I. and Spence, N. (1999) Potato virus Y from petunia can cause symptoms of potato tuber necrotic ringspot disease (PTNRD). *European Journal of Plant Pathology* 105, 617–621.

Boonham, N., Kreuze, J., Winter, S., van der Vlugt, R., Bergervoet, J., Tomlinson, J., and Mumford, R. (2014) Methods in virus diagnostics: From ELISA to next generation sequencing. *Virus Research* 186, 20–31.

Boonham, N., Walsh, K., Hims, M., North, J. and Barker, I. (2002) Biological and sequence comparisons of *Potato virus Y* isolates associated with potato tuber necrotic ringspot disease. *Plant Pathology* 51, 117–126.

Boonham, N., Walsh, K., Smith, P., Madagan, K., Graham, I. and Barker, I. (2003) Detection of potato viruses using microarray technology: towards a generic method for plant viral disease diagnosis. *Journal of Virological Methods* 108, 181–187.

Brattey, C., Badge, J.L., Burns, R., Foster, G.D., George, E., Goodfellow, H.A., *et al.* (2002) Potato latent virus: a proposed new species in the genus Carlavirus. *Plant Pathology* 51, 495–505.

Broadbent, L., Heathcote, G.D., McDermott, N. and Taylor, C.E. (1957) The effect of date of planting and of harvesting potatoes on virus infection and on yield. *Annals of Applied Biology* 45, 603–622.

Brucher, H. (1969) Observations on origin and expansion of Y^N-virus in South America. *Angewandte Botanik* 43, 241–249.

Calderoni, A. (1965) An unidentified virus of deforming type in potato varieties in Argentina. *American Potato Journal* 42, 257 (Abstract).

Canto, T., Aranda, M.A. and Fereres, A. (2009) Climate change effects on physiology and population processes of hosts and vectors that influence the spread of hemipteran-borne plant viruses. *Global Change Biology* 15, 1884–1894.

Carroll, M.W., Radcliffe, E.B., MacRae, I.V., Ragsdale, D.W., Olson, K.D. and Badibanga, T. (2009) Border treatment to reduce insecticide use in seed potato production: biological, economic, and managerial analysis. *American Potato Journal* 86, 31–37.

Chikh-Ali, M., Maoka, T., Natsuaki, T. and Natsuaki, K.T. (2010) PVY^{NTN-NW}, a novel recombinant strain of *Potato virus Y* predominating in potato fields in Syria. *Plant Pathology* 59, 31–41.

Chikh-Ali, M., Gray, S.M. and Karasev, A.V. (2013) An improved multiplex IC-RT-PCR assay distinguishes nine strains of *Potato virus Y*. *Plant Disease* 97, 1370–1374.

Chiko, A.W. and Guthrie, J.W. (1969) An hypothesis for selection of strains of potato leaf roll virus by passage through *Physalis floridana*. *American Potato Journal* 46, 155–167.

Chrzanowska, M. (1991) New isolates of the necrotic strain of potato virus Y (PVY^N) found recently in Poland. *Potato Research* 34, 179–182.
Cockerham, G. (1955) Strains of potato virus X. In: *Proceedings of the Second Conference on Potato Virus Diseases*. Veenham and Zohnen, Wageningen, the Netherlands, pp. 89–92.
Cockerham, G. (1970) Genetical studies on resistance to potato viruses X and Y. *Heredity* 25, 309–348.
Cordero, M., Ramos, P.L., Hernandez, L., Fernandez, A.I., Echemendia, A.L., Peral, R., *et al.* (2003) Identification of *Tomato mottle Taino begomovirus* in Cuban potato fields. *Phytoparasitica* 31, 479–489.
Cox, B.A. and Jones, R.A.C. (2010a) Genetic variability of the coat protein gene of *Potato virus S* and distinguishing its biologically distinct strains. *Archives of Virology* 155, 1163–1169.
Cox, B.A. and Jones, R.A.C. (2010b) Genetic variability of the coat protein gene of *Potato virus X*, and the current relationship between phylogenetic placement and resistance groupings. *Archives of Virology* 155, 1349–1356.
Cox, B.A. and Jones, R.A.C. (2012) Effects of tissue sampling position, primary and secondary infection, cultivar, and storage temperature and duration on the detection, concentration and distribution of three viruses within infected potato tubers. *Australasian Plant Pathology* 41, 197–210.
Crosslin, J.M. and Hamlin, L.L. (2010) First report of *Impatiens necrotic spot virus* infecting greenhouse-grown potatoes in Washington State. *Plant Disease* 94, 1507.
Crosslin, J.M., Eastwell, K.C., Davitt, C.M. and Abad, J.A. (2010) First report of seedborne *Cherry leaf roll virus* in wild potato, *Solanum acaule*, from South America. *Plant Disease* 94, 782.
de Bokx, J.A. and van der Want, J.P.H. (eds) (1987) *Viruses of Potatoes and Seed-Potato Production*, 2nd edn. Centre for Agricultural Publishing and Documentation, Wageningen, the Netherlands.
Djilani-Khouadja, F., Glais, L., Tribodet, M., Kerlan, C. and Fakhfakh, H. (2010) Incidence of potato viruses and characterisation of *Potato virus Y* variability in late season planted potato crops in northern Tunisia. *European Journal of Plant Pathology* 126, 479–488.
Donaire, L., Wang, Y., Gonzalez-Ibeas, D., Mayer, K.F., Aranda, M.A. and Llave, C. (2009) Deep-sequencing of plant viral small RNAs reveals effective and widespread targeting of viral genomes. *Virology* 392, 203–214.
Doring, T.F. (2011) Potential and limitations of plant virus epidemiology: lessons from the *Potato virus Y* pathosystem. *Potato Research* 54, 341–354.
Dullemans, A.M., Cuperus, C., Verbeek, M. and van der Vlught, R.A.A. (2011) Complete nucleotide sequence of a potato isolate of strain group C of *Potato virus Y* from 1938. *Archives of Virology* 156, 473–477.
Ellis, P., Stace-Smith, R. and de Villiers, G. (1977) Identification and geographic distribution of serotypes of potato virus Y. *Plant Disease* 81, 481–484.
Ellis, Y.P., Stace Smith, R., Bowler, G. and Mackenzie, D.J. (1996) Production of monoclonal antibodies for detection and identification of strains of potato virus Y. *Canadian Journal of Plant Pathology* 18, 64–70.
Fargette, D., Konate, G., Fauquet, C., Muller, E., Peterscmitt, M. and Thresh, J.M. (2006) Molecular ecology and emergence of tropical plant viruses. *Annual Review of Phytopathology* 44, 235–260.
Faria, J.C., Souza-Dias, J.A.C., Slack, S.A. and Maxwell, D.P. (1997) A new geminivirus associated with tomatoes in the state of São Paulo, Brazil. *Plant Disease* 81, 423.
Faurez, F., Baldwin, T., Tribodet, M. and Jacquot, E. (2012) Identification of new *Potato virus Y* (PVY) molecular determinants for the induction of vein necrosis in tobacco. *Molecular Plant Pathology* 13, 948–959.
Fletcher, J.D. and Lister, R.A. (2004) PVY^{NTN} in New Zealand potato crops. In: *Proceedings of the 12th European Association for Potato Research, Virology Section Meeting.* European Association for Potato Research, Rennes, France, p. 52 (Abstract).
Fortes, I.M. and Navas-Castillo, J. (2012) Potato, an experimental and natural host for the crinivirus *Tomato chlorosis virus*. *European Journal of Plant Pathology* 61, 994–1001.
Freitas, D.M.S., Nardin, I., Shimoyama, N., Souza-Dias, J.A.C. and Rezende, J.A.M. (2012) First report of *Tomato chlorosis virus* in potato in Brazil. *Plant Disease* 96, 593.
Fribourg, C.E. (1980) Historia y distribucion de los virus de la papa en America Latina. *Fitopatologia* 15, 3–24.
Fribourg, C.E., Jones, R.A.C. and Koenig, R. (1977a) Host plant reactions, physical properties and serology of three isolates of Andean potato latent virus from Peru. *Annals of Applied Biology* 86, 373–380.
Fribourg, C.E., Jones, R.A.C. and Koenig, R. (1977b) Andean potato mottle virus, a new member of the cowpea mosaic virus group. *Phytopathology* 67, 969–974.
Galvino-Costa, S.B.F., dos Reis Figueira, A., Camargos, V.V., Geraldino, P.S., Hu, X., Nikolaeva, O.V. , *et al.* (2011) A novel type of *Potato virus Y* recombinant genome, determined for the genetic strain PVY^E. *Plant Pathology* 61, 388–398.

Gebhart, G. and Valkonen, J.P.T. (2001) Organisation of genes controlling disease resistance in the potato genome. *Annual Review of Phytopathology* 39, 79–102.

Gildemacher, P.R., Demo, P., Barker, I., Kaguono, W., Woldegiorgis, G., Wagoire, W.W., *et al.* (2009) A description of seed potato systems in Kenya, Uganda and Ethiopia. *American Journal of Potato Research* 86, 373–382.

Glais, L., Tribodet, M., Gauthier, J.P., Astier-Manifacier, S., Robaglia, C. and Kerlan, C. (1998) RFLP mapping of the whole genome of potato viral isolates representative of different biological groups of potato virus Y. *Archives of Virology* 143, 2077–2091.

Glais, L., Tribodet, M. and Kerlan, C. (2002) Genomic variability in potato potyvirus Y (PVY): evidence that $PVY^{N}W$ and PVY^{NTN} variants are single to multiple recombinants between PVY^{O} and PVY^{N} isolates. *Archives of Virology* 147, 363–378.

Golnaraghi, A.R., Pourrahim, R., Farzadfar, S. and Ahoonmanesh, A. (2007) Identification and partial characterisation of a *Tospovirus* causing leaf and stem necrosis on potato. *Plant Pathology Journal* 6, 227–234.

Golnaraghi, A.R., Pourrahim, R., Ahoonmanesh, A., Zamsni-Zadeh, H.R. and Farzadfar, S. (2008) Detection and characterisation of a distinct isolate of *Tomato yellow fruit ring virus* from potato. *Plant Disease* 92, 1280–1287.

Goulden, M.G., Köhm, B.A., Santa-Cruz, S., Kavanagh, T.A. and Baulcombe, D.C. (1993) A feature of the coat protein of potato virus X affects both induced virus resistance in potato and viral fitness. *Virology* 197, 293–302.

Granval de Millán, N., Piccolo, R. and Gracia, O. (1998) Potato (*Solanum tuberosum* L.) tuber transmission of tospoviruses (TSWV, GRSV, TCSV) in Mendoza, Argentina. In: *Recent Progress in Tospovirus and Thrips Research*, 4th International Symposium on Tospoviruses and Thrips in Floral and Vegetable Crops, 2–6 May. Graduate Schools of Experimental Plant Sciences and Production Ecology, Wageningen, the Netherlands, pp. 46–47 (Abstract).

Gray, S., DeBoer, S., Lorenzen, J., Karasev, A.V., Whitworth, J., Nolte, P., *et al.* (2010) *Potato virus Y*: an evolving concern for potato crops in the United States and Canada. *Plant Disease* 94, 1384–1397.

Harrington, R. (2002) Insect pests and global climate change. In: Douglas, I. (ed.) *Encyclopaedia of Global Environmental Change*, *Vol 3, Causes and Consequences of Global Environmental Change*. John Wiley and Sons, Chichester, UK, pp. 381–386.

Harrington, R., Clark, S.J., Welham, S.J., Verrier, P.J., Denholm, C.H., Hullé, M., et al. and Consortium European Union Examine (2007) Environmental change and the phenology of European aphids. *Global Change Biology* 13, 1550–1564.

Harrison, B.D. and Jones, R.A.C. (1971) Factors affecting the development of spraing in potato tubers infected with potato mop-top virus. *Annals of Applied Biology* 68, 281–289.

Hinostroza-Orihuela, A.M. (1973) Some properties of potato virus S isolated from Peruvian potato varieties. *Potato Research* 16, 244–250.

Holland, M.B. and Jones, R.A.C. (2005) Benefits of virus testing in seed schemes. In: Pitt, A.J. and Donald, C. (eds) *Proceedings of 'Potato 2005'– Australian National Potato Conference*. Australian National Potato Conference, Cowes, Victoria, Australia, pp. 81–87.

Holmes, I.R. and Teakle, D.S. (1980) Incidence of potato viruses S, X and Y in potatoes in Queensland. *Australasian Plant Pathology* 9, 3–4.

Hooker, W.J. and Salazar, L.F. (1983) A new plant virus from the high jungle of the eastern Andes; Solanum apical leaf curling virus (SALCV). *Annals of Applied Biology* 103, 449–454.

Hu, X., Karasev, A.V., Brown, C.J. and Lorenzen, J. (2009a) Sequence characteristics of *Potato virus Y* recombinants. *Journal of General Virology* 90, 3033–3041.

Hu, X., Meacham, T., Ewing, L., Gray, S.M. and Karasev, A.V. (2009b) A novel recombinant strain of *Potato virus Y* suggests a new viral genetic determinant of vein necrosis in tobacco. *Virus Research* 143, 68–76.

Jeffries, C.J. (1998) Potato. In: *FAO/IPGRI Technical Guidelines for the Safe Movement of Germplasm*, No 19. FAO, Rome, pp. 62–63.

Jones, R.A.C. (1981a) The ecology of viruses infecting wild and cultivated potatoes in the Andean Region of South America. In: Thresh J.M. (ed.) *Pests, Pathogens and Vegetation*. Pitman, London, pp. 89–107.

Jones, R.A.C. (1981b) Use of continuous light to improve diagnosis of potato leaf roll virus in tuber indexing tests. *Plant Pathology* 30, 225–232.

Jones, R.A.C. (1982a) Tests for transmission of four potato viruses through potato true seed. *Annals of Applied Biology* 100, 315–320.

Jones, R.A.C. (1982b) Breakdown of resistance to potato virus X: production of a group four strain from strain group three. *Plant Pathology* 31, 325–331.
Jones, R.A.C. (1987) Problems associated with potyviruses in potato certification – field inspection and serological testing. *EPPO Bulletin* 17, 61–67.
Jones, R.A.C. (1988) Epidemiology and control of potato mop-top virus. In: Cooper, J.I. and Asher, M.J.C. (eds) *Viruses with Fungal Vectors*. Association of Applied Biologists, Wellesbourne, Warwick, UK, pp. 255–270.
Jones, R.A.C. (1990) Strain group specific and virus specific hypersensitive reactions to infection with potyviruses in potato cultivars. *Annals of Applied Biology* 117, 93–105.
Jones, R.A.C. (2006) Control of plant virus diseases. *Advances in Virus Research* 67, 205–244.
Jones, R.A.C. (2009) Plant virus emergence and evolution: origins, new encounter scenarios, factors driving emergence, effects of changing world conditions, and prospects for control. *Virus Research* 141, 113–130.
Jones, R.A.C. (2014) Trends in plant virus epidemiology: opportunities from new or improved technologies. *Virus Research* 186, 3–19.
Jones, R.A.C. and Barbetti, M.J. (2012) Influence of climate change on plant disease infections and epidemics caused by viruses and bacteria. *CAB Reviews* 7, No 022, 1–32. (online publication) http://www.cabi.org/cabreviews.
Jones, R.A.C. and Fribourg, C.E. (1979) Host plant reactions, some properties and serology of wild potato mosaic virus. *Phytopathology* 69, 446–449.
Jones, R.A.C., Koenig, R. and Lesemann, D.E. (1980) Pepino mosaic virus, a new potexvirus from pepino (*Solanum muricatum*). *Annals of Applied Biology* 94, 61–68.
Jones, R.A.C., Charkowski, A., Fribourg, C.E., Stevenson, W.R. and Slack, S.A. (2009) Potato virus and virus-like diseases. In: Barnet, O.W. (ed.) *Virus Diseases of Plants: Grape, Potato, and Wheat Image Collection and Teaching Resource CD-Rom*. APS Press, St Paul, Minnesota.
Jones, R.A.C., Salam, M.U., Maling, T., Diggle, A.J. and Thackray, D.J. (2010) Principles of predicting epidemics of plant virus disease. *Annual Review of Phytopathology* 48, 179–203.
Kaiser, W.J. (1980) Use of thermotherapy to free potato tubers of alfalfa mosaic, potato leaf roll and tomato black ring viruses. *Phytopathology* 70, 1119–1122.
Karasev, A.V. and Gray, S.N. (2013a) Continuous and emerging challenges of *Potato virus Y* in potato. *Annual Review of Phytopathology* 51, 571–586.
Karasev, A.V. and Gray, S.N. (2013b) Genetic diversity of *Potato virus Y* complex. *American Potato Journal* 90, 7–13.
Karasev, A.V., Nikolaeva, O.V., Hu, X., Sielaff, Z., Whitworth, J., Lorenzen, J.H., *et al.* (2010) Serological properties of ordinary and necrotic isolates of *Potato virus Y*: a case study of PVY^{N} misidentification. *American Journal of Potato Research* 87, 1–9.
Karasev, A.V., Hu, X., Brown, C.J., Kerlan, C., Nikolaeva, O.V., Crosslin, J.M., *et al.* (2011) Genetic diversity and origin of the ordinary strain of *Potato virus Y* (PVY) and origin of recombinant PVY strains. *Phytopathology* 101, 778–785.
Kassanis, B. (1949) Potato tubers freed from leaf-roll virus by heat. *Nature* 164, 881.
Kassanis, B. and Govier, D.A. (1971a) New evidence on the mechanism of aphid transmission of potato C and potato aucuba mosaic viruses. *Journal of General Virology* 10, 99–101.
Kassanis, B. and Govier, D.A. (1971b) The role of the helper virus in aphid transmission of potato aucuba mosaic virus and potato virus C. *Journal of General Virology* 13, 221–228.
Kavanagh, T., Goulden, M., Santa-Cruz, S., Chapman, S., Barker, I. and Baulcombe, D. (1992) Molecular analysis of a resistance-breaking strain of potato virus X. *Virology* 189, 609–617.
Kehoe, M.A. and Jones, R.A.C. (2011) A proposal to help resolve the disagreement between naming of *Potato virus Y* strain groups defined by resistance phenotypes and those defined by sequencing. *Archives of Virology* 156, 2273–2278.
Kerlan, C. and Moury, B. (2008) *Potato virus Y*. In: Granoff, A. and Webster, R. (eds) *Encyclopedia of Virology*, 3rd edn. Academic Press, New York, pp. 287–296.
Kerlan, C., Tribodet, M., Glais, L. and Guillet, M. (1999) Variability of potato virus Y in potato crops in France. *Journal of Phytopathology* 147, 643–651.
Kerlan, C., Nikolaeva, O.V., Hu, X., Meacham, T., Gray, S.M. and Karasev, A.V. (2011) Identification of the molecular make-up of the *Potato virus Y* strain PVY^{Z}: genetic typing of $PVY^{Z\text{-}NTN}$. *Phytopathology* 101, 1052–1060.

Kirchner S.M., Döring, T.F., Hiltunen, L.H., Virtanen, E. and Valkonen, J.P.T. (2011) Information-theory-based model selection for determining the main vector and period of transmission of *Potato virus Y*. *Annals of Applied Biology* 159, 414–427.

Koenig, R., Fribourg, C.E. and Jones, R.A.C. (1979) Symptomatological, serological and electrophoretic diversity of isolates of Andean potato latent virus from different regions of the Andes. *Phytopathology* 69, 748–752.

Kogovsek, P., Pompe-Novak, M., Baebler, S., Rotter, A., Gow, L., Gruden, K., *et al*. (2010) Aggressive and mild *Potato virus Y* isolates trigger different specific responses in susceptible potato plants. *Plant Pathology* 59, 1121–1132.

Komatsu, K., Kagiwada, S., Takahashi, S., Mori, T., Yamaji, Y., Hirata, H., *et al*. (2005) Phylogenetic characteristics, genomic heterogeneity and symptomatic variation of five closely related Japanese strains of *Potato virus X*. *Virus Genes* 31, 99–105.

Kreuze, J., Koenig, R., De Souza, J., Vetten, H.J., Muller, G., Flores, B., *et al*. (2013) The complete genome sequences of a Peruvian and a Colombian isolate of *Andean potato latent virus* and partial sequences of further isolates suggest the existence of two distinct potato-infecting tymovirus species. *Virus Research* 173, 431–435.

Kreuze, J.F., Perez, A., Untiveros, M., Quispe, D., Fuentes, S., Barker, I., *et al*. (2009) Complete viral genome sequence and discovery of novel viruses by deep sequencing of small RNAs: a generic method for diagnosis, discovery and sequencing of viruses. *Virology* 388, 1–7.

Lambert, D.H., Levy, L., Mavrodieva, V.A., Johnson, S.B., Babcock, M.J. and Vayda, M.E. (2003) First report of Potato mop-top virus on potato from the United States. *Plant Disease* 87, 872.

Le Romancer, M., Kerlan, C. and Nedellec, M. (1994) Biological characterization of various geographical isolates of *Potato virus Y* inducing superficial necrosis on potato tubers. *Plant Pathology* 43, 138–144.

Li, Y.-Y., Zhang, R.-N., Xiang, H.-Y., Li, Y.-Y., Abouelnasr, H., Li, D.-W., *et al*. (2013) Discovery and characterization of a novel Carlavirus infecting potatoes in China. *PLoS ONE* 8(6), e69255, doi:10.1371/journal.pone. 0069255.

Loebenstein, G. and Thottappilly, G. (eds) (2003) *Viruses and Virus-Like Diseases of Major Crops in Developing Countries*. Kluwer Academic Publishers, Dordrecht, the Netherlands.

Loebenstein, G., Berger, P.H., Brunt, A.A. and Lawson, R.H. (eds) (2001) *Virus and Virus-like Diseases of Potatoes and Production of Seed-Potatoes*. Kluwer Academic Publishers, Dordrecht, the Netherlands.

Lorenzen, J.H., Meacham, T., Berger, P.H., Shiel, P.J., Crosslin, J.M., Hamm, P.B., *et al*. (2006) Whole genome characterization of *Potato virus Y* isolates collected in the western USA and their comparison to isolates from Europe and Canada. *Archives of Virology* 151, 1055–1074.

Lorenzen, J.H., Nolte, P., Martin, D., Pasche, J.S. and Gudmestad, N.C. (2008) NE-11 represents a new strain variant class of *Potato virus Y*. *Archives of Virology* 153, 517–525.

McDonald, J.G. and Singh, R.P. (1996) Response of potato cultivars to North American isolates of PVY^{NTN}. *American Potato Journal* 73, 317–323.

MacKenzie, T.D., Fageria, M.S., Nie, X. and Singh, M. (2014) Effects of crop management practices on current season spread of *Potato virus Y*. *Plant Disease* 98, 213–222.

Malcuit, I., Marano, M.R., Kavanagh, T.A., De Jong, W., Forsyth, A. and Baulcombe, D.C. (1999) The 25-kDa movement protein of PVX elicits Nb-mediated hypersensitive cell death in potato. *Molecular Plant Microbe Interactions* 12, 536–543.

Malcuit, I., de Jong, W., Baulcombe, D.C., Shields, D.C. and Kavanagh, T.A. (2000) Acquisition of multiple virulence/avirulence determinants by potato virus X (PVX) has occurred through convergent evolution rather than through recombination. *Virus Genes* 20, 165–172.

Matoušek, J., Schubert, J., Ptáček, J., Kozlová, P. and Dědič, P. (2005) Complete nucleotide sequence and molecular probing of *Potato virus S* genome. *Acta Virologica* 49, 95–205.

Matros, A., Amme, S., Kettig, B., Buck-Sorlin, G.H., Sonnewald, U. and Mock, H.P. (2006) Growth at elevated CO_2 concentrations leads to modified profiles of secondary metabolites in tobacco cv. Samsun and to increased resistance against infection with potato virus Y. *Plant, Cell and Environment* 29, 126–137.

Mestre, P., Brigneti, G. and Baulcombe, D.C. (2000) An *Ry*-mediated resistance response in potato requires the intact active site of the Nia proteinase from *Potato virus Y*. *Plant Journal* 23, 653–661.

Mestre, P., Brigneti, G., Durrant, M.C. and Baulcombe, D.C. (2003) *Potato virus Y* NIa protease activity is not sufficient for elicitation of *Ry*-mediated disease resistance in potato. *Plant Journal* 36, 755–761.

Morales, F.J., Lastra, R., de Uzcategui, R.C. and Calvert, L. (2001) *Potato yellow mosaic virus*: a synonym of *Tomato yellow mosaic virus*. *Archives of Virology* 146, 2249–2253.

Moreira, A., Jones, R.A.C. and Fribourg, C.E. (1980) Properties of a resistance-breaking strain of potato virus X. *Annals of Applied Biology* 95, 93–103.

Mortimer-Jones, S.M., Jones, M.G.K., Jones, R.A.C., Thomson, G. and Dwyer, G.I. (2009) A single tube, quantitative real-time RT-PCR assay that detects four potato viruses simultaneously. *Journal of Virological Methods* 161, 289–296.

Moury, B., Caromel, B., Johansen, E., Simon, V., Chauvin, L., Jacquot, E., *et al.* (2011) The helper component proteinase cistron of *Potato virus Y* induces hypersensitivity and resistance in potato genotypes carrying dominant resistance genes on chromosome IV. *Molecular Plant Microbe Interactions* 24, 787–797.

Mumford, R.A. and Jones, R.A.C. (2005) *Pepino mosaic virus*. *Association of Applied Biologists Descriptions of Plant Viruses, No* 411, 11 pp.

Muniz, M. and Nombela, G. (2001) Differential variation in development of the B- and Q-biotypes of *Bemisia tabaci* on sweet pepper *Capsicum annnuum* L. at constant temperatures. *Environmental Entomology* 30, 720–727.

Nagata, T., de Avila, A.C., Barbosa, C.J., Melo, P.C.T., Juliatti, F.C. and Kitajima, E.W. (1995) Occurrence of different tospoviruses in six states in Brazil. *Fitopatologia Brasileira* 20, 90–95.

Navas-Castillo, J., Fiallo-Olivé, E. and Sánchez-Campos, S. (2011) Emerging virus diseases transmitted by whiteflies. *Annual Review of Phytopathology* 49, 219–248.

Nie, X. and Singh, R.P. (2002) A new approach for the simultaneous differentiation of biological and geographical strains of *Potato virus Y* by uniplex and multiplex RT-PCR. *Journal of Virological Methods* 104, 41–54.

Nie, X. and Singh, R.P. (2003a) Evolution on North American PVY^{NTN} strain Tu 660 from local PVY^{N} by mutation rather than recombination. *Virus Genes* 26, 39–47.

Nie, X. and Singh, R.P. (2003b) Specific differentiation of recombinant $PVY^{N:O}$ and PVY^{NTN} isolates by multiplex RT-PCR. *Journal of Virological Methods* 113, 69–77.

Nie, X., Singh, R.P. and Singh, M. (2004) Molecular and pathological characterization of N:O isolates of the *Potato virus Y* from Manitoba, Canada. *Canadian Journal of Plant Pathology* 26, 573–583.

Nisbet, C., Butzonich, I., Colavita, J., Daniels, J., Martin, J., Burns, R., *et al.* (2006) Charcterisation of *Potato rough dwarf virus* and *Potato virus P*: distinct strains of the same viral species in the genus *Carlavirus*. *Plant Pathology* 55, 803–812.

Nobrega, N.R. and Silberschmidt, K. (1944) Sobre uma provavel variante do virus “Y” da batatinha (Solanum virus 2, Orton) que tem a peculiaridade de provocar necroses em plantas de fumo. *Arquivos do Instituto Biológico, Sao Paulo* 15, 307–330.

Norse, D. and Gommes, R. (2003) Climate change and agriculture: physical and human dimensions. In: Bruinsma, J. (ed.) *World Agriculture Towards 2015/2030: An FAO Perspective*. Earthscan Publications Ltd, London, pp. 357–372.

Nyalugwe, E.P., Wilson, C.R., Coutts, B.A. and Jones, R.A.C. (2012) Biological properties of *Potato virus X* in potato: effects of mixed infection with *Potato virus S* and resistance phenotypes in cultivars from three continents. *Plant Disease* 96, 43–54.

Oshima, K., Sako, K., Hiraishi, C., Nakagawa, A., Matsuo, K., Ogawa, T., *et al.* (2000) Potato tuber necrotic ringspot disease occurring in Japan: its association with *Potato virus Y* necrotic strain. *Plant Disease* 84, 1109–1115.

Pappu, H.R., Jones, R.A.C. and Jain, R.K. (2009) Global status of tospovirus epidemics in diverse cropping systems: successes achieved and challenges ahead. *Virus Research* 141, 219–236.

Perry, K.L., Miller, L. and Williams, L. (2005) *Impatiens necrotic spot virus* in greenhouse-grown potatoes in New York State. *Plant Disease* 89, 340.

Piche, L.M., Singh, R.P., Nie, X. and Gudmestad, N.C. (2004) Diversity among *Potato virus Y* isolates obtained from potatoes in the United States. *Phytopathology* 94, 1368–1375.

Pourrahim, R., Farzadfar, S., Golnaraghi, A.R. and Ahoonmanesh, A. (2007) Incidence and distribution of important viral pathogens in some Iranian potato fields. *Plant Disease* 91, 609–615.

Pourrahim, R., Golnaraghi, A.R. and Farzadfar, Sh. (2012) Occurrence of *Impatiens necrotic spot virus* and *Tomato spotted wilt virus* on potatoes in Iran. *Plant Disease* 96, 771.

Ribeiro, S.G., Inoue-Nakata, A.K., Daniels, J. and Ávila, A.C. (2006) Potato deforming mosaic disease is caused by an isolate of *Tomato yellow vein streak virus*. *Plant Pathology* 55, 569.

Robles-Hernandez, L., Gonzalez-Franco, A.C., Hernandez-Huerta, J., Meacham, T.M., Nikolaeva, O.V. and Karasev, A.V. (2010) First identification of an unusual recombinant of *Potato virus Y* in potato in Mexico. *Plant Disease* 94, 1262.

Rodoni, B. (2005) *Potato virus Y* – a re-emerging problem. In: Pitt, A.J. and Donald, C. (eds) *Proceedings of 'Potato 2005'- Australian National Potato Conference*. Australian National Potato Conference, Cowes, Victoria, Australia, pp. 77–79.
Rolland, M., Lacroix, C., Blanchard, A., Baldwin, T., Kerlan, C. and Jacquot, E. (2008) *Potato virus Y* (PVY): from its discovery to the latest outbreaks. *Virologie* 12, 261–273.
Salazar, L.F. (2006) *Virus Diseases of Potatoes*. John Wiley and Sons Ltd, New York.
Salazar, L.F., Müller, G., Querci, M., Zapata, J.L. and Owens, R.A. (2000) *Potato yellow vein virus*: its host range, distribution in South America, and identification as a crinivirus transmitted by *Trialeurodes vaporiorum*. *Annals of Applied Biology* 137, 7–19.
Santa-Cruz, S. and Baulcombe, D.C. (1995) Analysis of potato virus X coat protein genes in relation to resistance conferred by the genes *Nx*, *Nb* and *Rx1* of potato. *Journal of General Virology* 76, 2057–2061.
Santala, J., Samuilova, O., Hannukkala, A., Latvala, S., Kortemaa, H., *et al.* (39 authors) (2010) Detection, distribution and control of *Potato mop-top virus*, a soil-borne virus, in northern Europe. *Annals of Applied Biology* 157, 163–178.
Santillan, F.W. (1979) *Estudio comparativo de 11 aislamientos de virus S de la region Andina*. MSc Thesis, National Agrarian University, La Molina, Lima, Peru.
Santillan, F.W., Fribourg, C.E. and Jones, R.A.C. (1980) Estudio comparativo de once aislamientos de virus S del la papa (PVS) de la region Andina. *Fitopatologia* 15, 42–43 (Abstract).
Saucke, H. and Doring, T.F. (2004) *Potato virus Y* reduction by straw mulch in organic potatoes. *Annals of Applied Biology* 144, 347–355.
Sawazaki, H.E., de Souza-Dias, J.A.C., Lorenzen, J.H., Jeffries, C., Madolo, D.G. and Machado, D.S. (2009) *Potato virus* Y^{NTN}: a coat and P1 protein sequences analysis of a Brazilian isolate. *Potato Research* 52, 379–392.
Schubert, J., Fomitcheva, V. and Sztangret-Wisniewska, J. (2007) Differentiation of *Potato virus Y* strains using improved sets of diagnostic PCR-primers. *Journal of Virological Methods* 140, 66–74.
Sigvald, R. (1986) Forecasting the incidence of potato virus Y^{O}. In: McLean, G.D., Garrett, R.G. and Ruesink, W.G. (ed.) *Plant Virus Epidemics: Monitoring, Modeling and Predicting Outbreaks*. Academic Press, Sydney, Australia, pp. 419–441.
Silvestre, R., Untiveros, M. and Cuellar, W.J. (2011) First Report of *Potato yellowing virus* (Genus Ilarvirus) in *Solanum phureja* from Ecuador. *Plant Disease* 95, 355.
Singh, R.P., Valkonen, J.P.T., Gray, S.M., Boonham, N., Jones, R.A.C., Kerlan, C., *et al.* (2008) Discussion Paper: the naming of *Potato virus Y* strains infecting potato. *Archives of Virology* 153, 1–13.
Soloman-Blackburn, R.M. and Barker, H. (2001a) A review of host major gene resistance to potato viruses X, Y, A and V in potato: genes, genetics and mapped locations. *Heredity* 86, 8–16.
Soloman-Blackburn, R.M. and Barker, H. (2001b) Breeding virus resistant potatoes (*Solanum tuberosum*): a review of traditional and molecular approaches. *Heredity* 86, 17–35.
Souza-Dias, J.A.C., Sawazaki, H.E., Pernambuco-Fo, P.C.A., Elias, L.M. and Maluf, H. (2008) *Tomato severe rugose virus*: another begomovirus causing leaf deformation and mosaic symptoms on potato in Brazil. *Plant Disease* 92, 487.
Souza-Richards, R., Adams, I.P., Kreuze, J.F., De Souza, J., Cuellar, W., Dullemans, A.M., *et al.* (2014) The complete genome sequences of two isolates of potato black ringspot virus and their relationship to other isolates and nepoviruses. *Archives of Virology* 159, 811–815.
Spetz, C., Taboada, A.M., Darwich, S., Ramsell, J., Salazar, L.F. and Valkonen, J.P.T. (2003) Molecular resolution of complexes of potyviruses infecting solanaceous crops at the centre of origin in Peru. *Journal of General Virology* 84, 2565–2578.
Steinbach, P. and Hamann, U. (1989) Symptome und ursache einer wenig bekannten ringnekrose der kartoffelknolle. *Archives of Phytopathology and Plant Protection* 25, 223–232.
Stevenson, W.R., Loria, R., Franc, G.D. and Weingartner, D.P. (eds) (2001) *Compendium of Potato Diseases,* 2nd edn. APS Press, St Paul, Minnesota.
Thompson, J.R., Perry, K.L. and De Jong, W. (2004) A new potato virus in a new lineage of picorna-like viruses. *Archives of Virology* 149, 2141–2154.
Tian, Y.-P. and Valkonen, J.P.T. (2013) Genetic determinants of *Potato virus Y* required to overcome or trigger hypersensivitve resistance to PVY strain group O controlled by the gene *Ny* in potato. *Molecular Plant–Microbe Interactions* 26, 297–305.
Todd, J.M. (1961) Tobacco veinal necrosis on potato in Scotland: control of the outbreak and some characters of the virus. In: *Proceedings of the Fourth Conference on Potato Virus Diseases*, Braunschweig, Germany, 1960. Veenham and Zohnen, Wageningen, the Netherlands, pp. 82–92.

Urbino, C., Polston, J.E., Patte, C.P. and Caruana, M.L. (2004) Characterization and genetic diversity of *Potato yellow mosaic virus* from the Caribbean. *Archives of Virology* 149, 417–424.

Usharani, K.S., Surendranath, B., Paul-Khurana, S.M., Gard, I.D. and Malathi, V.G. (2004) Potato leaf curl – a new disease of potato in northern India caused by a strain of *Tomato leaf curl New Delhi virus*. *Plant Pathology* 53, 235.

Valkonen, J.P.T. (1994) Natural genes and mechanisms for resistance to viruses in cultivated and wild potato species (*Solanum* spp.). *Plant Breeding* 112, 1–16.

Valkonen, J.P.T. (2007) Viruses: economical losses and biotechnological potential. In: Vreugdenhil, D. (ed.) *Potato Biology and Biotechnology: Advances and Perspectives*. Elsevier, Oxford, UK, pp. 619–641.

Valkonen, J.P.T., Pehu, E. and Watanabe, K. (1992) Symptom expression and seed transmission of alfalfa mosaic virus and potato yellowing virus (SB-22) in *Solanum brevidens* and *S. tuberosum*. *Potato Research* 35, 403–410.

Valkonen, J.P.T., Puurand, U., Slack, S.A., Makinen, K. and Saarma, M. (1995) Three strain groups of potato A potyvirus based on hypersensitive responses in potato, serological properties, and coat protein sequences. *Plant Disease* 79, 748–753.

Valkonen, J.P.T., Jones, R.A.C., Slack, S.A. and Watanabe, K.N. (1996) Resistance specificities to viruses in potato: standardization of nomenclature. *Plant Breeding* 115, 433–438.

Visser, J.C., Bellstedt, D.U. and Pirie, M.D. (2013) The recent recombinant evolution of a major crop pathogen, *Potato virus Y*. *PLoS ONE* 7(11), e50631, doi:10.1371/journal.pone.0050631.

Wagner, T.L. (1995) Temperature-dependent development, mortality, and adult size of sweetpotato whitefly biotype B (Homoptera: Aleyrodidae) on cotton. *Environmental Entomology* 24, 1179–1188.

Weidemann, H.L. and Maiss, E. (1996) Detection of the potato tuber necrotic ringspot strain of potato virus Y (PVY^{NTN}) by reverse transcription and immunocapture polymerase chain reaction. *Journal of Plant Diseases and Plant Protection* 103, 337–345.

Weilguny, H. and Singh, R.P. (1998) Separation of Slovenian isolates of PVY^{NTN} from the North American isolates PVY^{N} by a 3-primer PCR. *Journal of Virological Methods* 71, 57–68.

Wilson, C.R. (2001) Resistance to infection and translocation of *Tomato spotted wilt virus* in potatoes. *Plant Pathology* 50, 402–410.

Wilson, C.R. and Jones, R.A.C. (1990) Virus content of seed potato stocks produced in a unique seed potato production scheme. *Annals of Applied Biology* 116, 103–109.

Woolhouse, M.E.J., Haydon, D.T. and Antia, R. (2005) Emerging pathogens: the epidemiology and evolution of species jumps. *Trends in Ecology and Evolution* 20, 238–244.

Wright, N.S. (1977) The effect of separate infections by potato viruses X and S on Netted Gem potato. *American Potato Journal* 54, 147–149.

Wright, N.S., MacCarthy, H.R. and Cole, E.F. (1967) Detection and control of mild strains of potato leaf roll virus. *American Potato Journal* 44, 245–248.

Xu, H., DeHaan, T.-L. and De Boer, S.H. (2004) Detection and confirmation of *Potato mop-top virus* in potatoes produced in the United States and Canada. *Plant Disease* 88, 363–367.

Xu, H., Nie, J. and De Boer, S.H. (2005) Differentiation and molecular detection of Canadian necrotic strains of *Potato virus Y*. *Canadian Journal of Plant Pathology* 27, 125–131.

Yamamura, K. and Kiritani, K. (1998) A simple method to estimate the potential increase in the number of generations under global warming in temperate zones. *Applied Entomology and Zoology* 33, 289–298.

Yu, X.-Q., Wang, H.-Y., Lan, Y.-F., Zhu, X.-P., Li, X.-D., Fan, Z.-F., *et al.* (2008) Complete genome sequence of a Chinese isolate of *Potato virus X* and analysis of genetic diversity. *Journal of Phytopathology* 156, 346–351.

13 Weed Control Strategies for Potato (*Solanum tuberosum*)

Ian C. Burke[1]* and Wesley J. Everman[2]
[1]Department of Crop and Soil Science, Washington State University, Pullman, USA; [2]Department of Crop Science, North Carolina State University, Raleigh, USA

Weed management is critical for profitable potato production. Most growers utilize integrated management strategies that include cultural, mechanical, and chemical inputs to protect the crop from yield loss due to weeds. In 1992, the total cost of weed management inputs in conventional potato production were estimated at US$89 million (Bridges, 1992). Losses may have diminished in the past 30 years, but are now estimated conservatively at approximately US$54.8 million annually in the USA (not including tillage costs) (NASS, 2010). Losses from weeds on a worldwide basis are estimated to be an average of 8.3% of the 517.7 Mt of production (Oerke, 2006). The potential loss due to weeds exceeds that of any other pest of potato, becoming an equivalent loss of approximately 35% (Oerke, 2006).

13.1 Common and Troublesome Weeds in Potato

Troublesome weeds in potatoes consist primarily of broadleaf weed species, with the most predominant weeds being those in the nightshade (*Solanum* spp.) complex, common lambsquarters (*Chenopodium album*), and pigweed species (*Amaranthus* spp.) (Table 13.1).

Yellow and purple nutsedge can become troublesome weeds in potato production areas rotated with vegetable crops; concerns are lessened where potatoes are rotated with agronomic crops. Grass weeds, such as barnyardgrass (*Echinochloa crus-galli*), foxtail (*Setaria* spp.), and quackgrass (*Elytrigea repens*) can also be troublesome in potato production.

13.2 Integrated Weed Management

Integrated weed management is essential for potato production. Multiple tactics should be combined to form an integrated weed (or pest) management program. Integrated weed management can be divided into three phases: seedbank management, minimization of seedling establishment, and reduction of weed–crop interference (Gallandt *et al.*, 1998; Godfrey and Haviland, 2003).

Effective integrated weed management incorporates many factors including weeds present, soil properties and amendments, tillage, crop rotation, and chemical weed control. Integrating mechanical and chemical weed management inputs in potato production has long been a common practice. However, as practiced, mechanical and chemical inputs tend to

* E-mail: icburke@wsu.edu

Table 13.1. Common and troublesome species in potato in the USA, by state with significant potato hectarage. (Adapted from Bridges, 1992.)

State	Common	Troublesome
Colorado	Common sunflower	Common sunflower
Connecticut	Large crabgrass	Barnyardgrass
Delaware	Fall panicum	*Brassica* spp.
Florida	Pennsylvania smartweed	*Polygonum* spp.
Idaho	Barnyardgrass	Yellow nutsedge
Illinois	Redroot pigweed	Quackgrass
Louisiana	*Amaranthus* spp.	*Amaranthus* spp.
Massachusetts	Common lambsquarters	Common ragweed
Maryland	Eastern black nightshade	Velvetleaf
Maine	*Brassica* spp.	Quackgrass
Michigan	Barnyardgrass	Barnyardgrass
Minnesota	Green foxtail	Common ragweed
Montana	Canada thistle	Common sunflower
North Carolina	Common lambsquarters	*Cyperus* spp.
North Dakota	*Setaria* spp.	Redroot pigweed
Nebraska	Redroot pigweed	Redroot pigweed
New Jersey	Barnyardgrass	*Calystegia* spp.
New Mexico	*Amaranthus* spp.	*Solanum* spp.
Nevada	Fivehook bassia	*Solanum* spp.
New York	Redroot pigweed	Quackgrass
Ohio	*Amaranthus* spp.	*Amaranthus* spp.
Oregon	Hairy nightshade	Hairy nightshade
Pennsylvania	Common ragweed	Canada thistle
Rhode Island	Quackgrass	Barnyardgrass
South Dakota	Green foxtail	Canada thistle
Tennessee	*Ipomoea* spp.	*Ipomoea* spp.
Texas	Redroot pigweed	Johnsongrass
Utah	Redroot pigweed	Hairy nightshade
Washington	Common lambsquarters	Quackgrass

Notes: Barnyardgrass, *Echinochloa crus-galli* (L.) Beauv.; Canada thistle, *Cirsium arvense* (L.) Scop.; common lambsquarters, *Chenopodium album* L.; quackgrass, *Elymus repens* (L.) Gould; common ragweed, *Ambrosia artemisiifolia* L.; common sunflower, *Helianthus annuus* L.; eastern black nightshade, *Solanum ptychanthum* Dunal; fall panicum, *Panicum dichotomiflorum* Michx.; fivehook bassia, *Bassia hyssopifolia* (Pallas) Kuntze; green foxtail, *Setaria viridis* (L.) Beauv.; hairy nightshade, *Solanum physalifolium* Rusby; large crabgrass, *Digitaria sanguinalis* (L.) Scop.; Pennsylvania smartweed, *Polygonum pensylvanicum* L.; redroot pigweed, *Amaranthus retroflexus* L.; velvetleaf, *Abutilon theophrasti* Medik.; johnsongrass, *Sorghum halepense* (L.) Pers.

be tactical decisions, often made over a short period of time.

Growers would be well served to consider the long-term impacts of each input as important components of a holistic weed management system; for example, utilizing crop rotation as a critical component of integrated weed management by planting crops ahead of potato where effective herbicide options control nightshade species, yellow nutsedge, and quackgrass exist. The reduction of weed populations in preceding crops reduces early-season weed pressure, and the overall weed management inputs in the potato crop. Therefore, we will consider weed management strategically when practiced as a system, and then consider tactical inputs like herbicide applications.

Weed seedbank management

Persistent seedbanks remain one of the most important factors for the success of annual weeds by dispersing a weed population through time (Harper, 1977; Cavers and Benoit, 1989). Seedbanks decline over time due to a lack of additional inputs of new seeds. The rate of decline is dependent on many factors (Gallandt *et al.*, 1998), including species (Cook, 1980; Louda, 1989), environment (Burnside *et al.*, 1977; Donald and Zimdahl, 1987), and burial depth (Wilson, 1988; Thompson *et al.*, 1998; Benech-Arnold *et al.*, 2000). Seedbank decline occurs due to a host of factors including germination, predation, decay, and seed death. However, most of these processes are not well understood. The exception is germination processes, which are the mechanistic factors of germination usually available for important species.

Growers should have an underlying understanding of why cultivating conditions for weed seed germination, for example, is an important weed seedbank management strategy. Identifying and utilizing inputs that result in seedling death prior to planting potato, or other rotational crops, reduces weed pressure within the crop (Jordan *et al.*, 1995). Interestingly, seed persistence in the soil does not correlate well with the relative abundance of the species above ground (Roberts and Ricketts, 1979). Weeds are opportunistic, and even a very small number of seeds in the soil can produce a troublesome infestation.

An effective teaching tool is a simulation illustrating how a single failure or selection of a herbicide-resistant biotype can profoundly increase weed seed in the seedbank over a single season. These models also illustrate why managing the weed seedbank can be an effective strategy for the long-term sustainability of individual farms.

Minimization of seedling establishment

A small proportion (less than 20%) of the germinable seedbank actually produces seedlings in a given year (Roberts and Ricketts, 1979; Forcella, 1992; Gallandt *et al.*, 1998). Successful establishment requires that seeds are positioned in a site allowing them to avoid hazards and to experience conditions that promote germination, emergence, and growth (Harper, 1977). So-called "safe sites" are a function of seed depth, light, alternating temperatures, water availability, crop residues, and freedom from predators and diseases (Gallandt *et al.*, 1998). The safe site must function well past germination (Gallandt *et al.*, 1998). Inputs that reduce or eliminate safe sites, or deplete them of seed, are effective tools to minimize weed establishment.

Weed competition

Once germinated and established, weeds quickly begin to compete with potato plants. Weed–potato competition ultimately determines the yield loss of potatoes, as well as the total weed biomass and seed production. Weeds compete with potato for light, nutrients, water, and space, and can reduce potato yield (VanGessel and Renner, 1990; Boydston, 2010). Competition with weeds also affects tuber size and weight (Nelson and Thoreson, 1981; Wall and Friesen, 1990a), and can change the chemical composition of potatoes. Potato yield loss occurs at harvest, caused by the reduction in efficiency of harvest and increased likelihood of mechanical injury. Weeds cause physical injury by rhizome growth into the potato tuber, as is often the case with yellow nutsedge and quackgrass. Finally, weeds serve as hosts for numerous pests and pathogens (Ogg and Rogers, 1989; Eberlein *et al.*, 1991; WRIPM, 1992).

Potato is a relatively competitive crop, and thus less susceptible to yield loss from weeds compared to relatively non-competitive crops. Weeds that emerge with the crop and survive mechanical control inputs can cause significant yield loss and ultimately contribute seed to the weed seedbank. For example, yield loss in Russet Burbank due to competition with weeds that survived cultivation and hilling ranged from 7 to 43% when compared to weed-free conditions (Liebman *et al.*, 1996). Yield loss can vary by weed species, although direct comparisons of loss caused by individual species can be difficult.

A simple method of comparison is the Cousens rectangular hyperbola,

$$[Y_L = (ID)/[1 + (ID/A)],$$

where the relationship between weed density per meter of row and percent potato yield loss can be related (Cousens, 1988). Yield loss (Y_L) is based on percent reduction of weed-free yield. A is the asymptote for yield loss and, where appropriate, is constrained to 100% (Streibig *et al.*, 1989). D is the density per meter of crop row and I represents the yield loss per weed as weed density approaches zero. The parameter *I* allows for the approximate comparisons of weed competitiveness. Estimated *I* ranged from 0.06 to 2.68, where *I* was the percentage yield loss per unit weed population as density approached zero (Table 13.2).

Weed competition can vary considerably between years and among potato varieties. For example, green foxtail competitiveness at low densities ranged from 0.64 to 2.68% yield loss per unit weed, depending on the year (Wall and Friesen, 1990a,b). As a consequence, a very low weed population density is sufficient to cross an economic threshold and trigger a treatment. Green foxtail appears to be more competitive than the common and troublesome weeds in potato production, such as nightshade spp. and quackgrass.

The critical period of weed control in potato represents a large majority of the growing season. Canadian studies conducted in 1989 and 1990 revealed quackgrass caused a critical period of weed control that began between 3 and 15 days after emergence and lasted from 23 to as long as 68 days after emergence (Baziramakenga and Leroux, 1994). The researchers concluded that even a relatively low density of quackgrass should be controlled for the entire season to avoid a yield loss. The long critical period has been observed in potato production in Brazil (Costa *et al.*, 2008) and Angola (Monteiro *et al.*, 2011), where critical periods began approximately 21 days after treatment and lasted for longer than 40 days. Although considered very competitive, potato is still susceptible to yield loss and requires control inputs to minimize loss.

13.3 Management Tactics

Growers typically identify herbicides as the primary in-season weed management tactic. Some growers use a practice called "hilling" prior to canopy (foliage) row closure, for altering the size and shape of the potato hill, incorporating fertilizer and herbicides, and mechanical disruption of weeds. Herbicides are often applied early, although applications following hilling are not uncommon. To prevent phytotoxicity to the potato plant, some herbicides must be applied prior to potato emergence, or shortly thereafter. Later in the season, as herbicides dissipate to non-effective levels, the crop canopy contributes to weed control by competing for light. Utilizing good cultural practices, including seedling disease control and early-season insect control,

Table 13.2. Relative competitiveness of several weeds of potato.

Weed	Variety	Year	Estimated *i*[a]	Standard error	Estimated[b]	Source
Quackgrass	Superior	1989–1990	0.17	0.09	0.77	Baziramakenga and Leroux, 1994
Quackgrass	Russet Burbank	1987–1988	0.26	0.02	0.52	VanGessel and Renner, 1990
Green foxtail	Russet Burbank	1983	2.68	0.81	0.83	Wall and Friesen, 1990b
		1984	1.51	0.36	0.77	
		1985	0.64	0.17	0.71	
Redroot pigweed	Russet Burbank	1987–1988	0.54	0.02	0.41	VanGessel and Renner, 1990
Hairy nightshade	Russet Norkotah	2004–2005	0.25	0.02	0.37	Hutchinson *et al.*, 2011
Hairy nightshade	Russet Burbank	2004–2005	0.06	0.01	0.19	Hutchinson *et al.*, 2011

Notes: [a]Yield loss was estimated from figures and charts in various refereed journal articles and fit to the hyperbolic function with asymptote set to 100% maximum yield loss. Location effects were averaged where appropriate, and weed densities were converted to per meter of crop row. Estimated *i* values and standard errors are based on the hyperbolic yield loss function [$Y = iD/(1 + iD/A)$], where *D* is the weed density per meter of crop row and *i* is the percent yield loss per weed as weed density approaches zero.[b]Cousens, 1985.

helps establish a more vigorous crop to compete with weeds effectively. Mechanical and cultural practices, such as crop rotation and the use of cover crops, are important components in integrated weed management in potato.

Crop rotation

Crop rotation is an important cultural component of an integrated weed management program in potato. Liebman *et al.* (1996) noted the benefits of crop rotations, including increased yields and net profits. Well planned rotations to include both summer and winter annual crops, or the incorporation of cover crops prior to the potato rotation, minimize opportunities for weeds to germinate, emerge, and complete their life cycle (Boydston, 2010).

Potato production in the upper mid-west and eastern USA generally follows a rotation where excellent weed control can be achieved. One such crop planted prior to potato production is field maize (*Zea mays*). Other crops grown ahead of potato may include lucerne, soybean, and others based on herbicide options or the suppression of specific weeds (Westra *et al.*, 1995). Rotations in western US potato production regions are similar to the other regions, as most production occurs under irrigation. Soybean is seldom grown in a western US rotation, but other high-value crops, particularly onion, are added to the rotation.

Cover crops are becoming increasingly popular as a weed management tool in potato. In addition to acting as a scavenger for nutrients following the previous crop, cover crops compete with weeds for resources, reduce growth, and can even suppress weed seed germination. They also fix nitrogen, improve water infiltration, suppress diseases and nematodes, prevent soil erosion, add organic matter to the soil, and reclaim nutrients (Clark, 2007; Griffen *et al.*, 2009). Common cover crop options used in potato production include rye, clover, hairy vetch, and occasionally tillage radish (Boydston, 2010).

The most widely used cover crop is rye. Rye offers several advantages over other cover crops for use in potato (Björkman and Shail, 2010). Rye is a cold-tolerant grain that germinates in cool soil, making it an ideal winter cover crop, and has a fibrous root system that is effective at retaining soil nitrates. Rye has vigorous top growth and suppresses weeds successfully. To maximize cover crop competitiveness, growers can stimulate weed seed germination with irrigation, after which the weeds can be destroyed prior to planting the cover crop (Boydston, 2010). Rye can be controlled with a herbicide, or plowed into the soil with primary tillage (Björkman and Shail, 2010).

Clover species and hairy vetch are often used as cover crops because they are annual legumes and nitrogen fixers. Both are generally sown with a nurse crop, as they can be slow to establish. A good stand of either is effective for weed suppression; however, these crops are prone to winter kill in extreme conditions. For early-planted potatoes, clover and hairy vetch are not the best option. Fall weed suppression can be achieved with tillage radish, although spring weeds will not be affected as radish winterkills. Tillage radish is desirable as a cover crop due to its thick taproot, which can grow to a depth of 14 or more inches and breaks up compacted soils. A combination of cover crops may be more effective in meeting multiple needs, with rye being the most effective for weed suppression (Björkman and Shail, 2010).

Cover crops are important components of an integrated pest management system as well. Both sudangrass and rapeseed have reduced nematode populations when they are planted in the fall and incorporated as a green manure (Mojtahedi *et al.*, 1993a,b; Boydston and Hang, 1995). Mustard seed and foliage contain glucosinolate compounds that, when activated, produce isothiocyanates, which act as biofumigants for the management of soilborne diseases, and perhaps weeds (Brown and Morra, 1997).

Crop population and variety

Potato plant population ranges according to variety and region, and weed control is rarely, if ever, a consideration of the spatial requirements for potato. Optimal seed spacing and row spacing is determined by agronomic practices in the planting region, combined with the desired tuber size for a given variety. Colquhoun *et al.* (2009) reported significant variation among ten potato cultivars in the ability to tolerate weed competition and retain tuber yield when grown in similar conditions, yet growers seldom consider variety for its competitive ability. Instead, market demand and regional influences such as season length determine which varieties are planted. There may be an opportunity to increase potato competitiveness by incorporating population and variety decisions into integrated weed management plans.

Mechanical weed management

Hand pulling, chopping, and hoeing weeds are simple forms of weed control used on smaller farms where weed control options are limited, particularly those in less developed countries or those growing potatoes organically. Most commercial farms, however, use tractor tillage for mechanical weed control. Tillage is an important component of weed control in potatoes, regardless of region or production system. Before the development of herbicides, weeds in potatoes were controlled through cultivation (Wallace and Bellinder, 1989). Pre-emergence herbicides are used to supplement tillage, but many growers continue to cultivate from one to six times to control late-germinating weeds (Dallyn and Fricke, 1974). Multiple tillage operations are often employed by producers prior to and throughout the season; however, cultivation after emergence can injure potato roots and reduce yields (Nelson and Giles, 1986; Pierce and Burpee, 1995; Bailey *et al.*, 2001). Excessive tillage and improperly timed cultivations can affect potato yields adversely (Bailey *et al.*, 2001).

Tillage operations generally begin with primary tillage for seedbed preparation the fall before sowing if a crop with high residue has been produced. This is later followed by multiple tillage operations in the spring, prior to hilling and planting. Tillage practices vary by region, with tillage type, frequency, and timing determined by cropping system and producer preferences. Cultivation and hilling are very useful for weed control in season, but the operations can break the prophylactic layer of pre-emergence herbicides applied at planting. Consequently, some growers apply additional pre-emergence herbicides after hilling (Rioux *et al.*, 1979).

The first and most important tillage practice is primary tillage. Tillage in late fall is often necessary to reduce surface residue where potatoes are produced, and provides an opportunity for producers to conduct subsequent tillage operations in the spring before early planting. Primary tillage is accomplished by using a moldboard plow, disc, field cultivator, or field finisher to control early flushes of weeds while preparing the seedbed (Bohl and Johnson, 2010).

Secondary tillage is often used to control weeds that have emerged after planting and before hilling. As with other production practices, timing of hill formation varies by region and producer. Many producers plant their seed pieces under a small row hill, returning with cultivation and hilling once the shoots have emerged approximately 3 weeks after planting. A regional variation on this practice involves using a larger row hill at planting. Producers then wait to cultivate until the potatoes have emerged and grown for approximately 3 weeks. Soil is then added to the row as a consequence of hilling operations (Bohl and Johnson, 2010). As previously mentioned, late tillage may be undesirable due to the potential for root damage and subsequent yield loss.

A third option, popular in the Pacific Northwest USA, is used to minimize equipment passes through the field seedbed, loosen compacted soil, prevent erosion and water runoff, and disrupt weeds within the furrow. Following planting, producers use a Dammer Diker® (AG Engineering and Development Co, Kennewick, Washington, USA), which rips furrows between potato rows with a deep shank (Bohl and Johnson, 2010). Wings on the sides of each shank move and disrupt soil on the side of the potato hill, while paddle-type wheels implant small tillage reservoirs used to collect water and prevent erosion. The Dammer Diker® tillage operation typically occurs between planting and early post-emergence, and relies heavily on pre-emergence herbicides as the main weed control mechanism; post-emergence applications are used only where necessary.

Vine desiccation

Vine desiccation of potato is performed primarily to improve harvest efficiency and storability, with the added benefits of improved fresh market appearance and enhanced periderm development (Haderlie *et al.*, 1989; Renner, 1991). Desiccation of weeds and potato vines facilitates harvest, and may reduce the spread of viruses and diseases (Murphy, 1968). There are several factors that influence the efficacy of potato desiccants, with environment and cultivar being two of the most important (Mutch *et al.*, 1984). Hot, dry conditions will aid in desiccation, yet rapidly growing plants in the presence of high fertility or moisture will be difficult to manage with desiccants alone.

Mechanical vine desiccation can be effective when there are many late-season weeds and the potato vines are large and lush. Flail mowers and rollers are most commonly used in commercial production for non-chemical vine desiccation. It is important to avoid disturbing the hill, which can leave tubers exposed to the sun or, in more severe instances, cause damage to the tubers, affecting marketability or storage life.

Chemical vine desiccation has been practiced since the 1930s with the use of sulfuric acid. Several other products have been used as vine desiccants in potato, including carfentrazone, dinoseb, diquat, endothall, glufosinate, paraquat, pelargonic acid, and pyraflufen ethyl. Dinoseb is the only product not currently used for vine desiccation in the USA. All vine desiccation products are primarily contact herbicides that affect tissue above ground, with minimal translocation throughout the plant.

Chemical weed management

Managing weeds in potato production through the use of herbicides can be effective if managed properly, despite the limited options available. In 2011, 1.7 million pounds of herbicide-active ingredient were applied to potatoes in the USA (NASS, 2010). There are currently six herbicide families available for potato producers. Thus, limitations exist and herbicides with broad-spectrum control of both broadleaf and grass weeds are not available. Management of weeds with herbicides requires careful consideration when planning weed management programs.

Pre-emergence herbicides are critical for providing an optimal growing environment for potatoes early in the season. Pre-emergence herbicides labeled for use in potato include EPTC, linuron, metribuzin, *S*-metolachlor, pendimethalin, dimethenamid, rimsulfuron, flumioxazin, and fomesafen. These herbicides can be placed into four primary groups: root and shoot inhibitors, photosynthesis inhibitors, acetolactate synthesis inhibitors, and protoporphyrinogen oxidase inhibitors. Until recently, only root and shoot inhibitors and photosynthesis inhibitors were widely used for pre-emergence weed control. Metribuzin is currently the most widely used pesticide in potato production (Guenthner *et al.*, 1999).

Ethalfluralin, pendimethalin, and trifluralin are dinitroaniline herbicides that are typically applied to potato either pre-plant incorporated (triafluralin) or pre-emergence (ethalfluralin, pendimethalin). They are classified as seedling growth inhibitors (WSSA Herbicide Group 3/ HRAC Group K1).[1] A characteristic of dinitroaniline herbicides is that they inhibit lateral or secondary root formation (Parka and Soper, 1977), and consequently can increase yields compared to weed-free treatments not treated with a herbicide (Nelson and Giles, 1989). The three herbicides can essentially be used interchangeably (Wilcut *et al.*, 1995).

The labels for ethalfluralin, pendimethalin, and trifluralin specify the maximum waiting period between application and incorporation (24 h for trifluralin, 2 days for ethalfluralin, and 7 days for pendimethalin) (Senseman, 2007). The dinitroaniline herbicides control many annual grasses and some small-seeded broadleaf weeds including *Amaranthus* spp. and common lambsquarters. Use rates can be variable and are dependent on soil texture and organic matter content. In general, lower use rates are recommended in coarse-texture soils, and rates increase as soils become finer or as organic matter content increases.

Dimethenamid-P and *S*-metolachlor are chloroacetamide herbicides that are typically applied pre-emergence for the control of annual grasses including barnyardgrass, foxtails, and large crabgrass. *S*-metolachlor can also be applied pre-plant. The chloroacetamides are classified as root and shoot inhibitors (WSSA Group 15/HRAC Group K3). They can also control pigweed species, and *S*-metolachlor can suppress yellow nutsedge (Foy and Witt, 1997; Mueller and Hayes, 1997; Rabaey and Harvey, 1997; Tonks *et al.*, 1999). Dimethenamid-P provides better suppression of nightshade species than *S*-metolachlor (Hutchinson, 2012); however, *S*-metolachlor-based programs outperform dimethenamid-P-based systems on other weedy species, due to the much longer soil half-life of *S*-metolachlor (90 days) (Richardson *et al.*, 2004). Care must be taken when using *S*-metolachlor, as crops grown in rotation with potato could be injured, particularly wheat.

A single thiocarbamate herbicide continues to be used in potato. EPTC is used primarily in mixture with other pre-plant-incorporated herbicides. EPTC controls small-seeded broadleaf weeds and will suppress yellow nutsedge. However, EPTC has to be incorporated within 24 h or efficacy will diminish due to volatilization. EPTC can be applied pre-emergence or after hilling, and incorporated with specific incorporation equipment. In the Pacific Northwest potato producing region of the USA, EPTC can be incorporated with irrigation. Incorporation may be accomplished either by chemigation (herbicide injected into pressurized overhead irrigation pipes) or via a ground sprayer application that uses overhead irrigation to incorporate the herbicide into the soil. EPTC can also be applied in a hilling/layby application by directing the herbicide spray between the rows and incorporating using the aforementioned methods.

Flumioxazin, an *N*-phenylphthalimide herbicide, is new by comparison with the previously discussed chemistry. It controls weeds such as nightshades, common lambsquarters, redroot pigweed, and mustards (*Brassica* spp.). In contrast, fomesafen is the most recent pre-emergence herbicide to be labeled in potato. Fomesafen controls nightshade species, *Amaranthus* species, and suppresses yellow nutsedge. These herbicides are protoporphyrinogen oxidase inhibitors (or PPO, WSSA Group 14/HRAC Group E) (Wilson *et al.*, 2002). Fomesafen is most effective when applied pre-emergence to potatoes to avoid injury. Flumioxazin can only be applied between hilling and potato emergence, or injury will result. Unlike the previously discussed herbicides that require considerable incorporation, flumioxazin is most effective when concentrated in a narrower band of treated soil at the soil surface.

Metribuzin, an s-triazine or triazinone, is one of the oldest potato herbicides and is still widely used in potato production. Metribuzin controls a range of broadleaf weed species and suppresses many additional grasses. Resistance to triazine herbicides has developed in 69 species worldwide, limiting activity and requiring a combination of control practices for these weeds. Additionally, metribuzin has never provided effective control of weedy *Solanum* species.

Post-emergence herbicides are used primarily to maintain a weed-free crop after the crop has canopied. In extremely weedy fields, multiple post-emergence applications may be needed; however, most producers use secondary tillage and hilling to minimize post-emergence herbicide use. Herbicide options to control emerged weeds in potato are limited to metribuzin, rimsulfuron, sethoxydim, and clethodim, comprising three herbicide families. Several herbicides with residual activity only may be applied to newly emerged potatoes; however, emerged weeds will not be controlled.

Considerable research has been conducted to understand how best to incorporate and steward the post-emergence applied herbicide, rimsulfuron, which controls broadleaf weeds in potato (Eberlein *et al.*, 1994; Blackshaw *et al.*, 1995). As a sulfonylurea herbicide, rimsulfuron inhibits acetolactate synthase (ALS, WSSA Group 2/HRAC Group B). A considerable number of weeds have developed resistance to ALS inhibitors. As a consequence, rimsulfuron is usually applied in mixture with metribuzin. Rimsulfuron can be applied pre-emergence by chemigation, ground application, or by aircraft after hilling, drag-off, or Dammer Diker® operations. Post-emergence applications are often less successful, due to variation in the size of the weeds. Ideally, post-emergence applications should be timed when weeds are small, such as a cotyledon or in the two-leaf stage.

Metribuzin, as previously mentioned, is widely used in potato production for weed management. It is often applied pre-emergence; however, it controls many weed species post-emergence. Metribuzin should be applied when weeds are small and before potato row closure. Varietal sensitivity to metribuzin has been documented with visible symptoms consisting of chlorosis.

The only other post-emergence herbicides available for use in potato are clethodim and sethoxydim. Clethodim and sethoxydim are acetyl coenzyme A carboxylase inhibitors (ACCase, WSSA Group 1/HRAC Group A). They control emerged grass weeds. In general, clethodim is more effective on perennial grass species.

Varietal sensitivity is important to consider when selecting herbicides. Several documented cases of sensitivity to metribuzin, linuron, and rimsulfuron have occurred. Sensitivity to these herbicides varies by potato variety, but may include discoloration, shoot and root stunting, tuber loss, tuber malformation, and/or an overall reduction in tuber size and yield. Symptoms may be enhanced where potatoes are growing under stress due to moisture, disease, nutrient deficiency, or other environmental factors.

13.4 Summary

The potato is the leading vegetable in the world and is included with the major cereals as a primary food crop. Weeds affect potatoes negatively, and can cause considerable yield loss if they are not managed. Most growers utilize integrated weed management strategies that include cultural, mechanical, and chemical inputs to protect the crop from yield loss due to weeds.

Cultural methods of weed management are often synonymous with practices that increase productivity, and include inputs such as crop rotation. Tillage is an important component of weed control in potatoes, regardless of region or production system, and includes a wide variety of tactics ranging from simple hand weeding to complicated in-season tillage implements. Conventional potato weed management currently relies on inputs of herbicides. There are limited herbicide options for potato, and growers are encouraged to use them judiciously to mitigate the occurrence of herbicide resistance.

Finally, management of weeds in potato should be performed within a larger integrated pest management context that considers the long-term impacts and profitability of control inputs and the consequence of failures. Growers should always combine cultural, mechanical, and chemical methods of weed management with knowledge of weed biology in an integrated approach to manage weeds in potato successfully.

Note

[1] Herbicides are classified into groups based on their mode of action. The mode of action of a herbicide describes the effect the herbicide has on the plant at the tissue or cellular level. Two systems are commonly used; the Herbicide Resistance Action Committee system uses letters, while the Weed Science Society of America system uses numbers. For more information on herbicide modes of action and herbicide resistance, see the International Survey of Herbicide Resistant Weeds (www.weedscience.org).

References

Bailey, W.A., Wilson, H.P. and Hines, T.E. (2001) Influence of cultivation and herbicide programs on weed control and net returns in potato (*Solanum tuberosum*). *Weed Technology* 15, 654–659.

Baziramakenga, R. and Leroux, G.D. (1994) Critical period of quackgrass (*Elytrigia repens*) removal in potatoes (*Solanum tuberosum*). *Weed Science* 42, 528–533.

Benech-Arnold, R.L., Sánchez, R.A., Forcella, F., Kruck, B. and Ghersa, C.M. (2000) Environmental control of dormancy in weed seed banks in soil. *Field Crops Research* 67, 105–122.

Björkman, T. and Shail, J.W. (2010) Cornell cover crop guide for annual ryegrass (http://covercrops.cals.cornell.edu/annual-ryegrass.php, accessed 2 June 2014).

Blackshaw, R.E., Lynch, D.R. and Entz, T. (1995) Postemergence broadleaf weed control in potato (*Solanum tuberosum*) with rimsulfuron and HOE-075032. *Weed Technology* 9, 228–235.

Bohl, W.H. and Johnson, S.B. (2010) *Commercial Potato Production in North America: The Potato Association of America Handbook* (http://potatoassociation.org/documents/A_ProductionHandbook_Final_000.pdf, accessed 11 March 2013).

Boydston, R.A. (2010) Managing weeds in potato rotations without herbicides. *American Journal of Potato Research* 87, 420–427.
Boydston, R.A. and Hang, A. (1995) Rapeseed (*Brassica napus*) green manure crop suppresses weeds in potato (*Solanum tuberosum*). *Weed Technology* 9, 669–675.
Bridges, D.C. (1992) *Crop Losses Due to Weeds in the United States.* Weed Science Society of America, Champaign, Illinois.
Brown, P.D. and Morra, M.J. (1997) Control of soil-borne plant pests using glucosinolate-containing plants. *Advances in Agronomy* 61, 167–231.
Burnside, O.C., Wicks, G.A. and Fenster, C.R. (1977) Longevity of shattercane seed in soil across Nebraska. *Weed Research* 17, 139–143.
Cavers, P.B. and Benoit, D.L. (1989) Seed banks in arable land. In: Leck, M.A., Parker, V.T. and Simpson, R.L. (eds) *Ecology of Soil Seed Banks.* Academic Press, San Diego, California, pp. 309–328.
Clark, A. (2007) *Managing Cover Crops Profitably*, 3rd edn. Sustainable Agriculture Network, Beltsville, Maryland.
Colquhoun, J.B., Konieczka, C.M. and Rittmeyer, R.A. (2009) Ability of potato cultivars to tolerate and suppress weeds. *Weed Technology* 23, 287–291.
Cook, R. (1980) The biology of seeds in the soil. In: Solbrig, O.T. (ed.) *Demography and Evolution of Plant Populations*. University of California Press, Berkeley, California, pp. 107–129.
Costa, N.V., Cardosa, L.A., Rodrigues, A.C.P. and Martins, D. (2008) Weed interference periods in potato crop. *Planta Daninha* 26, 83–91.
Cousens, R. (1985) A simple model relating yield loss to weed density. *Annals of Applied Biology* 107, 239–252.
Cousens, R. (1988) Misinterpretations of results in weed research through inappropriate use of statistics. *Weed Research* 28, 281–289.
Dallyn, S.L. and Fricke, D.H. (1974) The use of minimum tillage plus herbicides in potato production. *American Potato Journal* 51, 177–184.
Donald, W.W. and Zimdahl, R.L. (1987) Persistence, germinability, and distribution of jointed goatgrass (*Aegilops cylindrical*) seed in soil. *Weed Science* 35, 149–154.
Eberlein, C.V., Barkdoll, A.W. and Davis, J.R. (1991) Pathogenicity of *Colletrotrichum coccodes* isolates to potato (*Solanum tuberosum*) and two nightshade (*Solanum spp.*) species. *Weed Technology* 5, 570–574.
Eberlein, C.V., Whitmore, J.C., Stranger, C.E. and Guttieri, M.J. (1994) Postemergence weed control in potatoes (*Solanum tuberosum*) with rimsulfuron. *Weed Technology* 8, 428–435.
Forcella, F. (1992) Prediction of weed seedling densities from buried seed reserves. *Weed Research* 32, 29–38.
Foy, C.L. and Witt, H.L. (1997) SAN 582, alachlor, and metolachlor control triazine-resistant (TR) smooth pigweed (*Amaranthus hybridus*) in to-till corn (*Zea mays*). *Weed Technology* 11, 623–625.
Gallandt, E.R., Liebman, M. and Huggins, D.R. (1998) Improving soil quality: implications for weed management. *Journal of Crop Production* 2, 95–121.
Godfrey, L.D. and Haviland, D.R. (2003) University of California Integrated Pest Management Guidelines: Potato, publication 3463 (http://www.ipm.ucdavis.edu/PMG/selectnewpest.potatoes.html, accessed 01 December 2012).
Griffen, T.S., Larkin, R.P. and Honeycut, C.W. (2009) Delayed tillage and cover crop effects in potato systems. *American Journal of Potato Research* 86, 79–87.
Guenthner, J.F., Wiese, M.V., Pavlista, S.D., Sieczka, J.B. and Wyman, J. (1999) Assessment of pesticide use in the US potato industry. *American Journal of Potato Research* 76, 25–29.
Haderlie, L.C., Halderson, J.L., Leino, P.W., Peterson, P.J. and Callihan, R.H. (1989) Chemical desiccation of potato vines. *American Potato Journal* 66, 53–62.
Harper, J.L. (1977) *Population Biology of Plants.* Academic Press, San Diego, California.
Hutchinson, P.J. (2012) Common lambsquarters and hairy nightshade control in potato with dimethenamid-P alone and in tank mixtures and comparison of control by dimethenamid-P with *S*-metolachlor and metolachlor. *Weed Technology* 26, 279–283.
Hutchinson, P.J.S., Beutler, B.R. and Farr, J. (2011) Hairy nightshade (*Solanum sarrachaoides*) competition with two potato varieties. *Weed Science* 59, 37–42.
International Survey of Herbicide Resistant Weeds (2013) www.weedscience.org (accessed 27 May 2013).
Jordan, N., Mortensen, D.A., Prentflow, D.M. and Cox, K.C. (1995) Simulation analysis of crop rotation effects on weed seedbanks. *American Journal of Botany* 82, 390–398.

Liebman, M., Drummon, F.A., Corson, S. and Zhang, J.X. (1996) Tillage and rotation crop effects on weed dynamics in potato production systems. *Agronomy Journal* 88, 18–26.

Louda, S.M. (1989) Predation in the dynamics of seed regeneration. In: Leck, M.A., Parker, V.T. and Simpson, R.L. (eds) *Ecology of Soil Seed Banks*. Academic Press, San Diego, California, pp. 25–51.

Mojtahedi, H., Santo, G.S. and Ingram, R.E. (1993a) Suppression of *Meloidogyne chitwoodi* with sudangrass cultivars as green manure. *Journal of Nematology* 25, 303–311.

Mojtahedi, H., Santo, G.S., Wilson, J.H. and Hang, A.N. (1993b) Managing *Meloidogyne chitwoodi* on potato with rapeseed as green manure. *Plant Disease* 77, 42–46.

Monteiro, A., Henriques, I. and Moreira, I. (2011) Critical period for weed control in potatoes in the Huambo Province (Angola). *Planta Daninha* 29, 351–362.

Mueller, T.C. and Hayes, R.M. (1997) Effect of tillage and soil-applied herbicides on broadleaf signalgrass (*Brachiaria platyphylla*) control in corn (*Zea mays*). *Weed Technology* 11, 698–703.

Murphy, H.J. (1968) Potato vine killing. *American Potato Journal* 45, 472–478.

Mutch, D.R., Penner, D., Roggenbuck, F. and Chase, R.W. (1984) The use of additives, temperature, and plant position to increase efficacy of dinoseb for potato (*Solanum tuberosum*) vine desiccation. *American Potato Journal* 61, 577–586.

NASS (National Agricultural Statistics Service) (2010) Quick Stats (http://quickstats.nass.usda.gov/results/43E98832-5F83-3949-A80B-37ED1C62AC53#DD56E66B-9D77-3DE5-BDD8-0BB83580724F, accessed 21 November 2012).

Nelson, D.C. and Giles, J.F. (1986) Implication of postemergence tillage on root injury and yields of potatoes. *American Potato Journal* 63, 445.

Nelson, D.C. and Giles, J.F. (1989) Weed management in two potato (*Solanum tuberosum*) cultivars using tillage and pendimethalin. *Weed Science* 37, 228–232.

Nelson, D.C. and Thoreson, M.C. (1981) Competition between potatoes (*Solanum tuberosum*) and weeds. *Weed Science* 29, 672–677.

Oerke, E.C. (2006) Crop losses to pests. *Journal of Agricultural Science* 144, 31–43.

Ogg, A.G. Jr and Rogers, B.S. (1989) Taxonomy, distribution, biology, and control of black nightshade (*Solanum nigrum*) and related species. *Reviews of Weed Science* 4, 25–58.

Parka, S.J. and Soper, O.F. (1977) The physiology and mode of action of the dinitroaniline herbicides. *Weed Science* 25, 79–87.

Pierce, F.J. and Burpee, C.G. (1995) Zone tillage effects on soil properties and yield and quality of potatoes (*Solanum tuberosum* L.). *Soil and Tillage Research* 35, 135–146.

Rabaey, T.L. and Harvey, G. (1997) Sequential applications control woolly cupgrass (*Erichloa villosa*) and wild-proso millet (*Panicum miliaceum*) in corn (*Zea mays*). *Weed Technology* 11, 537–542.

Renner, K.A. (1991) Chemical vine desiccation of two potato cultivars. *American Potato Journal* 68, 479–491.

Richardson, R.J., Whaley, C.M., Wilson, H.P. and Hines, T.E. (2004) Weed control and potato (*Solanum tuberosum*) tolerance with dimethenamid isomers and other herbicides. *American Journal of Potato Research* 81, 299–304.

Rioux, R., Comeau, J.E. and Genereux, H. (1979) Effect of cultural practices and herbicides on weed population and competition in potatoes. *Canadian Journal of Plant Science* 59, 367–374.

Roberts, H.A. and Ricketts, M.E. (1979) Quantitative relationship between the weed flora after cultivation and the seed population in the soil. *Weed Research* 19, 269–275.

Senseman, S.A. (2007) *Herbicide Handbook*. Weed Science Society of America, Lawrence, Kansas.

Streibig, J.C., Combellack, J.P., Pritchard, G.H. and Richardson, R.G. (1989) Estimation of thresholds for weed control in Australian cereals. *Weed Research* 29, 117–126.

Thompson, K., Bakker, J.P. and Bekker, R.M. (1998) Ecological correlates of seed persistence in the soil in the NW European flora. *Journal of Ecology* 86, 163–170.

Tonks, D.J., Eberlein, C.V., Guttieri, M.J. and Brinkman, B.A. (1999) SAN 582 efficacy and tolerance in potato (*Solanum tuberosum*). *Weed Technology* 13, 71–76.

VanGessel, M.J. and Renner, K.A. (1990) Redroot pigweed (*Amaranthus retroflexus*) and barnyardgrass (*Echinochloa crus-galli*) interference in potatoes (*Solanum tuberosum*). *Weed Science* 38, 338–343.

Wall, D.A. and Friesen, G.H. (1990a) Effect of duration of green foxtail (*Setaria viridis*) competition on potato (*Solanum tuberosum*) yield. *Weed Technology* 4, 539–542.

Wall, D.A. and Friesen, G.H. (1990b) Green foxtail (*Setaria viridis*) competition in potato (*Solanum tuberosum*). *Weed Science* 38, 396–400.

Wallace, R.W. and Bellinder, R.R. (1989) Potato (*Solanum tuberosum*) yields and weed populations in conventional and reduced tillage systems. *Weed Technology* 3, 590–595.

WRIPM (Western Regional Integrated Pest Management Project) (1992) *Integrated Pest Management for Potatoes in the Western United States*, 2nd edn. Publication 3316. University of California, Oakland, California.

Westra, J.V., Boyle, K.J. and Porter, G.A. (1995) Net revenues of potatoes rotated with other crops. *American Potato Journal* 72, 99–117.

Wilcut, J.W., York, A.C. and Jordan, D.L. (1995) Weed management systems for oil seed crops. In: Smith, A.E. (ed.) *Handbook of Weed Management Systems*. Marcel-Dekker, New York, pp. 343–400.

Wilson, D.E., Nissen, S.J. and Thompson, A. (2002) Potato (*Solanum tuberosum*) variety and weed response to sulfentrazone and flumioxazin. *Weed Technology* 16, 567–574.

Wilson, R.G. (1988) Biology of weed seeds in the soil. In: Altieri, M.A. and Liebman, M. (eds) *Weed Management in Agroecosystems: Ecological Approaches*. CRC Press, Boca Raton, Florida, pp. 25–39.

14 Tuber Physiological Disorders

Loretta Mikitzel*
New Brunswick Department of Agriculture, Aquaculture and Fisheries, Potato Development Centre, Wicklow, Canada

Defects in potato tubers that are brought about by unfavorable environmental factors that alter normal plant and tuber growth are called physiological disorders. These disorders are not due to pathogens, insects, or nematodes, and are often referred to as abiotic, non-infectious or non-pathogenic disorders. As these disorders are not caused by pathogenic microorganisms, they should not be referred to as diseases. The major causes of physiological disorders include improper cultural, storage or handling practices, high or low temperatures, and inadequate, excessive or uneven soil moisture or nutrient levels. Often, a disorder is caused by the combined effects of environment and cultural practices.

Physiological disorders affect the external and internal appearance of tubers. External disorders, such as changes in tuber size, shape, or skin color are readily apparent, but internal disorders are often undetected until the tuber is cut. These defects reduce the value of the crop because the tubers no longer meet market grade standards. Losses due to physiological disorders may be minor or very serious. Some cultivars are more susceptible than others to particular disorders. Physiological disorders occur erratically from year to year and location to location. Since they are often detected only after harvest, it is difficult to identify the patterns in the field that may aid in diagnosing the cause of the disorder. Also, physiological disorders are slow to develop, and manifestation of symptoms may not occur for some time after the disorder is initiated.

Losses from many physiological disorders can be minimized by using good cultural practices that promote uniform plant and tuber growth over the course of the growing season, and by maintaining proper storage conditions. The cause of some physiological disorders is unavoidable or unknown, and they are therefore uncontrollable.

14.1 Tuber Malformations

Tuber malformations include knobby (a type of second growth), dumb-bell, bottleneck, and pointed-end tubers. Tuber malformations are also referred to as misshapen or rough tubers.

Knobby tubers are formed due to a loss of apical dominance, resulting in secondary growth in one or more lateral eyes (Krauss, 1985). The growth results in protuberances that range in size from a raised, swollen bud to very large knobs, sometimes larger than the original tuber (Fig. 14.1). Knobs vary in size and shape, depending on the stage of tuber development in which the disorder began. The plant and the interior of the tuber show no symptoms.

* E-mail: loretta.mikitzel@gnb.ca

Fig. 14.1. Knobby tubers result from secondary growth from lateral eyes. (Photo courtesy of M.J. Pavek, Washington State University.)

Pointed-end, dumb-bell, and bottleneck tubers result from irregular radial growth and are characterized by a constriction along the long axis of the tuber. The type of symptom is determined by the time at which the growth interruption occurred. Dumb-bell shaped tubers have a constriction in the middle of the tuber due to midseason growth interruption. Early-season, growth-restricting stress leads to bottleneck or pointed stem-end tubers, and late-season stress leads to pointed apical (bud) end tubers that have a narrower bud end than basal (stem) end.

Cultivars with long tubers tend to develop malformations more readily than those with round tubers; Russet Burbank is particularly susceptible. No cultivar is completely free of tuber malformations. Tubers with pointed ends tend to have high reducing sugar content and may develop a translucent end (Iritani and Weller, 1973).

Conditions or stresses that modify plant growth and temporarily slow or stop tuber growth lead to tuber malformations. When tuber growth resumes with favorable conditions, growth and starch synthesis resume, but only in certain areas of the tuber, commonly next to buds where cell division is most active (Reeve *et al.*, 1973). Abnormally shaped tubers result. Tuber malformations are associated with wide fluctuations in soil and air temperatures, moisture stress, defoliation due to hail, frost or insects, an imbalance of fertilizer, or a combination of these factors (Lugt, 1960; Bodlaender *et al.*, 1964; Hiller *et al.*, 1985; Krauss, 1985). Sudden or severe stress usually results in a higher incidence of tuber malformations than continuous or gradually increasing stress (Thornton, 2001a).

Factors that contribute to the incidence and severity of tuber malformations include single-stemmed plants that produce few tubers per hill, poor stands, reduced tuber set caused by Rhizoctonia or other reasons, and excessive vine growth (Hiller and Thornton, 1993).

It is difficult to avoid this disorder completely because the unfavorable conditions that cause tuber malformations often cannot be controlled. Good cultural practices that promote uniform stand, proper plant population, balanced nutrition and even moisture will reduce the incidence and severity of tuber malformations. When markets allow, growers should plant cultivars that are less susceptible.

14.2 Tuber Cracking

There are two types of physiological tuber cracking: growth cracks and thumbnail cracks (also called air checks).

Growth cracks are shallow to deep fissures in the surface of the tuber. They are caused when internal pressure exceeds the tensile strength of the surface tissues during tuber enlargement and the outer periderm bursts

(Jefferies and MacKerron, 1987b; Thornton, 2001a). High internal cellular tension develops from tissue expansion during rapid tuber growth. Growth cracks will suberize, or wound heal, and eventually become covered with normal-looking skin. They occur more often at the bud end and tend to run longitudinally on the tuber. There may be one or more cracks per tuber. Cracks formed early in the season will be flatter at harvest than those formed later (van der Zaag, 1996). Cultivars vary greatly in susceptibility to growth cracking.

Growth cracks occur when there are rapid changes in tuber growth rate, as when a long, dry period, or period of high temperature stress, is followed by rain (Jefferies and MacKerron, 1987b). Rapid water uptake results in large changes in tuber turgidity and growth rate (Thornton, 2001a). Susceptibility to growth cracks is increased by wide plant spacing and low tuber set. If fertilizer placement or timing cause uneven tuber growth, growth cracks may develop (Strand, 2006).

Growth cracks may also develop when plants are infected by *Potato yellow dwarf virus*, *Potato mop-top virus*, *Potato spindle tuber viroid*, or when plants are exposed to certain herbicides (Thornton, 2001a). Carnegie and McCreath (2010) indicated that *Potato virus A* (PVA) and *Potato virus Y* (PVY) might also cause growth cracking in some potato cultivars. Growth cracks caused by virus are difficult to distinguish from those caused by physiological conditions. Identifying the cause of such growth cracks requires observing foliar symptoms and/or laboratory analysis.

To reduce the incidence of growth cracks, when possible, one should plant cultivars that are less susceptible, maintain uniform growing conditions by maintaining adequate soil moisture and fertility throughout the growing season, and adjust seed piece spacing to ensure uniform plant spacing and adequate plant population.

Thumbnail cracks or air checks are random, shallow, arch-shaped breaks in the skin caused by handling cold, very turgid tubers. The cracks become visible after the tuber is exposed to drying conditions. These cracks are due to mild physical impacts and are generally superficial. Freshly harvested tubers exposed to the air may also develop thumbnail cracks without an impact. Similar cracks may develop in storage. Breaks in the skin of tubers are often related to high tuber hydration and temperatures below 7.2°C (Strand, 2006). Thumbnail cracks heal slowly, especially at low relative humidity, leading to tuber dehydration and increased potential for tuber rot.

To minimize thumbnail cracks, do not harvest tubers when pulp temperatures are below 10°C, adjust machinery so drop heights are minimized to reduce impacts, and wound heal at high relative humidity. In addition, tuber pulp temperature should be maintained above 10°C when handling tubers that have been removed from storage.

14.3 Enlarged Lenticels

Lenticels are small, inconspicuous pores scattered (one per square centimeter of surface) on the tuber surface. They are essential for gas exchange and oxygen diffusion into the tuber for respiration (Wigginton, 1973). When tubers lack oxygen, as in excessively wet, waterlogged soils or in very compacted dry soils, the lenticels swell and become prominent. The lenticels become enlarged (hypertrophied or proliferated) and protrude beyond the periderm, forming raised masses of corky tissue approximately 0.5 mm in diameter on the tuber surface (Lulai, 2001). Enlarged lenticels, which resemble miniature kernels of popped maize, can look like scab lesions, but are smaller and lighter in color. They detract from the tuber appearance and are sites of entry for decay organisms. It is important to note that some potato cultivars appear to be more susceptible to enlarged lenticels than others.

Enlarged lenticels, often referred to as water spot or water scab (Thornton, 2001a), may also form after harvest or in storage when free moisture is allowed to remain on tuber surfaces.

To prevent enlarged lenticels, avoid over-irrigation during the growing season. Provide good field drainage, especially in low-lying areas that may stay wet for long periods of time. Do not allow free moisture to form on tubers after harvest.

14.4 Greening

Tuber greening is the light-induced formation of green color just below the skin of the tuber due to chlorophyll accumulation in the leucoplasts. When tubers, or parts of tubers, become green due to exposure to direct or indirect sunlight in the field, the disorder is known as sunburn or sun green. Greening due to postharvest exposure to artificial light is often referred to as lightburn, or the tubers have been light-struck. The higher the light intensity and longer the exposure, the more intense the green color. The skin and flesh, from 2 mm to 10 mm in depth, are affected (O'Brien and Rich, 1976; Reeves, 1988). The outer tissues become deep green, while the inner tissues may be yellowish-green or deep yellow. Long exposure to low light levels in storage, at market or at home, produces a lighter, more diffuse green color over the entire tuber. Relatively low light intensities, 3–11 W/m^2 for as little as 24 h, will induce greening (Gull and Isenberg, 1958). Greening occurs more rapidly at temperatures above 13°C (Yamaguchi *et al.*, 1960; Thornton, 2001a). All potato cultivars are susceptible to greening, but white-skinned cultivars are more prone to greening than red or russet-type tubers (Reeves, 1988; Strand, 2006). Sunburn does not kill the affected tissue.

Chlorophyll itself is tasteless and harmless, but green tubers also accumulate glycoalkaloids (such as solanine), which are bitter, mildly toxic to humans, and tend to make tubers unpalatable. The production of chlorophyll and glycoalkaloids in green tubers occur independently of each other (Edwards *et al.*, 1998). Green tubers are unacceptable for fresh market or processing, but green seed tubers may be planted (van der Zaag, 1996).

Tubers protruding from the hill or growing close to the soil surface become green from direct sunlight or light penetrating through cracks in the soil. Contributing factors to greening in the field are shallow planting of seed pieces or incorrect hilling (Bohl and Love, 2005), hills eroded by rain, cracks in dry soils, and tubers forced to the surface due to high plant densities and crowding (Wale *et al.*, 2008). Cultivars that naturally set tubers near the soil surface are more prone to greening. In the marketplace, transparent packaging and indoor lights increase greening, especially of immature tubers, which green more rapidly than mature tubers (Yamaguchi *et al.*, 1960).

Tubers exposed to intense light and high temperatures develop sunscald. The damaged external tissues have a blistered appearance and metallic color. Internal underlying tissues are watery and turn brown. Affected tissues may be invaded by secondary bacteria, or may dry and become granular or leathery (O'Brien and Rich, 1976).

To prevent tuber greening, plant seed pieces sufficiently deep at the correct plant population, and form hills of proper shape and size to ensure tubers remain covered by soil. In some cases, fields are rolled to fill in soil cracks. After harvest, store tubers in complete darkness and avoid long exposure to artificial lights. Tubers accumulate chlorophyll at a slower rate under fiber optic lights than under fluorescent lights (Olsen and Brandt, 2005).

14.5 Elephant Hide

Also known as alligator hide in russet potatoes, elephant hide is the thick, coarse russeting, furrowing or cracking of the periderm. The cracks are shallow, appear corky or scaly, and may affect part of the tuber or the entire surface (Okazawa and Iriuda, 1980). Interior tissues are not affected. On smooth-skinned cultivars, the disorder is known as fishy skin or turtle back (Zimmerman-Gries and Blodgett, 1963).

The exact cause of elephant hide is unknown. Elephant hide has been associated with salts/fertilizer near the tuber surface (Hodgson *et al.*, 1974), decaying organic matter in contact with the tuber, excessive soil moisture, and improper timing of field-applied, sprout-inhibiting chemicals (Hiller *et al.*, 1985). Tubers exposed to high temperatures produce increased numbers of cork layers (Okazawa and Iriuda, 1980; Ginzberg *et al.*, 2009). When these cells do not slough off, due to increased pectin and hemicellulose levels, the thick russeting develops. Ginzberg *et al.* (2009) postulated that periderm thickening due to high temperatures was a defense mechanism, with the heavy russeting a secondary outcome when the inflexible skin cracked as the tuber grew.

There is no known control for elephant hide. Culling of affected tubers, especially for the fresh market, is necessary.

14.6 Second Growth

There are three distinct types of second growth expressed by potato plants in the field: heat sprouts, also known as aerial sprouts or heat runners; chain tubers, also called tuber chaining, gemmation, or chain tuberization; and little tuber, also called kinder tuber, little potato, sprout tuber, or no top.

Heat sprouts and chain tubers (Fig. 14.2) are caused by high soil temperatures (28°C and higher), with or without drought (Bodlaender *et al.*, 1964). Water stress alone does not cause these types of second growth. The longer the period of elevated temperature and the higher the temperature, the more second-growth tubers will be found (Bodlaender *et al.*, 1964; Lugt *et al.*, 1964). Fluctuating nitrogen nutrition is also known to result in second growth (Krauss, 1985).

Heat sprouts arise from stolon or tuber buds. They grow as short sprouts underground that elongate, emerge from the soil, and develop into leafy aboveground stems. Heat sprouts may develop before tubers are set, or from stolon reformation from the bud end of tubers at any stage of tuber development. Heat sprouts often develop when tubers are close to the soil surface.

Chain tuber refers to multiple, small tubers forming on a single stolon. High temperatures favor stolon branching (Struik *et al.*, 1989), and a tuber forms on each branch or bud. In some cases, a stolon develops from primary tuber eyes and a secondary tuber forms at the end of the stolon (Lugt, 1960; Lugt *et al.*, 1964; Okazawa and Iriuda, 1980). Small, young tubers have a greater tendency to produce stolons at the bud end (Krauss, 1985). In severe cases, secondary tubers form directly from the primary tuber eye.

Heat sprouts and chain tubers result when tuber growth is renewed after development has been interrupted during extended exposure to warm soil temperatures or excessive nitrogen. Second growth is related to reduced apical dominance of primary tubers, a temporary alteration in tuberization response, or disruption of source–sink relationships (Bodlaender *et al.*, 1964; Lugt *et al.*, 1964; Krauss, 1985). Gibberellin (GA), the phytohormone believed to be the dominant regulator of tuber formation, is high in elongating stolon tips and low in developing tubers (Xu *et al.*, 1998). Low GA levels promote tuber growth and high GA activity inhibits tuber growth. Xu *et al.* (1998) demonstrated that by alternating between low and high GA-containing media, chain tubers could be induced to form on cuttings *in vitro*. With high soil temperature or high nitrogen endogenous GA levels increase, tuber growth ceases and regrowth (i.e. new stolon growth) occurs (Krauss, 1985). When temperature decreases or nitrogen is removed, the plants revert back to a tuberizing state.

When chain tubers occur, the secondary tuber becomes dominant and the primary tuber

Fig. 14.2. Heat sprouts and chain tubers, two types of second growth. (Photo courtesy or L. van Marion, McCain Foods Europe.)

loses its sink properties and stops growing (Lugt *et al.*, 1964; Krauss, 1985). After vine killing, carbohydrates are translocated to the secondary tuber from the primary tuber, which will have a reduced specific gravity and become glassy (Lugt, 1960; van der Zaag, 1996). Secondary tubers are physiologically younger than primary or normal tubers, and when planted, will differ in emergence and stem and tuber numbers produced (Jefferies and MacKerron, 1987a).

Little tuber describes the phenomenon of daughter tubers forming directly from an eye, or a very short stolon from an eye of a seed tuber, without sprout or foliage development. They may form in storage prior to planting, or in the field. Cultivars differ in susceptibility to little tuber. Little tuber is associated with physiologically old seed tubers, with tubers that have been stored at warm temperatures (20°C or higher), then planted into cold (less than 10°C), dry soils, or sprouted tubers that have been transferred from warm storage to cold storage, then planted at a later date (Hiller *et al.*, 1985).

Cultural practices that encourage uniform vine and tuber growth will help to minimize second growth. Ensure uniform growth with proper plant spacing and population, correct fertilizer rate and placement, and uniform soil moisture levels.

To control little tuber, store seed tubers in cool storage at 3–4°C, avoid using seed tubers stored for an excessively long time or aged at high temperatures, and do not plant physiologically old seed into cold, dry soil. Storing seed tubers in the light can slow physiological aging (van der Zaag, 1996).

14.7 Skinning and Scald

Skinning, or feathering, and scald are skin defects of immature tubers. Skinning occurs when immature tubers are harvested or handled. The skin of the tuber is not “set” and is only loosely held in place and easily scuffed or slipped off. The skin may be completely or partially removed, exposing the tissues beneath. Loose shreds of skin adhering to the surface of the tuber will dry and become thin and papery.

The potato periderm is composed of three layers: phellem (cork), the outermost layer, which is suberized and protective; phellogen (cork cambium) just beneath the phellem; and phelloderm, the innermost layer just beneath the phellogen. The phellem and phelloderm layers are derived from the mitotic activity of the phellogen, which is a single layer of meristematic cells. The phellem is comprised of several layers of cells (approximately six to ten) and is the portion of the periderm commonly referred to as the skin of the potato. In immature tubers, the active meristematic cells of the phellogen have thin radial walls that fracture easily with mechanical pressure, allowing the phellem or skin to separate or scuff off (Lulai and Freeman, 2001), thus skinning the tuber. As the phellogen becomes inactive as the tuber matures, its cell walls thicken, become resistant to fracture, and the skin becomes tightly bound to the tuber (Lulai and Freeman, 2001).

Skinned tubers are highly susceptible to dehydration and invasion by pathogens (Lulai and Orr, 1995). The skin of the bud end of the tuber is more fragile and prone to skinning than the stem end (Lulai and Orr, 1993). Under proper wound-healing conditions, skinned areas of the tuber will heal.

Scald (not to be confused with sunscald, see “Greening” above) may develop when skinned tubers are exposed to sun and warm winds of high velocity and low relative humidity after harvest (Whiteman and Lutz, 1954; O'Brien and Rich, 1976). Relatively short exposure (less than 4 h) is needed for damage to appear. The skinned areas turn brown, due to rapid superficial drying, and have a scalded appearance. In storage, affected tubers often decay and suffer weight loss.

To control skinning and scald, avoid harvesting immature tubers. Handle tubers carefully during harvest. Harvest tubers 14–21 days after vine kill to ensure proper tuber maturity and skin set. Do not expose freshly harvested tubers to sun, wind, or low relative humidity for extended periods during or after harvesting.

14.8 Pink Eye

Initially believed to be caused by a pathogen, current research suggests pink eye is a physiological disorder (Lulai *et al.*, 2006; Sabba *et al.*, 2008). It is characterized by puffy, pink-colored areas around the eyes and on the periderm of tissues between the eyes, initially on the bud end of the tuber. As the disorder progresses, water-soaked or dried and cracked corky patches develop on the tuber (Sabba *et al.*, 2008). The corky patches develop late in the season or during storage and are known as "corky patch syndrome" or "bull hide" (Secor and Rouse, 1992). The thickened corky patches may extend up to 3 mm into the tuber flesh (Nolte *et al.*, 1993). Internal symptoms of pink eye progress from erratic browning of cortical cell walls to necrotic zones, often surrounded by an internal, pink eye-related periderm (Sabba *et al.*, 2008). On exposure to ultraviolet light, tissues directly beneath the areas affected by pink eye exhibit intense autofluoresence (Nolte *et al.*, 1993), due to the presence of suberin poly(phenolic(s)) (Lulai *et al.*, 2006).

Pink eye has been implicated as a precursor to many tuber diseases such as bud end decay (Secor and Rouse, 1992). Suberized cells of the native periderm of pink eye surfaces are compromised or deteriorated, increasing tuber susceptibility to rot organisms (Lulai *et al.*, 2006). It is hypothesized that very wet and hot soil conditions contribute to low oxygen levels in the soil, resulting in cell damage and death in the tuber periderm and underlying cortical cells, and the development of pink eye (Sabba *et al.*, 2008).

Pink eye is associated with excessive moisture late in the season in conjunction with high soil and air temperatures (Secor and Rouse, 1992; Copas *et al.*, 2008). Pink eye is most severe after periods of saturated soil conditions or prolonged periods of free water followed by high temperatures. Copas *et al.* (2008) found pink rot symptoms appeared 7–10 days after excessive irrigation or precipitation events. Tubers from fields with a long history of potato production tend to develop pink eye (Secor and Rouse, 1992).

Control of pink eye is difficult. Avoid excessive moisture prior to and during tuber harvest. There is no control in storage, but it is important to provide cool, dry conditions to prevent the development of soft rot.

14.9 Brown Center and Hollow Heart

Brown center (also known as incipient hollow heart, brown heart, or sugar center) and hollow heart are considered to be two different phases of the same internal disorder, and are likely induced by similar conditions. Brown center and hollow heart, although related, can occur independently of each other. Brown center is believed to be a precursor to hollow heart, although hollow heart does not always develop (Hiller *et al.*, 1985). Affected tubers and plants show no external symptoms.

Brown center is a small, firm, brown discoloration of pith tissue near the center of the tuber. Tubers are most susceptible to brown center when they are very small, from the tuber initiation stage up to the time when tubers weigh 56 g (Thornton, 2001b). The discoloration arises from damage to cell membranes and organelles, and necrosis of affected cells (Van Denburgh *et al.*, 1986). Cool soil temperatures, for example 10–15°C, around the time of tuber initiation are reported to induce brown center (Hiller *et al.*, 1979; Van Denburgh *et al.*, 1980). Damage to pith cell membranes begins to appear after 5 days at soil temperatures of 10°C, with brown center visible to the naked eye after 16 days (Van Denburgh *et al.*, 1986). High soil moisture increases the severity of brown center (Hiller and Koller, 1981). If tuber growth is slow and uniform after discoloration develops, growth of living cells spreads and disperses the dead ones, dissipating the brown color by the end of the growing season (Thornton, 2001b).

If, however, the tuber grows rapidly, the affected tissues may tear or split apart and form a cavity due to growth of the perimedullary region of the tuber outpacing that of the pith region (Rex and Mazza, 1989). The result is the disorder known as hollow heart. In the past, the only way to detect hollow heart was to cut the tuber open. Equipment that detects hollow heart via X-rays is commercially available and in use in North America. Other non-destructive means of detecting hollow heart that have been studied include acoustic impact (Elbatawi, 2008), ultrasonics (Jivanuwong, 1998), and nuclear magnetic resonance imaging (Wang *et al.*, 1997).

Hollow heart refers to a cavity in the central pith of a potato (Fig. 14.3). It may be lens-, star-, or irregularly shaped, and range in size from a small gap of a few millimeters long to a hole that occupies almost the entire tuber pith

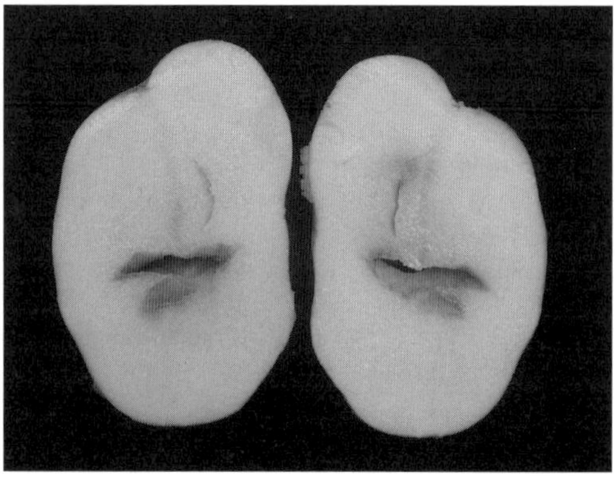

Fig. 14.3. Hollow heart in Yukon Gold tuber. (Photo courtesy of S.B. Johnson, University of Maine.)

area. There may be one or more cavities in the tuber that are transverse or longitudinal to the pith. The cavity may develop a suberin lining similar in structure to the periderm, giving it a tan or dark brown color (Dean *et al.*, 1977). Depending on when the cavity is formed, it may be near the center of the tuber or towards the bud end or stem end. On occasion, fungi or bacteria invade the cavity and rot the tuber from the inside (Wale *et al.*, 2008).

Cavities formed near the stem end of the tuber are called stem-end hollow heart. Stem-end hollow heart is formed early in the growing season, shortly after tuber initiation, and is preceded by a brown center (Levitt, 1942). It is often associated with a period of restricted growth due to cool soil temperatures, followed by accelerated growth rates as the soil warms up. A cavity near the bud end of the tuber is called bud-end hollow heart. It is not associated with brown center and is initiated late in the tuber bulking period. Bud-end hollow heart is commonly caused by tuber growth stoppage due to water or nutrient stress late in the season (Rex and Mazza, 1989). Cultivars vary in susceptibility to hollow heart (Rex and Mazza, 1989; Jansky and Thompson, 1990). Russet Burbank and Yukon Gold are very prone to hollow heart. Large tubers tend to develop hollow heart more frequently than smaller tubers (Nelson and Thoreson, 1986; Jansky and Thompson, 1990), but hollow heart may be found in small tubers during a period of exceptionally rapid growth (Hiller *et al.*, 1985). Hollow heart does not affect seed tuber quality, nor affect incidence of the disorder in the subsequent crop.

Tuber growth rate is an important factor in the development of hollow heart. Rapidly growing tubers are more likely to develop hollow heart. Hollow heart typically appears when stress conditions affect the plant and reduces tuber bulking, followed by a relief from the stress and a rapid increase in tuber bulking rate, such as when rainfall follows a moisture deficit. Nelson (1970) observed over a period of 4 years that "hollow heart is particularly serious when stands are poor, moisture is adequate or heavy, temperatures are below normal and plants are vigorous enough to maintain a rapid rate of tuber growth". Growing season or year of production strongly influences the development of hollow heart.

Wide plant spacing promotes larger tubers and favors hollow heart (Rex *et al.*, 1987; Creamer *et al.*, 1999). High rates of nitrogen fertilizer especially late in the season promote the disorder (Wilson *et al.*, 2009), while potassium fertilization tends to reduce hollow heart (Nelson, 1970; Panique *et al.*, 1997). Nitrogen applied near tuber initiation may stimulate tuber growth rate and enhance hollow heart formation (Hiller and Thornton, 1993). A connection exists between tuber calcium concentration and hollow heart. Tubers with non-periderm calcium levels of 100 $\mu g\ g^{-1}$ dry weight or less have a greater incidence of hollow heart (Kleinhenz *et al.*, 1999).

Controlling hollow heart and brown center completely is very difficult, if not impossible. However, some cultivars are less susceptible to both conditions and should be chosen in regions known to exacerbate their formation. Other than cultivar choice, managing hollow heart and brown center involves using cultural practices that promote uniform steady plant and tuber growth rates and reduced plant stress. Delay planting to avoid cool soil temperatures at tuber initiation. Establish uniform plant stands by planting at the proper depth and spacing. The use of large seed pieces ensures a good stand and several stems per hill. Adequate plant competition and a full tuber set help reduce periods of rapid tuber growth. Avoid excessive irrigation and fertilization. Maintain uniform soil moisture and fertility levels for steady, uniform plant growth.

14.10 Bruise

There are three types of bruising: blackspot bruise and shatter bruise, caused by physical impacts during harvest and handling, and pressure bruise, which results from the weight and pressure of other tubers or container surfaces during storage, when tubers lose weight due to moisture loss.

Blackspot (also called internal black spot or blue spot) bruise appears as a small to large bluish-gray or black discolored area just beneath the skin of the tuber in the perimedullary tissue (Fig. 14.4). There are no external symptoms and damage is visible only when tubers are peeled. When tubers bounce against hard surfaces with enough force, membranes rupture and allow the enzyme, polyphenol oxidase (PPO), to come into contact with the phenol tyrosine in the intercellular space. The PPO oxidizes the phenols to form the black pigment, melanin (Dean *et al.*, 1993; Dean, 1996). The damaged area turns pink to reddish-brown first, then grayish-black as the melanin forms. At 21°C, the discoloration begins to appear in 6 h and is complete by 24 h (Strand, 2006).

Cultivars vary in susceptibility to blackspot bruise, and blackspot bruise susceptibility has been correlated with tuber tyrosine level (Dean *et al.*, 1993; McNabnay *et al.*, 1999) and tuber cell size (Hudson, 1975). The stem end of the tuber (Kunkel and Gardner, 1959), large tubers, tubers with high specific gravity, dehydrated or non-turgid tubers, and very mature tubers are more susceptible to blackspot bruise (Brook, 1996; Corsini *et al.*, 1999; Karlsson *et al.*, 2006). Inadequate potassium levels increase tuber blackspot bruise susceptibility (McNabnay *et al.*, 1999). Tubers grown on sandy or loamy soils tend to have more blackspot bruise due to the lower water-holding capacity of the soil (Corsini *et al.*, 1999). Immature tubers harvested from green vines are more resistant to blackspot bruise than more mature tubers (Pavek *et al.*, 1985). Blackspot bruise susceptibility increases during storage (Corsini *et al.*, 1999).

Karlsson *et al.* (2006) studied tuber calcium levels and blackspot bruise incidence. Two cultivars with consistently low tuber calcium concentrations, Atlantic and Snowden, had high incidence of bruise, while Dark Red Norland and Superior had low incidence of bruise and

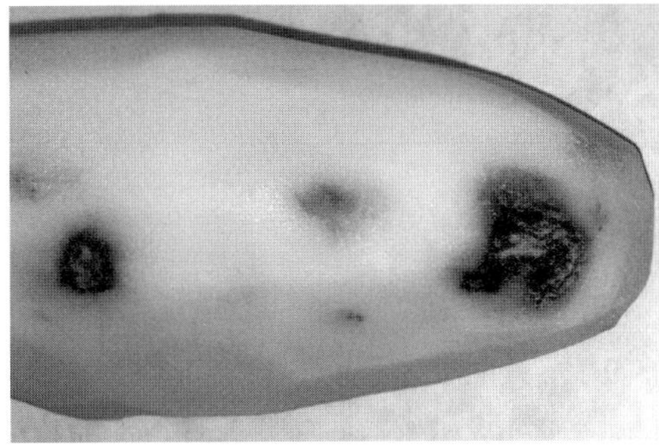

Fig. 14.4. Blackspot bruise damage on a peeled potato. (Photo courtesy of L. Mikitzel, New Brunswick Department of Agriculture, Aquaculture and Fisheries.)

consistently high tuber calcium concentrations. They found blackspot bruise incidence was minimized when medullary calcium concentrations were 200–250 ppm on a dry-weight basis. Tuber calcium concentration can be increased by supplemental calcium applications during the growing season.

Tuber temperature during harvest, grading, and transport influence blackspot development. Tubers are more susceptible to blackspot bruise at low temperatures (less than 12°C) than they are at higher temperatures (Smittle *et al.*, 1974; van der Zaag, 1996; Corsini *et al.*, 1999).

In some cultivars, tuber damage from physical impacts is manifested as a brownish patch with a hard, grayish white starch deposit in the center (van der Zaag, 1996; Christ, 1998; Thornton, 2001b). The starch deposit is comprised of an abundance of densely packed, large, intracellular starch granules, two to four times larger than those in normal cells (Keenan *et al.*, 1989). It is located several millimeters beneath the skin, usually between the periderm and vascular ring (Christ, 1998). This disorder is referred to as white knot, starch spot, or white spot bruise, and is visible only on cutting the tuber. The cultivar, Atlantic, is prone to white knot and it is usually found when specific gravity is high (Keenan *et al.*, 1989; Christ, 1998). White knot forms a hard spot when tubers are processed.

Shatter bruise occurs when a significant impact produces cracks in the outer skin that extend into the tuber flesh below. The tuber has a shattered appearance at the point of impact. As the tissues dry out and separate, the cracks become apparent. Shatter bruise increases with high tuber turgidity and temperatures below 10°C (Smittle *et al.*, 1974).

Pressure bruises develop in storage and are flattened or slightly sunken areas on the tuber where it has been in contact with other tubers or solid surfaces. When relative humidity is below 95%, the external pressure causes the affected tissue to lose water and become flattened. The depression will persist if storage relative humidity remains below 90%. Internally, tissues may exhibit gray to black discoloration, usually more intense in the vascular region (Olsen *et al.*, 2003). Warm storage temperatures, excessive potato pile height, and long storage favor pressure bruise.

Bruise damage is minimized by careful harvest of tubers that are not too cold and of the proper hydration level. To ensure adequate hydration at harvest, follow good cultural practices to promote proper vine and root growth. Supply adequate potassium nutrition. If the soil is dry, and irrigation is available, apply a light irrigation to soften soil clods, reduce abrasion, and increase tuber hydration. Tuber pulp temperature should be between 10°C and 15°C. Avoid harvesting when pulp temperature is below 7°C and above 18°C. Adjust harvesting equipment to minimize drop heights and add padding to areas known to damage tubers. During harvest, keep conveyors full of tubers to help reduce roll back. The use of a canvas sling in bulk trucks helps to reduce drop height from the boom into the truck.

Control of pressure bruise includes maintaining storage relative humidity above 90%, piling tubers not more than 5.5 m high, and holding tubers at the optimum storage temperature.

14.11 Blackheart

The term "blackheart" describes a sharply delineated, purplish-gray to black area in the center (pith) and closely associated parenchyma cells of the tuber (Fig. 14.5). Occasionally, the discoloration radiates to the skin. The black discoloration of blackheart is due to melanin, the same pigment responsible for blackspot bruise (Reeve, 1968). The affected tissue is usually firm and has no odor. A cavity may form if the affected tissue shrinks. This cavity is not lined with wound periderm, as in hollow heart, but with gray or black layers, and is referred to as a "cat's eye" (Strand, 2006). Only in extreme cases are there visible external symptoms of moist, discolored patches (Snowden, 1992).

Blackheart is the result of inadequate oxygen supply. It develops when internal tuber tissues do not receive enough oxygen to support respiration (Davis, 1926). Oxygen deficiency may be due to inadequate ventilation in closed storage bins or transport vehicles, or exposure of tubers to near freezing or greater than 32°C temperatures. At high and low temperatures, gas diffusion is too slow to reach the internal tuber tissues. At lower temperatures, longer times are required for blackheart to develop (Stewart and Mix, 1917). In the field, high soil temperatures and waterlogged soils contribute to blackheart development. Tubers will not recover once blackheart has developed. Seed tubers that have blackheart should not be planted, due to increased susceptibility to soft rot and poor emergence problems (Wale *et al.*, 2008).

Some potato cultivars tolerate conditions of low oxygen better than others, but precautions must be taken with all potatoes. Avoid planting potatoes in poorly drained areas. Harvest tubers from hot, dry soils as soon as the vines are dead. After digging, do not leave tubers lying on hot, dry soils (O'Brien and Rich, 1976). Provide good ventilation in storage and transit. Avoid exposing tubers to temperatures greater than 32°C or prolonged near-freezing storage.

14.12 Translucent End/Jelly End

The disorder, translucent end, is also known as sugar end, glassy end, incipient jelly end, or dark end. In affected tubers, the stem end contains higher reducing sugars (glucose and fructose) and lower dry-matter content than the bud end (Iritani and Weller, 1973; Sowokinos *et al.*, 2000). The tissue appears translucent or water soaked. Usually, less than 5 cm of the stem end is affected (Strand, 2006). When processed into chips or French fries, heat during processing causes the reducing sugars to react with amino acids and form undesirable dark compounds. This disorder is sometimes referred to as dark end, because one end of the French fries made

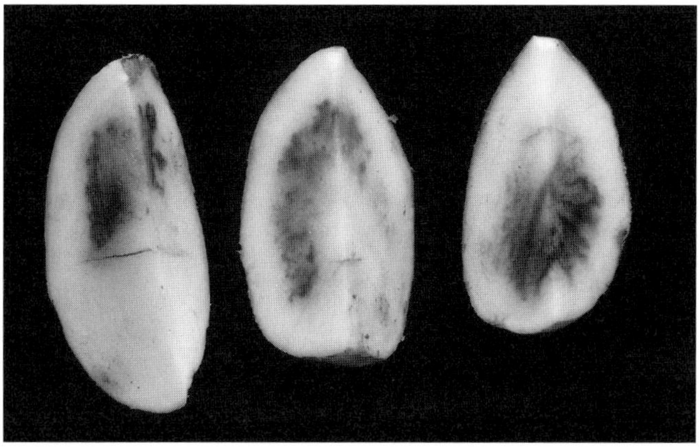

Fig. 14.5. Blackheart. (Photo courtesy of K.I. Al-Mughrabi, New Brunswick Department of Agriculture, Aquaculture and Fisheries.)

from affected tubers is dark and the other end is light. Certain cultivars, especially those with long-type tubers like Russet Burbank, tend to be more prone to translucent end, and it is often associated with tubers exhibiting second growth or pointed stem ends (Hiller and Thornton, 1993).

When translucent end tissue becomes soft and jelly-like, it is referred to as jelly end. A condition called jelly end rot develops when jelly end tuber tissues break down and collapse in storage (Iritani and Weller, 1973). There is a distinct line between the affected and healthy tissues (Thornton, 2001b). The rot does not usually spread in storage (Hodgson *et al.*, 1974).

Moisture stress and heat stress during tuber initiation and early tuber bulking result in translucent end (Iritani and Weller, 1973; Kleinkopf *et al.*, 1988; Shock *et al.*, 1993). Shock *et al.* (1993) found that a single transitory water deficit was sufficient to induce translucent end. They also found that reducing sugar content of the stem end of the affected tubers did not increase during or just after the moisture stress, but several weeks later. When stressed, immature tubers and tubers stored at low temperatures (5°C) immediately following harvest are more prone to translucent end (Iritani and Weller, 1973, 1978). Inadequate fertility that leads to premature vine death has been associated with high reducing sugars in the stem ends of tubers (Iritani and Weller, 1978).

Sugar accumulation in translucent end tissue is due to stress-induced alteration of the enzymes responsible for sucrose and starch metabolism. Geigenberger *et al.* (1997) demonstrated that, with moderate water stress, sucrose phosphate synthase, the enzyme responsible for sucrose synthesis, was activated in intact tubers. With more extreme water stress, they also found inhibition of the enzymes involved in the terminal reactions of starch synthesis. Sowokinos *et al.* (2000) compared enzyme activity in translucent end and normal tissues. In translucent end tissue, compared with normal tissue, adenosine diphosphate (ADP)-glucose pyrophosphorylase and uridine diphosphate (UDP)-glucose pyrophosphorylase activity were decreased and starch phosphorylase activity was increased, all indicative of starch mobilization. In addition, activity of acid invertase, the vacuolar enzyme that breaks down sucrose into glucose and fructose, was twice as high in translucent end tissue as in normal tissue mid-growing season, and four times as high after several months in storage. Increased inorganic phosphate (Pi) (Sowokinos *et al.*, 2000) and decreased pH (Iritani and Weller, 1973; Sowokinos *et al.*, 2000) are also characteristic of translucent end tissues. In translucent end tubers, there is a shift in carbohydrate metabolism away from starch synthesis and toward the accumulation of reducing sugars. A recent review by Thompson *et al.* (2008) revealed that, once translucent end was triggered, reducing sugar accumulation could not be reversed.

To reduce the occurrence of translucent ends, use cultural practices that promote good plant stands and uniform crop growth. Irrigation should match water requirements of the crop. Avoid moisture stress, especially during early tuber growth. If irrigation is not possible, use cultural practices that help to retain soil moisture, such as proper spacing, hilling, and cultivation. Grow cultivars that are less susceptible to translucent end.

14.13 Internal Brown Spot and Heat Necrosis

Non-pathogenic necrosis in potato tubers has been referred to by many different names: internal brown spot, internal heat necrosis, internal browning, internal rust spot, internal brown fleck, physiological internal necrosis, and chocolate spot (see Hiller *et al.*, 1985, and references therein). These names more than likely refer to the same disorder or similar disorders that are difficult to distinguish from one another. In the western USA, internal brown spot is the accepted term for this disorder, in the eastern USA, internal heat necrosis is the term used, while in Europe, internal rust spot is customarily used. Internal brown spot (IBS) will be used here. For an in-depth review of this disorder, see Yencho *et al.* (2008).

Symptoms include reddish-brown or rust-colored groups of necrotic parenchyma cells that appear spongy and suberized and are lacking in starch (Baruzzini *et al.*, 1989). The blotches or spots of necrotic cells may be small, round, or irregularly shaped. In IBS, the blotches are scattered throughout the flesh of affected tubers. The spots can occur anywhere in the tuber, but are more commonly found inside the vascular ring. The lesions seldom occur outside of the vascular ring in the cortex. When the spots are concentrated along or just inside the vascular ring, some authors characterize the disorder as internal heat necrosis; however, internal heat necrosis may occur anywhere in the tuber (Yencho *et al.*, 2008). It has also been reported that internal heat necrosis occurs in tubers growing near the soil surface (Thornton, 2001b), while IBS can be found in tubers throughout the profile of the potato hill (Hiller *et al.*, 1985). IBS differs from brown center in that the necrotic spots do not concentrate in the pith or center of the tuber. There are normally no external plant or tuber symptoms with IBS; the disorder is seen only after the tubers are cut open. Several publications contain color plates of tubers affected by IBS and heat necrosis (Hiller *et al.*, 1985; Hiller and Thornton, 1993; Stevenson *et al.*, 2001). Large tubers (Iritani *et al.*, 1984; Sterrett *et al.*, 1991a) and tubers with high specific gravity (Iritani *et al.*, 1984) have a higher incidence of IBS. Cultivars vary greatly in IBS susceptibility (Collier *et al.*, 1980; Sterrett and Henninger, 1991; Davies, 1998); the cultivars, Yukon Gold and Atlantic, are highly susceptible, whereas Desiree is very resistant.

IBS may develop soon after tuber initiation (Olsen *et al.*, 1996), or during periods of uneven or rapid tuber growth (Hiller *et al.*, 1985; Hiller and Thornton, 1993). Initially, the necrotic spots occur toward the bud end of the tuber, and over time, the color intensity and area of the tuber affected increase (Sterrett *et al.*, 1991a). Symptoms may increase in storage, particularly in tubers stored at higher temperatures (Iritani *et al.*, 1984; Hiller and Thornton, 1993).

The exact cause of IBS is poorly understood. Localized calcium deficiencies have been implicated as initiating cell death and tissue necrosis associated with IBS (Collier *et al.*, 1980; Davies, 1998). With restricted calcium (due to environmental stress), the loss of membrane integrity is accompanied by oxidative damage, which, when severe, leads to cell death (Davies, 1998). In resistant cultivars, activity of enzymatic antioxidants is significantly higher than in susceptible cultivars (Davies, 1998). Tubers with IBS have lower calcium concentrations than unaffected tubers (Collier *et al.*, 1978, 1980; Sterrett and Henninger, 1991; Olsen *et al.*, 1996). However, there is no clear threshold for calcium below which the disorder is induced (Davies, 1998). Ozgen *et al.* (2006) found IBS incidence to decrease as tuber calcium content increased from 100 to 250 μg g^{-1} dry weight. When calcium fertility or availability is low and soil temperatures are high, the incidence of IBS increases. Recently, Sterrett *et al.* (2006) found that higher tuber manganese and sulfur levels and lower phosphorus levels were associated with lower IBS incidence. Tuber mineral status may be more important to IBS susceptibility, and not the concentration of a single mineral.

The frequency and severity of IBS are increased with high day and night temperatures early in the growing season, together with low rainfall (few rain events) (Sterrett *et al.*, 1991a,b; Yencho *et al.*, 2008). IBS may be more common on sandy soils (O'Brien and Rich, 1976), which have low cation exchange capacity and great heat conductance. Growing location also influences the incidence and severity of IBS (Sterrett and Henninger, 1991; Sterrett *et al.*, 1991a).

To reduce the incidence of IBS, when possible, plant less susceptible cultivars. Ensure adequate calcium level in the stolon and tuber zone, especially on sandy or low-calcium soils. Delaying planting reduces the incidence of IBS (Iritani *et al.*, 1984; Sterrett *et al.*, 1991b), possibly by mediating tuber size. In warmer climates, time planting to mature the crop before excessive heat is expected. Prevent exposing tubers to high soil temperatures at harvest. Closer plant spacing may be used to limit the production of large tubers and reduce IBS (Creamer *et al.*, 1999). Maintain adequate soil moisture through irrigation, mulching, and full canopy, and use cultural practices that promote uniform, unstressed vine and tuber growth.

14.14 Low-Temperature Injury

Low-temperature injury ranges from slight damage caused by exposure to non-freezing temperatures to outright freezing and death of tuber tissues. Low-temperature injury may occur in the field, during storage, or in transit. Synonyms for low-temperature injury are freezing necrosis, freezing injury, frost necrosis, frost injury, chilling injury, and mahogany browning. The severity of low-temperature injury depends on temperature, length of exposure, and cultivar. Within any lot of tubers, at a given temperature, the degree of tissue damage and the range of injuries found may vary considerably.

Ice crystals form in the tissues of tubers exposed to freezing temperatures, causing irreversible damage. Potato tissues freeze at temperatures of −1.9°C to −1.4°C, and when exposed to temperatures at or below −2.8°C, extensive mortality occurs (Boydston *et al.*, 2006). Damage due to prolonged freezing or very low temperatures is only visible once the tubers have warmed or thawed. The damage that occurs on thawing is from intracellular ice crystals rupturing cell membranes. Tubers killed by freezing become soft, watery, and exude much liquid. These tubers disintegrate rapidly in a watery rot, and are often colonized by bacteria. In a cold or dry environment, frozen tubers thaw into a mealy, granular, chalky mass (O'Brien and Rich, 1976). In the field, freezing usually occurs on the end or side of the tuber closest to the soil surface, usually the bud end. In partially frozen tubers, there is a distinct purplish to brownish line of corky tissue between healthy and damaged tissues.

Shorter exposures to temperatures just at freezing or brief exposure below the freezing point are marked by several types of internal discoloration. There are no external symptoms, and tubers must be cut open to see the damage. Tissues at the stem end and vascular ring are most susceptible to this low-temperature or chilling injury (Thornton, 2001b). A net type of necrosis may develop where the phloem is blackened and dark strands and necrotic spots are scattered throughout the tuber tissue. Blue-gray or black discolorations or blotches may be present in the vascular ring and associated tissues. The vascular ring may break down. Irregularly shaped, opaque bluish-gray to black patches or diffuse smoky areas may develop anywhere in the tissues. Tubers with low-temperature injury, even those that appear to be only slightly injured, should not be used for seed, as germination may be delayed or very poor.

Exposure to cold, non-freezing (0–3°C) temperatures may also lead to tuber injury. Starch is converted to sugars at these temperatures, causing the tubers to be sweet. Mahogany browning, where affected tissues turn gray or reddish-brown, is a disorder induced when tubers are exposed to temperatures slightly below freezing (Thornton, 2001b). Tubers held close to 0°C in long-term storage may develop external dark brown, slightly sunken areas that are predisposed to surface mold. Areas under the surface lesions vary from reddish-brown to dark brown. In some cases, discolored areas may be found in the center of the tuber as well (van der Zaag, 1996).

The vascular necrosis in tubers associated with low-temperature injury may resemble that caused by *Potato leaf roll virus*. Unlike the diseased tubers, tissues damaged by low temperature will fluoresce on exposure to ultraviolet light (Hruschka *et al.*, 1969).

Potato tubers are able to be supercooled to −7°C, and remain unharmed if allowed to warm up (Thornton, 2001b; Boydston *et al.*, 2006). However, if these tubers are jarred or handled, ice crystals will form and cause cell death.

To minimize low-temperature injury, ensure adequate soil depth over tubers, with proper hilling, and harvest before the danger of severe frost. In storage and in transit, maintain temperature above 3°C. Intermittent warming during long-term storage at low temperatures can prevent low-temperature injury (Hruschka *et al.*, 1969). If tubers with low-temperature injury must be stored, use high airflow rates to minimize losses.

14.15 Vascular Discoloration

Vascular discoloration is known by several names – vascular necrosis, stem-end discoloration, stem-end browning, stem-end necrosis, vascular browning, phloem necrosis, and xylem ring discoloration. It is characterized by discoloration of varying intensity in the vascular ring at the stem end of the tuber. Symptoms, visible after the tuber is cut, include necrotic, light tan, reddish-brown or dark brown streaks at the point of stolon attachment. The discoloration is usually confined to within 1 cm of the stem end, but in severe cases may extend the length of the tuber. Vascular discoloration affects the marketability of table-stock and processing potatoes. This vascular discoloration often causes concern, because it resembles vascular discoloration caused by Verticillium or Fusarium wilt.

Sometimes, the cells at the point of stolon attachment will collapse to form a necrotic brown lesion (van der Zaag, 1996; Wale *et al.*, 2008). This lesion may be invaded by fungi or bacteria. Vascular discoloration has no effect on seed quality, and affected tubers are viable as long as no pathogen invades the lesion.

Vascular discoloration occurs when plant haulms are killed rapidly under dry conditions, particularly when the plants are still green and the weather is hot (Sieczka and Thornton, 1993). The disorder is most noticeable when quick-acting chemical desiccants are used, but may also occur when the haulm is destroyed mechanically or by heavy frost. Growing location appears to contribute to the appearance of this disorder (Halderson *et al.*, 1985). The tendency for vascular discoloration is increased when vines are less mature at the time of top killing (Halderson *et al.*, 1985). Vascular discoloration may be present at harvest or may increase in intensity during the first few months of storage.

To control vascular discoloration, use chemicals that kill the haulm gradually. Avoid rapid vine-killing methods during hot weather, especially if the soil is dry.

14.16 Internal Anthocyanin Discoloration

On occasion, atypical anthocyanin synthesis occurs in the flesh of cultivars where normally there is none. Some cultivars are prone to internal anthocyanin discoloration, while others are rarely, if ever, affected. Abnormal pigmentation has been seen in many potato cultivars; Record and Ulster Sceptre in Great Britain, Roseval in France (Storey and Davies, 1992), and Nooksack, Hilite Russet, Frontier Russet, and Russet Norkotah in the USA and Canada (Olsen *et al.*, 2003).

When affected tubers are cut, purplish red stripes and flame-shaped discolorations can be seen in the perimedullary region. The purple or red color occurs when the natural balance between anthocyanin precursors and inhibitors is disrupted (Storey and Davies, 1992). This disorder appears to be associated with light, as it is observed mainly in green tubers (van der Zaag, 1996). It is more prevalent in tubers found near the soil surface. Flesh pigmentation has also been associated with low temperatures (below 10°C) in the field or storage (Olsen *et al.*, 2003). When cooked, the discoloration becomes slightly less apparent, or may disappear if pale.

To reduce the development of anthocyanin discoloration, ensure adequate hilling in the field.

References

Baruzzini, L., Ghirardelli, L.A. and Honsell, E. (1989) Ultrastructural changes in tubers of potato (*Solanum tuberosum* L.) affected by rust spot. *Potato Research* 32, 405–410.

Bodlaender, K.B.A., Lugt, C. and Marinus, J. (1964) The induction of second-growth in potato tubers. *European Potato Journal* 7, 57–71.

Bohl, W.H. and Love, S.L. (2005) Effect of planting depth and hilling practices on total, U.S. No. 1, and field greening tuber yields. *American Journal of Potato Research* 82, 441–450.

Boydston, R.A., Seymour, M.D., Brown, C.R. and Alva, A.K. (2006) Freezing behavior of potato (*Solanum tuberosum*) tubers in soil. *American Journal of Potato Research* 83, 305–315.

Brook, R.C. (1996) *Potato Bruising: How and Why Emphasizing Black Spot Bruise*. Running Water Publishing, Haslett, Michigan.

Carnegie, S.F. and McCreath, M. (2010) Mosaic virus symptoms in potato crops and the occurrence of growth cracking in tubers. *Potato Research* 53, 17–24.

Christ, B.J. (1998) *Identifying Potato Diseases in Pennsylvania*. The Pennsylvania State University, University Park, Pennsylvania.

Collier, F.G., Wurr, D.C.E. and Huntington, V.C. (1978) The effect of calcium nutrition on the incidence of internal rust spot in the potato. *Journal of Agricultural Science Cambridge* 91, 241–243.

Collier, F.G., Wurr, D.C.E. and Huntington, V.C. (1980) The susceptibility of potato varieties to internal rust spot. *Journal of Agricultural Science Cambridge* 94, 407–410.

Copas, M.E., Bussan, A.J. and Drilias, M.J. (2008) Influence of compaction and subsoil tillage on soil conditions and pink eye. *American Journal of Potato Research* 85, 342–352.

Corsini, D., Stark, J. and Thornton, M. (1999) Factors contributing to the blackspot bruise potential of Idaho potato fields. *American Journal of Potato Research* 76, 221–226.

Creamer, N.G., Crozier, C.R. and Cubeta, M.A. (1999) Influence of seedpiece spacing and population on yield, internal quality, and economic performance of Atlantic, Superior and Snowden potato varieties in Eastern North Carolina. *American Journal of Potato Research* 76, 257–261.

Davies, H.V. (1998) Physiological mechanisms associated with the development of internal necrotic disorders of potato. *American Journal of Potato Research* 75, 37–44.

Davis, W.B. (1926) Physiological investigation of black heart of potato tuber. *Botanical Gazette* 81, 323–338.

Dean, B. (1996) The chemical nature of blackspot bruising. In: Brook, R.C. (ed.) *Potato Bruising; How and Why; Emphasizing Black Spot Bruise*. Running Water Publishing, Haslett, Michigan, pp. 29–38.

Dean, B.B., Kolattukudy, P.E. and Davis, R.W. (1977) Chemical composition and ultrastructure of suberin from hollow heart tissue of potato tubers (*Solanum tuberosum*). *Plant Physiology* 59, 1008–1010.

Dean, B.B., Jackowiak, N., Nagel, M., Pavek, J. and Corsini, D. (1993) Blackspot pigment development of resistant and susceptible *Solanum tuberosum* L. genotypes at harvest and during storage measured by three methods of evaluation. *American Potato Journal* 70, 201–217.

Edwards, E.J., Saint, R.E. and Cobb, A.H. (1998) Is there a link between greening and light-enhanced glycoalkaloid accumulation in potato (*Solanum tuberosum* L.) tubers? *Journal of the Science of Food and Agriculture* 76, 327–333.

Elbatawi, I.E. (2008) An acoustic impact method to detect hollow heart of potato tubers. *Biosystems Engineering* 100, 206–213.

Geigenberger, P., Reimholz, R., Geiger, M., Merlo, L., Canale, V. and Stitt, M. (1997) Regulation of sucrose and starch metabolism in potato tubers in response to short-term water deficit. *Planta* 201, 502–518.

Ginzberg, I., Barel, G., Ophir, R., Tzin, E., Tanami, Z., Muddarangappa, T., *et al.* (2009) Transcriptomic profiling of heat-stress response of potato periderm. *Journal of Experimental Botany* 60, 4411–4421.

Gull, D.D. and Isenberg, F.M. (1958) Lightburn and off-flavor development in potato tubers exposed to florescent lights. *Proceedings of the American Society for Horticultural Science* 71, 446–454.

Halderson, J.L., Corsini, D.L. and Haderlie, L.C. (1985) Potato vine kill: stem-end discoloration effects on Russet Burbank. *American Potato Journal* 62, 273–279.

Hiller, L.K. and Koller, D.C. (1981) Brown center/hollow heart of potatoes – What do we know? *Proceedings of the Washington State Potato Conference* 20, 73–80.

Hiller, L.K. and Thornton, R.E. (1993) *Management of physiological disorders*. In: Rowe, R.C. (ed.) *Potato Health Management*. Phytopathological Society Press, St Paul, Minnesota, pp. 87–94.

Hiller, L.K., Koller, D.C. and Van Denburgh, R.W. (1979). Brown center of potatoes – What have we learned? *Proceedings of the Washington State Potato Conference* 18, 21–27.

Hiller, L.K., Koller, D.C. and Thornton, R.E. (1985) Physiological disorders of potato tubers. In: Li, P.H. (ed.) *Potato Physiology*. Academic Press, Inc, Orlando, Florida, pp. 389–455.

Hodgson, W.A., Pond, D.D. and Munro, J. (1974) *Diseases and Pests of Potatoes*. Agriculture Canada Publication No 1492.

Hruschka, H.V., Smith, W.L. Jr and Baker, J.E. (1969) Reducing chilling injury of potatoes by intermittent warming. *American Potato Journal* 46, 38–53.

Hudson, D.E. (1975) The relationship of cell size, intercellular space, and specific gravity on bruise depth in potatoes. *American Potato Journal* 52, 9–14.

Iritani, W.M. and Weller, L. (1973) The development of translucent end tubers. *American Potato Journal* 50, 223–233.

Iritani, W.M. and Weller, L. (1978) Influence of low fertility and vine killing on sugar development in apical and basal portions of Russet Burbank potatoes. *American Potato Journal* 55, 239–246.

Iritani, W.M., Weller, L.D. and Knowles, N.R. (1984) Factors influencing incidence of internal brown spot in Russet Burbank potatoes. *American Potato Journal* 61, 335–343.

Jansky, S.H. and Thompson, D.M. (1990) Expression of hollow heart in segregating tetraploid potato families. *American Potato Journal* 67, 695–703.

Jefferies, R.A. and MacKerron, D.K.L. (1987a) Effect of second growth on the quality of tubers as seed in the cultivar Record. *Potato Research* 30, 337–340.

Jefferies, R.A. and MacKerron, D.K.L. (1987b) Observations on the incidence of tuber growth cracking in relation to weather patterns. *Potato Research* 30, 613–623.

Jivanuwong, S. (1998) Nondestructive detection of hollow heart in potatoes using ultrasonics. MSc thesis. Virginia Polytechnic Institute and State University, Blacksburg, Virginia.

Karlsson, B.H., Crump, P.M. and Palta, J.P. (2006) Enhancing tuber calcium concentration may reduce incidence of blackspot bruise injury in potatoes. *HortScience* 41, 1213–1221.

Keenan, M.K., Baker, K.W., Coffin, R. and Yada, R.Y. (1989) A histological examination of 'white knot' disorder in tubers of the cultivar Atlantic. *American Potato Journal* 66, 489–493.

Kleinhenz, M.D., Palta, J.P., Gunter, C. and Kelling, K.A. (1999) Impact of source and timing on calcium and nitrogen applications on 'Atlantic' potato tuber calcium concentrations and internal quality. *Journal of the American Society of Horticultural Science* 24, 498–506.

Kleinkopf, G.E., Westermann, D.T. and Wille, M.J. (1988) Soil temperature effects on sugar end development. *Proceedings of the University of Idaho Winter Commodity Schools* 20, 180–181.

Krauss, A. (1985) Interaction of nitrogen nutrition, phytohormones, and tuberization. In: Li, P.H. (ed.) *Potato Physiology*. Academic Press, Inc, Orlando, Florida, pp. 209–230.

Kunkel, R. and Gardner, W.H. (1959) Blackspot of Russet Burbank potatoes. *Proceedings of the American Society of Horticultural Science* 73, 439–444.

Levitt, J. (1942) A histological study of hollow heart of potatoes. *American Potato Journal* 19, 134–143.

Lugt, C. (1960) Second-growth phenomena. *European Potato Journal* 3, 307–324.

Lugt, C., Bodlaender, K.B.A. and Goodijk, G. (1964) Observations on the induction of second-growth in potato tubers. *European Potato Journal* 7, 219–227.

Lulai, E.C. (2001) Tuber periderm and disease resistance. In: Stevenson, W.R., Loria, R., Franc, G.D., and Weigartner, D.P. (eds) *Compendium of Potato Diseases*, 2nd edn. Phytopathological Society Press, St Paul, Minnesota, pp. 3–6.

Lulai, E.C. and Freeman, T.P. (2001) The importance of phellogen cells and their structural characteristics in susceptibility and resistance to excoriation in immature and mature potato tuber (*Solanum tuberosum* L.) periderm. *Annals of Botany* 88, 555–561.

Lulai, E.C. and Orr, P.H. (1993) Determining the feasibility of measuring genotypic differences in skin-set. *American Potato Journal* 70, 599–609.

Lulai, E.C. and Orr, P.H. (1995) Porometric measurements indicate wound severity and tuber maturity affect the early stages of wound-healing. *American Potato Journal* 72, 225–241.

Lulai, E.C., Weiland, J.J., Suttle, J.C., Sabba, R.P. and Bussan, A.J. (2006) Pink eye is an unusual periderm disorder characterized by aberrant suberization: a cytological analysis. *American Journal of Potato Research* 83, 409–421.

McNabnay, M., Dean, B.B., Bajema, R.W. and Hyde, G.M. (1999) The effect of potassium deficiency on chemical, biochemical and physical factors commonly associated with blackspot development in potato tubers. *American Journal of Potato Research* 76, 53–60.

Nelson, D.C. (1970) Effect of planting date, spacing, and potassium on hollow heart in Norgold Russet potatoes. *American Potato Journal* 47, 130–135.

Nelson, D.C. and Thoreson, M.C. (1986) Relationships between tuber size and time of harvest to hollow heart initiation in dryland Norgold Russet potatoes. *American Potato Journal* 63, 155–161.

Nolte, P., Secore, G.A., Gudmestad, N.C. and Henningson, P.J. (1993) Detection and identification of fluorescent compounds in potato tuber tissue with corky patch syndrome. *American Potato Journal* 70, 649–666.

O'Brien, M.J. and Rich, A.E. (1976) Abiotic diseases. In: *Potato Diseases*. Agricultural Research Service Handbook 474. United States Department of Agriculture, Washington, DC, pp. 61–73.

Okazawa, Y. and Iriuda, N. (1980) On occurrence of defected potato tubers with rough russeted skin. *Japanese Journal of Crop Science* 49, 58–65.

Olsen, N. and Brandt, T. (2005) The affect of light source on greening and other quality attributes of 'Russet Burbank' potatoes (http://extension.uidaho.edu/kimberly/files/2014/05/Greeningwocontact.pdf, accessed 19 May 2014).

Olsen, N.L., Hiller, L.K. and Mikitzel, L.J. (1996) The dependence of internal brown spot development upon calcium fertility in potato tubers. *Potato Research* 39, 165–178.

Olsen, N., Kleinkopf, G.E. and Stark, J.C. (2003) Physiological disorders. In: Stark, J.C. and Love, S.L. (eds) *Potato Production Systems*. University of Idaho Agricultural Communications, Moscow, Idaho, pp. 309–327.

Ozgen, S., Karlsson, B.H. and Palta, J.P. (2006) Response of potatoes (cv Russet Burbank) to supplemental calcium applications under field conditions: tuber calcium, yield, and incidence of internal brown spot. *American Journal of Potato Research* 83, 195–204.

Panique, E., Kelling, K.A., Schulte, E.E., Hero, D.E., Stevenson, W.R. and James, R.V. (1997) Potassium rate and source effects on potato yield, quality, and disease interaction. *American Potato Journal* 74, 379–398.

Pavek, J., Corsini, D. and Nissley, F. (1985) A rapid method for determining blackspot susceptibility of potato clones. *American Potato Journal* 62, 511–517.

Reeve, R.M. (1968) Further histological comparisons of black spot, physiological internal necrosis, black heart, and hollow heart in potatoes. *American Potato Journal* 45, 391–401.

Reeve, R.M., Timm, H. and Weaver, M.L. (1973) Parenchyma cell growth in potato tubers. II. Cell division vs. cell enlargement. *American Potato Journal* 50, 71–78.

Reeves, A.F. (1988) Varietal differences in potato tuber greening. *American Potato Journal* 65, 651–658.

Rex, B.L. and Mazza, G. (1989) Cause, control and detection of hollow heart in potatoes: a review. *American Potato Journal* 66, 165–183.

Rex, B.L., Russell, W.A. and Wolfe, H.R. (1987) The effect of spacing of seedpieces on yield, quality and economic value for processing of Shepody potatoes in Manitoba. *American Potato Journal* 64, 177–189.

Sabba, R.P., Bussan, A.J. and Lulai, E.C. (2008) Relationship between pink eye symptoms and cell damage in the tuber periderm and cortex. *American Journal of Potato Research* 85, 466–476.

Secor, G. and Rouse, D. (1992) Proceedings of the second conference on pink eye of potatoes. *American Potato Journal* 69, 149–154.

Shock, C.C., Holmes, Z., Stieber, T.D., Eldredge, E.P. and Zhang, P. (1993) The effect of timed water stress on quality, total solids and reducing sugar content of potatoes. *American Potato Journal* 70, 227–241.

Sieczka, J.B. and Thornton, R.E. (1993) *Commercial Potato Production in North America. Potato Association of America Handbook*, Volume 57. University of Maine, Orono, Maine.

Smittle, D.A., Thornton, R.E., Peterson, C.L. and Dean, B.B. (1974) Harvesting potatoes with minimum damage. *American Potato Journal* 51, 152–164.

Snowden, A.L. (1992) Temperate roots and tubers. In: *Color Atlas of Post-harvest Diseases and Disorders of Fruits and Vegetables Volume 2: Vegetables*. CRC Press, Inc, Boca Raton, Florida, pp. 263–342.

Sowokinos, J.R., Shock, C.C., Stieber, T.D. and Eldredge, E.P. (2000) Compositional and enzymatic changes associated with the sugar-end defect in Russet Burbank potatoes. *American Journal of Potato Research* 77, 47–56.

Sterrett, S.B. and Henninger, M.R. (1991) Influence of calcium on internal heat necrosis of Atlantic potato. *American Potato Journal* 68, 467–477.

Sterrett, S.B., Henninger, M.R. and Lee, G.S. (1991a) Relationship of internal heat necrosis of potato to time and temperature after planting. *Journal of the American Society for Horticultural Science* 116, 697–700.

Sterrett, S.B., Henninger, M.R. and Lee, G.S. (1991b) Predictive model for onset and development of internal heat necrosis of 'Atlantic' potato. *Journal of the American Society for Horticultural Science* 116, 701–705.

Sterrett, S.B., Haynes, K.G., Yencho, G.C., Henninger, M.R. and Vinyard, B.T. (2006) 4x-2x potato clones with resistance or susceptibility to internal heat necrosis differ in tuber mineral status. *Crop Science* 46, 1471–1478.

Stevenson, W.R., Loria, R., Franc, G.D. and Weigartner, D.P. (2001) *Compendium of Potato Diseases*, 2nd edn. Phytopathological Society Press, St Paul, Minnesota.

Stewart, F.C. and Mix, A.J. (1917) Blackheart and the aeration of potatoes in storage. *New York Agricultural Experiment Station Bulletin* 436, 321–362.

Storey, R.M.J. and Davies, H.V. (1992) Tuber quality. In: Harris, P. (ed.) *The Potato Crop: The Scientific Basis for Improvement*. Chapman and Hall, London, pp. 507–569.

Strand, L.L. (2006) *Integrated Pest Management for Potatoes in the Western United States*, 2nd edn. University of California, Oakland, California.

Struik, P.C., Geertsema, J. and Custers, C.H.M.G. (1989) Effects of shoot, root and stolon temperature on the development of the potato (*Solanum tuberosum* L.) plant. III. Development of tubers. *Potato Research* 32, 151–158.

Thompson, A.L., Love, S.L., Sowokinos, J.R., Thornton, M.K. and Shock, C.C. (2008) Review of the sugar end disorder in potato (*Solanum tuberosum*, L.). *American Journal of Potato Research* 85, 375–386.

Thornton, R.E. (2001a) Physiological disorders of tubers: external symptoms. In: Stevenson, W.R., Loria, R., Franc, G.D. and Weigartner, D.P. (eds) *Compendium of Potato Diseases*, 2nd edn. Phytopathological Society Press, St. Paul, MN, pp. 84-88.

Thornton, R.E. (2001b) Physiological disorders of tubers: internal symptoms. In: Stevenson, W.R., Loria, R., Franc, G.D., and Weigartner, D.P. (eds) *Compendium of Potato Diseases*, 2nd edn. Phytopathological Society Press, St Paul, Minnesota, pp. 88–92.

Van Denburgh, R.W., Hiller, L.K. and Koller, D.C. (1980) The effect of soil temperatures on brown center development in potatoes. *American Potato Journal* 57, 371–375.

Van Denburgh, R.W., Hiller, L.K. and Koller, D.C. (1986) Ultrastructural changes in potato tuber pith cells during brown center development. *Plant Physiology* 81, 167–170.

van der Zaag, D.E. (1996) *Potato Diseases: Diseases, Pests and Defects*. The NIVAA, Den Haag, the Netherlands.

Wale, S., Platt, H.W. and Cattlin, N. (2008) *Diseases,Pests and Disorders of Potatoes: A Color Handbook*. Academic Press, San Diego, California.

Wang, P.C., Wang, C.Y., Song, H.F. and Yan, Z.J. (1997) Nuclear magnetic resonance imaging detects internal defects in potatoes. [Abstract] *HortScience* 32, 488.

Whiteman, T.M. and Lutz, J.M. (1954) Sunken scald spot field injury evident in stored potatoes. *American Potato Journal* 31, 43–49.

Wigginton, M.J. (1973) Diffusion of oxygen through lenticels in potato tuber. *Potato Research* 16, 85–87.

Wilson, M.L., Rosen, C.J. and Moncrief, J.F. (2009) Potato response to a polymer-coated urea on an irrigated, coarse-textured soil. *Agronomy Journal* 101, 897–905.

Xu, X., van Lammeren, A.A.M., Vermeer, E. and Vreugdenhil, D. (1998) The role of gibberellin, abscisic acid, and sucrose in the regulation of potato tuber formation *in vitro*. *Plant Physiology* 117, 575–584.

Yamaguchi, M., Hughes, D.L. and Howard, F.D. (1960) Effect of season, storage temperature, and temperature during light exposure on chlorophyll accumulation in White Rose potatoes. *Proceedings of the American Society for Horticultural Science* 75, 529–536.

Yencho, G.C., McCord, P.H., Haynes, K.G. and Sterrett, S.B.R. (2008) Internal heat necrosis – a review. *American Journal of Potato Research* 85, 69–76.

Zimmerman-Gries, S. and Blodgett, E.G. (1963) Some skin defects of potatoes in Israel. *European Potato Journal* 6, 191–200.

15 Postharvest Storage and Physiology

Paul C. Bethke*
USDA Agricultural Research Service and University of Wisconsin Department of Horticulture, Madison, USA

Potato storage is an important component of the potato production system in many parts of the world. This chapter begins by describing briefly the purpose and value of potato storage, and some of the many options available for storing the crop.

Physiological defects that develop in storage are described, including pressure bruise, cold-induced sweetening, sugar-end defect, and stem-end chip defect. The central importance of the reducing sugars, glucose and fructose, to tuber processing quality is introduced in the context of physiological defects that cause dark-colored fried products. The general features of carbohydrate metabolism in stored potato tubers are described, with details provided for the subcellular formation and utilization of starch, sucrose, glucose, and fructose. These biochemical details are the foundation for understanding the section on the sugar management of potatoes in storage.

Tuber respiration is discussed, with an emphasis on how storage duration and temperature impact respiration rates. The effects of respiration on crop value and potato storage management are illustrated. The mechanism of tuber water loss is described, and the relative importance of water loss through the periderm, lenticels, wounds, and sprouts is considered. Sprout control is essential to maintain crop value, and as such, options for residue-free, organic, and synthetic sprout control are presented. Finally, guidelines for the storage of potatoes in bulk bins are provided, in order to give a general understanding of how potatoes in large, ventilated storage facilities are managed.

15.1 The Purpose of Potato Storage

Potato storage makes the crop available for consumption or sale over an extended period of time. Insulated storage structures protect the crop from freezing in regions where postharvest temperatures fall below 0°C. In regions where a warm or wet period follows harvest, ventilated storage structures protect the crop from disease by keeping water off the tubers. In regions where most potatoes are harvested during a short period of time, prices tend to be low immediately after harvest, due to an abundant supply. Storage allows the potatoes to be sold when supplies are reduced and prices are higher. Where tubers are used for processing, local storage allows for tubers to be available for many months. Local availability minimizes shipping costs and reduces idle time at processing plants. In situations where storage conditions can be controlled precisely, a well-designed storage facility can facilitate the

* E-mail: paul.bethke@ars.usda.gov

maintenance of crop quality. Proper storage is essential to maintain the quality and viability of seed tubers being held until the next production season.

It is difficult to assess how much of the world's potato crop goes into storage, because data are not available for many locations. One estimate is that approximately 25% of the world-wide potato crop destined for human consumption is stored for some length of time (van der Zaag, 1989). A few examples demonstrate the relative importance of storage in different parts of the world. In the USA, 59% of the 2010 crop was in storage for 3 months or more (USDA National Agricultural Statistics Service, 2011). Likewise, approximately 58% of the crop in Great Britain is commercially stored prior to being utilized, half of which is used for the fresh market (Cunnington, 2008). In India, 40% of total production is put into refrigerated storage; of this, approximately 25% is for consumption and 15% for seed (CIP, n.d.; Fuglie *et al.*, 2000). In Kenya, harvest occurs regionally every 2–3 months. Consequently, storage of ware potatoes in Kenya is not routine (CIP, n.d.). Similarly, in Ecuador, seasonal climatic variation is minimal and potato cultivation occurs continually throughout the year. Few potatoes are put into storage, since recently harvested potatoes are always available (CIP, n.d.).

15.2 Types of Potato Storage

Potatoes are grown worldwide. The way that potatoes are stored depends on cultural expectations for the quality of tubers coming out of storage, the desired duration of the storage period, and local environmental conditions. Potato storage methods are highly diverse, because each of these parameters varies widely between production regions. For example, in most of North America and Europe, harvest occurs in the fall and is followed by an extended cool or cold period, whereas harvest in India occurs in the spring and is followed by an extended warm period. The methods used to store tubers and the duration of the storage period reflects these very different climatic conditions, as well as the intended market.

Ground storage is the simplest type of potato storage. In this case, mature potatoes are left unharvested and kept in the ground for an extended period of time. Ground storage requires moderate soil temperatures and even soil moisture following senescence of the potato vines. Freezing temperatures and waterlogged soils can be disastrous for the crop, as one increases the risk of physical breakdown and the other of pathogenic breakdown. The advantage of ground storage is that infrastructure cost is very low. Ground storage also eliminates the increased risk of disease in storage that comes from damage during harvest, transportation, and handling. Ground storage is used in Ethiopia for half of the seed crop (Gildemacher *et al.*, 2009).

A potato clamp is another form of simple storage. It is suitable for locations where winter temperatures do not fall much below freezing. Clamps are linear, steep-sided piles of potatoes that are covered with straw and a layer of soil. The straw and soil provide insulation. The straw functions to keep the soil out of the pile, thereby maximizing the air volume next to the tubers. In some cases, plastic sheeting is placed over the straw a few weeks after harvest, and this is covered with soil. The sheets are arranged to provide protection from precipitation while allowing for ventilation.

Within the clamp, evaporation from tubers and the surrounding soil creates an environment of high relative humidity (RH), which helps to maintain tuber hydration. Heat produced by respiration keeps the tubers warmer than their surroundings, but it can also result in an excessively high level (0.88%) of carbon dioxide (CO_2), which is detrimental to tuber quality (Crook and Watson, 1950). One benefit of using clamps is that one can use material readily available on the farm. As recently as the late 1940s, clamps were the predominant form of potato storage in England (Samuel and Wilson, 1949). Due to advances in technology, only 5% of the potatoes in Great Britain were stored outdoors 40 years later (Statham, 1989; van der Zaag, 1989).

Excessive heat during storage leads to an increase in disease. Locations with warm climates like India rely on large-scale refrigerated storages to hold potatoes for an extended period (Fuglie *et al.*, 2000); however, these storages are prohibitively expensive to growers in poorer subtropical regions. As an inexpensive alternative, small landholders may store ware potatoes in aboveground cribs that allow air movement through permeable tops and bottoms. In these regions, seed tubers may be kept in small storages that allow diffuse light to reach the tubers, promoting the growth of strong, short sprouts that are less susceptible to damage during planting than sprouts that have developed in the dark.

In regions where subsurface soils routinely freeze during the winter, ground storage or aboveground clamps and cribs may be impractical. In this case, tubers must be stored in a well-insulated facility that prevents freezing. The simplest of these are belowground cellars that maintain temperatures above 0°C via walls insulated with soil, straw, or synthetic material. The RH in a belowground cellar depends on the rate of evaporation from the potatoes and on how well the walls, floor, and roof are sealed to prevent air and water vapor movement. Belowground potato storage is implemented most easily for small volumes of tubers, but some designs have capacities of 25 t (Pringle *et al.*, 2009).

Large volumes of potatoes can be stored successfully in bags, boxes, or bulk piles. Each method has advantages, some of which are related to climatic factors, storage facility, and the intended market. When tubers are stored in bags made of burlap, the bags are typically stacked to a depth of 2 m or less in buildings where air temperature and RH are controlled to suit the needs of the crop.

Large-scale storage of potatoes in pallet boxes is common in the UK (Cunnington, 2008). Ventilation air, with or without supplemental humidification, can be supplied through floor vents or from side walls. In sophisticated storage systems, boxes are designed to form air ducts when stacked in place, and this facilitates efficient air movement throughout the entire storage. Box storage allows tracking of small lots of tubers from individual plots or fields, since each box can be labeled and followed independently.

Large-scale storage of potatoes in bulk piles inside a climate-controlled, insulated building is typical in the USA (Brook *et al.*, 1995) and the Netherlands, and for processing potatoes in Great Britain (Cunnington, 2008). Tubers are typically piled 4.5–5.5 m high into rooms or large bins on either side of a central plenum. Fans at one end of the plenum force temperature-controlled air through moist fiber pads and past supplemental humidification systems. Lateral air ducts located on the bottom of the potato pile distribute the humidified air throughout the storage. These ducts can be made of corrugated steel, with perforations to distribute the air, or can be in-floor ducts covered with wooden slats. Air that passes through the pile is drawn back toward the fans to be recirculated. Fresh air, as needed to prevent excessive CO_2 build-up and regulate temperature, is pulled through adjustable louvers in the outside wall and blended with the return air before passing through the humidification system. In many cases, supplemental refrigeration augments the cooling capacity available from introducing cool outside air and the evaporating of water in the humidification system.

15.3 Physiological Defects Occurring During Storage

Potatoes in storage may develop quality defects that decrease the value of the stored crop. Several of these are physiological defects that are independent of pathogen activity.

The most universal storage-induced defect is pressure bruising of tubers. Pressure bruise results when tubers within a pile lose their ability to resist the downward force applied by the weight of the tubers above. In its mildest manifestation, this results in pressure flattening, which is apparent as flat spots or indentations on tubers coming out of storage. When the force of overhead pressure exceeds the ability of the tubers to withstand compression forces, pressure bruise may be observed (Castleberry and Jayanty, 2012). Bruising results when the cells under pressure-flattened areas sustain physical damage that results in the formation of melanin pigments (Lulai *et al.*, 1996). Pressure bruising is thought to be promoted by tuber water loss before harvest or in storage that results in the reduction of cellular turgor pressure. Susceptibility to pressure bruise varies with cultivar (Castleberry and Jayanty, 2012).

Several tuber quality defects in processing potatoes arise from the accumulation of the reducing sugars, glucose and fructose, in storage. During high-temperature frying, the carbonyl group on a reducing sugar reacts with the alpha- amino group of amino acids in a non-enzymatic, Maillard reaction (Shallenberger *et al.*, 1959). Products of the Maillard reaction include dark-colored melanoidin pigments and numerous other compounds that contribute positively or negatively to cooked product taste and aroma. Tubers with elevated amounts of reducing sugars produce unacceptably dark-colored chips or fries that may have a bitter flavor that is undesirable to consumers (Kumar *et al.*, 2004). Processors may reject tubers if the reducing sugar content exceeds the commercially acceptable range. Acrylamide is another product of the Maillard reaction, and it is formed when reducing sugars react with free asparagine. Acrylamide consumption is currently being investigated as a potential health risk; maintaining low reducing sugars in process and chip potatoes reduces acrylamide, which, in turn, reduces the potential health risk (Bethke and Bussan, 2013).

Cold-induced sweetening (CIS), also referred to as low-temperature sweetening, typically develops when potato tubers are stored at temperatures at or below 10°C (Sowokinos, 2001b; Malone *et al.*, 2006). The effects of CIS on reducing sugar accumulation are typically observed 1–3 weeks after tubers are transferred to excessively cool temperatures, with higher accumulations developing at lower storage temperatures. Cultivars differ widely in their susceptibility to CIS. Some chip and fry processing cultivars resist CIS and can therefore be stored successfully under 5.5°C. Storage temperatures of less than 4°C almost inevitably result in unacceptable chip and fry color, regardless of cultivar.

Sugar-end defect is a serious physiological defect and quality concern of fry processing cultivars (Thompson *et al.*, 2008). The defining characteristic of sugar-end defect appears when the tubers are cut into fries and cooked. One end of the fry will be darker in color than the other end. It is usually the basal or stem end of tubers that fries dark because it has high concentrations of reducing sugars relative to the apical or bud end (Sowokinos *et al.*, 2000; Matsuura-Endo *et al.*, 2004). Sugar-end defects are typically not observed at harvest, but develop following weeks of storage. Unlike CIS, sugar-end defects are not caused by postharvest conditions and do not require low-temperature storage to develop. Instead, sugar-end defects result from transient periods of heat or water stress during early tuber development (Iritani *et al.*, 1973; Sowokinos *et al.*, 1985; Shock *et al.*, 1992, 1993; Kincaid *et al.*, 1993; Eldredge *et al.*, 1996). Sugar-end defects may not become apparent until tubers have been in storage for several weeks, after which time defects become increasingly severe. Unlike CIS, sugar- end defects cannot be removed by reconditioning at warm temperatures (Thompson *et al.*, 2008). The significance of this is that once tubers develop sugar-end defects, their value as fry processing material may be reduced substantially.

Stem-end chip defect is similar to sugar-end defect in that products made from defective tubers have dark color, and reducing sugars are high in tissues located at the stem-end of the tuber. However, in stem-end chip defect, the defect is more specifically localized at or near the vascular regions (Wang *et al.*, 2012). Stem-end chip defects tend to develop in storage, but low temperatures are not required for defect formation. Unlike sugar-end defects, stem-end chip defects for many potato cultivars become less severe following extended storage periods. Stem-end defect is thought to be caused by environmental or pathogen stress during tuber development. Specifically, high temperatures during growth and infection with the vascular pathogen, *Verticillium dahliae*, are thought to contribute to stem-end chip defect formation (Wang *et al.*, 2012; Wang and Bethke, 2013).

Senescence sweetening describes a developmentally controlled accumulation of reducing sugars that occurs several months after harvest (Burton, 1966). Senescence sweetening is characterized by a blush of dark color in the center of chips and fries that appears after they are fried. Pronounced end-to-end asymmetry is not a feature of senescence sweetening, and this helps to differentiate it from sugar-end defect or stem-end chip defect. Cultivars vary widely in how long it takes before they begin senescence sweetening; the sweetening begins to occur anywhere from a few months to 9 months or more after harvest. Conditions in storage that promote higher rates of respiration tend to promote earlier senescence sweetening (Herrman *et al.*, 1996). Senescence sweetening is an irreversible process that can develop rapidly in storage; warming the tubers will not improve chip or fry processing quality. Moreover, increasing the storage temperature in an attempt to recondition tubers (decrease reducing sugars) may actually exacerbate the issue by increasing sugar accumulation in the tuber.

15.4 Carbohydrate Metabolism in Stored Potatoes

The major value of potatoes as a food stock is due largely to the tuber's ability to accumulate massive amounts of starch. The typical tuber is 15–19% starch by weight. Starch is a dynamic compound, and during the storage period, it is broken down to sugars that are consumed by respiration. Starch contributes to pools of sugar and sugar phosphates in the cytosol and vacuole, which are used for the synthesis of carbon-containing compounds such as suberin and additional starch. Carbohydrate metabolism as it relates to the accumulation of reducing sugars is a critical determinant of stored tuber quality for chip and fry processing.

A simplified diagram illustrating carbohydrate metabolism in potato tubers is shown in Fig. 15.1. Indicated in the diagram are four subcellular locations in which key metabolic activities take place: the amyloplast, cytosol, vacuole, and mitochondria. Carbon intermediates are shuttled between these locations, where enzyme activities, specific to each, contribute to overall changes in starch content, sugar content, and respiration rate observed at the level of whole tubers.

Starch synthesis and storage occur in amyloplasts. Glucose-6-phosphate (G-6-P) is transported from the cytosol into amyloplasts via a transmembrane transport protein. Once inside, the enzyme, plastidic phosphoglucomutase (pPGM), converts G-6-P into glucose-1-phosphate (G-1-P). ADP-glucose pyrophosphorylase (AGPase) then converts G-1-P to the immediate substrate for starch synthesis, ADP-glucose. The reaction catalyzed by AGPase requires ATP imported from the cytosol through the ATP/ADP translocator, and both the enzyme and the transporter exert strong control over the rate of starch synthesis (Sowokinos, 1976; Tjaden *et al.*, 1998; Geigenberger, 2003; Geigenberger *et al.*, 2004). Starch, which is composed of branched amylopectin and linear amylose domains, is synthesized by several enzymes including granule bound and soluble starch synthases (SS), starch branching enzymes (SBE), and debranching enzymes (Takaha *et al.*, 1993; Larsson *et al.*, 1996; Lloyd *et al.*, 2004; Zeeman *et al.*, 2007).

The details of starch breakdown in potato tuber amyloplasts are poorly understood (Smith *et al.*, 2005; Rathore *et al.*, 2009; Zeeman *et al.*, 2010). There is evidence for two enzymatic routes of breakdown: phosphorolytic, using starch phosphorylase (SP), and hydrolytic, using amylase.

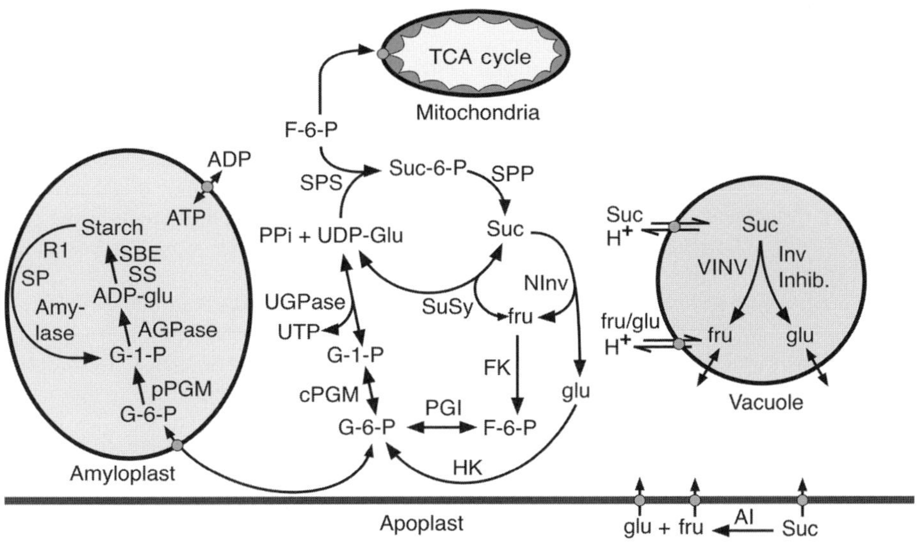

Fig. 15.1. Simplified diagram of carbohydrate metabolism in potato tubers showing key enzymes and metabolites in the amyloplast, cytosol, apoplast, mitochondria, and vacuole. See text for details and explanation of abbreviations.

Starch breakdown is promoted by the R1 enzyme, an alpha-glucan water dikinase that phosphorylates amylopectin at the C6 position (Ritte *et al.*, 2002). Transgenic plants with reduced amounts of R1 mRNA are less prone to sugar accumulation at low temperatures, and these data support the involvement of R1 in starch breakdown during CIS (Rommens *et al.*, 2006). Starch granules are degraded into branched and linear polyglucans by enzymes that have not been identified clearly in potato tubers. Branched polymers are reduced to linear glucans by debranching enzymes and limit-dextrinase. The linear glucans produced may be degraded into neutral sugars by L-type starch phosphorylase (Smith *et al.*, 2005). This enzyme catalyzes the reversible conversion of alpha-1, 4-glucan and inorganic phosphate into G-1-P.

A second route for starch degradation likely involves hydrolysis by amylases. Starch-associated alpha-amylase activity has been identified in potato tubers, and the isolated enzyme, alpha-amylase, was able to hydrolyze potato starch (Witt and Sauter, 1996). Since alpha-amylase produces small oligomers of glucose, it is likely that beta-amylase and alpha-glucosidase participate in the complete breakdown of starch to simple sugars (Cochrane *et al.*, 1991; Nielsen *et al.*, 1997). The activity of beta-amylase in tubers has been shown to increase as storage temperature declines (Cochrane *et al.*, 1991; Nielsen *et al.*, 1997).

The identity of the metabolites transported out of amyloplasts during starch degradation is unknown, but maltose and glucose are likely candidates based on research done primarily with chloroplasts (Zeeman *et al.*, 2007; Rathore *et al.*, 2009). Once in the cytosol, maltose can be converted to G-6-P by a glucosyltransferase and hexokinase, or to G-1-P by cytosolic H-type phosphorylase (Rathore *et al.*, 2009). Cytosolic phosphoglucomutase (cPGM) readily converts G-6-P to G-1-P, and phosphoglucoisomerase (PGI) converts G-6-P to fructose-6-phosphate (F-6-P). Both reactions are freely reversible. Cytosolic G-1-P, G-6-P, and F-6-P constitute the hexose-phosphate pool and are used to supply carbon for numerous biochemical reactions and to provide substrates for respiration.

Sucrose and invertase activity

Sucrose formation is the first committed step in the accumulation of reducing sugars by potato tubers, and the enzymes responsible for sucrose synthesis have been identified (Sowokinos, 2001b). The most important is UDP-glucose pyrophosphorylase (UGPase), which synthesizes UDP-glucose and pyrophosphatase (PPi) from UTP and G-1-P. Isoenzymes of UGPase with different enzymatic properties have been characterized (Sowokinos, 2001a; Gupta and Sowokinos, 2003). Those with reduced affinity (higher K_m) for G-1-P form sucrose less readily in stored tubers than those with greater affinity. In many potato cultivars, favorable isoenzymes of UGPase contribute to the maintenance of low sucrose and reducing sugars during storage (Sowokinos, 2001a).

Sucrose is synthesized in the cytosol by the combined activities of sucrose phosphate synthase (SPS), which produces sucrose-6-phosphate (Suc-6-P) from UDP-glucose and F-6-P, and sucrose phosphate phosphatase (SPP), which dephosphorylates sucrose-6-phosphate to sucrose.

Sucrose is partitioned between the cytosol and vacuole. Transport through the tonoplast is likely to be facilitated by a sucrose-H^+ antiporter. In two reports, the concentration of sucrose in the cytosol of developing potato tubers was found to be approximately equal to that in the vacuole (Farre *et al.*, 2001, 2008). Comparable data are not available for mature tubers. Vacuolar acid invertase (VINV) cleaves sucrose into the reducing sugars, glucose and fructose, in a non-reversible reaction. The expression of the vacuolar acid invertase gene and activity of the vacuolar acid invertase enzyme are strongly upregulated by low-temperature storage of tubers (Pressey and Shaw, 1966; Matsuura-Endo *et al.*, 2004; Bhaskar *et al.*, 2010). Rates of invertase activity vary widely between cultivars (Matsuura-Endo *et al.*, 2004; McKenzie *et al.*, 2005).

When invertase activity has been strongly reduced through biotechnology, dramatic reductions in reducing sugar accumulation have been observed in tubers stored at low temperatures (Bhaskar *et al.*, 2010; Ye *et al.*, 2010; Liu *et al.*, 2011; Wu *et al.*, 2011). Hence, vacuolar acid invertase is a key determinant of reducing sugar accumulation during low-temperature storage. Interestingly, in lines of potato with very low invertase activity, sucrose accumulation depended on the UGPase isoforms present (Wu *et al.*, 2011). In cultivars containing A-II isoforms, such as Snowden, Dakota Pearl, and Atlantic, sucrose accumulation in low-invertase tubers was comparable to that in wild-type tubers. In cultivars lacking A-II isoenzymes, such as MegaChip, sucrose accumulation was much greater in the transgenic lines than in wild type (Wu *et al.*, 2011).

A proteinaceous invertase inhibitor has also been localized to the vacuole in mature potato tubers, and changes in its expression and activity contribute to differences between cultivars in sensitivity to low-temperature storage (Liu *et al.*, 2010, 2011; Brummell *et al.*, 2011). When RNA interference was used to decrease the activity of sucrose-phosphate phosphatase, invertase activity was inhibited as well. It was proposed that sucrose-6-phosphate might play a role in regulating the expression of vacuolar acid invertase (Chen *et al.*, 2008).

It is important to note that the glucose and fructose formed in the vacuole as a result of invertase activity are not part of the cytosolic pool of free sugars and sugar phosphates. As such, they are not immediately available for renewed starch synthesis or for respiration in mitochondria. It is likely that this partitioning of reducing sugars into a discrete subcellular location is one reason that removing them through respiratory activity is slow or ineffective, such as in the case of sugar-end defect and senescence sweetening.

Sucrose synthase (SuSy) can also metabolize cytosolic sucrose, and this reaction produces fructose and UDP-glucose. Sucrose synthase activity is very high during potato tuber bulking and is an important determinant of the ability of developing tubers to attract assimilates (sink strength). In mature tubers, however, only a low level of activity is detectable. Neutral invertase (NInv) can cleave cytosolic sucrose to glucose and fructose, but little is known about the activity or regulation of this enzyme in mature tubers. Cytosolic glucose and fructose are readily phosphorylated by hexokinase (HK) and fructokinase (FK), respectively, to produce G-6-P and F-6-P. Sucrose in the apoplastic space can be cleaved by cell wall resident acid invertase (AI) (Li *et al.*, 2005), which also affects tuber quality traits.

Sugar management of stored potatoes

Sugar management is an essential component of storage management for fry and chip processing potatoes. The goal of sugar management is to keep tuber reducing sugar concentrations below established limits for acceptable processing quality.

Sugar management of potatoes occurs during three important phases of storage: preconditioning, cooling to set point, and monitoring during long-term storage. During the preconditioning phase, tubers are stored at relatively warm temperatures (12–13°C) to allow excess sugars to dissipate through respiration or starch synthesis and to allow tubers with harvest-inflicted surface wounds to heal. Successful preconditioning minimizes the adverse effects of non-optimal irrigation, under- or over-maturity of tubers at harvest, vine desiccant application, mechanical disturbance, and damage during harvest, transport, and loading into storage. Tubers that are not preconditioned adequately prior to cooling often exhibit a larger increase in tuber reducing sugars as temperatures are dropped compared with conditioned tubers (Pritchard and Adam, 1992).

Changes in reducing sugar often lag behind changes in sucrose; therefore, tuber sucrose levels are typically used to guide management decisions during preconditioning. Specific targets for stored tuber sugars vary by cultivar and are often reported; however, general guidelines also exist. For chip potatoes, where sugar management is critical, sucrose should be less than 0.7 mg g^{-1} fresh weight (FW) and glucose should be less than 0.035 mg g^{-1} FW at the end of preconditioning. For fry processing potatoes, a wider range of sugars may be acceptable, depending on cultivar. Sucrose contents of 1–2 mg g^{-1} FW, stable glucose contents, and acceptable fry color may be used as a starting point for fry processing cultivars that have little or no storage data. Depending on the growing season and cultivar, it may take several weeks, or even months, before tubers are preconditioned properly. The advantages of preconditioning need to be balanced against the potential disadvantages in terms of increased risk from disease and decreased time until the onset of senescence sweetening.

The second stage of sugar management occurs as tubers are cooled to final temperature set points. During this phase, storage managers try to prevent the development of CIS by slowly decreasing the temperature of the storage. Storage temperature is typically reduced between 0.12 and 0.24°C day^{-1} (temperature ramping), and as such, it may take several weeks to cool processing and chip potatoes to their holding temperature. Tuber responses to decreasing temperatures are monitored by quantifying tuber sugars. An uptick in tuber sucrose indicates that reducing sugar contents are likely to increase. If an increase in sucrose content is observed, temperature ramping may be suspended temporarily to allow sucrose contents to drop to acceptable amounts.

The final stage of sugar management occurs during long-term storage at a constant holding temperature. The goal during this stage is to identify problems in processing quality just as they develop, so that appropriate action can be taken to prevent raw product rejection by the chip or fry processor. This is done by tracking sucrose and reducing sugar content during storage. An increase in sucrose may indicate that tubers have experienced a stress, such as a period of high CO_2 due to inadequate ventilation, or that they are beginning to undergo senescence sweetening. Corrective action may be possible to reduce stress-induced sugar accumulation, but senescence sweetening is irreversible.

Sugar monitoring at holding temperatures is used as an essential guide for subsequent management decisions. Prevention management of tuber sugar accumulation includes changes in the rate or duration of ventilation, a decrease in allowable CO_2 content of the storage air, temperature adjustment, and the timely removal and processing of tubers. Because of the irreversible nature of senescence sweetening, stored tubers should be monitored frequently, and cultivar-specific guidelines should be followed if available.

15.5 Potato Respiration

Potato tubers are living entities, and like other eukaryotic organisms, they produce energy via aerobic respiration in mitochondria to maintain cellular function. Respiration consumes oxygen and produces oxidation by-products: heat, CO_2, and water. As described below, excessive tuber respiration is often undesirable, as it leads to weight loss via a reduction of stored carbon, a build-up of CO_2 in storage, and an increase in vital heat.

Respiration rates, and corresponding heat production, are often highest immediately after harvest, especially when tubers are harvested during warm periods. Without cooling air, heat produced by respiration can increase tuber temperature in storage easily by more than 0.5°C day^{-1} shortly after harvest. In addition to elevating CO_2 in the storage atmosphere and contributing to weight loss, excessive heat in storage also aggravates disease problems by promoting the growth of bacterial and fungal pathogens.

An abundant supply of stored energy is available to tubers in the form of starch. This is dissociated enzymatically into simple sugars and sugar phosphates through the combined activities of enzymes in the amyloplast and cytosol (see Fig. 15.1). Sugar phosphates enter glycolysis and the mitochondrial tricarboxylic acid (TCA) cycle, and are used to produce high-energy compounds including ATP. This process generates CO_2 and water.

A simplified equation for the conversion of carbohydrate to CO_2 and water during respiration is:

$$C_6H_{12}O_6 + 6O_2 \rightarrow 6CO_2 + 6H_2O + 2872\ \text{kJ mol}^{-1}\ \text{heat.}$$

Although this equation does not capture the biochemical complexity of the respiratory process, it is adequate for understanding the significance of respiration for potato storage. For healthy potatoes in storage, the respiratory quotient is close to unity in most cases (Burton, 1966; Isherwood and Burton, 1975). Hence, potato tuber respiration is fully aerobic, and from each mole of a 6-carbon sugar, 6 moles of O_2 are consumed and 6 moles of CO_2 and 6 moles of water are produced. As a result, water produced in respiration partially offsets water lost through evaporation. The specific energy of glucose is approximately 2872 kJ mol^{-1}, and approximately this amount of energy is released as heat (Green *et al.*, 1941).

Respiration rates

The respiration rate of potato tubers has been measured on many occasions (Burton, 1964, 1966; Schippers, 1977a,b; Dwelle and Stalknecht, 1978; Bethke and Busse, 2010). A typical seasonal pattern is shown for cultivar Russet Burbank in Fig. 15.2 (Dwelle and Stalknecht, 1978), which

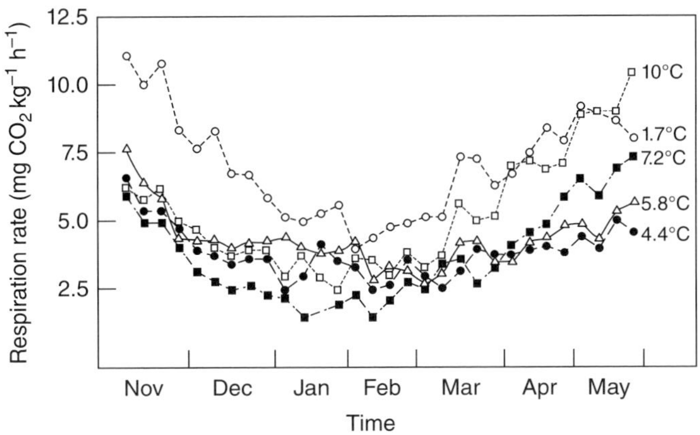

Fig. 15.2. A typical seasonal pattern for the respiration rate of Russet Burbank tubers stored from November to May at five temperatures from 1.7°C to 10°C. (From Dwelle and Stalknecht, 1978; reprinted with permission from Springer Science and Business Media.)

also indicates how the respiration rate is influenced by storage temperature. Immature tubers have very high respiration rates, and these decrease as tubers mature (Burton, 1964; Bethke and Busse, 2010).

Respiration rates decline most rapidly during the first 2 months after harvest, and reach a minimum approximately 4 months after harvest. Tubers of Russet Burbank harvested in early October have been shown to reach their minimum respiration rate around 1 February (Fig. 15.2). Later in storage, respiration rates gradually increase. This increase is observed regardless of whether or not sprouting occurs, but the magnitude of the increase is less in cases where sprouting is prevented (Isherwood and Burton, 1975).

The temperature for minimum tuber respiration is approximately 5–7°C (Schippers, 1977a; Burton, 1978). Respiration rates increase as temperature increases, and the differences between tubers stored at 4.4, 5.8, 7.2, and 10°C may increase as storage duration increases (Fig. 15.2). Storage below approximately 4°C also results in an increase in tuber respiration rate compared with those stored warmer (5–7°C). In Fig. 15.2, a large difference in respiration rate is apparent between tubers stored at 1.7°C and 4.4°C, regardless of time in storage. A similar dependence of respiration rate on temperature has been observed for many cultivars (Schippers, 1977a).

Factors influencing respiration rates

Harvest operations and harvest damage can have a large effect on tuber respiration rate during the first weeks of storage, and as such can increase dramatically the need for cooling air during the initial storage period (Fig. 15.3). Respiration rates for cultivar Katahdin were measured 1 or 11 days after tubers were hand harvested, or after they had experienced one or more mechanical harvest operations. Tubers that had been windrowed, picked up with a mechanical harvester, and piled into storage had respiration rates twice those of tubers that had been hand harvested. Each step in the harvest operation was additive: each contributing to the final respiration rate (Fig. 15.3). Bruising, skinning, dropping, and other mechanical stresses increase tuber respiration rates, which persist for days or weeks (Schippers, 1977a; Burton, 1978; Pisarczyk, 1982; Bethke and Busse, 2010).

Infection with pathogenic organisms leads to an increase in the net respiration of stored potatoes (Gwinn *et al.*, 1989; Fennir *et al.*, 2005). The increase in tuber respiration is likely due to the production of ethylene, exacerbated by the infection (Creech *et al.*, 1973; Gwinn *et al.*, 1989). Ethylene is a volatile plant growth regulator that at very low atmospheric concentrations (0.15 µl l^{-1} or less) stimulates potato tuber respiration (Huelin and Barker, 1939; Reid and Pratt, 1972). At higher concentrations, ethylene can

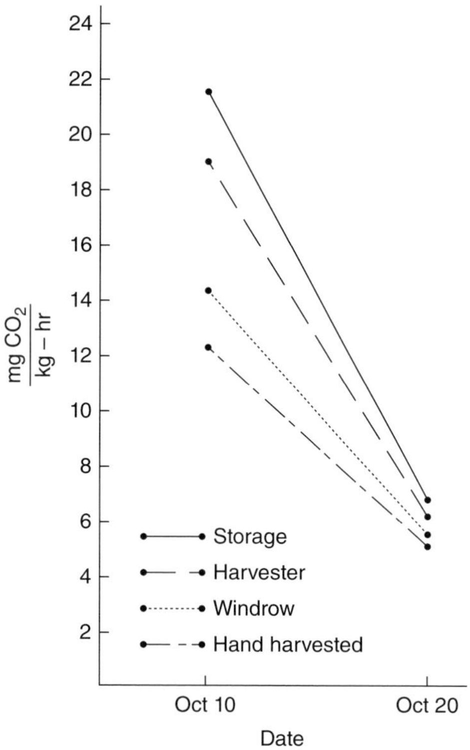

Fig. 15.3. Respiration rate of Katahdin tubers 1 or 11 days after harvesting by hand, or after one or more mechanical harvest operations. (From Pisarczyk, 1982; reprinted with permission from Springer Science and Business Media.)

also cause an increase in tuber reducing sugars and a darkening of fried product color (Daniels-Lake *et al.*, 2005b).

Computer-controlled ventilation systems within many modern potato storages function to minimize the adverse effects of tuber respiration. Ventilation rates are adjusted to maintain a constant small temperature differential across the tubers, such that tubers farthest from the ventilation source are slightly warmer than those close to the source. For bulk storage with a potato pile 6 m deep, this temperature differential is typically 0.25–0.8°C. Fresh, outside air is introduced through external access doors as needed, to flush out harmful concentrations of CO_2. Maximum allowable thresholds for CO_2 vary with the market class of potato. They may be as high as 3000 $\mu l\ l^{-1}$ for seed and fresh market potatoes, or as low as 1200 $\mu l\ l^{-1}$ for chip processing potatoes. Supplemental refrigeration systems are used along with cool ambient air and evaporative cooling to maintain temperature set points despite ongoing heat production from the tubers.

15.6 Water Loss from Stored Potatoes

Potato tubers are 75–80% water by weight. Water loss from stored tubers decreases product value by decreasing the weight of tubers available for sale. Water loss can decrease tuber quality by increasing the incidence of pressure flattening and pressure bruise that occurs in bulk storage, as well as increasing tuber specific gravity, which influences processing quality.

Water is lost from tubers by evaporation, and the rate of water loss depends on the vapor pressure difference (VPD) between the moist environment within the tuber and the ambient air. Free water is contained within the cell wall matrix in tubers, and in many cases it can safely be assumed that the RH at the site of water evaporation from the tuber is 100%. The outer layer of the skin, or phellem, is the multicellular barrier that reduces the diffusion of water from the tuber to the atmosphere. Suberin, which is a mixture of soluble waxes, poly(aliphatic) domains and poly(phenolic) domains, is synthesized in the cell wall matrix of phellem cells as they mature. The aliphatic waxes of the phellem are primarily responsible for reducing the diffusion coefficient for water vapor of the skin. Removal of the waxes by chemical means increased water permeability by 100 times (Schreiber *et al.*, 2005). Burton (1966) estimated that the presence of the mature skin reduced the rate of diffusion from a potato tuber by 300- to 500-fold.

Cuts, scrapes, and physical forces that result in the skinning of tubers substantially increase tuber water loss. Moderate skinning, less than 5% of tuber surface area, can double water loss in the short term. Subsequent wound healing and formation of a secondary periderm is required to re-establish an effective resistance to the diffusion of water vapor. This process depends on the deposition of soluble waxes and seems much less dependent on the synthesis and deposition of the polymeric aliphatic and phenolic components of suberin (Lulai and Orr, 1995). It is for this reason that all potatoes going into storage need to be stored initially under conditions that promote final maturation of the native periderm and the formation of wound periderm.

Favorable conditions for wound healing and periderm maturation include an environment with oxygen and CO_2 concentration similar to those in fresh air, RH above 80%, and temperature of 10–13°C (Wigginton, 1974). This temperature range is low enough to slow the replication of soft rot bacterial and other pathogenic organisms, but warm enough for wounds to heal and native periderm to mature within approximately 2 weeks (Lulai and Orr, 1995; Lulai and Corsini, 1998; Schreiber *et al.*, 2005). Wound periderm may never achieve the same level of water permeability as native periderm (Sabba and Lulai, 2002; Schreiber *et al.*, 2005). Consistent with this observation are data showing that the amount of weight lost by tubers in storage increases with the severity of harvest injury (Misener, 1994).

Potato tubers contain lenticels, specialized pores that are used to promote gas exchange. Lenticels are particularly apparent when tubers are removed from wet soils, as they swell open under these conditions. Although lenticels are a pathway for relatively unrestricted diffusion of water vapor from the tuber, water loss from lenticels is a small fraction of the total water loss. This reflects the large difference in area between the lenticels and the remainder of the tuber surface. Most of the water lost from tubers diffuses through the periderm.

Rate of water loss

The rate of water loss from tubers depends on the resistance to water vapor diffusion through the periderm, the surface area of the tuber, and the water VPD between the tuber and the atmosphere. The resistance to water vapor diffusion and the surface area of the tuber are under genetic control and are influenced strongly by field management practices. Since the RH within the tuber is always close to 100%, the magnitude of the VPD depends on the management of temperature and RH in the potato storage atmosphere.

The water-holding capacity of cool air is less than that of warm air. The significance of this is that cooling air, which is always at a lower temperature than the tubers, will inevitably remove moisture from tubers. Tubers with high rates of respiration produce heat at higher rates than tubers with lower rates of respiration. The increased amount of cooling required to maintain a constant temperature will lead to greater water loss from rapidly respiring tubers relative to tubers with low respiration rates. Poorly insulated and poorly sealed storage buildings have a greater requirement for cooling air when ambient temperatures are high. As a result, tubers in these storages will have greater weight loss compared to those in well-insulated storages.

Humidification

To minimize water loss from potatoes in storage, the storage air may be humidified. Various kinds of humidification systems exist, including wet fiber pads, centrifugal humidifiers, high-pressure water spray nozzles, and air-assisted spray nozzles (Brook *et al.*, 1995). Evaporation from tubers adds moisture to the storage atmosphere at the expense of tuber weight loss (Hunter, 1986). To minimize tuber dehydration, humidity sensors used to control a humidification system need to be positioned in the ventilation air upstream of the tubers.

Minimizing tuber water loss in storage through humidity management necessitates the ability to measure storage temperature and humidity. Temperature is measured easily with temperature probes accurate to <0.5°C. Relative humidity is more difficult to measure, especially above 90% RH. The accuracy of humidity probes is usually no better than ±2–3% RH in the range of 90–100% RH. This lack of accuracy for RH measurements can have a significant effect on tuber weight loss when sensors consistently report values 1–3% higher or lower than the actual RH.

15.7 Sprout Control

Sprouted tubers are undesirable, because sprouts promote rapid dehydration, but more importantly, sprouted potatoes are unacceptable to consumers and cannot be sold to processors. Sprouts on tubers are primary sites for water loss. Sprout area is directly proportional to rate of water loss (Burton, 1966); this indicates that the resistance to water movement through the epidermis of sprouts is much less than the resistance of the periderm.

Stored tubers must be managed to prevent sprouting if tuber quality and grower profitability are to be maintained. For fresh market potatoes, sprouting can be reduced through the use of cold storage temperatures. Storing tubers at 3–4°C maintains dormancy for an extended period and slows sprout growth after dormancy is lost (Wiltshire and Cobb, 1996). This method is less desirable for tubers intended for processing, as it is likely to result in dark-colored fries and chips due to CIS. Moreover, low-temperature storage is often insufficient to prevent sprouting of cultivars that have short dormancy, or for tubers stored for long periods of time. With careful cultivar selection, cold storage can be expected to be an effective means of sprout control for several months.

Sprout inhibitors

Chemical or physical sprout control is necessary for long-term storage of fresh market potatoes, and for chip and fry processing potatoes that will be stored at 9–13°C for more than a few weeks. A wide variety of materials are available for sprout control. The suitability of each depends on customer requirements and product performance needs. Where tubers must be free of chemical residues when sold, ethylene can be used as a sprout inhibitor. Ethylene is mixed with the storage air to a concentration of 4 $\mu l\ l^{-1}$ or more (Daniels-Lake *et al.*, 2005a). When used with processing potatoes, initial ethylene concentrations are applied at low rates and then increased slowly over several weeks (Daniels-Lake *et al.*, 2007). This minimizes the effect of ethylene on reducing sugar accumulation and fried product darkening (Prange *et al.*, 1998). Another form of residue-free sprout control is irradiation with high-energy electrons (Kleinkopf *et al.*, 2003).

Several sprout inhibitors have been produced from natural products and are appropriate for use on organic potatoes, including clove oil, caraway oil, mint oil, and others (Vaughn and Spencer, 1993; Kleinkopf *et al.*, 2003). Most of these products volatilize slowly from tubers during storage and must be reapplied several times during the storage season to maintain efficacy. Other natural products approved for use as a potato sprout inhibitor are substituted naphthalenes and hydrogen peroxide. Both require multiple applications to achieve full season sprout control.

Synthetic sprout inhibitors have been used for many years. In most growing regions, chloropropam (CIPC, isopropyl N-(3-chlorphenyl) carbamate), has replaced maleic hydrazide as the preferred sprout inhibitor. CIPC interferes with mitotic cell division and must therefore be applied after wound healing is complete. In ventilated potato storages, CIPC is applied through the storage ventilation system as a fog. In many cases, a single application is sufficient for season-long control of sprouting. Although used most often in modern potato storages, CIPC is also effective at inhibiting sprout growth and maintaining tuber quality when used in small-scale heaps and pits (Mehta *et al.*, 2010). Health concerns related to CIPC residues on potato led to the imposition of maximum residue levels of 10 mg kg^{-1} in Europe and 30 mg kg^{-1} in the USA (Kleinkopf *et al.*, 2003; EFSA, 2012).

15.8 Potato Storage Management Summary

Successful storage management results in maintenance of stored crop quality, and in many cases, an increase in crop value. Achieving this requires planning and attention to detail in how activities at harvest and in storage are conducted. Below are guidelines for the management of potatoes in bulk storage. These are meant to illustrate the general approach toward potato storage as practiced in the developed world. Specific details for tuber management will necessarily vary by growing region, yearly climate differences, and experience with individual storage types.

Guidelines for storage of potatoes in bulk bins

- Sanitize storage bins and cool them to preconditioning temperature. Humidify to ~93% RH prior to harvest.

- Harvest the crop when skin set and chemical maturity are appropriate for the desired end use.
- Harvest chip and fry processing tubers before crop is exposed to cold temperatures that can lead to CIS.
- Minimize tuber skinning and damage during harvest. Skinning and bruising facilitate pathogen entry and can increase respiration rate and tuber sugars.
- Avoid bringing soil into the storage. Adhering soil makes it more difficult to dry the tubers in the short term and may prevent uniform airflow in the long term.
- The ideal pulp temperature at harvest is 10–16°C. Harvest operations should be suspended when the pulp temperature of tubers arriving at the potato storage is over 18°C. Warm temperatures increase the risk of pathogenic spoilage of the crop.
- Observe all tubers moving into storage and make note of potential storage issues such as excessive shatter bruising, disease/rot, and moisture.
- Pile tubers to a uniform depth of 5–5.5 m. Operate the bin piler such that tubers drop no more than 6 inches when leaving the conveyer. The top of the pile should be level, which encourages uniform airflow through the pile and greatly facilitates the detection of wet spots and sunken areas later in the year.
- Avoid piling cool potatoes on top of warm potatoes. The interface is a site where condensation is likely to occur, and this promotes disease development.
- Calibrated temperature probes should be placed in tubers on the bottom and near the top of the pile to monitor pulp temperatures.
- Humidify to 95% RH. Reduce humidity slightly if disease problems develop.
- Cool to 14–15.5°C as quickly as possible.
- Initial air temperature settings should be based on measured pulp temperatures. Set to 0.8°C less than the pulp temperature near the top of the pile.
- Remove field heat in about 3–4 days using cool air circulation.
- Ramp temperature down to 13°C at 0.3–0.5°C day^{-1} or less to reduce evaporative loss and refrigeration required.
- Adjust airflow rate or duration to establish a uniform pile temperature and top-to-bottom of pile temperature differential (delta T). A delta T of 0.6–0.8°C works well; for example, 13°C tuber pulp temperature on the bottom of the pile and 13.8°C on the top of the pile for a 5.5 m high pile.
- Keep records for each storage bin: cultivar, field, pulp temperature during loading, extent of skinning and bruising, and chemical maturity of the crop. Improvements in storage management accrue when past successes and failures can be evaluated in light of reliable records.
- Hold potatoes at approximately 13°C for wound healing and preconditioning. Monitor tuber sugars to assess progress in preconditioning.
- Preconditioning of fry processing potatoes requires 2–3 weeks, but can be longer. Glucose should be 0.8 mg g^{-1} FW or less at bud end for good color in processing.
- Precondition chip stock so that sucrose is ~0.7 mg g^{-1} FW and glucose is less than 0.035 mg g^{-1} FW. This can take from several weeks to 2–3 months.
- For fresh market potatoes, allow wound healing to proceed for at least 3 days at 10–13°C, then ramp to final storage temperature at a rate of 0.25°C day^{-1}.
- Monitor atmospheric CO_2 in the potato storage bin. Add fresh air when needed to keep CO_2 below thresholds.
- Ramp chip and fry processing cultivars to temperature set point at a rate of 0.1–0.16°C day^{-1}. Slow cooling can reduce the potential for CIS.
- General guidelines for temperature set points are 7–10°C for fry processing, 9–10°C for chips, 3–5°C for fresh market, and 3°C for seed.
- Ramping may take 6–8 weeks for processing potatoes.
- Maintain pile temperature differential (delta T) by adjusting fan speed or by using intermittent ventilation.
- Apply CIPC or other sprout inhibitor after suberization is complete. A minimum of 3–4 weeks in storage is recommended; follow label recommendations.
- Monitor sucrose for advance notice that processing materials are likely to fry dark because of responses to environment or senescence sweetening.

References

Bethke, P.C. and Bussan, A.J. (2013) Acrylamide in processed potato products. *American Journal of Potato Research* 90, 403–424.
Bethke, P.C. and Busse, J.S. (2010) Vine-kill treatment and harvest date have persistent effects on tuber physiology. *American Journal of Potato Research* 87, 299–309.
Bhaskar, P.B., Wu, L., Bethke, P.C., Vaillancourt, B., Jansky, S.H., Busse, J.S., *et al.* (2010) Suppression of the vacuolar invertase gene prevents cold-induced sweetening in potato. *Plant Physiology* 154, 939–948.
Brook, R.C., Fick, R.J. and Forbush, T.D. (1995) Potato storage design and management. *American Journal of Potato Research* 72, 463–480.
Brummell, D.A., Chen, R.K.Y., Harris, J.C., Zhang, H., Hamiaux, C., Kralicek, A.V., *et al.* (2011) Induction of vacuolar invertase inhibitor mRNA in potato tubers contributes to cold-induced sweetening resistance and includes spliced hybrid mRNA variants. *Journal of Experimental Botany* 62, 3519–3534.
Burton, W.G. (1964) The respiration of developing potato tubers. *European Potato Journal* 7, 90–101.
Burton, W.G. (1966) *The Potato*, 2nd edn. H. Veenman and Zonen, Wageningen, the Netherlands.
Burton, W.G. (1978) Post-harvest behaviour and storage of potatoes. In: Coaker, T.H. (ed.) *Applied Biology*, Volume 3. Academic Press, London, pp. 86–228.
Castleberry, H.C. and Jayanty, S.S. (2012) An experimental study of pressure flattening during long-term storage in four Russet potato cultivars with differences in at-harvest tuber moisture loss. *American Journal of Potato Research* 89, 269–276.
Chen, S., Hajirezaei, M.R., Zanor, M.I., Hornyik, C., Debast, S., Lacomme, C., *et al.* (2008) RNA interference-mediated repression of sucrose-phosphatase in transgenic potato tubers (*Solanum tuberosum*) strongly affects the hexose-to-sucrose ratio upon cold storage with only minor effects on total soluble carbohydrate accumulation. *Plant, Cell and Environment* 31, 165–176.
CIP (International Potato Center) (n.d.) The World Potato Atlas (https://research.cip.cgiar.org/confluence/display/wpa/Home, accessed 18 June 2012).
Cochrane, P.M., Duffus, C.M., Allison, M.J. and Mackay, G.R. (1991) Amylolytic activity in stored potato tubers. 2. The effect of low-temperature storage on the activities of α-and β-amylase and α-glucosidase in potato tubers. *Potato Research* 34, 333–341.
Creech, D.L., Workman, M. and Harrison, M.D. (1973) Influence of storage factors on endogenous ethylene production by potato tubers. *American Potato Journal* 50, 145–150.
Crook, E.M. and Watson, D.J. (1950) Studies on the storage of potatoes. III. The composition of the atmosphere in a potato clamp. Journal of Agricultural Science 40, 227–232.
Cunnington, A.C. (2008) Developments in potato storage in Great Britain. *European Potato Journal* 51, 403–410.
Daniels-Lake, B.J., Prange, R.K., Nowak, J., Asiedu, S.K. and Walsh, J.R. (2005a) Sprout development and processing quality changes in potato tubers stored under ethylene: I. Effects of ethylene concentration. *American Journal of Potato Research* 82, 389–397.
Daniels-Lake, B.J., Prange, R.K. and Walsh, J. (2005b) Carbon dioxide and ethylene: a combined influence on potato fry color. *HortScience* 40, 1824–1828.
Daniels-Lake, B.J., Prange, R.K., Kalt, W. and Walsh, J.R. (2007) Methods to minimize the effect of ethylene sprout inhibitor on potato fry colour. *European Potato Journal* 49, 303–326.
Dwelle, R.B. and Stalknecht, G.F. (1978) Respiration and sugar content of potato tubers as influenced by storage temperature. *American Potato Journal* 55, 561–571.
Eldredge, E.P., Homes, Z.A., Mosley, A.R., Shock, C.C. and Stieber, T.D. (1996) Effects of transitory water stress on potato tuber stem-end reducing sugar and fry color. *American Potato Journal* 73, 517–530.
EFSA (European Food Safety Authority) (2012) Review of the existing maximum residue levels (MRLs) for chlorpropham according to Article 12 of Regulation (EC) No 396/2005. *EFSA Journal* 10, 1–53.
Farre, E.M., Tiessen, A., Roessner, U., Geigenberger, P., Trethewey, R.N. and Willmitzer, L. (2001) Analysis of the compartmentation of glycolytic intermediates, nucleotides, sugars, organic acids, amino acids, and sugar alcohols in potato tubers using a nonaqueous fractionation method. *Plant Physiology* 127, 685–700.
Farre, E.M., Fernie, A.R. and Willmitzer, L. (2008) Analysis of subcellular metabolite levels of potato tubers (*Solanum tuberosum*) displaying alterations in cellular or extracellular sucrose metabolism. *Metabolomics* 4, 161–170.
Fennir, M., Landry, J. and Raghavan, G. (2005) Respiration rate of potatoes (*Solanum tuberosum* L.) as affected by soft rot (*Erwinia carotovora*) and determined at various storage temperatures. *West Indian Journal of Engineering* 27, 62–71.

Fuglie, K.O., Khatana, V.S., Ilangantileke, S.G., Scott, G.J. and Singh, J. (2000) Economics of potato storage in Northern India. *Quarterly Journal of International Agriculture* 39, 131–148.
Geigenberger, P. (2003) Regulation of sucrose to starch conversion in growing potato tubers. *Journal of Experimental Botany* 54, 457–465.
Geigenberger, P., Stitt, M. and Fernie, A.R. (2004) Metabolic control analysis and regulation of the conversion of sucrose to starch in growing potato tubers. *Plant, Cell and Environment* 27, 655–673.
Gildemacher, P.R., Demo, P., Barker, I., Kaguongo, W., Woldegiorgis, G., Wagoire, W.W., *et al.* (2009) A description of seed potato systems in Kenya, Uganda and Ethiopia. *American Potato Journal* 86, 373–382.
Green, W., Hukill, W. and Rose, D. (1941) Calimetric measurements of the heat of respiration of fruits and vegetables. United States Department of Agriculture, Technical Bulletin No 771, Washington, DC.
Gupta, S. and Sowokinos, J.R. (2003) Physicochemical and kinetic properties of unique isozymes of UDP-Glc pyrophosphorylase that are associated with resistance to sweetening in cold-stored potato tubers. *Journal of Plant Physiology* 160, 589–600.
Gwinn, K., Stelzig, D. and Bhatia, S. (1989) Differential ethylene production by potato tuber tissue inoculated with a compatible or an incompatible race of *Phytophthora infestans*. *American Potato Journal* 66, 417–423.
Herrman, T.J., Love, S.L., Shafii, B. and Dwelle, R.B. (1996) Chipping performance of three processing potato cultivars during long-term storage at two temperature regimes. *American Potato Journal* 73, 411–425.
Huelin, F. and Barker, J. (1939) The effect of ethylene on the respiration and carbohydrate metabolism of potatoes. *New Phytologist* 38, 85–104.
Hunter, J.H. (1986) Heat of respiration and weight loss from potatoes in storage. In: Cargill, B.F. (ed.) *Engineering for Potatoes*. Michigan State University and American Society of Agricultural Engineers, East Lansing, Michigan, pp. 511–550.
Iritani, W.M., Weller, L. and Russell, T.S. (1973) Relative differences in sugar content of basal and apical portions of Russet Burbank potatoes. *American Potato Journal* 50, 24–31.
Isherwood, F.A. and Burton, W.G. (1975) The effect of senescence, handling, sprouting and chemical sprout suppression upon the respiratory quotient of stored potato tubers. *Potato Research* 18, 98–104.
Kincaid, D.C., Westermann, D.T. and Trout, T.J. (1993) Irrigation and soil-temperature effects on Russet Burbank quality. *American Potato Journal* 70, 711–723.
Kleinkopf, G.E., Oberg, N.A. and Olsen, N.L. (2003) Sprout inhibition in storage: current status, new chemistries and natural compounds. *American Journal of Potato Research* 80, 317–327.
Kumar, D., Singh, B.P. and Kumar, P. (2004) An overview of the factors affecting sugar content of potatoes. *Annals of Applied Biology* 145, 247–256.
Larsson, C.T., Hofvander, P., Khoshnoodi, J., Ek, B., Rask, L. and Larsson, H. (1996) Three isoforms of starch synthase and two isoforms of branching enzyme are present in potato tuber starch. *Plant Science* 117, 9–16.
Li, L., Strahwald, J., Hofferbert, H.R., Lubeck, J., Tacke, E., Junghans, H., *et al.* (2005) DNA variation at the invertase locus invGE/GF is associated with tuber quality traits in populations of potato breeding clones. *Genetics* 170, 813–821.
Liu, X., Song, B., Zhang, H., Li, X., Xie, C. and Liu, J. (2010) Cloning and molecular characterization of putative invertase inhibitor genes and their possible contributions to cold-induced sweetening of potato tubers. *Molecular Genetics and Genomics* 284, 147–159.
Liu, X., Zhang, C., Ou, Y., Lin, Y., Song, B., Xie, C., *et al.* (2011) Systematic analysis of potato acid invertase genes reveals that a cold-responsive member, StvacINV1, regulates cold-induced sweetening of tubers. *Molecular Genetics and Genomics* 286, 109–118.
Lloyd, J.R., Blennow, A., Burhenne, K. and Kossman, J. (2004) Repression of a novel isoform of disproportionating enzyme (stDPE2) in potato leads to inhibition of starch degradation in leaves but not tubers stored at low temperature. *Plant Physiology* 134, 1347–1354.
Lulai, E.C. and Corsini, D.L. (1998) Differential deposition of suberin phenolic and aliphatic domains and their roles in resistance to infection during potato tuber (*Solanum tuberosum* L.) wound-healing. *Physiological and Molecular Plant Pathology* 53, 209–222.
Lulai, E.C. and Orr, P.H. (1995) Porometric measurements indicate wound severity and tuber maturity affect the early stages of wound-healing. *American Potato Journal* 72, 225–241.
Lulai, E.C., Glynn, M.T. and Orr, P.H. (1996) Cellular changes and physiological responses to tuber pressure-bruising. *American Potato Journal* 73, 197–209.
McKenzie, M.J., Sowokinos, J.R., Shea, I.M., Gupta, S.K., Lindlauf, R.R. and Anderson, J.A.D. (2005) Investigations on the role of acid invertase and UDP-glucose pyrophosphorylase in potato clones with varying resistance to cold-induced sweetening. *American Journal of Potato Research* 82, 231–239.

Malone, J.G., Mittova, V., Ratcliffe, R.G. and Kruger, N.J. (2006) The response of carbohydrate metabolism in potato tubers to low temperature. *Plant and Cell Physiology* 47, 1309–1322.
Matsuura-Endo, C., Kobayashi, A., Noda, T., Takigawa, S., Yamauchi, H. and Mori, M. (2004) Changes in sugar content and activity of vacuolar acid invertase during low-temperature storage of potato tubers from six Japanese cultivars. *Journal of Plant Research* 117, 131–137.
Mehta, A., Singh, B., Ezekiel, R. and Kumar, D. (2010) Effect of CIPC on sprout inhibition and processing quality of potatoes stored under traditional storage systems in India. *Potato Research* 53, 1–15.
Misener, G.C. (1994) Relationship between damage index and weight loss of potatoes during storage. *Drying Technology* 12, 1735–1741.
Nielsen, T.H., Deiting, U. and Stitt, M. (1997) A β-amylase in potato tubers is induced by storage at low temperature. *Plant Physiology* 113, 503–510.
Pisarczyk, J. (1982) Field harvest damage affects potato tuber respiration and sugar content. *American Potato Journal* 59, 205–211.
Prange, R.K., Kalt, W., Daniels-Lake, B.K., Liew, C.L. and Page, R.T. (1998) Using ethylene as a sprout control agent in stored 'Russet Burbank' potatoes. *Journal of the American Society for Horticultural Science* 123, 463–469.
Pressey, R. and Shaw, R. (1966) Effect of temperature on invertase, invertase inhibitor and sugars in potato tubers. *Plant Physiology* 41, 1657–1661.
Pringle, R.T., Bishop, C.F.H. and Clayton, R.C. (2009) *Potatoes Postharvest*. CAB International, Wallingford, UK.
Pritchard, M.K. and Adam, L.R. (1992) Preconditioning and storage of chemically immature Russet Burbank and Shepody potatoes. *American Potato Journal* 69, 805–815.
Rathore, R.S., Garg, N., Garg, S. and Kumar, A. (2009) Starch phosphorylase: role in starch metabolism and biotechnological applications. *Critical Reviews in Biotechnology* 29, 214–224.
Reid, M.S. and Pratt, H.K. (1972) Effects of ethylene on potato tuber respiration. *Plant Physiology* 49, 252–255.
Ritte, G., Lloyd, J.R., Eckermann, N., Rottman, A. and Kossman, J. (2002) The starch-related R1 protein is an alpha-glucan, water dikinase. *Proceedings of the National Academy of Sciences* 99, 7166–7171.
Rommens, C.M., Ye, J., Richael, C. and Swords, K. (2006) Improving potato storage and processing characteristics through all-native DNA transformation. *Journal of Agricultural and Food Chemistry* 54, 9882–9887.
Sabba, R.P. and Lulai, E.C. (2002) Histological analysis of the maturation of native and wound periderm in potato (*Solanum tuberosum* L.) tuber. *Annals of Botany* 90, 1–10.
Samuel, G.G. and Wilson, A.R. (1949) A summary of investigations on clamp storage of potatoes in England. *Tijdschrift Over Plantenziekten* 55, 179–186.
Schippers, P. (1977a) Rate of respiration of potato-tubers during storage 1. Review of literature. *Potato Research* 20, 173–188.
Schippers, P. (1977b) Rate of respiration of potato-tubers during storage 2. Results of experiments in 1972 and 1973. *Potato Research* 20, 189–206.
Schreiber, L., Franke, R. and Hartmann, K. (2005) Wax and suberin development of native and wound periderm of potato (*Solanum tuberosum* L.) and its relation to peridermal transpiration. *Planta* 220, 520–530.
Shallenberger, R.S., Smith, O. and Treadway, R.H. (1959) Food color changes, role of the sugars in the browning reaction in potato chips. *Journal of Agricultural and Food Chemistry* 7, 274–277.
Shock, C., Zalewski, J.C., Stieber, T.D. and Burnett, D.S. (1992) Impact of early-season water deficits on Russet Burbank plant development, tuber yield and quality. *American Potato Journal* 69, 793–803.
Shock, C.C., Holmes, Z.A., Stieber, T.D., Eldredge, E.P. and Zhang, P. (1993) The effect of timed water stress on quality, total solids and reducing sugar content of potatoes. *American Potato Journal* 70, 227–241.
Smith, A.M., Zeeman, S.C. and Smith, S.M. (2005) Starch degradation. *Annual Review of Plant Biology* 56, 73–98.
Sowokinos, J. (1976) Pyrophosphorylase in *S. tuberosum*. *Plant Physiology* 57, 63–68.
Sowokinos, J.R. (2001a) Allele and isozyme patterns of UDP-glucose pyrophosphorylase as a marker for cold-sweetening resistance in potatoes. *American Journal of Potato Research* 78, 57–64.
Sowokinos, J.R. (2001b) Biochemical and molecular control of cold-induced sweetening in potatoes. *American Journal of Potato Research* 78, 221–236.
Sowokinos, J.R., Lulai, E. and Knoper, J. (1985) Translucent tissue defects in *Solanum tuberosum* L. I. Alterations in amyloplast membrane integrity, enzyme activities, sugars, and starch content 1. *Plant Physiology* 78, 489–494.

Sowokinos, J.R., Shock, C.C., Stieber, T.D. and Eldredge, E.P. (2000) Compositional and enzymatic changes associated with the sugar-end defect in Russet Burbank potatoes. *American Journal of Potato Research* 77, 47–56.
Statham, O.J.H. (1989) A review of potato storage practices trends and developments in Great Britain. In: Cargill, B.F., Brook, R.C. and Forbush, T.D. (eds) *Potato Storage: Technology and Practice*. American Society of Agricultural Engineers, St Joseph, Michigan, pp. 9–37.
Takaha, T., Yanase, M., Okada, S. and Smith, S.M. (1993) Disproportionating enzyme (4-alpha-glucanotransferase; EC 2.4.1.25) of potato. Purification, molecular cloning, and potential role in starch metabolism. *Journal of Biological Chemistry* 268, 1391–1396.
Thompson, A.L., Love, S.L., Sowokinos, J.R., Thornton, M.K. and Shock, C.C. (2008) Review of the sugar end disorder in potato (*Solanum tuberosum,* L.). *American Journal of Potato Research* 85, 375–386.
Tjaden, J., Möhlmann, T. and Kampfenkel, K. (1998) Altered plastidic ATP/ADP-transporter activity influences potato (*Solanum tuberosum* L.) tuber morphology, yield and composition of tuber starch. *The Plant Journal* 16, 531–540.
USDA National Agricultural Statistics Service (2011) Potatoes 2010 Summary (September 2011) (www.nass.usda.gov/Publications/Todays_Reports/reports/pots0911.pdf, accessed 30 May, 2014).
van der Zaag, D.E. (1989) Some thoughts about the situation on storage of potatoes in the world. In: Cargill, B.F., Brook, R.C. and Forbush, T.D. (eds) *Potato Storage: Technology and Practice*. American Society of Agricultural Engineers, St Joseph, Michigan, pp. 1–7.
Vaughn, S.F. and Spencer, G.F. (1993) Naturally-occurring aromatic-compounds inhibit potato-tuber sprouting. *American Potato Journal* 70, 527–533.
Wang, Y. and Bethke, P.C. (2013) Effects of *Verticillium dahliae* infection on stem-end chip defect development in potatoes. *Crop Science* 53, 595–601.
Wang, Y., Bussan, A.J. and Bethke, P.C. (2012) Stem-end defect in chipping potatoes (*Solanum tuberosum* L.) as influenced by mild environmental stresses. *American Journal of Potato Research* 89 392–399.
Wigginton, M.J. (1974). Effects of temperature, oxygen tension and relative humidity on the wound-healing process in the potato tuber. *European Potato Journal* 17, 200–214.
Wiltshire, J.J.J. and Cobb, A.H. (1996) A review of the physiology of potato tuber dormancy. *Annals of Applied Biology* 129, 553–569.
Witt, W.W. and Sauter, J.J. (1996) Purification and properties of a starch granule-degrading α-amylase from potato tubers. *Journal of Experimental Botany* 47, 1789–1795.
Wu, L., Bhaskar, P.B., Busse, J.S., Zhang, R., Bethke, P.C. and Jiang, J. (2011) Developing cold-chipping potato varieties by silencing the vacuolar invertase gene. *Crop Science* 51, 981–990.
Ye, J., Shakya, R., Shrestha, P. and Rommens, C.M. (2010) Tuber-specific silencing of the acid invertase gene substantially lowers the acrylamide-forming potential of potato. *Journal of Agricultural and Food Chemistry* 58, 12162–12167.
Zeeman, S.C., Delatte, T., Messerli, G., Umhang, M., Stettler, M., Mettler, T., *et al.* (2007) Starch breakdown: recent discoveries suggest distinct pathways and novel mechanisms. *Functional Plant Biology* 34, 465–473.
Zeeman, S.C., Kossmann, J. and Smith, A.M. (2010) Starch: its metabolism, evolution, and biotechnological modification in plants. *Annual Review of Plant Biology* 61, 209–234.

16 Traditional Breeding and Cultivar Development

Richard Novy*
USDA-ARS, Aberdeen, Idaho, USA

The history and origin of cultivated potato have been thoroughly detailed by Brown and Henfling (see Chapter 1, this volume), with Bradshaw (2007a,b) also having reviewed the domestication, evolution, and germplasm resources of cultivated potato. Both book chapters are recommended for their detailed reviews and provide the reader with the necessary background on the history and ascent of potato to its current role as the world's most important non-grain food crop. This chapter details the contributions that potato breeding and cultivar development have made to the global acceptance of potato as a major food crop, the unique characteristics of potato that impact its breeding, the emerging issues of importance in cultivar development, and the methodologies used in traditional potato breeding.

16.1 History of Potato Breeding and Cultivar Development

The first reports of directed breeding in potato with hybridizations made between cultivars were made in the early 19th century and were attributed to Knight (1807) and Knott in about 1810 (Glendinning, 1983). Potato cultivars at this time appeared to consist of an assemblage of landrace potato cultivars from the Andean regions of South America and the lowlands of south-central Chile, respectively (Ames and Spooner, 2008; also see Chapter 2, this volume). The catastrophic late blight epidemic of the 1840s spurred the introduction of new germplasm and the development of new cultivars by primarily farmers, hobby breeders, and seedsmen (Bradshaw, 2007a). One of the more notable breeders of the time was Chauncey Goodrich, a minister from Utica, New York. Goodrich received eight cultivars from Panama in 1851, the best of which he named Rough Purple Chile, believing that the cultivar's origin had originally been Chilean. The impact of the introduction of Rough Purple Chile is profound, with 100% of 44 "prominent" North American potato cultivars (minimum of 200 acres of US seed) having this cultivar represented in their ancestry; it is also represented in the ancestry of European cultivars such as Imperator, Richter's Jubel, and Busola (Love, 1999). Garnet Chile was a progeny of Rough Purple Chile, and was successful both as a cultivar and as a parent in breeding; Early Rose being a direct descendant. It was from a single fruit of Early Rose that Luther Burbank obtained a seedling with high merit that he sold to a Massachusetts seedsman for US$150. Using this money, Luther Burbank moved and settled in California, taking ten tubers of his Burbank

* E-mail: rich.novy@ars.usda.gov

cultivar with him, as allowed by the seedsman. The cultivar Burbank soon became popular in California and from there spread to other western potato production states with a russet- skinned genetic sport from Montana, becoming the cultivar Russet Burbank, which is still the predominant cultivar grown in the USA (Bethke *et al.*, 2014).

Breeders of the 19th century, such as Goodrich and Burbank, have left a lasting legacy. Their cultivars are present in the pedigrees of most modern-day potato cultivars—in fact, prior to the release of the cultivar Katahdin by the US Department of Agriculture in 1932, all potato cultivars in the USA had previously been developed by such private breeders (Hoopes and Plaisted, 1987). The extent of progress in potato breeding since the first fledgling attempts by Knight and Knott in the early 19th century is exemplified in the *World Catalogue of Potato Varieties* (Pieterse and Hils, 2009), with over 4500 cultivars listed.

16.2 Attributes of Potato and Their Significance for the Breeding of Potato

Every crop has unique attributes that must be taken into consideration by breeders in formulating the approach to be taken for its genetic enhancement. This section details the important characteristics of potato that guide the approach taken by breeders for its genetic improvement.

Dual reproduction (sexual and asexual)

Potato cultivars have "fixed genotypes" in that they are propagated via asexual means, specifically by the cutting of tubers into seed pieces, which are planted in the field. Thus, plants of any given cultivar are genetically identical to one another, with the exception of occasional genetic variants. Selections can be made from these genetic variants, as was the case with strain (line) selections of Russet Norkotah (Norkotah) made by the state breeding programs of Colorado and Texas (Miller *et al.*, 1999). Strain selections of Norkotah have met with commercial success. Janksy and Miller (2010) reported that there were more certified seed acres of strain selections of Norkotah being grown in the USA in 2009 than of the original Norkotah cultivar.

Although strain or line selection is preferable from the standpoint of maintaining the desirable attributes of a cultivar while improving the less desirable attributes, it requires that the trait for improvement be readily identifiable, as is the case with the initial selection of giant hill variants of Norkotah having more robust vine growth to withstand hail and early dying disease better, or tuber skin variants such as the Russet Burbank originating from the white-skinned Burbank. Otherwise, it literally can be like looking for "a needle in a haystack" to identify the infrequent genetic variant of a cultivar.

Traditional potato breeding focuses on sexual rather than asexual reproduction in the earliest stage of the breeding process. Superior parents are identified and subsequently hybridized, allowing for the production of botanical seed, also called "true potato seed", as distinguished from "potato seed"—a tuber piece with an eye (sprout). Each seedling that arises from the germinated true potato seed represents a unique individual with new combinations of genes. The process of selection and evaluation then begins, and 10 or more years after the seedling is first produced, a new cultivar may be released. Additional background and greater detail regarding the sexual component of potato breeding is presented later in this chapter in the section on traditional breeding methodologies. Also detailed in this section is how asexual reproduction is again employed by breeders to maintain the unique combination of genes of superior seedlings for their release as potato cultivars.

Tetraploid genetics

Cultivated potato, found in fields worldwide, is an autotetraploid with four sets of 12 chromosomes ($2n = 4\times = 48$), allowing the pairing of up to four homologous chromosomes during meiosis. At the tetraploid level, a locus can be represented by up to four different alleles (variants) of a gene, whereas a diploid would have a maximum of two divergent alleles. More complex interactions between genes at different loci (inter-allelic), as well as between up to four genes at the same locus (intra-allelic), are possible in tetraploids than in diploids.

Cultivated potato also has diploid derivatives and wild relatives that are useful in genetic and molecular analyses. For example, sequencing of the potato genome utilized two diploid clones that were identified as having a relatively high frequency of presence/absence variants, and other deleterious gene mutations, that likely contributed to observations of inbreeding depression in potato (Potato Genome Sequencing Consortium, 2011).

The maintenance of desirable gene interactions in highly heterozygous, outcrossing potato, while minimizing homozygosity which contributes to the expression of deleterious genes, requires breeders to hybridize parents that are not closely related. The backcross method, as practiced in self-pollinated crops, is not normally utilized in potato for the incorporation of simply inherited traits, due to the inherent inbreeding depression associated with the use of a recurrent parent. Instead, a modified backcross technique whereby a different parent is used as the recurrent parent in each backcross generation is commonly employed in potato (Hoopes and Plaisted, 1987).

Relative to diploids, the segregation of traits becomes much more complex. For example, in a simple gene model with two alleles, "A" (dominant) and "a" (recessive) at a single locus, diploid genotypes would consist of two homozygote classes (AA and aa) and one heterozygote class (Aa). In tetraploid potato, two homozygote classes are also represented: AAAA and aaaa, defined as quadruplex and nulliplex genotypes, respectively. However, rather than there being just one heterozyote genotype, triplex (A^3a), duplex (A^2a^2), and simplex (Aa^3) heterozygote genotypes are now represented. The ratio of phenotypes obtained from the crossing of the three heterozygote classes to a nulliplex where the presence of the A allele confers one phenotypic class (dominant gene effect) with the other phenotypic class being nulliplex (recessive) can vary considerably (Table 16.1). Under a chromosome segregation model whereby no crossing over is expected between the locus and the centromere, only dominant phenotypes would be expected with triplex heterozygotes, whereas recessive phenotypes are represented in the progeny of duplex and simplex heterozygotes (Table 16.1). A greater frequency of recessive phenotypes is obtained if the locus is at a sufficient distance from the centromere for crossing over to occur on a regular basis—termed chromatid segregation, with recessive phenotypes now also being represented in the progeny of triplex heterozygotes where none were found previously with chromosome segregation (Table 16.1).

The complex genetics associated with the higher ploidy of potato has limited the understanding of the inheritance of genetic traits in potato relative to diploid crops. Even the advent of molecular biology and its associated molecular markers has not alleviated the difficulty of working with a tetraploid crop. As discussed by van Eck (2007), most molecular marker techniques currently available do not allow full classification of multiple alleles at a locus in tetraploids—thereby limiting their usefulness in associating traits of interest with molecular markers. Currently in potato, the development and use of marker-assisted selection in potato breeding has been primarily for traits conferred by a single major gene, such as resistance to viruses or nematodes (Simko *et al.*, 2007), with less progress having been made for polygenic or quantitative traits.

The recent sequencing of the 844 Mb potato genome will aid in identifying allelic variants of genes contributing to important quantitative traits in potato (Potato Genome Sequencing

Table 16.1. Heterozygote genotypes in potato and their expected chromosome and chromatid phenotypic segregation in progeny following crosses to a nulliplex (a^4). The "A" allele conferring complete dominance over the recessive "a" allele, with two phenotypic classes therefore being possible.

	Expected phenotypic classes	
Heterozygote cross	Chromosome segregation	Chromatid segregation
$A^3a \times a^4$	All A phenotype	27A:1a
$A^2a^2 \times a^4$	5A:1a	3.7A:1a
$Aa^3 \times a^4$	1A:1a	0.87A:1a

Consortium, 2011), thereby facilitating the continued genetic improvement of potato. Readers are encouraged to see Veilleux and Boluarte-Medina (Chapter 17, this volume) for more detail regarding the importance of molecular breeding and how sequencing of the potato genome can facilitate breeding.

Diversity of wild relatives

Early potato breeding efforts were highlighted by the importation and utilization of new sources of germplasm such as Rough Purple Chile. Germplasm with desirable traits not found in cultivated potato are as important to contemporary potato breeders as they were to Chauncy Goodrich over 161 years ago. Fortunately, potato has a diversity of wild relatives with genetic traits important to the continued enhancement of cultivated potato. Hawkes (1990) identified 217 potato relatives, but more recent taxonomical assessments have reduced that number to approximately 100 (Spooner, 2009). These wild relatives have a wide geographic distribution, stretching from south-western USA to the central regions of Chile and Argentina (Hijmans and Spooner, 2001). Collections or gene banks of these wild potato species exist in the USA, the UK, Peru, the Netherlands, Germany, Argentina, and Russia. These germplasm collections truly represent banks in which genes are deposited, to be withdrawn to address emerging problems in potato production. These gene banks also play a vital role in conserving species diversity, as increasing human encroachment on native habitats is associated with the loss of unique genotypes.

Unfortunately, the successful utilization of wild potato species for the development of potato cultivars has been infrequent. Love (1999) identified 11 species and subspecies in the pedigrees of prominent North American cultivars—approximately 11% of the known wild relatives of cultivated potato. While the numbers of wild relatives used successfully in the development of potato cultivars is low, 34 of 44 (77%) potato cultivars have "exotic germplasm" in their pedigrees—defined by Love as the previously mentioned 11 wild species, four South and Central American cultivars, and three accessions derived from *Solanum demissum*. Love stated that the use of exotic germplasm in potato breeding appeared to be a recent phenomenon, with the only meaningful introgression prior to 1966 occurring from the use of *S. demissum* for its late blight resistance in the late 1930s and 1940s.

Why the relatively sparse utilization of the diversity of wild potato relatives by potato breeders? Bradshaw (2007a,b) identified ploidy differences (the majority of wild species being diploid), endosperm balance number (EBN) mismatches, pollen–pistil incompatibilities, and nuclear-cytoplasmic male sterilities as contributing factors. However, Bradshaw also stated that the use of chromosome doubling techniques or unreduced $2n$ gametes from wild species, or the development of haploids (dihaploids) of cultivated potato following crosses with *Solanum phureja* as the male parent could allow for the successful utilization of virtually any potato species by breeders. Somatic hybridization or cell fusion is also another tool for overcoming hybridization barriers and allowing the introgression of desirable traits, such as virus, fungal, nematode, and insect resistances, from wild potato species into cultivated potato (Novy and Helgeson, 1994a,b; Brown *et al.*, 1995, 2006; Novy *et al.*, 2002a, 2007; Thompson *et al.*, 2007; Zhang *et al.*, 2007; Butler *et al.*, 2011).

The primary barrier to the utilization of wild species is often the undesirable attributes that accompany the desired genes of interest. For example, late blight resistance is often associated with later maturity, which is not desirable in the development of early- to mid-season potato cultivars with late blight resistance (Nowicki *et al.*, 2012). Market classes of potato, especially for processing, have very specific commercial requirements with respect to tuber size and length, starch and sugar concentrations, and extended tuber dormancy, to allow for processing from long-term storage. The use of wild species tends to cause substantial deviations in attributes important to the potato industry, and as stated by Brown (2011): "The inconvenient accompaniment of introducing exotic genetic variation is that the breeding products are often outside of the targeted market niche."

While wild species use by breeders can be problematic, continued genetic enhancement of potato requires their successful utilization. An example is the use of the wild species *Solanum etuberosum* and *Solanum berthaultii*. Both species have desirable traits, with multiple virus resistances identified in *S. etuberosum*—a diploid, non-tuber-bearing species that cannot be hybridized with tetraploid or diploid forms of cultivated potato. Sexual incompatibilities were overcome by the use of somatic hybridization with the generation of somatic hybrids between *S. etuberosum* and a Gp. Tuberosum dihaploid × *S. berthaultii* hybrid clone (Novy and Helgeson, 1994a). The tri-species somatic hybrids had very vigorous foliar growth in the field, with low yields of tubers having very poor type (Fig. 16.1)— not totally unexpected, with approximately 75% of the genome originating from wild species.

However, following two generations of hybridization of the somatic hybrids to cultivated potato, breeding clones having round, white-skinned tubers with the appearance desired for use by the chipping industry were obtained (Fig 16.1). Following an additional two generations of crossing to cultivated potato (four generations removed from the somatic hybrid progenitors), breeding clones with the long tuber type and russeted skin desired by the western US potato industry also were obtained (Fig. 16.1)—indicative of how species can be used successfully in a breeding program. The breeding clone with the long, russeted tubers referenced in Fig. 16.1 has potato leaf roll (PLRV) resistance derived from *S. etuberosum* and the necessary processing attributes for its continued evaluation and advancement for possible release as a potato cultivar. In addition to PLRV resistance, the somatic hybrids and their progenies have been shown to have resistances to *Potato virus Y*, green peach aphid, pink rot, *Pythium* leak, and potato psyllid (Novy and Helgeson, 1994b; Novy *et al.*, 2002a, 2007; Thompson *et al.*, 2007; Butler *et al.*, 2011), with unpublished data suggesting resistance to wireworm and Colorado potato beetle as well. A major gene for PLRV resistance derived from *S. etuberosum*, *Rlretb*, has also been mapped (Gillen and Novy, 2007; Kelley *et al.*, 2009). The numerous resistances identified in the progenies of *S. etuberosum* and *S. berthaultii* highlight the contributions that wild species can make to the genetic improvement of potato.

16.3 Emerging Issues of Importance in Potato Breeding

Potato has a myriad of diseases, pests, and physiological disorders that are problematic for its production, the extent of which are detailed in this volume (see Waters and Jensen, Chapter 9; Palomares-Rius *et al.*, Chapter 10; Kirk and

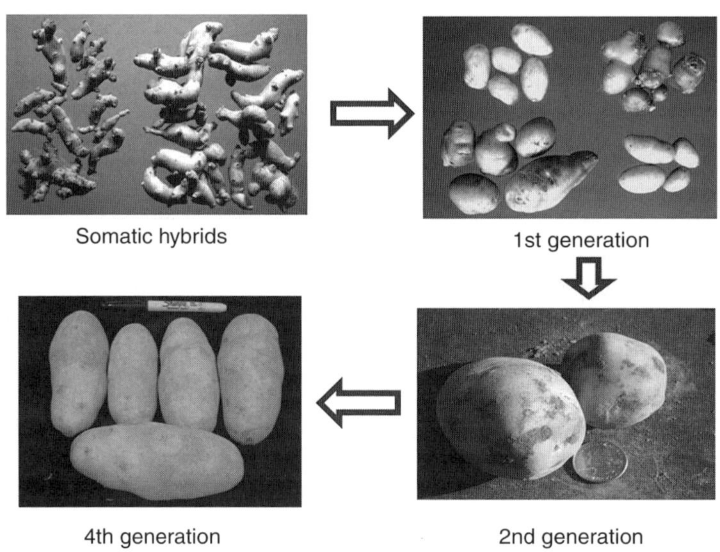

Fig. 16.1. Tubers of somatic hybrids of *Solanum etuberosum* and a Gp. Tuberosum dihaploid × *Solanum berthaultii* hybrid and the improvement in tuber type over generations with successive hybridizations to cultivated potato.

Wharton, Chapter 11; Jones, Chapter 12; and Mikitzel, Chapter 14). Potato breeders must actively breed and select for genetic resistance to those that impact yield and quality seriously in the production regions for which they develop new potato cultivars. In addition, they must ensure during the development of a new cultivar that it has no greater susceptibility to lesser pests and diseases than those cultivars currently used by the industry. Agronomic characteristics such as yield, tuber size distribution and dormancy, starch content, sensory attributes, and processing and storage characteristics (see Bethke, Chapter 15, this volume) must also meet industry standards with respect to market class (see Bond, Chapter 3, this volume). Enhanced nutritional value (biofortification) and flavor have also come to the forefront in potato breeding, with enhancement of these traits also being discussed in Chapter 18 (Navarre *et al.*) and Chapter 19 (Taylor), this volume. Bradshaw (2007a) provides a thorough review of resistance and agronomic traits important in potato breeding, with Simko *et al.* (2007) and van Eck (2007) providing detailed analyses of the genetics of resistance and morphological and tuber traits, respectively. Rather than restate the well-written summaries provided by the aforementioned authors, this section will discuss the food safety, disease, and sustainability issues of potato that have most recently emerged as important breeding objectives.

Acrylamide

Acrylamide is a chemical compound that is a neurotoxin, a carcinogen in rodents, and a suspected carcinogen in humans. In 2002, moderate levels of acrylamide were reported in heated protein-rich foods, with higher concentrations found in carbohydrate-rich foods such as potato and their fried products (Tareke *et al.*, 2002). Formation of acrylamide has subsequently been shown to be a product of the Maillard reaction and is primarily a consequence of a chemical reaction occurring between the free amino acid, asparagine (Asn), and reducing sugars (glucose/fructose) at high temperatures. Recent reviews of acrylamide formation in foods (Lineback *et al.*, 2012) and potato specifically (Vinci *et al.*, 2012; Bethke and Bussan, 2013) are recommended for an understanding of the factors contributing to acrylamide formation and strategies for its mitigation. The effects of acrylamide on human health at the amounts found in food are not yet clear, with no country having yet used regulatory action to set limits on the acrylamide content in foods or in the diet (Lineback *et al.*, 2012). However, acrylamide in foods is of concern, with the US potato industry proactively seeking to reduce its levels in processed potato products several years prior to the US Food and Drug Administration issuing draft guidance to the food industry in November 2013 on approaches for reducing the levels of acrylamide in certain foods.

Potato cultivars with low levels of reducing sugars and/or free asparagine in tubers would be a means of reducing acrylamide in processed potato products (Vinci *et al.*, 2012). For several decades, potato breeders have bred for reduced tuber sugar levels to facilitate the development of processing cultivars having lighter fry or chip color following frying. Fortuitously, breeding for reduced sugar levels in potato has also contributed to reducing acrylamide. However, efforts in reducing the concentration of asparagine in potato were not a breeding objective until the recent elucidation of its role in acrylamide formation. In potato, asparagine concentrations are in excess compared to reducing sugar content, with the conclusion being that acrylamide formation will be determined largely by reducing sugar levels (Vinci *et al.*, 2012). However, while the majority of published data show stronger correlations between reducing sugars and acrylamide formation than for asparagine, cold-induced sweetening in tubers following storage can change the balance, with reducing sugars no longer being limiting and asparagine content now becoming critical in acrylamide formation (Matsuura-Endo *et al.*, 2006). A combined breeding approach for reducing acrylamide formation by lowering concentrations of both tuber reducing sugars and free asparagine appears warranted. Support for such an approach was voiced by Shepherd *et al.* (2010), with the authors making the following statement based on data from their analysis of acrylamide formation in a tetraploid breeding population segregating for tuber sugars and asparagine: “In conclusion our data indicate that conventional breeding approaches need to select for both low reducing sugar levels and low asparagine levels, in order to produce new improved processing cultivars that will deliver reduced acrylamide levels in the processed products."

Zebra chip disease

First described in Mexico in 1994 and subsequently reported in Central America, the USA (Secor *et al.*, 2006; Gudmestad and Secor, 2007) and New Zealand (Liefting *et al.*, 2008), zebra chip (ZC) has emerged as a devastating disease of potato. Substantial reductions in yield as well as tuber quality have resulted in great economic losses to this disease in potato. The primary symptoms of ZC in the tuber include browning of vascular tissue, flecking of internal tissues, and discoloration of the medullary ray tissues. These tuber symptoms become more pronounced following frying of tubers for chips or French fries, with dark striping and streaks—thus, the designation, zebra chip disease. In New Zealand, up to 60% losses in harvestable yield and downgrading of the potato crop due to reduced starch content in tubers has been reported (Liefting *et al.*, 2008).

The disease has been associated with a bacterium, *Candidatus* Liberibacter solanacearum (syn. Ca. L. psyllaurous) (Hansen *et al.*, 2008; Secor *et al.*, 2009), which is transmitted by the potato psyllid, *Bactericera cockerelli* (Šulc) (Munyaneza *et al.*, 2007). Being a relatively new disease of potato, little was known regarding resistance in cultivars. Muyaneza *et al.* (2011) evaluated nine potato cultivars commonly grown in the USA for resistance to ZC over a 2-year period in field-cage evaluations using liberibacter-infected potato pysllids. Cultivars included chip and fry processing types, as well as an early russeted cultivar used for fresh consumption. In these field-cage evaluations, none of the nine cultivars exhibited resistance to ZC, with almost 100% of the inoculated plants developing severe ZC foliar and tuber symptoms, as well as yield reductions of 50% or greater. An evaluation of a breeding clone and 11 potato cultivars widely grown in New Zealand for ZC incidence under full, reduced, and no insecticide treatments (for control of potato psyllid) identified the fresh market cultivar, Nadine, as having reduced ZC symptoms in raw tubers (Anderson *et al.*, 2013). However, 50% and 60% reductions in marketable yield were observed between full and no insecticide treatments for Nadine over the 2 years of evaluations. The authors concluded (with the exception of reduced ZC symptoms in raw tubers of Nadine) that "there was no real indication that any of the other cultivars in the trial had any effective resistance to TPP (tomato–potato psyllid) or ZC".

Butler *et al.* (2011) evaluated 22 potato breeding clones and cultivars to assess whether host genotype impacted adult potato psyllid behavior and transmission of liberibacter. The authors reported observing significant differences among potato germplasm for the occurrence and duration of potato psyllid probing, the duration of psyllid cleaning and resting, and the amount of time psyllids spent off the potato leaflet, suggesting that resistance to the insect vector of liberibacter did exist. Reduced transmission of the liberibacter responsible for ZC was also observed in psyllid-resistant germplasm, as well as in three breeding clones with no apparent resistance to the insect vector—indicative of possible resistance to the liberibacter pathogen itself.

Improved sustainable production

Improved sustainability in production has become progressively more important to the potato industry, with increasing efforts in reducing production inputs such as pesticides and nutrient applications. Since 2009, following the voiced concerns of shareholders, McDonald's, the largest global food-service retailer, has made sustainability a priority, with their website detailing their efforts and progress in improved sustainability.

Potato cultivars with host plant resistance and a reduced need for nutrients can contribute to reduced pesticide and fertilizer applications during the production of the crop. As an example of host plant resistance contributing to greater sustainability in potato production, the late blight-resistant processing cultivars, Defender and Palisade Russet (Novy *et al.*, 2006, 2012) can reduce significantly the number of applications of fungicides needed for the control of the late blight disease. On the basis of their data, Stevenson *et al.* made the following conclusion regarding the cultivar, Defender:

> Even if only 20% of Washington and Wisconsin potato growers adopted this new cultivar,

> savings of several million dollars due to lower fungicide costs and fewer losses in storage could result. Defender additionally presents a viable alternative for organic potato producers. All evidence to date suggests that its improved resistance to late blight and early blight is a long-lasting type of resistance that promises to reduce fungicide use.
>
> (Stevenson *et al.*, 2007)

Improved efficiencies in nutrient use by potato cultivars have the capacity to reduce potential groundwater contamination with nitrates and mitigate water-quality degradation associated with phosphate runoff. Trehan (2009a,b) found the Indian cultivar, Kufri Puhraj, to be the most efficient in nitrogen, phosphorus, and potassium use. The author concluded that the root system of this cultivar was more efficient in uptake of nitrogen. Observed variability in phosphorus and potassium efficiency among ten Indian potato cultivars appeared to relate to both root uptake and plant utilization (Trehan, 2009a). A substantial reduction in nitrogen relative to the North American industry standard, Russet Burbank, also has been reported for the following potato cultivars, with percent reduction in nitrogen relative to Russet Burbank given in parentheses: Alta Russet (33%), Alturas (40%), and Bannock Russet (40%) (Novy *et al.*, 2002b, 2003; Bizimungu *et al.*, 2011).

16.4 Traditional Breeding Methodologies

As discussed previously in the chapter, potato has the unique ability for both sexual and asexual (clonal) reproduction. Once a breeding clone with enhanced characteristics relative to commercially available potato cultivars is identified, that genotype with its desirable attributes can be propagated clonally via the tuber, allowing the crop uniformity necessary in commercial potato production. However, in order to develop a new potato cultivar using traditional breeding methods, sexual reproduction is the starting point for any breeding program. Potato flowers, which make potato such a beautiful crop when in full bloom in the field, are also the instruments for the creation of new potato cultivars by the breeder via sexual hybridization. This section presents the methodology used by the breeder in the development of new potato cultivars, with the Aberdeen (Idaho) potato breeding program being referenced, with variance among programs generally occurring in the number of breeding clones evaluated, selection intensity in each field generation, and the size of field plots and their number of replicates in trials (Hoopes and Plaisted, 1987).

Parental selection and hybridization

A breeder will put time and thought into the annual selection of parents for hybridization in the "crossing block"—this term is used to describe the assemblage of parents for that year's hybridizations. The value of the sexual progeny resulting from the intercrossing of parents relies heavily on the merit of the parents and the directed crosses among them. The selection of parents is based on a combination of attributes, such as desirable agronomics, processing and sensory qualities, as well as pest and disease resistances; no parent has all the desired attributes, thus its selection and hybridization with other parents that complement its known weaknesses. Parental choices include breeding clones/cultivars that produce higher frequencies of desirable offspring with certain other parents (specific combining ability), or when hybridized with an array of parents (general combining ability). Observations on parental combining ability may be based on breeders' personal experience over years or through statistical evaluations of progeny performance (Brown and Dale, 1998; Ruiz De Galarreta *et al.*, 2006). Parents can also comprise breeding clones and cultivars from other breeding programs, germplasm representing hybrids between adapted and non-adapted germplasm, and potato wild species.

Parental crossing sheets that detail the strengths and weaknesses of each parent along with parentage are then generated and used in decision making regarding the crosses to be conducted, as well as a means of tracking hybridizations to ensure unique and non-redundant hybridizations are made. To maximize heterosis and avoid inbreeding depression, hybridizations among closely related parents are generally avoided.

Potato flowers used in hybridizing are perfect with both male (anthers) and female (carpels) sexual organs borne on the same flower.

Cultivars or breeding clones used in hybridizations can, with few exceptions, be used as the female parent. However, this is not the case with respect to male fertility, with sterility often being an impediment with approximately 40–50% of parents displaying male sterilities. If a parent is male fertile, an inflorescence—consisting of an average of ten flowers—will be tagged on that plant for the collection of pollen into glass vials (Fig. 16.2a). Pollen, if kept dry, can be stored for up to a month at 2.5°C or 11–12 months at −20 to −24°C (Howard, 1958; Blomquist and Lauer, 1962). Other inflorescences on the same plant can be used as female flowers, with their respective anthers being removed (emasculated), allowing for easier pollination of the stigma with pollen from another male-fertile parent. Crossing tags, placed by the breeder on the female inflorescence, detail the female and male parents used in the cross and the date the tag was placed on the inflorescence. Pollination of flowers typically is done in the morning to early afternoon, after flowers have opened and prior to the corolla closing by mid- to late afternoon (Fig. 16.2b). If successful, a pollinated flower will swell at its base—ovary—and develop into a fruit—also called a berry—that looks like a small, green tomato (Fig. 16.2c), otherwise the flower will wither and drop from the plant within 3–5 days following pollination. Plastic bags with air holes are tied around each developing fruit cluster, with an accompanying crossing tag to ensure fruits do not drop, with the parental background of the cross subsequently being lost. Additional detail regarding hybridization of potato is also presented by Plaisted (1993).

True potato seed

Potato fruits that form from successful hybridizations contain true potato seed—to distinguish from the more common reference to potato seed that comprises a piece of a tuber containing an eye. True potato seed looks very much like tomato seed (Fig. 16.2d), and is extracted from mature potato fruits that have changed from green to gray-green in color and have softened, generally 5–7 weeks following hybridization. True potato seeds in fruits are extracted in the Aberdeen program using a blender with dulled blades, with seeds then being sieved and dried. Dried seed is stored in coin envelopes, with each envelope representing a unique hybridization

Fig. 16.2. (a) Pollen is collected in a glass vial (arrow) and used for pollination of emasculated flowers (b). If successful, potato fruits develop (c), with true potato seed (TPS) extracted from mature fruit (d). Potato seedlings from germinated TPS (e) develop into plants and tuberize, with harvested seedling tubers (f) being planted in the field as single hills for selection the following year.

event between two parents, with progeny seed of the unique hybridization being termed a "family". Plaisted (1993) reported that seed dried and stored at room temperature in a cabinet containing a room dehumidifier germinated following 10 years of storage. Storage in sealed conditions with a dessicant or in metalized polyester film can further extend the viability of seed beyond 10 years. True potato seed has an average dormancy of approximately 6 months; soaking for 24 h in 2000 ppm gibberellic acid (GA_3) can be used to overcome dormancy in those instances when germination prior to 6 months is desired (Lam and Erickson, 1966). For rapid generation and cycling of true potato seed into seedlings, Janksy *et al.* (2012) recommended the harvest of fruits 3 weeks following initial pollination, with an additional 2 weeks of seed maturation in the harvested fruit. Freshly extracted seed from the fruit can then be soaked in GA_3 and planted immediately.

Greenhouse seedling tubers

Following the generation of true potato seed, the next step in the breeding process is the germination of the true potato seed to produce potato seedlings and their associated tubers. True potato seed representing a family of individuals is planted in a small tray and allowed to germinate (Fig. 16.2e). Due to the tetraploid nature of potato and its associated heterozygosity at loci, each germinated seedling is a unique genotype that differs from its parents and siblings. Approximately 2–3 weeks following planting as true potato seed, potato seedlings begin emerging, with another 2 weeks of growth prior to their transplanting into individual cells or pots (size varies among breeding programs), with subsequent growth into mature plants. Seedling tubers are harvested (Fig. 16.2f) approximately 2.5–3 months following transplanting. The largest tuber (A-size) of each seedling plant is retained and bulked with the largest tubers of its siblings in that family. Oftentimes, seedlings will produce one to two additional tubers (termed B- and C-size) that are smaller than the A-tubers retained by the breeding programs. These smaller seedling tubers are also saved and bulked as families that historically have been exchanged among US public breeding programs. This reciprocity in the exchange of seedling tubers among breeding programs bolsters the numbers and diversity of seedling tubers evaluated in potato production regions of the USA, with obvious benefits in developing new potato cultivars for the US potato industry. Harvested seedling tubers are stored at cold temperatures (3–5°C) over the winter months to preserve tuber dormancy and quality until planting in the spring. The production of seedling tubers is an important transition point in preserving the seedling genotypes generated via sexual reproduction. The diversity of genotypes generated subsequently advance in the breeding program as clonally propagated breeding clones.

Single-hill selection (1st field generation)

Seedling tubers harvested from the greenhouse in the previous year and stored under refrigeration are planted in the spring as single-hill selections. Single hill describes the unique genotype derived from a planted seedling tuber, represented by one plant in the first field generation. A spacing of 91 cm is used between seedling tubers to prevent the intermixing of tubers among single-hill clones at harvest, ensuring a selection is "true-to-type" and is representative of only one genotype. This is critical in that from this point forward, a selected breeding clone will be propagated asexually as it advances through a breeding program.

Single-hill selections are made based on the market class the family represents (e.g. processing for fries or chips, specialties with unique skin or flesh color), with emphasis at this early stage on an acceptable visual appearance of tubers for the corresponding market class. Love *et al.* (1997) assessed 27 tuber traits involving skin, eyes, size, shape, yield, and defects such as growth cracks and second growth (knobs) that are commonly used by breeders in visual selection in the field. Twenty-four (88%) tuber traits were identified as having sufficiently high heritabilities and consistency in expression over years to validate their effectiveness in visual selection. Only three traits involving the uniformities of skin russeting, tuber shape, and size were determined to be sufficiently variable in their expression that the authors recommended they not be applied in early-generation visual selection. The findings of this study validate the effectiveness of visual selection in early-generation breeding, surprisingly even for yield attributes. Brown (1988) had previously shown that the non-genetic component of seedling tuber size unduly influenced single-hill breeding clone yields, with the recommendation being that little emphasis be placed on yield at this stage of selection. Yield and its respective components are selected more efficiently using field-grown tubers in subsequent clonal generations (Caligari *et al.*, 1986; Maris, 1986; Brown, 1988; Kumar and Gopal, 2006). Selection against obvious negative attributes such as tuber malformations are recommended during single-hill selection, with literature supporting this approach for improving breeding efficiencies in the earliest stages of the selection process (Maris, 1988; Ruiz De Galarreta *et al.*, 2006).

Intense selection pressure is applied at this stage, with typically 1–4% of evaluated single hills being selected and advanced based on a survey of US potato breeding programs (Kathleen Haynes, personal communication). Bradshaw (2007a,b) cited literature that stated intense, early-generation visual selection was inefficient and discussed an alternative approach using mid-parent values and seedling progeny tests to improve efficiencies in the Scottish Crop Research Institute breeding program. Additional details of this breeding strategy for increasing efficiencies in combining processing qualities and disease and pest resistances are described by Bradshaw *et al.* (2003). The impression of inefficiency in the breeding process is also supported by Wenzel *et al.* (1983), with the authors calculating that 500,000–2,000,000 seedling tubers must be evaluated for the selection and release of one cultivar. Hoopes and Plaisted (1987) stated greater efficiencies than did Wenzel, with a calculation of approximately 200,000 seedlings evaluated for every cultivar named and released.

An evaluation of the Northwest Potato Variety Development Program (NPVDP), comprised of researchers in the USA in Idaho, Oregon, and Washington, challenges the belief that intense early-generation selection (1–2% of single hills being selected in the NPVDP) is inefficient in the identification of new potato cultivars. Since the inception of the NPVDP in 1985, 41 cultivars have been released, with six of the fifteen most widely grown cultivars grown in the USA in 2012 having originated from this program (USDA, National Agricultural Statistics Service, Crop Production, December 2012). Based on an annual evaluation of approximately 210,000 seedling tubers in the NPVDP and the release of 41 cultivars over 27 years, one seedling tuber in approximately 138,000 will be released as a cultivar. Brown (2011) cited a figure of one cultivar being identified and released per 60,000 seedlings evaluated in a period from 2004 to 2012. This more recent data on the numbers of seedlings evaluated per cultivar released would indicate much higher efficiencies in traditional breeding programs than previous citations would indicate.

12-hill selection (2nd field generation)

On the basis of 1–2% of single hill clones being selected in the field and retained, approximately 2000–2250 breeding clones are planted the following year in the Aberdeen potato breeding program as non-replicated plots consisting of 12 genetically identical plants in what is termed a 12-hill (12H) plot. This field generation is the first for which field data on flower color, vine maturity and size, and resistance to diseases such as foliar early blight (*Alternaria solani*) and verticillium wilt (*Verticillium dahliae*) are collected.

During the field season, 12H plots are also visually evaluated for the presence of virus—most notably *Potato virus Y* (PVY), which has become an increasingly difficult pathogen for breeding programs due to the introduction of newer strains of PVY with less obvious foliar symptoms of infection than the previously prevalent PVY^{O} common strain. Removal or roguing of virus-infected plots is necessary to prevent the spread of the virus by aphids throughout the 12H plots; the advent of enzyme-linked immunosorbent assay (ELISA) tests that can be conducted in the field have been adopted by breeding programs to confirm the presence or absence of virus in "real time" to allow the removal of infected plants that are asymptomatic or with weakly expressed symptoms.

At the end of the field season, 12H plots are dug and laid on the ground for visual evaluation. Similar to the single-hill generation, selection against negative tuber attributes is applied. However, with plots consisting of 12 plants rather than a single plant, yield now factors into the decision making to a much greater degree than in the previous single-hill generation, as does positive selection for tuber type. The larger number of tubers in the 12H plots also allows for tubers to be cut in the field to allow evaluations for internal defects such as brown center/hollow heart, and vascular and stem-end discolorations; with plots being left in the field, where a high percentage of internal defects are noted. In the single hill, with almost all tubers needed for seed for 12H plots, selection against internal defects is not readily applicable during the selection process.

Annually, 20–25% of 12H plots are selected in the field, with all tubers of a plot being harvested. Yield obtained from the plots is adequate for the first time to allow the collection of data on specific gravity (dry matter content), as well as processing attributes (fry or chip color) following approximately 5 months storage at 4.4 and 7.2°C. Following the obtainment of processing data, tubers of all field-selected 12H plots are inspected visually in the cellar for final disposition, with processing data and storage characteristics such as tuber dormancy and shrinkage, stem-end/vascular discoloration, skin color retention (in the case of red- or purple-skinned selections), and storage rots (dry and soft) all being factored into the final determination of which breeding clones advance or are discarded in the breeding program. This final cellar disposition in March results in the discard of an additional 50–60% of the 12H breeding clones originally selected in the field the previous fall.

Seed increase and replicated trials (3rd field generation and higher)

Ten tubers of 12H breeding clones retained for advancement in the breeding program are assayed for the presence of the three major potato viruses, PVY, PVX, and PLRV, using ELISA. Those tubers free from all viruses are planted in a seed increase plot at a field site isolated from any nearby commercial production to decrease the possibility of current-season infection by viruliferous aphids. The number of individuals represented in previous field generations does not allow for the cost-effective evaluation for virus until numbers are made more manageable at the end of the 2nd field season. Virus-free seed is crucial to the success of a potato breeding program in allowing for accurate assessments of the merit of breeding clones without the confounding presence of viruses.

Replicated trials in the third field season in the Aberdeen program are the first opportunity for statistical inferences to be made among breeding clones relative to check cultivars, especially in the context of yield, which up to this point has not been a primary determinant in the advancement of a breeding clone due to the first and second field generations being non-replicated. Most potato breeding programs use a randomized complete-block design. Data are collected on total yield, as well as on marketable yield (US No 1); US No 1 yield comprising all tubers greater than 4.8 cm diameter (approximately 113.4 g) and with cull tubers, with defects such as second growth and growth cracks having been removed, as well as any tubers showing wet rot. In addition, tubers of a breeding clone with the potential for processing or fresh market russet usage are separated by size into weight categories of less than 113.4, 113.4–170.0, 170.1–283.5, and >283.5 g. Tuber size distribution is important to the fresh pack and processing industries, with market usage defining desirable tuber size. For example, the 170.1–283.5 g (6.4–8.3 cm diameter) is of importance for processing into French fries and other frozen potato products, whereas smaller size tubers with diameters of 1.9–6.4 cm (≤170.1 g) are desired for fresh pack of specialty cultivars having unique skin or flesh colors, where smaller size commands a premium price from consumers. Tubers ranging in size from 226.8 to 396.9 g are the most valuable for the russet fresh pack industry (Guenthner, 2003), with those sizes being used in "count cartons" of varying tuber numbers based on tuber size (e.g. a 60-count carton would contain 60 tubers ranging in weight from 283.5 to 453.6 g).

Besides yield data, other characteristics of breeding clones evaluated for the first time in the third field season include blackspot bruise susceptibility and storage disease response following inoculation with *Erwinia* and *Fusarium* for soft and dry rot, respectively. The ability to store for prolonged durations in storage is important to the industry of the western USA, with tubers of the long dormancy cultivar, Russet Burbank, being stored for periods as long as 9–10 months following harvest; increased susceptibility to storage rots relative to industry standards is therefore not desirable, and susceptibilities are therefore best identified early in the breeding program as soon as adequate tubers are available for controlled inoculations.

As a clone advances in a breeding program, with more manageable numbers and greater seed amounts than in earlier stages, the number of replicates in trials increases, as does the number of trial sites. This allows an assessment of a breeding clone's merit in differing environments and changes in its performance relative to other trial entries. The changes in relative performance due to environmental influences are termed genotype × environment interactions, with Bradshaw providing a comprehensive review of its impact in potato breeding (Bradshaw 1994, 2007a). Bradshaw (2007b) indicates that early-generation selection is best practiced in environments as closely similar to those of commercial production environments where a new cultivar would be grown. Based on this assessment, Bradshaw advocated that potato producing countries, when possible, should have their own potato breeding programs to facilitate the breeding and development of new cultivars. Additional data on breeding clones' responses to diseases and pests, such as viruses, late blight, and nematodes, are also now collected during field years 4–6; this information being useful in determining whether a breeding clone will be retained in the breeding program or discarded.

Advanced regional trials

Dependent on there being adequate seed amounts, breeding clones saved following the sixth field year of evaluation in the Aberdeen program are advanced to the Tri-State Potato Variety Trials (TSPVT), with sites in the states of Idaho, Oregon, and Washington. Sites in the three states allow a more comprehensive evaluation of a breeding clone's merit for the north-western US potato industry, which produced 57% of the US potato crop in 2012 (USDA, National Agricultural Statistics Service, Crop Production Summary, 2012). A breeding clone having merit for release as a cultivar is generally retained in the Tri-State regional trial for 2 years, although clones showing exceptional merit have been retained for just 1 year prior to advancing to the Western Regional Potato Variety Trial (WRPVT). This trial represents the final 3 years of evaluation prior to the release of a meritorious breeding clone as a potato cultivar, with trial sites in the states of California, Colorado, Idaho, Oregon, Texas, and Washington. Similar regional trials are also conducted by breeding programs of the north-central, south-western, and eastern USA, with the data collected being used by breeders in making final determinations regarding the release of a breeding clone as a new cultivar.

It is during evaluations in the TSPVT and WRPVT that detailed postharvest data on processing and sensory attributes are collected, as well as economic analyses for the fresh and processing value of breeding clones relative to industry standards. Data are presented in formal reports and at meetings with potato industry representatives present, and introduces them to the merit of a breeding clone for consideration for their company's use and evaluation. During the final 3 years of evaluation in the WRPVT, concurrent production management trials of promising breeding clones are also being conducted by researchers in the NPVDP. These trials establish guidelines for the management of potential new cultivars with respect to nutrients, irrigation—western US production typically being irrigated—and storage. The adoption of a new cultivar by the industry is improved if management regimes to optimize performance are also included in the cultivar release, in extension publications, and in presentations given at grower meetings.

Seeds of a promising clone for industry evaluations are initially obtained from breeding programs, concomitant with the clone being established in tissue culture under aseptic laboratory conditions. *In vitro* plantlets tested and confirmed as being free of viruses, viroids, and bacteria become the starting point for the production of certified seed, as outlined by Love *et al.* (2003). Certified seed originating from pathogen-free, *in vitro* plantlets is generally available for larger-scale field evaluations by industry collaborators prior to a final decision being made regarding the release of a breeding clone as a cultivar.

Cultivar release

Following the third year of evaluation in the WRPVT, a decision is made regarding the merit of a breeding clone for release as a cultivar. The total time elapsed from the generation of true potato seed until the determination of whether to release a breeding clone as a cultivar generally ranges from 12 to 14 years. This time frame is necessary for the collection of agronomic and disease/pest response data, the development of management guidelines, and the production of sufficient certified seed. While seemingly a long duration, an established breeding program has breeding clones at all stages of development, with cultivar releases occurring more frequently than 12–14 years. For example, the NPVDP annually released one to four cultivars from 2007 to 2012.

Key to the decision making regarding the release of a new cultivar is industry feedback. As mentioned in the previous section, processing companies, as examples, will be collecting their own agronomic, processing, and storage data on advanced breeding clones relative to industry standards even prior to release. The larger-scale field, processing, and storage evaluations conducted by industry may reveal weaknesses that may not be observed in the smaller-scale evaluations conducted by breeding programs—for example, it is difficult for a breeding program to mimic the large potato piles and the environment typically found in commercial storage structures, as described by Kleinkopf and Olsen (2003).

On the basis of research data and industry feedback, a name is selected for a new potato cultivar, and in the case of new cultivars to be released in Idaho, the state's Foundation Seed Stock Committee is approached with a release document that details the new cultivar and its merit for release. Committee members review the submitted release, and following discussion a formal vote is taken for approval of the release of the cultivar, with a formal document detailing the new cultivar being released to the public.

Prior to the amendment of the Plant Variety Protection Act in 1995, allowing for potato to be included among crops eligible for protection, potato cultivars were public releases, with no stipulations concerning their use. Plant Variety Protection (PVP) for potato allowed US breeding programs to collect research fees or royalties for the use of their protected cultivars, with all programs now applying for PVP for newly released cultivars. The release of a new cultivar by a potato breeding program was not generally accompanied by any promotion or marketing. However, within the last decade, breeding programs such as in the state of Colorado, as well as the Tri-State program, have formed collaborations with grower-supported marketing entities to promote cultivars released from those programs. A primary benefit to the breeding programs is a greater return of cultivar royalties to help support continued research efforts—important with the continued erosion of federal and state research funds in support of potato breeding.

16.5 Conclusion

Traditional breeding allows for the genetic "reshuffling" of genes and their recombination into new genotypes that may carry the desired assemblage of resistance and agronomic traits necessary for release as a new cultivar. While molecular biology techniques can be useful for improving on a weakness in a cultivar, such as improving resistance to blackspot bruise and cold-induced sweetening (Rommens *et al.*, 2006) or late blight (Halterman *et al.*, 2008), such technology cannot currently be used in the development of a wholly new cultivar, due to the inability to pyramid all the necessary traits required by industry, many of which are quantitative with no clear understanding of the genes involved in their expression. Traditional breeding therefore remains the primary source of new potato cultivars, with a greater acceptance and adoption of them by the potato industry. For example, while Russet Burbank is still the predominant potato cultivar grown in the USA, that cultivar has seen its acreage steadily replaced by newer cultivars. Idaho's acreage of Russet Burbank from 1992 to 2012 decreased from 87.8% to 52.5%, with the cultivars, Russet Norkotah (released in 1987) and Ranger Russet (released in 1991), contributing to the decrease in Russet Burbank acreage. Two of the top ten cultivars grown in 2012 in the fall-production states of the USA—these states representing 76% of total US potato acreage—represent cultivars released within the last 15 years—Umatilla Russet (1998) and Alturas (2002) (USDA, National Agricultural Statistics Service, Crop Production, December 2012).

For potato breeders, greater efficiencies in the development and mapping of molecular markers and the recent completion of the sequencing of the potato genome hold promise in providing insights into gene expression and interactions in potato—especially useful in this tetraploid crop with its complex genetics. Marker-assisted selection in potato breeding will also likely expand beyond its current use with monogenic traits to encompass also polygenic traits such as yield, and tuber starch and sugar content. In the future, greater consumer and industry acceptance of genetically modified potato cultivars using intragenic technology as proposed by Rommens (2007) also has the potential to add to the potato breeder's toolbox. It is an exciting time for the potato breeding community, with many new technologies having the potential to improve breeding efficiencies. None the less, potato breeders should not lose sight of the "art" component inherent with the breeding process, whereby experience gained over years of selecting in the field needs to be recognized as an important contributor to successful cultivar development. This sentiment is expressed succinctly by Brown (2011), who states, "a breeder's mind and intuition have something important to contribute in the midst of gene chips and genome sequences".

References

Ames, M. and Spooner, D.M. (2008) DNA from herbarium specimens settles a controversy about origins of the European potato. *American Journal of Botany* 95, 252–257.

Anderson, J.A.D., Walker, G.P., Alspach, P.A., Jeram, M. and Wright, P.J. (2013) Assessment of susceptibility to zebra chip and *Bactericera cockerelli* of selected potato cultivars under different insecticide regimes in New Zealand. *American Journal of Potato Research* 90, 58–65.

Bethke, P.C. and Bussan, A.J. (2013). Acrylamide in processed potato products. *American Journal of Potato Research* 90, 403–424.

Bethke, P.C., Nassar, A.M.K., Kubow, S., Leclere, Y.N., Li, X.Q, Haroon, M., Molen, T., Bamberg, J., Martin, M. and Donnelly, D.J. (2014) History and origin of Russet Burbank (Netted Gem) a sport of Burbank. *American Journal of Potato Research*. DOI 10.1007/s12230-014-9397-5.

Bizimungu, B., Holm, D.G., Kawchuk, L.M., Konschuh, M., Schaupmeyer, C., Wahab, J., *et al.* (2011) Alta Crown: a new Russet potato cultivar with resistance to common scab and a low incidence of tubers deformities. *American Journal of Potato Research* 88, 72–81.

Blomquist, A.W. and Lauer, F.I. (1962) A simplified technique for handling and storing potato pollen. *American Potato Journal* 39, 340–343.

Bradshaw, J. (2007a) Breeding potato as a major staple crop. In: Kang, M.S. and Priyadarshan, P.M. (eds) *Breeding Major Food Staples*. Blackwell Publishing Professional, Ames, Iowa, pp. 277–332.

Bradshaw, J. (2007b) Potato-breeding strategy. In: Vreugdenhil, D. (ed.) *Potato Biology and Biotechnology, Advances and Perspectives*. Elsevier Ltd, Oxford, UK, pp. 157–177.

Bradshaw, J.E. (1994) Quantitative genetics theory for tetrasomic inheritance. In: Bradshaw, J.E. and Mackay, G.R. (eds) *Potato Genetics*. CAB International, Wallingford, UK, pp. 71–99.

Bradshaw, J.E., Dale, M.F.B. and Mackay, G.R. (2003) Use of mid-parent values and progeny tests to increase the efficiency of potato breeding for combined processing quality and disease and pest resistance. *Theoretical and Applied Genetics* 107, 36–42.

Brown, C.R. (2011) The contribution of traditional potato breeding to scientific potato improvement. *Potato Research* 54, 287–300.

Brown, C.R., Mojtahedi, H. and Santo, G.S. (1995) Introgression of resistance to Columbia and Northern root-knot nematodes from *Solanum bulbocastanum* into cultivated potato. *Euphytica* 83, 71–78.

Brown, C.R., Mojtahedi, H., James, S., Novy, R.G. and Love, S. (2006) Development and evaluation of potato breeding lines with introgressed resistance to Columbia root-knot nematode (*Meloidogyne chitwoodi*). *American Journal of Potato Research* 83, 1–8.

Brown, J. (1988) The effect of the weight of the seedling-derived tuber on subsequent clonal generations in a potato breeding programme. *Annals of Applied Biology* 113, 69–78.

Brown, J. and Dale, M.F.B. (1998) Identifying superior parents in a potato breeding program using cross prediction techniques. *Euphytica* 104, 143–149.

Butler, C.D., Gonzalez, B., Manjunath, K.L., Lee, R.F., Novy, R.G., Miller, J.C., *et al.* (2011) Behavioral responses of adult potato psyllid, *Bactericera cockerelli* (Hemiptera: Triozidae), to potato germplasm and transmission of Candidatus *Liberibacter psyllaurous*. *Crop Protection* 30, 1233–1238.

Caligari, P.D.S., Brown, J. and Abbott, R.J. (1986) Selection for yield and yield components in the early generations of a potato breeding programme. *Theoretical and Applied Genetics* 73, 218–222.

Gillen, A.M. and Novy, R.G. (2007) Molecular characterization of the progeny of *Solanum etuberosum* identifies a genomic region associated with resistance to potato leafroll virus. *Euphytica* 155, 403–415.

Glendinning, D.R. (1983) Potato introductions and breeding up to the early 20th century. *New Phytologist* 94, 479–505.

Gudmestad, N.C. and Secor, G.A. (2007) Zebra chip: a new disease of potato. Nebraska Potato Eyes 19, PHREC 07-05a. University of Nebraska–Lincoln, Lincoln, Nebraska.

Guenthner, J.F. (2003) Principles of economics and marketing. In: Stark, J.C. and Love, S.L. (eds) *Potato Production Systems*. University of Idaho Agricultural Communications, Moscow, Idaho, pp. 383–397.

Halterman, D., Kramer, L.C., Weilgus, S. and Jiang, J. (2008) Performance of transgenic potato containing the late blight resistance gene *RB*. *Plant Disease* 92, 339–343.

Hansen, A.K., Trumble, J.T., Stouthamer, R. and Paine, T.D. (2008) A new huanglongbing species, '*Candidatus* Liberibacter psyllaurous' found to infect tomato and potato, is vectored by the psyllid *Bactericera cockerelli* (Sulc). *Applied and Environmental Microbiology* 74, 5862–5865.

Hawkes, J.G. (1990) *The Potato: Evolution, Biodiversity and Genetic Resources*. Belhaven Press, Oxford, UK.

Hijmans, R.J. and Spooner, D.M. (2001) Geographic distribution of wild potato species. *American Journal of Botany* 88, 2101–2112.

Hoopes, R.W. and Plaisted, R.L. (1987) Potato. In: Fehr, W.R. (ed.) *Principles of Cultivar Development, Volume 2*. Macmillan Publishing Company, New York, pp. 385–436.

Howard, H.W. (1958) The storage of potato pollen. *American Potato Journal* 35, 676–678.

Jansky, S.H. and Miller, J.C. Jr (2010) Evaluation of Verticillium wilt resistance in Russet Norkotah and six strain selections. *American Journal of Potato Research* 87, 492–496.

Jansky, S.H., Hamernik, A. and Cai, X. (2012) Rapid cycling with true potato seed. *Seed Science and Technology* 40, 43–50.

Kelley, K.B., Whitworth, J.L. and Novy, R.G. (2009) Mapping of the potato leafroll virus resistance gene, Rlr_{etb}, from *Solanum etuberosum* identifies interchromosomal translocations among its E-genome chromosomes 4 and 9 relative to the A-genome of *Solanum* L. sect. *Petota*. *Molecular Breeding* 23, 489–500.

Kleinkopf, G.E. and Olsen, N. (2003) Storage management. In: Stark, J.C. and Love, S.L. (eds) *Potato Production Systems*. University of Idaho Agricultural Communications, Moscow, Idaho, pp. 363–381.

Knight, T.A. (1807) On raising new and early varieties of the potatoe. *Transactions of the Horticultural Society of London* 1, 57–59.

Kumar, R. and Gopal, J. (2006) Repeatability of progeny mean, combining ability, heterosis and heterobeltiosis in early generations of a potato breeding programme. *Potato Research* 49, 131–141.

Lam, S.L. and Erickson, H.T. (1966) Interaction of light and gibberellin on potato seed germination. *American Potato Journal* 43, 443–449.

Liefting, L.W., Perez-Egusquiza, Z.C., Clover, G.R.G. and Anderson, J.A.D. (2008) A new 'Candidatus Liberibacter' species in *Solanum tuberosum* in New Zealand. *Plant Disease* 92, 1474.

Lineback, D.R., Coughlin, J.R. and Stadler, R.H. (2012) Acrylamide in foods: a review of the science and future considerations. *Annual Review of Food Science and Technology* 3, 15–35.

Love, S.L. (1999) Founding clones, major contributing ancestors, and exotic progenitors of prominent North American potato cultivars. *American Journal of Potato Research* 76, 263–272.

Love, S.L., Werner, B.K. and Pavek, J.J. (1997) Selection for individual traits in the early generations of a potato breeding program dedicated to producing cultivars with tubers having long shape and russet skin. *American Potato Journal* 74, 199–213.

Love, S.L., Nolte, P., Corsini, D.L, Whitmore, J.C., Ewing, L.L. and Whitworth, J.L. (2003) Seed production and certification. In: Stark, J.C. and Love, S.L. (eds) *Potato Production Systems*. University of Idaho Agricultural Communications, Moscow, Idaho, pp. 49–69.

Maris, B. (1986) The effect of seed tuber weight on characters in the first and the second clonal generation of potato populations. *Euphytica* 35, 465–482.

Maris, B. (1988) Correlations within and between characters between and within generations as a measure for the early generation selection in potato breeding. *Euphytica* 37, 205–224.

Matsuura-Endo, C., Ohara-Takada, A., Chuda, Y., Ono, H., Yada, H., Yoshida, M., *et al.* (2006) Effects of storage temperature on the contents of sugars and free amino acids in tubers from different potato cultivars and acrylamide in chips. *Bioscience Biotechnology and Biochemistry* 70, 1173–1180.

Miller J.C. Jr, Scheuring, D.C., Miller, J.P. and Fernandez, G.C.J. (1999) Selection, evaluation, and identification of improved Russet Norkotah strains. *American Journal of Potato Research* 76, 161–167.

Munyaneza, J.E., Crosslin, J.M. and Upton, J.E. (2007) Association of *Bactericera cockerelli* (Homoptera: Psyllidae) with "zebra chip", a new potato disease in southwestern United States and Mexico. *Journal of Economic Entomology* 100, 656–663.

Munyaneza, J.E., Buchman, J.L., Sengoda, V.G., Fisher, T.W. and Pearson, C.C. (2011) Susceptibility of selected potato varieties to zebra chip potato disease. *American Journal of Potato Research* 88, 435–440.

Novy, R.G. and Helgeson, J.P. (1994a) Somatic hybrids between *Solanum etuberosum* and diploid, tuber-bearing *Solanum* clones. *Theoretical and Applied Genetics* 89, 775–782.

Novy, R.G. and Helgeson, J.P. (1994b) Resistance to potato virus Y in somatic hybrids between *Solanum etuberosum* and *S. tuberosum* × *S. berthaultii* hybrid. *Theoretical and Applied Genetics* 89, 783–786.

Novy, R.G., Nasruddin, A., Ragsdale, D.W. and Radcliffe, E.B. (2002a) Genetic resistances to potato leafroll virus, potato virus Y, and green peach aphid in progeny of *Solanum etuberosum*. *American Journal of Potato Research* 79, 9–18.

Novy, R.G., Corsini, D.L., Love, S.L., Pavek, J.J., Mosley, A.R., James, S.R., *et al.* (2002b) Bannock Russet: a dual purpose, russet potato cultivar with high U.S. No. 1 yield and multiple disease resistances. *American Journal of Potato Research* 79, 147–153.

Novy, R.G., Corsini, D.L., Love, S.L., Pavek, J.J., Mosley, A.R., James, S.R., *et al.* (2003) Alturas: a multi-purpose, russet potato cultivar with high yield and tuber specific gravity. *American Journal of Potato Research* 80, 295–301.

Novy, R.G., Love, S.L., Corsini, D.L., Pavek, J.J., Whitworth, J.L., Mosley, A.R., *et al.* (2006) Defender: a high-yielding, processing potato cultivar with foliar and tuber resistance to late blight. *American Journal of Potato Research* 83, 9–19.

Novy, R.G., Gillen, A.M. and Whitworth, J.L. (2007) Characterization of the expression and inheritance of potato leafroll virus (PLRV) and potato virus Y (PVY) resistance in three generations of germplasm derived from *Solanum etuberosum*. *Theoretical and Applied Genetics* 114, 1161–1172.

Novy, R.G., Whitworth, J.L., Stark, J.C., Charlton, B.A., Yilma, S., Knowles, N.R., *et al.* (2012) Palisade Russet: a late blight resistant potato cultivar having a low incidence of sugar ends and high specific gravity. *American Journal of Potato Research* 89, 89–101.

Nowicki, M., Foolad, M.R., Nowakowska, M. and Kozik, E.U. (2012) Potato and tomato late blight caused by *Phytophthora infestans*: an overview of pathology and resistance breeding. *Plant Disease* 96, 4–17.

Pierterse, L. and Hils, U. (2009) *World Catalogue of Potato Varieties 2009/10*. Agrimedia GmbH, Bergen/Dumme, Germany, 326 pp.

Plaisted, R.L. (1993) Potato. In: Fehr, W.R. and Hadley, H.H. (eds) *Hybridization of Crop Plants, 3rd printing*. American Society of Agronomy, Inc and Crop Science Society of America, Inc, Madison, Wisconsin, pp. 483–494.

Potato Genome Sequencing Consortium (2011) Genome sequence and analysis of the tuber crop potato. *Nature* 475, 189–195.

Rommens, C.M. (2007) Intragenic crop improvement: combining the benefits of traditional breeding and genetic engineering. *Journal of Agricultural and Food Chemistry* 55, 4281–4288.

Rommens, C.M., Ye, J., Richael, C. and Swords, K. (2006) Improving potato storage and processing characteristics through all-native DNA transformation. *Journal of Agricultural and Food Chemistry* 54, 9882–9887.

Ruiz De Galarreta, J.I., Ezpeleta, B., Pascualena, J. and Ritter, E. (2006) Combining ability and correlations for yield components in early generations of potato breeding. *Plant Breeding* 125, 183–186.

Secor, G.A., Lee, I. and Bottner, K. (2006) First report of a defect of processing potatoes in Texas and Nebraska associated with a new Phytoplasma. *Plant Disease* 90, 377.

Secor, G.A., Rivera, V.V., Abad, J.A., Lee, I.M., Clover, G.R.G., Liefting, L.W., *et al.* (2009) Association of '*Candidatus* Liberibacter solanacearum' with the zebra chip disease of potato established by graft and psyllid transmission, electron microscopy and PCR. *Plant Disease* 93, 574–583.

Shepherd, L.V., Bradshaw, J.E., Dale, M.F., McNicol, J.W., Pont, S.D., Mottram, D.S., *et al.* (2010) Variation in acrylamide producing potential in potato: segregation of the trait in a breeding population. *Food Chemistry* 123, 568–573.

Simko, I., Jansky, S., Stephenson, S. and Spooner, D. (2007) Genetics of resistance to pests and disease. In: Vreugdenhil, D. (ed.) *Potato Biology and Biotechnology, Advances and Perspectives*. Elsevier Ltd, Oxford, UK, pp. 91–115.

Spooner, D.M. (2009) DNA barcoding will frequently fail in complicated groups: an example in wild potatoes. *American Journal of Botany* 96, 1177–1189.

Stevenson, W.R., James, R.V., Inglis, D.A., Johnson, D.A., Schotzko, R.T. and Thornton, R.E. (2007) Fungicide spray programs for Defender, a new potato cultivar with resistance to late blight and early blight. *Plant Disease* 91, 1327–1336.

Tareke, E., Rydberg, P., Karlsson, P., Eriksson, S. and Tornqvist, M. (2002) Analysis of acrylamide, a carcinogen formed in heated foodstuffs. *Journal of Agricultural and Food Chemistry* 50, 4998–5006.

Thompson, A.L., Taylor, R.J., Pasche, J.S., Novy, R.G. and Gudmestad, N.C. (2007) Resistance to *Phytophthora erythroseptica* and *Pythium ultimum* in a potato clone derived from *S. berthaultii* and *S. etuberosum*. *American Journal of Potato Research* 84, 149–160.

Trehan, S.P. (2009a) Improving nutrient use efficiency by exploiting genetic diversity of potato. *Potato Journal* 36, 121–135.

Trehan, S.P. (2009b) Mechanism of high nitrogen efficiency in potato cultivars. *Advances in Horticultural Science* 23, 179–184.

USDA, National Agricultural Statistics Service, Crop Production (2012) http://www.nass.usda.gov/ (accessed 14 November 2013).

van Eck, H.J. (2007) Genetics of morphological and tuber traits. In: Vreugdenhil, D. (ed.) *Potato Biology and Biotechnology, Advances and Perspectives*. Elsevier Ltd, Oxford, UK, pp. 91–115.

Vinci, R.M., Mestdagh, F. and De Meulenaer, B. (2012) Acrylamide formation in fried potato products – present and future, a critical review on mitigation strategies. *Food Chemistry* 133, 1138–1154.

Wenzel, G., Bapat, V.A. and Uhrig, H. (1983) New strategy to tackle breeding problems of potato. In: Sen, S.K and Giles, K.L. (eds) *Plant Cell Culture in Crop Improvement*. Plenum Press, New York, pp. 337–349.

Zhang, L., Mojtahedi, H., Kuang, H., Baker, B. and Brown, C.R. (2007) Marker-assisted selection of Columbia root-knot nematode resistance introgressed from *Solanum bulbocastanum*. *Crop Science* 47, 2021–2026.

17 Molecular Breeding of Potato in the Postgenomic Era

Richard E. Veilleux* and T. Boluarte-Medina
Department of Horticulture, Virginia Tech, Blacksburg, USA

Since the potato genome was published in 2011 (Potato Genome Sequencing Consortium, 2011), with all of its associated tools, including the genome browser, blast engine, transcriptome with links to other plant genomes, the Infinium 8303 Potato Array, new possibilities have become available for potato breeding that complement the traditional approach described in Chapter 16 of this volume. Amino acid sequences of genes for which a function has been assigned in model plants can be blasted against the translated nucleotide sequence database for potato to identify genes in potato with the most similar structure to the query. Genetic maps can be constructed in a matter of days by processing DNA samples from a segregating population on the Infinium 8303 Potato Array, yielding a dense map with thousands of markers in a fraction of the time that was previously required for constructing maps using simple sequence repeats (SSRs), amplified fragment length polymorphisms (AFLPs), or other markers (Felcher *et al.*, 2012). Because of the distribution of single-nucleotide polymorphism (SNP) markers across the linkage groups, the SNP chip presents a convenient option for the identification of candidate genes through association genetics. The first genome available for potato has been assembled for an obscure cultigen, DM 1-3 516 R44 (hereafter DM for doubled monoploid), derived by anther culture of a diploid ($2n = 2x = 24$) selection of *Solanum tuberosum* Group Phureja, followed by chromosome doubling of an anther-derived monoploid plant. Its fingerling tubers, with little or no keeping quality, are a far cry from commercial potatoes that comprise the major world crop. However, it is unmistakably a potato, and its homozygosity, a rarity for potato, enabled sequence assembly from short reads using available technology, whereas previous attempts to assemble the genome of a heterozygous diploid selection had failed. So DM, despite its limitations as a cultivated potato, serves as the draft potato genome, providing a scaffold for other potato genomes and a set of some 39,000 genes, most of which are highly similar to those in potato cultivars, enabling the transcriptome derived from sequencing cultivar expressed sequence tags (ESTs) to be aligned to the DM genome. The potato genome assembly represents 86% of the 844 Mb genome (Potato Genome Sequencing Consortium, 2011).

17.1 SNP Chip Development for Assisting Mapping

Gene mapping (genome mapping) can be accomplished by assigning DNA fragments to plant

* E-mail: potato@vt.edu

chromosomes based either on recombination in a segregating population to generate a genetic map or on physical distance between markers to generate a physical map of an organism. Genotyping is the identification of alleles, alternative forms of a gene that differentiate individuals; differences in the DNA sequences between alleles comprise any of several types of molecular markers that can be associated with phenotypes. New methods of high-throughput DNA purification, PCR amplification, DNA sequencing, and the availability of bioinformatics tools for handling large data sets facilitate the development of genetic maps and molecular genotyping. Molecular markers on genetic maps enable the identification of plants with desirable genotypes and phenotypes at the seedling stage to hasten breeding programs. SNPs are a powerful type of molecular marker that differentiate two alleles by a single nucleotide change, which may be associated with an amino acid difference (non-synonymous SNP) that can alter the protein product of a gene.

A recent addition to the molecular breeding toolkit for cultivated potato is the Illumina Infinium 8303 Potato Array (Hamilton *et al.*, 2011). RNASeq reads based on cDNA extracted from three commercial cultivars (Atlantic, Premier Russet, and Snowden) were used to generate a 28- to 29-Mb sequence for each cultivar (Hamilton *et al.*, 2011). From more than 2 million SNPs derived from the transcriptome data of the three cultivars, a set of more than 69,000 high-confidence SNPs was identified after the application of various filters. The participation of the potato research community was solicited to aid in the selection of SNPs for inclusion on the chip in order to feature SNPs within genes considered valuable to various breeding and genetics programs. Other SNPs were selected to provide adequate distribution across the genome. An initial application of the potato array was used on a panel of 248 genotypes comprising processing cultivars, table stocks, diploid breeding lines, and a few non-cultivated potato species used in breeding; the SNP chip separated these four classes of germplasm easily (Hamilton *et al.*, 2011). The broad genome coverage was verified in a mapping study where two diploid populations that shared a common parent were mapped based on SNP chip application (Felcher *et al.*, 2012).

17.2 Marker-Assisted Selection

The application of marker-assisted selection in potato breeding was initially limited by the paucity of molecular markers, which only numbered some 350 or so in 2004 (Barone, 2004). The difficult breeding structure of potato, such that backcross breeding cannot be readily applied to recover the highly heterozygous genotype of a recurrent parent, is another limitation. However, some progress has been made in the utilization of molecular markers for breeding and selection in populations, especially for late blight resistance (Rickert *et al.*, 2003; Bormann *et al.*, 2004; Costanzo *et al.*, 2005; Park *et al.*, 2005; Colton *et al.*, 2006; Malosetti *et al.*, 2007; Wickramasinghe *et al.*, 2009; DeBolt, 2010; Sliwka *et al.*, 2010; Tan *et al.*, 2010; Busch *et al.*, 2011; Mori *et al.*, 2011; Gyetvai *et al.*, 2012), virus resistance (Hämäläinen *et al.*, 1997; Gebhardt *et al.*, 2006; Witek *et al.*, 2006; Velásquez *et al.*, 2007; Valkonen *et al.*, 2008; Kelley *et al.*, 2009; Whitworth *et al.*, 2009; Mori *et al.*, 2011; Ortega and Lopez-Vizcon, 2012), nematode resistance (Pajerowska *et al.*, 2005; Zhang *et al.*, 2007; Achenbach *et al.*, 2009; Rüping *et al.*, 2010; Milczarek *et al.*, 2011; Mori *et al.*, 2011; Schultz *et al.*, 2012), chip color (Kawchuk *et al.*, 2008), dry matter (Bhering *et al.*, 2009), cold-induced sweetening (Baldwin *et al.*, 2011), and processing quality (Li, 2008). An online Sol*R*gene database has been compiled with relevant information about resistance genes in tuber-bearing potato germplasm (Ferreira *et al.*, 2010). Yao *et al.* (2011) combined marker-assisted selection specifically for late blight resistance in conjunction with background AFLP marker selection to maintain diversity in their breeding population while increasing the frequency of resistant individuals. Analysis of data provided by the potato genome browser promises to speed up the advance of potato breeding in the molecular era. Jupe *et al.* (2012) identified and localized on the potato genome the nucleotide-binding and leucine-rich repeat domain (NB-LRR) gene family that comprised a vast majority of resistance (*R*) genes. These data elucidate the evolution of the potato *R* genes and facilitate identification of functional NB-LRR genes for further applications (Agrawal *et al.*, 2008).

Most of the above work was limited by the availability of genetic and physical maps of potato that tended to be constructed from less than ideal populations derived from segregation of as many as four different alleles per locus from crosses between heterozygous diploid selections. The development of an ultradense AFLP genetic map (van Os *et al.*, 2006) was an arduous task, which resulted in a map populated with markers that needed subsequent conversion to PCR markers in order to be used easily (Lo *et al.*, 2003). A potato consensus map was constructed based on quantitative trait loci (QTLs) reported in independent studies on potato and related species for late blight resistance and for maturity, since resistance to this pathogen was often associated with maturity. QTLs were projected and clustered into meta-QTLs for both late blight and maturity, and used for meta-analysis to refine genomic regions of interest. The authors suggested that anchoring these meta-QTLs to the potato genome sequence would improve candidate gene selection (Mueller *et al.*, 2005).

However, the genomic era has given rise to an unprecedented abundance of molecular markers (Hamilton *et al.*, 2011). In addition to the SNPs developed from the transcriptome sequencing of three tetraploid cultivars that form two tracks on the potato genome browser, the Infinium High-Confidence SNP (69K) and its subset, the SolCAP 8303 Infinium SNP array, a third track comprising genome-wide SNPs that differentiate either allele of heterozygous RH (a diploid hybrid of dihaploid *S. tuberosum* Group Tuberosum and *S. tuberosum* Group Phureja) from DM (the RH SNP track) presents some 3.67 million SNPs identified from aligning RH illumina reads to the PGSC v2.1.11 DM Pseudomolecules (Fig. 17.1). Populations derived by crossing homozygous DM to either of two heterozygous diploid selections were used to develop linkage maps based on the segregation of SNP markers included on the SolCAP 8303 Infinium Array (Felcher *et al.*, 2012). The array facilitated a high-throughput mapping scheme with wide genome

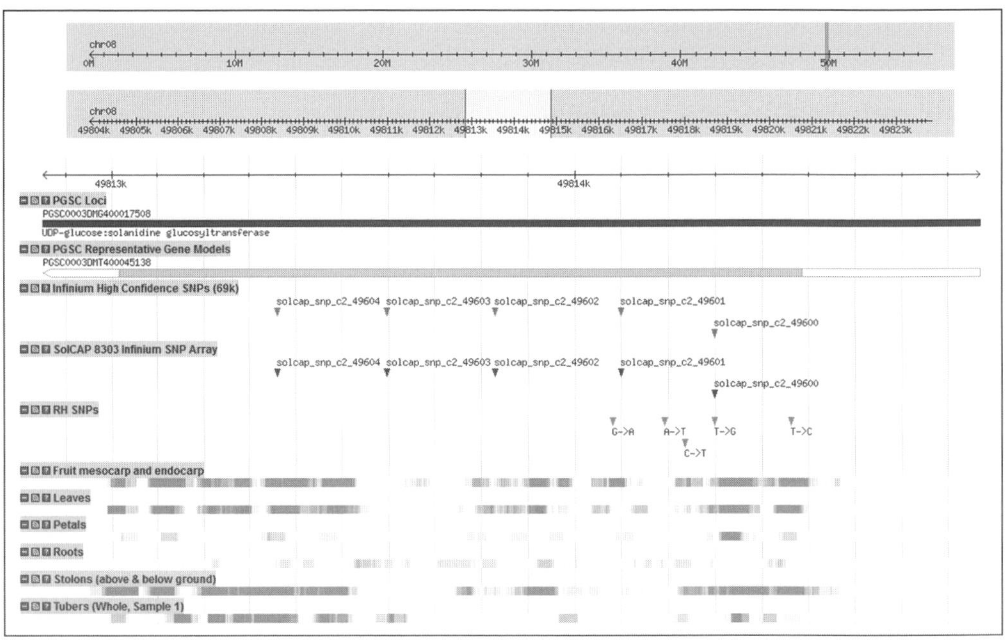

Fig. 17.1 Screen shot of the Potato Genome Browser v4.03 showing the gene model for UDP-glucose: solanidine glucosyltransferase (PGSC0003DMG400017508) with three single-nucleotide polymorphism (SNP) tracks: five Infinium high-confidence SNPs, the same five SNPs that were included on the SolCAP 8303 Infinium SNP Array, and five SNPs of which only one coincided with the Infinium SNPs that were identified from aligning the RH genome sequence against the DM reference genome. RNASeq tracks show relative expression of the gene in six tissues, with greater expression indicated by darker shading.

coverage, demonstrating the possibilities for QTL analysis, genome-wide association studies, and map-based gene cloning (Felcher *et al.*, 2012).

Ortega and Lopez-Vizcon (2012) discuss the current state of marker-assisted selection in potato breeding from the practical perspective of application in a commercial breeding program. Among the reasons they present for the limited impact of marker-assisted selection in potato breeding to date are the paucity of markers in adapted tetraploid germplasm, the high cost of screening a large population of seedlings, and the poor adaptation of clones in which markers have been developed. The availability of new resources should help to facilitate the adoption of marker-assisted selection.

17.3 Transcriptomics

The age of transcriptomics in potato began with the sequencing of EST libraries from various tissues of the plant (Crookshanks *et al.*, 2001; Ronning *et al.*, 2003; Armstrong *et al.*, 2005; Nielsen *et al.*, 2005) during its interaction with a pathogen, such as *Phytophthora infestans* (Birch *et al.*, 2003; Evers *et al.*, 2004), or subjected to abiotic stress (Rensink *et al.*, 2005b; Rotter *et al.*, 2007). Comparative analysis of solanaceous species revealed some 23% of ESTs unique to potato compared to 55–81% of ESTs with significant similarity among six of the species examined (Rensink *et al.*, 2005c). Transciptomic analysis of an isogenic series of potato differing by ploidy demonstrated only subtle expression changes related to ploidy differences among 1*x*, 2*x*, and 4*x* lines (Stupar *et al.*, 2007).

Once sufficient EST data had been generated, Kloosterman *et al.* (2008) developed a 44K 60-mer oligo array (the Potato Oligo Chip Initiative) that allowed facile identification of differentially expressed genes among tissues or stress challenges, thereby facilitating functional annotation of potato genes. This breakthrough was subsequently implemented to describe differential gene expression patterns as potato was subjected to drought (Watkinson *et al.*, 2006; Mane *et al.*, 2008; Evers *et al.*, 2010), cold (Oufir *et al.*, 2008; Evers *et al.*, 2012), salinity (Legay *et al.*, 2009; Evers *et al.*, 2012), and heat stress (Ginzberg *et al.*, 2009). A bioinformatics *in silico* approach was used to identify 217 cumulative abiotic stress genes from multiple microarrays (Barozai and Wahid, 2012). Transcriptome analysis of potato cultivar Kennebec exposed to various stress treatments including salt and drought provided an array of candidate genes for investigation of abiotic stress response (Rensink *et al.*, 2005a,b) and the possibility of introducing these genes transgenically into potato cultivars. Genes regulated by day length and light quality were identified by Rutitzky *et al.* (2009), who found conserved as well as species-specific photoperiodically regulated genes between potato and tobacco. A potato microarray dedicated to food safety, with emphasis on nutrient-related pathways, basal physiology, and stress-related metabolic pathways, was designed to assess the relative importance of the cultivar or environment on the transcriptomics of two cultivars grown under two different fertilizer regimes and two different plant protection regimes; genetic differences between the two cultivars were more pronounced than those due to different cultural practices (van Dijk *et al.*, 2009, 2010). In a subsequent study of a single cultivar grown under organic and conventional agricultural practices, van Dijk *et al.* (2012) used the microarray to identify differing expression patterns of genes and pathways under the different environmental conditions.

Microarray analysis of a transgenic potato cultivar carrying the *trehalose-6-phosphate synthase 1* gene of yeast to improve stress tolerance revealed the concomitant up- and downregulation of 74 and 25 native genes (Kondrák *et al.*, 2011). Stushnoff *et al.* (2010) examined pigmented and non-pigmented tuber tissues from potato clones with differential expression of anthocyanins with a microarray to identify 27 genes likely to have affected anthocyanin accumulation. D'Ippólito *et al.* (2010) used a macroarray to demonstrate the upregulation of several hundred cDNA clones resulting from the inoculation of potato with *Fusarium solani* f. sp. *eumartii*, the cause of dry rot in potato. Ballvora *et al.* (2007) used microarrays and real-time reverse transcriptase polymerase chain reaction (RT-PCR) on a wild relative of cultivated potato and an advanced tetraploid clone to study gene expression induced by inoculation with *P. infestans*; they found some unique genes in the defense pathway for each genotype, or that the same genes were activated at different times during disease development.

An alternative instrument for transcriptome analysis is the cDNA-AFLP method, where cDNA rather than genomic DNA is used in conjunction with AFLP to generate transcriptome-derived fragments (TDFs), which allow the monitoring of differentially expressed genes among tissues, genotypes, or treatments applied (Ritter *et al.*, 2008). Among *in vitro* plantlets, mature leaves, and dormant tubers of potato, most primer combinations yielded one or two segregating TDFs (Ritter *et al.*, 2008). Henriquez and Daayf (2010) combined this technique with subtractive hybridization to remove constitutively expressed genes and common transcripts between treatments, allowing the identification of 41 differentially expressed TDFs from the interaction between potato and *P. infestans*.

The publication of the potato genome (Potato Genome Sequencing Consortium, 2011) along with the potato genome browser (http://solanaceae.plantbiology.msu.edu/integrated_searches.shtml) has expanded the possibility of potato transcriptomics by providing a reference genome along with gene-specific transcriptome tracks derived from RNASeq data generated from multiple tissues and treatments (Fig. 17.1) (Massa *et al.*, 2011). The amount of data generated by RNASeq overshadows what can be obtained from microarrays. Novel candidate genes for plant–pathogen interactions have been identified by sequencing the transcriptome of compatible and incompatible potato–*P. infestans* interactions (Gao *et al.*, 2011). Of course, cDNA prepared from RNA isolation does not discriminate between translated and non-translated RNA. Use of RNA derived from isolated polysomes will provide a more accurate picture of the translatome (Mustroph *et al.*, 2009). Another methodology for differentiating classes of RNA has been to separate total RNA using polyacrylamide gel electrophoresis in order to select small RNA (20–30 nt), ligate to adapters, then use reverse transcription and amplification to provide a "DNA Colony Template Library" for deep sequencing (Hwang *et al.*, 2011c). Application of this methodology to potato plants exposed to drought stress allowed the identification of several families of microRNAs associated with the stress response (Hwang *et al.*, 2011a,b,c). SAGE (serial analysis of gene expression) or DeepSAGE represents another transcriptomic approach using next-generation sequencing (Velculescu *et al.*, 1995). Gyetvai *et al.* (2012) applied SAGE technology to capture the transcriptomes of compatible and incompatible interactions between potato and *P. infestans* to identify transcriptomic changes that revealed candidate genes for plant host–pathogen interactions.

17.4 Proteomics

Two-dimensional protein gel electrophoresis has been used to compare the proteomes of potato under various environmental conditions, in reaction to disease agents, and to examine possible unintended changes resulting from transgene introduction. Initially, tuber proteins were classified and characterized to determine cultivar differences (Bauw *et al.*, 2006), those localized to the amyloplast (Stensballe *et al.*, 2008), changes over the life cycle of potato tubers during growth and development (Lehesranta *et al.*, 2006; Agrawal *et al.*, 2008) and during the postharvest aging process (Delaplace *et al.*, 2009). Proteins involved in plant defense have been demonstrated to be localized primarily in the skin of potato (Barel and Ginzberg, 2008). Only fertility management among a three-factorial experiment comparing two cover crops in organic and conventional production schemes significantly affected the proteome of potato cultivar Santé (Lehesranta *et al.*, 2007). A catalog of differentially expressed proteins in the petioles of *Solanum andigena* exposed to long or short day conditions has been compiled and categorized (Shah *et al.*, 2011). Proteins extracted from vacuoles of field-grown tubers of potato cultivar Kuras have been catalogued and classified by Jørgensen *et al.* (2011).

Proteomic analyses have also been conducted on potato in conjunction with pathogen interactions and abiotic stress. Liu and Halterman (2009) identified 12 protein spots that differentiated susceptible from transgenic cultivar Katahdin after both were inoculated with the late blight pathogen, *P. infestans*. Katahdin had been transformed with a single copy of the *RB* gene (a late blight resistance gene cloned from *Solanum bulbocastanum*) to confer resistance. Similarly, Barzic and Com (2012) observed that 13 proteins were upregulated in a resistant cultivar after exposure to bacterial soft rot (*Pectobacterium atrosepticum*). Thirty proteins were found to be differentially expressed in plants within a population segregating for after-cooking darkening of the tubers, with a greater wound response exhibited by individuals more prone to this tuber defect (Murphy *et al.*, 2010). Proteomic analysis of ten potato cultivars differing in susceptibility to tuber bruising revealed a class III lipase as a putative factor contributing to this postharvest handling problem (Urbany *et al.*, 2012). Reducing cold-induced sweetening during potato storage, a problem that requires "reconditioning" of stored potatoes at room temperature for 2 weeks before processing, has been tackled by proteomics, with the finding that some 46 of 4463 potato proteins were differentially expressed between samples stored for 0 or 5 months at 5°C (Yang *et al.*, 2011b); some of the proteins identified were enzymes involved in starch/sugar conversion. Aghaei *et al.* (2008) examined the proteomes of salt-tolerant and sensitive cultivars of potato to determine that 47 proteins of more than 300 observed were differentially expressed under NaCl stress in both cultivars; upregulation of defense-related proteins was determined by the relative intensity of protein spots after 2-D polyacrylamide gel electrophoresis.

Lehesranta *et al.* (2005) examined 1077 protein spots obtained from various potato cultivars and genetic stocks, as well as transgenic cultivars harboring either an empty vector or various alien genes, several of which resulted in stunted growth; nine proteins of 730 differentiated genetically modified potatoes from non-transgenic controls—many fewer than required to distinguish the varieties and land races. Transgenic potatoes with tuber-specific expression of a seed protein (amaranth albumin I) exhibited a 60% increase in tuber protein, as well as greater expression of native proteins that contributed to a rebalancing of the proteome (Chakraborty *et al.*, 2010). In contrast, Khalf *et al.* (2010) observed little, if any, difference in the proteomes of potatoes transformed with tomato cathepsin D inhibitor intended to convey insect resistance, compared to non-transgenic controls. Slight differences between control and transgenic potato carrying the Arabidopsis *DREB1A* gene intended to convey stress tolerance were identified through a proteomic approach (Nakamura *et al.*, 2010); the authors concluded that the transgenics may have expressed slightly greater levels of allergens.

Two recent technical improvements to proteome analysis in potato have provided better differentiation of spots on protein blots. These include differential centrifugation to separate leaf tissue into cell wall and cytoplasmic fractions (Lim *et al.*, 2012) and Mass Accuracy Precursor Alignment, allowing computer-aided measurements for peptide identification (Hoehenwarter *et al.*, 2011). As methodology for proteome analysis improves and becomes implemented in plant science research, proteomic research will become less descriptive and better able to target peptides associated with response to various treatments. In human genetics, proteomics has been used to distinguish marker SNPs from causal SNPs using proteome-wide analysis of SNPs (PWAS), in essence to identify allele-specific binding to transcription factors to understand gene regulation in susceptibility to disease such as type 1 diabetes (Butter *et al.*, 2012). As the technology improves, we can expect similar approaches to plant molecular breeding. In potato, a combination of association mapping with proteomics has been used to identify novel candidate genes that affect tuber starch and sugar content as they influence cold-induced sweetening in processing cultivars (Fischer *et al.*, 2013).

17.5 Candidate Gene Approach

With the candidate gene approach, genotyping is used to target functional and positional genes expected to be associated with the expression of a trait of interest (Pflieger *et al.*, 2001). Functional candidates are genes that have been shown or are suspected to have a functional role in the phenotype of interest. Allelic variants of such genes may be causal for the observed natural trait variation. In this case, DNA polymorphisms located within the candidate gene or physically close to it will be associated with trait variation (Huitema *et al.*, 2004).

Gene sequences must be tested for linkage to the trait of interest by molecular mapping in order to select the most promising candidates from many functional candidate genes (Pajerowska *et al.*, 2005). The candidate gene approach is facilitated greatly by plant genomic resources such as EST databases and analysis of gene function in model organisms. The candidate gene approach depends on and takes advantage of the knowledge available in the literature and databases on the physiology, biochemistry, and molecular genetics of a trait of interest. It is biased, in that it makes assumptions about the identity and function of the genes underlying the trait, but does not require a sequenced genome and is feasible with limited human and financial resources (Huitema *et al.*, 2004).

In the postgenome era, the candidate gene approach is facilitated by the possibility of identifying DNA variation within specific domains of interest in a specific gene. Scientific publications, gene banks, and genome browsers are available to find information on a trait of interest in other plant species and to provide lists of possible candidate genes and their DNA or protein sequences, as well as an identification code to find the gene model in a genome browser or gene bank. The availability of the potato genome browser enables us to use an amino acid sequence to find orthologous sequences and determine if these genes have similar functions in potato.

The gene function can be further analyzed with programs such as "Smart Domain" (Schultz *et al.*, 1998, 2012). With Smart Domain, it is possible to identify important protein domains within the query sequence and determine a smaller stretch within the gene, in which a SNP causing an amino acid replacement could affect the trait of interest. By designing primers to amplify such specific domains of a gene in potato, it is possible to identify meaningful SNPs.

The candidate gene approach can now be even more specific with the availability of information from RNASeq technology (Wang *et al.*, 2009); RNASeq provides expression profiles for genes from samples extracted from different tissues or organs of the plant, or from the same tissues under different environmental conditions. These data allow a viewer to determine if a candidate gene is suitably expressed in the organ of interest; for example, meiosis-specific gene expression in pollen mother cells.

The success of the candidate gene approach to identify genes of interest is therefore based on the reliable phenotypic characterization of a population, to determine phenotypic extremes for the trait of interest. The selected extremes can serve as the base for sequencing the specific domain in order to identify allelic differences in SNPs. If a particular SNP produces a change in the amino acid sequence (i.e. a non-synonymous SNP), and this change affects the protein structure and possibly its function, this SNP could be used for further applications, such as SNP genotyping for large population characterization or genetic engineering.

To apply the candidate gene approach in potato, due to its heterozygosity it is necessary to clone multiple amplicons of the domain for each selection for reliable sequencing of all possible allelic versions of the SNP. Once a true SNP is determined, this SNP is used to design allele-specific primers and screen the whole population by allelic discrimination and determine if the SNP is linked significantly to the trait of interest. In an optimal case, the gene of interest might bear SNPs that were included in the Illumina SNP chip for potato (Hamilton *et al.*, 2011); if that is the case, a phenotypically characterized population can be screened directly using the potato SNP chip to determine if the SNP is meaningful for the trait of interest. If so, the SNP could also be used to design allele-specific primers for allelic discrimination in larger populations. A third possibility is that a candidate gene can also be physically near a SNP that is in the Infinium 8303 Potato Array. If the gene of interest is close to a SNP within the SNP chip that is linked significantly to the trait of interest, then linkage analysis of the candidate gene would determine if the SNP is informative for screening a population for the trait of interest by using the SNP chip.

17.6 Transgenics

While molecular markers and other molecular applications have been successful in characterizing existing genetic variation within species, plant biotechnology generates new genetic diversity that often extends beyond species boundaries (Shewry *et al.*, 2008). The introduction of single genes with the hope of delivering improved traits to potato through transgenic research has been attempted with a range of genes from various organisms (Veilleux and De Jong, 2007). The first wave of commercially available transgenic potatoes with Colorado potato beetle resistance, PVY, and *Potato leaf roll virus* resistance was released in the mid-1990s, but was subsequently removed from the market due to international discrimination against genetically modified potato (Kaniewski and Thomas, 2004). In the interim, scientific research has continued into the possibility of transgenic approaches to improve various traits. Drought tolerance has been studied transgenically by insertion of the yeast *trehalose-6-phosphate synthase 1* gene (Kondrák *et al.*, 2011), the Arabidopsis *dehydroascorbate reductase gene* (Eltayeb *et al.*, 2011), a *betaine aldehyde dehydrogenase* (*BADH*) gene from spinach (Zhang *et al.*, 2011), or constitutive expression of the potato *StMYB1R-1* transcription factor (Shin *et al.*, 2011). In addition, Waterer *et al.* (2010) investigated the effect of wheat mitochondrial *Mn superoxide dismutase* (SOD3:1), barley *dehydrin* 4 (DHN 4), a canola transcriptional factor (*DREB/CBF1*), or a bromegrass stress-inducible *ROB5* gene in potato cultivar Desiree. The transgenes were placed under the control of either the constitutive P35S promoter or a stress-inducible Arabidopsis COR78 promoter. The promoter/transgene combinations, COR78:DHN4 and COR78:ROB5, were the most likely to enhance tuber yield under drought stress. Kim *et al.* (2011) inserted into potato cultivar Atlantic a peroxiredoxin gene (*2-Cys Prx*) from Arabidopsis controlled by either the constitutive P35S promoter or the stress-inducible PSWPA2 promoter from sweet potato. The transgenics were more tolerant to heat shock than untransformed controls, and the stress-inducible promoter resulted in greater expression of the transgene than P35S. Stacking antioxidant genes may provide greater protection against multiple stress conditions than deploying single genes (Ahmad *et al.*, 2010). The strategy of using stress-inducible promoters restricts expression of the transgene to conditions when resistance is needed, with the intention of economizing physiological processes and avoiding the undesirable effects of transgene expression during non-stress conditions.

Improvement of salt stress tolerance of potato has also been the target of considerable transgenic research in recent years. Bayat *et al.* (2010) reported improved *in vitro* salt tolerance for potato cultivars transformed with a barley antiporter gene, *HvNHX2*. Overexpression of the *GalUR* gene, an ascorbic acid pathway enzyme, provided potato cultivar Taedong Valley with enhanced tolerance to salt stress, as measured by growth and microtuberization of *in vitro* plantlets (Upadhyaya *et al.*, 2011). The Arabidopsis transcription factor, *AtDREB1A*, under the control of a stress-inducible Arabidopsis promoter (*rd294*) in potato cultivar Desiree, conferred salinity tolerance relative to the level of expression of *AtDREB1A* under salt stress conditions (Celebi-Toprak *et al.*, 2005; Watanabe *et al.*, 2011). These initial studies of transgenics using stress response genes, identified through the application of transcriptomic and proteomic tools in model plants as well as in potato, suggest a new wave of commercial transgenic potatoes with wide environmental adaptation in the near future.

Resistance against biotic stress has been emphasized in transgenic studies in recent years as well. Defense against fungal pathogens has been illustrated by the insertion of thionin genes from brassicaceous species into potato, which was reported to express increased resistance to gray mold (*Botrytis cinerea*) (Hoshikawa *et al.*, 2012). The *StoVe1* gene from an aubergine species was overexpressed to reduce the infection rate of *Verticillium dahliae* on potato cultivar Desiree (Liu *et al.*, 2012). Likewise, expression of both chitinase (chiA) and ribosome inactivating protein (rip30) was combined to enhance the resistance of potato to *Rhizoctonia solani* (M'Hamdi *et al.*, 2012). In another study, the *Rpi-bt1* resistance gene from *S. bulbocastanum* under the control of a potato ubiquitin promoter conferred resistance to late blight (Oosumi *et al.*, 2009). A new source of transgenic resistance to PVY was found in the construction of a recombinant heavy chain variable region (VH antibody) with added hydrophobic residues targeting cytosolic expression (Bouaziz *et al.*, 2009). Stacking of resistance genes by inserting a double gene construct (*Nicotiana tabacum* AP24 osmotine and *Phyllomedusa sauvagii* dermaseptin) into potato, followed by retransformation with a *Gallus gallus* lysozyme construct, yielded plants that were highly resistant to *Erwinia carotovora* and inhibited fungal growth (Rivero *et al.*, 2012).

Pest resistance as well as pathogen resistance has been the focus of several studies on transgenic potato. The different crystal (cry) endotoxins produced by the soil bacterium, *Bacillus thuringiensis*, have been used to convey transgenic resistance to several insect pests of potato (Pardo-López *et al.*, 2013). Various *cry* genes, including *cry1Ab* under the control of a tuber-specific, granule-bound starch synthase (GBSSi) promoter (Kumar *et al.*, 2010), or a light-inducible phosphoenolpyruvate carboxylae (PEPC) promoter (Hagh *et al.*, 2009), and *cryIIa1* with a constitutive promoter (Douches *et al.*, 2011), have all been demonstrated to diminish damage by potato tuberworm (*Phthorimaea operculella*). The *Amaranthus caudatus* agglutinin (*ACA*) gene has been demonstrated to convey partial resistance to aphid attack in transgenic potato in both greenhouse and field trials (Yang *et al.*, 2011a). Suppression of root invasion by the potato cyst nematode (*Globodera pallida*) was achieved in transgenic potato by tissue-specific expression in the root tips, using the Arabidopsis MDK4-20 promoter to drive the expression of a disulfide-constrained 7-mer peptide (nAChRbp) to inhibit chemoreception of the nematode pest without affecting the incidence of non-invasive nematodes (Green *et al.*, 2012).

Miscellaneous other potato transgenics have emerged in recent years, including the introduction of a uridine diphosphate (UDP)-glucose flavoloid-3-*O*-glucosyltransferase gene (Wei *et al.*, 2012) to enhance the tuber skin color of Desiree, and overexpressing the sucrose synthase (*SuSy*) gene in tubers to increase starch accumulation and yield (Baroja-Fernández *et al.*, 2009). A molecular farming approach was used to harvest a viral protein gene (glycoprotein (GP) 5) from the virus that causes porcine reproductive and respiratory syndrome, in order to develop an inexpensive vaccination for swine (Chen and Liu, 2011). Increased tuber size and greater tuber yield were reported for a Korean potato cultivar overexpressing the Arabidopsis jasmonic acid carboxyl methyltransfease gene (Sohn *et al.*, 2011). A reduction in enzymatic browning of potato tubers was achieved by transformation with a hairpin construct designed to silence native polyphenol oxidase (*PPO*) genes (Llorente *et al.*, 2011). An unintended consequence was that the transgenic, non-browning potatoes maintained their aroma in storage longer than the controls (Llorente *et al.*, 2010).

The emergence of potato genomics has opened many new possibilities for the improvement of potato using transgenic approaches. An obvious strategy is to identify potato orthologs of genes identified in other organisms in order to develop politically less objectionable cisgenic rather than transgenic approaches to potato transformation. Overexpression of such orthologs may result in similarly desirable traits in the cisgenic plants, as would have been obtained in transgenic lines using a heterologous gene. The genome also offers the possibility of promoter analysis to identify regulatory sequences in order to target the expression of transgenes to the specific tissue where they can be most effective, while reducing or eliminating the presence of alien proteins in the edible tubers.

A recent breakthrough technology has been described whereby whole pathways can be introduced into a crop plant through multitransgenic binary vector construction using zinc finger nucleases and homing endonucleases (Zeevi *et al.*, 2012). The initial study described the insertion of up to nine genes concurrently into Arabidopsis using a single vector. Likewise, targeted mutagenesis through the regulated expression of zinc finger nucleases offers new opportunities for determining the function of previously unannotated genes through homologous recombination by a knockout technique without the need for the development of extensive mutant populations (Armstrong *et al.*, 2005). Li *et al.* (2012) demonstrated the potential of this TALEN (transcription activator-like effector nuclease) technology by modifying a susceptibility gene of rice to *Xanthomonas oryzae*, thereby conveying resistance to the pathogen. It is important to note that the possibility of undesirable somaclonal variation is ever-present during the application of any of these technologies, so that it is always essential to regenerate many different plants from unique events in order to select for normal phenotypes (Barrell and Conner, 2011).

17.7 Comparative Genomics

Comparative genomics is the study of the relationship of genome structure and function across different biological species or strains (Goodstein *et al.*, 2012). Researchers use comparative genomics to understand the function and evolutionary processes that act on genomes. Similarities and differences in the proteins, RNA, and regulatory regions of different organisms are important components of comparative genomics. Genome elements that are responsible for similarities between different species should be conserved through time, while elements responsible for differences among species should be divergent. Elements that are unimportant to the evolutionary success of the organism will not be conserved. A comparison of potato genes with those of 11 other plant genomes revealed more than 8000 unique genes in some 1200 gene families (Fig. 17.2a) (Potato Genome Sequencing Consortium, 2011).

While comparative genomics is crucial to identify mechanisms of evolution, it is also a tool for molecular geneticists to identify genes coding for functional proteins important for crop improvement. Comparative genomics can be the base for gene identification in the postgenomic era. The analysis of crop genomes and genetic maps is crucial in understanding plant origin and evolution, revealing ancestral rearrangements and polyploidization events through evidence of genome duplication. Scientists can address fundamental questions about species proliferation, adaptation, and functional modulations (Ballvora *et al.*, 2007).

Species within the Solanaceae harbor diverse phenotypes that have been exploited for different agronomic purposes. Potato has been bred for tubers (modified stems), while tomato, pepper, and aubergine have been bred for enhanced fruit production. Likewise, petunia has been bred and selected for floral phenotypes, while tobacco has been bred for leaf size and secondary metabolism (Rensink *et al.*, 2005c). Genomes within the Solanaceae have undergone relatively few rearrangements and duplications, and therefore have similar gene content and order. This exceptionally high level of conservation of genome organization at the macro and micro levels makes this family a model to explore the basis of phenotypic diversity and adaptation to natural and agricultural environments (Mueller *et al.*, 2005).

Rensink *et al.* (2005c) documented a genomic-scale comparison of the available coding sequences (ESTs and ETs) from six solanaceous species, comparing nucleotide sequences for potato, tomato, pepper, aubergine, tobacco, and *Nicotiana benthamiana*. Including ortholog analysis, they confirmed a high level of sequence conservation. Phylogenetic and comparative genomic analyses with Arabidopsis, rice, and 21 other gene indices revealed sequence divergence during speciation, as evidenced by transcripts likely unique among the Solanaceae and unique to individual Solanaceae species (Rensink *et al.*, 2005c).

Arabidopsis is often the model of choice for anchoring comparative genomic studies at a more detailed level, because it has been characterized more completely, and is the first plant species with its genome completely sequenced. The Arabidopsis system, though, has its limitations for studying certain biological processes. The genome of potato has been compared with those of Arabidopsis and grape, revealing various degrees of syntenic blocks (Fig. 17.2c) (Potato Genome Sequencing Consortium, 2011). Moore *et al.* (2005) examined the ripening of fleshy fruits based in tomato microarrays. By analyzing different stages of fruit development in closely related heterologous species, they identified candidate ESTs and genes implicated in fruit ripening in the Solanaceae (Moore *et al.*, 2005). Most studies on comparative genomics of solanaceous species have been performed generally by analyzing two species at a time (Ballvora *et al.*, 2007; Guyot *et al.*, 2012).

Wu and Tanksley (2010) used comparative genomics based on a large set of single-copy conserved orthologous markers (COSII) from studies in tomato, potato, aubergine, pepper, and diploid *Nicotiana* species, to deduce the broad features and outcomes of chromosomal evolution in this family over the past 30 million years. The authors concluded that a rate of 0.03–0.12 rearrangements per chromosome per million years caused the chromosomal changes present in the family. They determined that a higher frequency of inversions rather than translocations occurred, and identified hotspots of chromosomal breakage. They reconstructed the most likely genome configuration for the ancestors of the Solanaceae species.

Molecular markers, and more recently, high-throughput genome sequencing efforts, have increased dramatically the knowledge of and ability to characterize genetic diversity in the germplasm pool for essentially any crop species (see Chapter 2, this volume). The availability of full genome sequences for several crop species allows comparison between many accessions within species, providing new keys for biodiversity exploitation and interpretation. High-throughput genotyping by re-sequencing is enabling the identification of a wealth of SNPs to produce the haplotype maps necessary to associate molecular variation to phenotype accurately. A genome sequence provides a useful starting point for insight into genetic variation. Genome-wide molecular tools can be used to

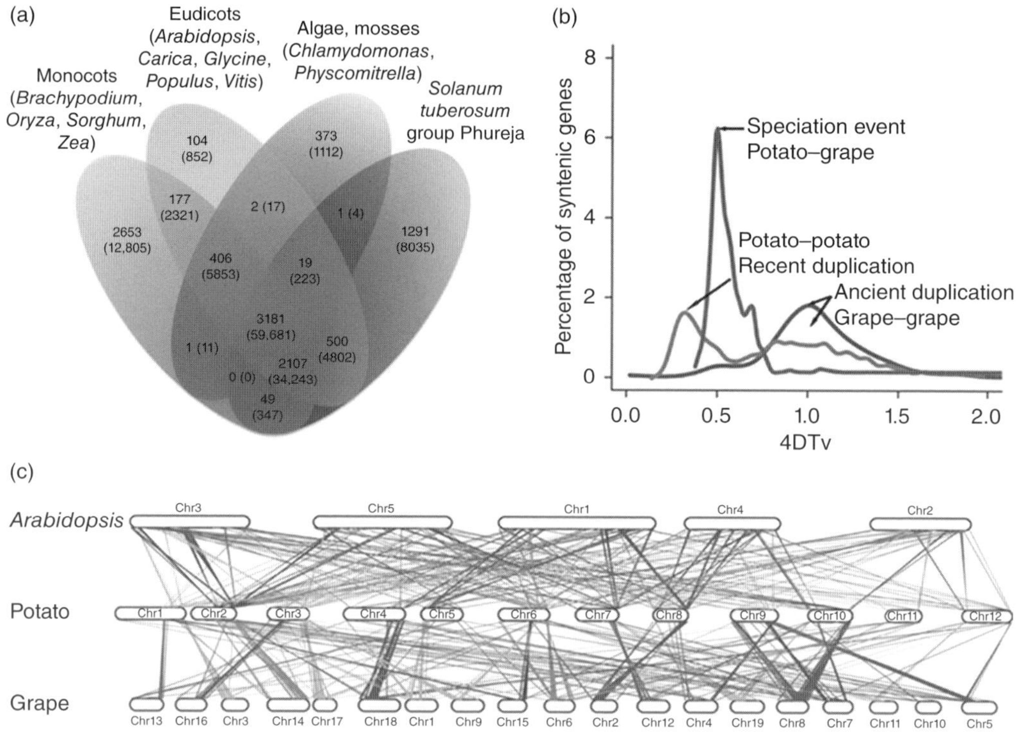

Fig. 17.2. Comparative analyses and evolution of the potato genome. (a) Clusters of orthologous and paralogous gene families in 12 plant species as identified by OrthoMCL33. Gene family number is listed in each of the components; the number of genes within the families for all of the species within the component is noted in parentheses. (b) Genome duplication in dicot genomes as revealed through 4DTv analyses. (c) Syntenic blocks between *Arabidopsis thaliana*, potato, and *Vitis vinifera* (grape), demonstrating a high degree of conserved gene order between these taxa (Potato Genome Sequencing Consortium, 2011).

assist crop improvement, explore genome organization, and study the evolution of gene families. In fact, detailed analyses of repetitive sequences, transposable elements (TEs), and comparative genomics of genes including intergenic regions shed new light on the mechanisms of genome evolution, speciation, and domestication (Barabaschi *et al.*, 2012).

Crop genomes show tremendous diversity in both size and structure, much of which has likely been created by whole-genome duplication (WGD) events. In the past 10 years, the availability of complete or draft genome sequences has allowed the reconstruction of a more detailed picture of plant genome evolution from the early angiosperm ancestor, to the origin of present plants characterized by segment and genome duplications, fragment rearrangements, and genome size reductions (Hancock, 2012).

Potato is the first sequenced genome of an asterid, a clade within eudicots that encompasses nearly 70,000 species characterized by unique morphological, developmental, and compositional features. The sequencing effort identified 2642 high-confidence asterid-specific genes and 3372 potato lineage-specific genes. Genes encoding transcription factors, self-incompatibility, and defense-related proteins were evident, and presumably contributed to the unique characteristics of asterids (Potato Genome Sequencing Consortium, 2011). Based on genomics 4DTv results analysis, the authors suggest two WGD events (Fig. 17.2b). The same events were identified in three rosid genomes (*V. vinifera*, *A. thaliana*, and *Populus trichocarpa*). The ancient WGD corresponds to the ancestral hexaploidization event in grape, consistent with a previous report based on EST analysis that the two main branches of eudicots, the asterids and rosids, may share the same palaeo-hexaploid duplication event (Tang *et al.*, 2008). The consortia suggest the second event occurred after the divergence between dicots and monocots (185 ± 55 million years ago) (Jaillon *et al.*, 2007). Therefore, the recent duplication could have occurred ~67 million years ago, consistent with the WGD that occurred near the Cretaceous–Tertiary boundary ~65 million years ago (Busch *et al.*, 2011). The divergence of potato and grape occurred ~89 million years ago, which was likely to represent the split between the rosids and asterids (Potato Genome Sequencing Consortium, 2011). Barabaschi *et al.* (2012) compiled recent data in monocots (Salse *et al.*, 2009) and in eudicots (Paterson *et al.*, 2009) to support the idea that "duplication-resistant" gene families corresponded to transcriptional regulators or disease resistance genes that were usually maintained after WGD (Huang *et al.*, 2005; Guyot *et al.*, 2012)

Completely sequenced genomes are indispensable for comparative genomics; they offer the possibility of utilizing the transcriptomes of different species or different cultivars with contrasting phenotypes to study a trait of interest, to support crop improvement or scientific research. An example of a downstream application of comparative transcriptomics is to evaluate inadvertent gene expression associated with a transgenic; for example, in cultivars transformed with a gene for resistance to *P. infestans* (Gyetvai *et al.*, 2012). With the release of the potato and tomato genome sequences, the efforts of mapping and cloning disease-resistance genes from *Solanum* species will increase. Resistance genes belonging to the NBS-LRR class have a conserved protein motif that enables the generation of resistance gene analogs (RGA), polymerase chain reaction fragments derived from *R* genes. RGAs from *S. bulbocastanum* and other *Solanum* species were generated to perform comparative genomics meta-analysis of nearly 800 NBS-LRR gene sequences defining *R* gene diversity bins to reflect evolutionary relationships and DNA cross-hybridization. The results demonstrate that all tested *R* gene lineages are of ancient origin (Bedoya-Reina and Barrero, 2010).

Comparative genetics was used to study the formation of tubers (the actual potatoes) by modification of a stolon that was unique for potato. Tomato is closely related to potato, but does not produce stolons or modified stems for tuber production. Transcript data from both potato and tomato were analyzed to address genetic regulation of the formation of stolons and the transition of stolons to tubers. The formation of stolons and tubers was found to coincide with the upregulation of genes associated with starch biosynthesis and storage proteins, and Kunitz protease inhibitor genes associated with pests and pathogens (Moore *et al.*, 2005).

Now that sequenced plant genomes are available, computer programs are under development to exploit the information they contain, as well as gene family data for analysis to aid comparative genomic studies for diverse applications. Such programs combine plant genome databases in web portals that are available as an aid to link model systems with other plants of economic and ecological importance for comparative translational genomic studies (Goodstein *et al.*, 2012). The SOL Genomics Network collects and curates information about Solanaceae-specific genes across a range of species, and makes the information available through a public portal (Menda *et al.*, 2008).

17.8 Conclusion

Potato breeding has not had the huge impact on agricultural production that has been realized in so many other crops where genetic improvement has been the major contributor to agricultural productivity. Yield improvements in potato over the past century have been due more to improved cultural practices and supporting technologies. However, potato breeders have now been equipped with an enviable array of genomic technologies that can be deployed for cultivar improvement. It is likely that the genetic gains that have already been realized in other crops with more tractable breeding systems can finally become a reality for improvements to potato during the 21st century.

References

Achenbach, U., Paulo, J., Ilarionova, E., Lübeck, J., Strahwald, J., Tacke, E., *et al.* (2009) Using SNP markers to dissect linkage disequilibrium at a major quantitative trait locus for resistance to the potato cyst nematode *Globodera pallida* on potato chromosome V. *Theoretical and Applied Genetics* 118, 619–629.

Aghaei, K., Ehsanpour, A.A. and Komatsu, S. (2008) Proteome analysis of potato under salt stress. *Journal of Proteome Research* 7, 4858–4868.

Agrawal, L., Chakraborty, S., Jaiswal, D.K., Gupta, S., Datta, A. and Chakraborty, N. (2008) Comparative proteomics of tuber induction, development and maturation reveal the complexity of tuberization process in potato (*Solanum tuberosum* L.). *Journal of Proteome Research* 7, 3803–3817.

Ahmad, R., Kim, Y.H., Kim, M.D., Kwon, S.Y., Cho, K., Lee, H.S., *et al.* (2010) Simultaneous expression of choline oxidase, superoxide dismutase and ascorbate peroxidase in potato plant chloroplasts provides synergistically enhanced protection against various abiotic stresses. *Physiologia Plantarum* 138, 520–533.

Armstrong, M.R., Whisson, S.C., Pritchard, L., Bos, J.I.B., Venter, E., Avrova, A.O., *et al.* (2005) An ancestral oomycete locus contains late blight avirulence gene *Avr3a*, encoding a protein that is recognized in the host cytoplasm. *Proceedings of the National Academy of Sciences* 102, 7766–7771.

Baldwin, S.J., Dodds, K.G., Auvray, B., Genet, R.A., Macknight, R.C. and Jacobs, J.M.E. (2011) Association mapping of cold-induced sweetening in potato using historical phenotypic data. *Annals of Applied Biology* 158, 248–256.

Ballvora, A., Jöcker, A., Viehöver, P., Ishihara, H., Paal, J., Meksem, K., *et al.* (2007) Comparative sequence analysis of *Solanum* and *Arabidopsis* in a hot spot for pathogen resistance on potato chromosome V reveals a patchwork of conserved and rapidly evolving genome segments. *BMC Genomics* 8.

Barabaschi, D., Guerra, D., Lacrima, K., Laino, P., Michelotti, V., Urso, S., *et al.* (2012) Emerging knowledge from genome sequencing of crop species. *Molecular Biotechnology* 50, 250–266.

Barel, G. and Ginzberg, I. (2008) Potato skin proteome is enriched with plant defence components. *Journal of Experimental Botany* 59, 3347–3357.

Baroja-Fernández, E., Muñoz, F.J., Montero, M., Etxeberria, E., Sesma, M.T., Ovecka, M., *et al.* (2009) Enhancing sucrose synthase activity in transgenic potato (*Solanum tuberosum* L.) tubers results in increased levels of starch, ADPglucose and UDPglucose and total yield. *Plant and Cell Physiology* 50, 1651–1662.

Barone, A. (2004) Molecular marker-assisted selection for potato breeding. *American Journal of Potato Research* 81, 111–117.

Barozai, M.Y.K. and Wahid, H.A. (2012) *In silico* identification and characterization of cumulative abiotic stress responding genes in potato (*Solanum tuberosum* L.). *Pakistan Journal of Botany* 44, 57–69.

Barrell, P.J. and Conner, A.J. (2011) Facilitating the recovery of phenotypically normal transgenic lines in clonal crops: a new strategy illustrated in potato. *Theoretical and Applied Genetics* 122, 1171–1177.

Barzic, M.R. and Com, E. (2012) Proteins involved in the interaction of potato tubers with *Pectobacterium atrosepticum*: a proteomic approach to understanding partial resistance. *Journal of Phytopathology* 160, 561–575.

Bauw, G., Nielsen, H.V., Emmersen, J., Nielsen, K.L., Jørgensen, M. and Welinder, K.G. (2006) Patatins, Kunitz protease inhibitors and other major proteins in tuber of potato cv. Kuras. *FEBS Journal* 273, 3569–3584.

Bayat, F., Shiran, B., Belyaev, D.V., Yur'eva, N.O., Sobol'kova, G.I., Alizadeh, H., *et al.* (2010) Potato plants bearing a vacuolar Na^{+}/H^{+} antiporter HvNHX2 from barley are characterized by improved salt tolerance. *Russian Journal of Plant Physiology* 57, 696–706.

Bedoya-Reina, O.C. and Barrero, L.S. (2010) Preliminary assessment of COSII gene diversity in lulo and a relative species: initial identification of genes potentially associated with domestication. *Gene* 458, 27–36.

Bhering, L.L., Pinto, C., Benites, F.R.G., Leite, M.E. and da Silva, F.L. (2009) Marker assisted selection for dry matter and reducing sugars in potato tubers. *Ciência Rural* 39, 38–44.

Birch, P.R.J., Avrova, A.O., Armstrong, M., Venter, E., Taleb, N., Gilroy, E.M., *et al.* (2003) The potato – *Phytophthora infestans* interaction transcriptome. *Canadian Journal of Plant Pathology* 25, 226–231.

Bormann, C.A., Rickert, A.M., Ruiz, R.A.C., Paal, J., Lübeck, J., Strahwald, J., *et al.* (2004) Tagging quantitative trait loci for maturity-corrected late blight resistance in tetraploid potato with PCR-based candidate gene markers. *Molecular Plant–Microbe Interactions* 17, 1126–1138.

Bouaziz, D., Ayadi, M., Bidani, A., Rouis, S., Nouri-Ellouz, O., Jellouli, R., *et al.* (2009) A stable cytosolic expression of VH antibody fragment directed against PVY NIa protein in transgenic potato plant confers partial protection against the virus. *Plant Science* 176, 489–496.

Busch, B.L., Schmitz, G., Rossmann, S., Piron, F., Ding, J., Bendahmane, A., *et al.* (2011) Shoot branching and leaf dissection in tomato are regulated by homologous gene modules. *Plant Cell* 23, 3595–3609.

Butter, F., Davison, L., Viturawong, T., Scheibe, M., Vermeulen, M., Todd, J.A., *et al.* (2012) Proteome-wide analysis of disease-associated SNPs that show allele-specific transcription factor binding. *PLoS Genetics* 8, e1002982.

Celebi-Toprak, F., Behnam, B., Serrano, G., Kasuga, M., Yamaguchi-Shinozaki, K., Naka, H., *et al.* (2005) Tolerance to salt stress of the transgenic tetrasomic tetraploid potato, *Solanum tuberosum* cv. Desirée appears to be induced by the *DREB1A* gene and *rd29A* promoter of *Arabidopsis thaliana*. *Breeding Science* 55, 311–319.

Chakraborty, S., Chakraborty, N., Agrawal, L., Ghosh, S., Narula, K., Shekhar, S., *et al.* (2010) Next-generation protein-rich potato expressing the seed protein gene *AmA1* is a result of proteome rebalancing in transgenic tuber. *Proceedings of the National Academy of Sciences* 107, 17533–17538.

Chen, X. and Liu, J.L. (2011) Generation and immunogenicity of transgenic potato expressing the GP5 protein of porcine reproductive and respiratory syndrome virus. *Journal of Virological Methods* 173, 153–158.

Colton, L.M., Groza, H.I., Wielgus, S.M. and Jiang, J.M. (2006) Marker-assisted selection for the broad-spectrum potato late blight resistance conferred by gene RB derived from a wild potato species. *Crop Science* 46, 589–594.

Costanzo, S., Simko, I., Christ, B.J. and Haynes, K.G. (2005) QTL analysis of late blight resistance in a diploid potato family of *Solanum phureja* × *S. stenotomum*. *Theoretical and Applied Genetics* 111, 609–617.

Crookshanks, M., Emmersen, J., Welinder, K.G. and Nielsen, K.L. (2001) The potato tuber transcriptome: analysis of 6077 expressed sequence tags. *FEBS Letters* 506, 123–126.

DeBolt, S. (2010) Copy number variation shapes genome diversity in Arabidopsis over immediate family generational scales. *Genome Biology and Evolution* 2, 441–453.

D'Ippólito, S., Martin, M.L., Salcedo, M.F., Atencio, H.M., Casalongué, C.A., Godoy, A.V., *et al.* (2010) Transcriptome profiling of *Fusarium solani* f. sp. *eumartii* – infected potato tubers provides evidence of an inducible defense response. *Physiological and Molecular Plant Pathology* 75, 3–12.

Delaplace, P., Fauconnier, M.L., Sergeant, K., Dierick, J.F., Oufir, M., van der Wal, F., *et al.* (2009) Potato (*Solanum tuberosum* L.) tuber ageing induces changes in the proteome and antioxidants associated with the sprouting pattern. *Journal of Experimental Botany* 60, 1273–1288.

Douches, D.S., Coombs, J.J., Lacey, L.A., Felcher, K.J. and Pett, W.L. (2011) Evaluations of transgenic potatoes for resistance to potato tuberworm in the laboratory and field. *American Journal of Potato Research* 88, 91–95.

Eltayeb, A.E., Yamamoto, S., Habora, M.E.E., Yin, L.N., Tsujimoto, H. and Tanaka, K. (2011) Transgenic potato overexpressing Arabidopsis cytosolic *AtDHAR1* showed higher tolerance to herbicide, drought and salt stresses. *Breeding Science* 61, 3–10.

Evers, D., Gigliotti, S., Nicot, N., Solinhac, L., Herrera, C., Bartolini, I., *et al.* (2004) Late blight infection of potato: transcriptomics approaches. *Acta Physiologiae Plantarum* 26, 268.

Evers, D., Lefèbre, I., Legay, S., Lamoureux, D., Hausman, J.F., Rosales, R.O.G., *et al.* (2010) Identification of drought-responsive compounds in potato through a combined transcriptomic and targeted metabolite approach. *Journal of Experimental Botany* 61, 2327–2343.

Evers, D., Legay, S., Lamoureux, D., Hausman, J.F., Hoffmann, L. and Renaut, J. (2012) Towards a synthetic view of potato cold and salt stress response by transcriptomic and proteomic analyses. *Plant Molecular Biology* 78, 503–514.

Felcher, K.J., Coombs, J.J., Massa, A.N., Hansey, C.N., Hamilton, J.P., Veilleux, R.E., *et al.* (2012) Integration of two diploid potato linkage maps with the potato genome sequence. *PLoS ONE* 7, e36347.

Ferreira, S.J., Senning, M., Sonnewald, S., Keβling, P.M., Goldstein, R. and Sonnewald, U. (2010) Comparative transcriptome analysis coupled to X-ray CT reveals sucrose supply and growth velocity as major determinants of potato tuber starch biosynthesis. *BMC Genomics* 11.

Fischer, M., Schreiber, L., Colby, T., Kuckenberg, M., Tacke, E., Hofferbert, H.R., *et al.* (2013) Novel candidate genes influencing natural variation in potato tuber cold sweetening identified by comparative proteomics and association mapping. *BMC Plant Biology* 13, 113.

Gao, L., Tu, Z., Katagiri, F. and Bradeen, J.M. (2011) RNA-seq analysis of potato tuber transcriptome dynamics in response to the late blight pathogen *Phytophthora infestans*. *Phytopathology* 101, S58.
Gebhardt, C., Bellin, D., Henselewski, H., Lehmann, W., Schwarzfischer, J. and Valkonen, J.P.T. (2006) Marker-assisted combination of major genes for pathogen resistance in potato. *Theoretical and Applied Genetics* 112, 1458–1464.
Ginzberg, I., Barel, G., Ophir, R., Tzin, E., Tanami, Z., Muddarangappa, T., *et al.* (2009) Transcriptomic profiling of heat-stress response in potato periderm. *Journal of Experimental Botany* 60, 4411–4421.
Goodstein, D.M., Shu, S.Q., Howson, R., Neupane, R., Hayes, R.D., Fazo, J., *et al.* (2012) Phytozome: a comparative platform for green plant genomics. *Nucleic Acids Research* 40, D1178–D1186.
Green, J., Wang, D., Lilley, C.J., Urwin, P.E. and Atkinson, H.J. (2012) Transgenic potatoes for potato cyst nematode control can replace pesticide use without impact on soil quality. *Plos One* 7, e30973.
Guyot, R., Lefebvre-Pautigny, F., Tranchant-Dubreuil, C., Rigoreau, M., Hamon, P., Leroy, T., *et al.* (2012) Ancestral synteny shared between distantly-related plant species from the asterid (*Coffea canephora* and *Solanum* sp.) and rosid (*Vitis vinifera*) clades. *BMC Genomics* 13, 103.
Gyetvai, G., Sønderkær, M., Göbel, U., Basekow, R., Ballvora, A., Imhoff, M., *et al.* (2012) The transcriptome of compatible and incompatible interactions of potato (*Solanum tuberosum*) with *Phytophthora infestans* revealed by DeepSAGE analysis. *Plos One* 7, e31526.
Hagh, Z.G., Rahnama, H., Panahandeh, J., Rouz, B.B.K., Jafari, K.M.A. and Mahna, N. (2009) Green-tissue-specific, C4-PEPC-promoter-driven expression of *Cry*1Ab makes transgenic potato plants resistant to tuber moth (*Phthorimaea operculella*, Zeller). *Plant Cell Reports* 28, 1869–1879.
Hämäläinen, J.H., Watanabe, K.N., Valkonen, J.P.T., Arihara, A., Plaisted, R.L., Pehu, E., *et al.* (1997) Mapping and marker-assisted selection for a gene for extreme resistance to potato virus Y. *Theoretical and Applied Genetics* 94, 192–197.
Hamilton, J., Hansey, C., Whitty, B., Stoffel, K., Massa, A., Van Deynze, A., *et al.* (2011) Single nucleotide polymorphism discovery in elite north american potato germplasm. *BMC Genomics* 12, 302.
Hancock, J.F. (2012) *Plant Evolution and the Origin of Crop Species (3rd edn)*. CAB International, Wallingford, UK.
Henriquez, M.A. and Daayf, F. (2010) Identification and cloning of differentially expressed genes involved in the interaction between potato and *Phytophthora infestans* using a subtractive hybridization and cDNA-AFLP combinational approach. *Journal of Integrative Plant Biology* 52, 453–467.
Hoehenwarter, W., Larhlimi, A., Hummel, J., Egelhofer, V., Selbig, J., van Dongen, J.T., *et al.* (2011) MAPA distinguishes genotype-specific variability of highly similar regulatory protein isoforms in potato tuber. *Journal of Proteome Research* 10, 2979–2991.
Hoshikawa, K., Ishihara, G., Takahashi, H. and Nakamura, I. (2012) Enhanced resistance to gray mold (*Botrytis cinerea*) in transgenic potato plants expressing thionin genes isolated from Brassicaceae species. *Plant Biotechnology* 29, 87–93.
Huang, S., Van Der Vossen, E.A.G., Kuang, H., Vleeshouwers, V.G.A.A., Zhang, N., Borm, T.J.A., *et al.* (2005) Comparative genomics enabled the isolation of the *R3a* late blight resistance gene in potato. *The Plant Journal* 42, 251–261.
Huitema, E., Bos, J.I.B., Tian, M.Y., Win, J., Waugh, M.E. and Kamoun, S. (2004) Linking sequence to phenotype in *Phytophthora*-plant interactions. *Trends in Microbiology* 12, 193–200.
Hwang, E.W., Shin, S.J. and Kwon, H.B. (2011a) Identification of microRNAs and their putative targets that respond to drought stress in *Solanum tuberosum*. *Journal of the Korean Society for Applied Biological Chemistry* 54, 317–324.
Hwang, E.W., Shin, S.J., Park, S.C., Jeong, M.J. and Kwon, H.B. (2011b) Identification of miR172 family members and their putative targets responding to drought stress in *Solanum tuberosum*. *Genes and Genomics* 33, 105–110.
Hwang, E.W., Shin, S.J., Yu, B.K., Byun, M.O. and Kwon, H.B. (2011c) miR171 family members are involved in drought response in *Solanum tuberosum*. *Journal of Plant Biology* 54, 43–48.
Jaillon, O., Aury, J.M., Noel, B., Policriti, A., Clepet, C., Casagrande, A., *et al.* (2007) The grapevine genome sequence suggests ancestral hexaploidization in major angiosperm phyla. *Nature* 449, 463–468.
Jørgensen, M., Stensballe, A. and Welinder, K.G. (2011) Extensive post-translational processing of potato tuber storage proteins and vacuolar targeting. *FEBS Journal* 278, 4070–4087.
Jupe, F., Pritchard, L., Etherington, G.J., MacKenzie, K., Cock, P.J.A., Wright, F., *et al.* (2012) Identification and localisation of the NB-LRR gene family within the potato genome. *BMC Genomics* 13.
Kaniewski, W.K. and Thomas, P.E. (2004) The potato story. *AgBioForum* 7, 41–46.

Kawchuk, L.M., Lynch, D.R., Yada, R.Y., Bizimungu, B. and Lynn, J. (2008) Marker assisted selection of potato clones that process with light chip color. *American Journal of Potato Research* 85, 227–231.
Kelley, K.B., Whitworth, J.L. and Novy, R.G. (2009) Mapping of the potato leafroll virus resistance gene, Rlr_{etb}, from *Solanum etuberosum* identifies interchromosomal translocations among its E-genome chromosomes 4 and 9 relative to the A-genome of *Solanum* L. sect. *Petota*. *Molecular Breeding* 23, 489–500.
Khalf, M., Goulet, C., Vorster, J., Brunelle, F., Anguenot, R., Fliss, I., *et al.* (2010) Tubers from potato lines expressing a tomato Kunitz protease inhibitor are substantially equivalent to parental and transgenic controls. *Plant Biotechnology Journal* 8, 155–169.
Kim, M.D., Kim, Y.H., Kwon, S.Y., Jang, B.Y., Lee, S.Y., Yun, D.J., *et al.* (2011) Overexpression of 2-cysteine peroxiredoxin enhances tolerance to methyl viologen-mediated oxidative stress and high temperature in potato plants. *Plant Physiology and Biochemistry* 49, 891–897.
Kloosterman, B., De Koeyer, D., Griffiths, R., Flinn, B., Steuernagel, B., Scholz, U., *et al.* (2008) Genes driving potato tuber initiation and growth: identification based on transcriptional changes using the POCI array. *Functional and Integrative Genomics* 8, 329–340.
Kondrák, M., Marincs, F., Kalapos, B., Juhász, Z. and Bánfalvi, Z. (2011) Transcriptome analysis of potato leaves expressing the *trehalose-6-phosphate synthase 1* gene of yeast. *PLoS One* 6, e23466.
Kumar, M., Chimote, V., Singh, R., Mishra, G.P., Naik, P.S., Pandey, S.K., *et al.* (2010) Development of Bt transgenic potatoes for effective control of potato tuber moth by using *cry1Ab* gene regulated by GBSS promoter. *Crop Protection* 29, 121–127.
Legay, S., Lamoureux, D., Hausman, J.F., Hoffmann, L. and Evers, D. (2009) Monitoring gene expression of potato under salinity using cDNA microarrays. *Plant Cell Reports* 28, 1799–1816.
Lehesranta, S.J., Davies, H.V., Shepherd, L.V.T., Nunan, N., McNicol, J.W., Auriola, S., *et al.* (2005) Comparison of tuber proteomes of potato varieties, landraces, and genetically modified lines. *Plant Physiology* 138, 1690–1699.
Lehesranta, S.J., Davies, H.V., Shepherd, L.V.T., Koistinen, K.M., Massat, N., Nunan, N., *et al.* (2006) Proteomic analysis of the potato tuber life cycle. *Proteomics* 6, 6042–6052.
Lehesranta, S.J., Koistinen, K.M., Massat, N., Davies, H.V., Shepherd, L.V.T., McNicol, J.W., *et al.* (2007) Effects of agricultural production systems and their components on protein profiles of potato tubers. *Proteomics* 7, 597–604.
Li, T., Liu, B., Spalding, M.H., Weeks, D.P. and Yang, B. (2012) High-efficiency TALEN-based gene editing produces disease-resistant rice. *Nature Biotechnology* 30, 390–392.
Li, X.Q. (2008) Molecular characterization and biotechnological improvement of the processing quality of potatoes. *Canadian Journal of Plant Science* 88, 639–648.
Lim, S., Chisholm, K., Coffin, R.H., Peters, R.D., A-Mughrabi, K.I., Wang-Pruski, G., *et al.* (2012) Protein profiling in potato (*Solanum tuberosum* L.) leaf tissues by differential centrifugation. *Journal of Proteome Research* 11, 2594–2601.
Liu, S.P., Zhu, Y.P., Xie, C., Jue, D.W., Hong, Y.B., Chen, M., *et al.* (2012) Transgenic potato plants expressing *StoVe1* exhibit enhanced resistance to *Verticillium dahliae*. *Plant Molecular Biology Reporter* 30, 1032–1039.
Liu, Z.Y. and Halterman, D. (2009) Analysis of proteins differentially accumulated during potato late blight resistance mediated by the *RB* resistance gene. *Physiological and Molecular Plant Pathology* 74, 151–160.
Llorente, B., Rodríguez, V., Alonso, G.D., Torres, H.N., Flawiá, M.M. and Bravo-Almonacid, F.F. (2010) Improvement of aroma in transgenic potato as a consequence of impairing tuber browning. *Plos One* 5, e14030.
Llorente, B., Alonso, G.D., Bravo-Almonacid, F., Rodriguez, V., Lopez, M.G., Carrari, F., *et al.* (2011) Safety assessment of nonbrowning potatoes: opening the discussion about the relevance of substantial equivalence on next generation biotech crops. *Plant Biotechnology Journal* 9, 136–150.
Lo, H.S., Wang, Z., Hu, Y., Yang, H.H., Gere, S., Buetow, K.H., *et al.* (2003) Allelic variation in gene expression is common in the human genome. *Genome Research* 13, 1855–1862.
Malosetti, M., van der Linden, C.G., Vosman, B. and van Eeuwijk, F.A. (2007) A mixed-model approach to association mapping using pedigree information with an illustration of resistance to *Phytophthora infestans* in potato. *Genetics* 175, 879–889.
Mane, S.P., Robinet, C.V., Ulanov, A., Schafleitner, R., Tincopa, L., Gaudin, A., *et al.* (2008) Molecular and physiological adaptation to prolonged drought stress in the leaves of two Andean potato genotypes. *Functional Plant Biology* 35, 669–688.
Massa, A.N., Childs, K.L., Lin, H.N., Bryan, G.J., Giuliano, G. and Buell, C.R. (2011) The transcriptome of the reference potato genome *Solanum tuberosum* Group Phureja clone DM1-3 516R44. *Plos One* 6, e26801.

Menda, N., Buels, R.M., Tecle, I. and Mueller, L.A. (2008) A community-based annotation framework for linking Solanaceae genomes with phenomes. *Plant Physiology* 147, 1788–1799.

M'Hamdi, M., Chikh-Rouhou, H., Boughalleb, N. and de Galarreta, J.I.R. (2012) Enhanced resistance to *Rhizoctonia solani* by combined expression of chitinase and Ribosome Inactivating Protein in transgenic potatoes (*Solanum tuberosum* L.). *Spanish Journal of Agricultural Research* 10, 778–785.

Milczarek, D., Flis, B. and Przetakiewicz, A. (2011) Suitability of molecular markers for selection of potatoes resistant to *Globodera* spp. *American Journal of Potato Research* 88, 245–255.

Moore, S., Payton, P., Wright, M., Tanksley, S. and Giovannoni, J. (2005) Utilization of tomato microarrays for comparative gene expression analysis in the Solanaceae. *Journal of Experimental Botany* 56, 2885–2895.

Mori, K., Sakamoto, Y., Mukojima, N., Tamiya, S., Nakao, T., Ishii, T., *et al.* (2011) Development of a multiplex PCR method for simultaneous detection of diagnostic DNA markers of five disease and pest resistance genes in potato. *Euphytica* 180, 347–355.

Mueller, L.A., Solow, T.H., Taylor, N., Skwarecki, B., Buels, R., Binns, J., *et al.* (2005) The SOL Genomics Network. A comparative resource for Solanaceae biology and beyond. *Plant Physiology* 138, 1310–1317.

Murphy, J.P., Kong, F.M., Pinto, D.M. and Wang-Pruski, G. (2010) Relative quantitative proteomic analysis reveals wound response proteins correlated with after-cooking darkening. *Proteomics* 10, 4258–4269.

Mustroph, A., Zanetti, M.E., Jang, C.J.H., Holtan, H.E., Repetti, P.P., Galbraith, D.W., Girke, T., *et al.* (2009) Profiling translatomes of discrete cell populations resolves altered cellular priorities during hypoxia in *Arabidopsis*. *Proceedings of the National Academy of Sciences* 106, 18843–18848.

Nakamura, R., Satoh, R., Nakamura, R., Shimazaki, T., Kasuga, M., Yamaguchi-Shinozaki, K., *et al.* (2010) Immunoproteomic and two-dimensional difference gel electrophoresis analysis of Arabidopsis dehydration response element-binding protein 1A (DREB1A)-transgenic potato. *Biological and Pharmaceutical Bulletin* 33, 1418–1425.

Nielsen, K.L., Grønkjaer, K., Welinder, K.G. and Emmersen, J. (2005) Global transcript profiling of potato tuber using LongSAGE. *Plant Biotechnology Journal* 3, 175–185.

Oosumi, T., Rockhold, D.R., Maccree, M.M., Deahl, K.L., McCue, K.F. and Belknap, W.R. (2009) Gene *Rpi-bt1* from *Solanum bulbocastanum* confers resistance to late blight in transgenic potatoes. *American Journal of Potato Research* 86, 456–465.

Ortega, F. and Lopez-Vizcon, C. (2012) Application of molecular marker-assisted selection (MAS) for disease resistance in a practical potato breeding programme. *Potato Research* 55, 1–13.

Oufir, M., Legay, S., Nicot, N., Van Moer, K., Hoffmann, L., Renaut, J., *et al.* (2008) Gene expression in potato during cold exposure: changes in carbohydrate and polyamine metabolisms. *Plant Science* 175, 839–852.

Pajerowska, K.M., Parker, J.E. and Gebhardt, C. (2005) Potato homologs of *Arabidopsis thaliana* genes functional in defense signaling—identification, genetic mapping, and molecular cloning. *Molecular Plant–Microbe Interactions* 18, 1107–1119.

Pardo-López, L., Soberón, M. and Bravo, A. (2013) *Bacillus thuringiensis* insecticidal three-domain Cry toxins: mode of action, insect resistance and consequences for crop protection. *FEMS Microbiology Reviews* 37, 3–22.

Park, T.H., Gros, J., Sikkema, A., Vleeshouwers, V.G.A.A., Muskens, M., Allefs, S., *et al.* (2005) The late blight resistance locus *Rpi-blb3* from *Solanum bulbocastanum* belongs to a major late blight *R* gene cluster on chromosome 4 of potato. *Molecular Plant–Microbe Interactions* 18, 722–729.

Paterson, A.H., Bowers, J.E., Bruggmann, R., Dubchak, I., Grimwood, J., Gundlach, H., *et al.* (2009) The *Sorghum bicolor* genome and the diversification of grasses. *Nature* 457, 551–556.

Pflieger, S., Lefebvre, V. and Causse, M. (2001) The candidate gene approach in plant genetics: a review. *Molecular Breeding* 7, 275–291.

Potato Genome Sequencing Consortium (2011) Genome sequence and analysis of the tuber crop potato. *Nature* 475, 189–195.

Rensink, W., Hart, A., Liu, J., Shu, O.Y., Zismann, V. and Buell, C.R. (2005a) Analyzing the potato abiotic stress transcriptome using expressed sequence tags. *Genome* 48, 598–605.

Rensink, W.A., Iobst, S., Hart, A., Stegalkina, S., Liu, J. and Buell, C.R. (2005b) Gene expression profiling of potato responses to cold, heat, and salt stress. *Functional and Integrative Genomics* 5, 201–207.

Rensink, W.A., Lee, Y., Liu, J., Iobst, S., Ouyang, S. and Buell, C.R. (2005c) Comparative analyses of six solanaceous transcriptomes reveal a high degree of sequence conservation and species-specific transcripts. *BMC Genomics* 6, 124.

Rickert, A.M., Kim, J.H., Meyer, S., Nagel, A., Ballvora, A., Oefner, P.J., *et al.* (2003) First-generation SNP/InDel markers tagging loci for pathogen resistance in the potato genome. *Plant Biotechnology Journal* 1, 399–410.

Ritter, E., de Galarreta, J.I.R., van Eck, H.J. and Sánchez, I. (2008) Construction of a potato transcriptome map based on the cDNA-AFLP technique. *Theoretical and Applied Genetics* 116, 1003–1013.

Rivero, M., Furman, N., Mencacci, N., Picca, P., Touma, L., Lentz, E., *et al.* (2012) Stacking of antimicrobial genes in potato transgenic plants confers increased resistance to bacterial and fungal pathogens. *Journal of Biotechnology* 157, 334–343.

Ronning, C.M., Stegalkina, S.S., Ascenzi, R.A., Bougri, O., Hart, A.L., Utterbach, T.R., *et al.* (2003) Comparative analyses of potato expressed sequence tag libraries. *Plant Physiology* 131, 419–429.

Rotter, A., Usadel, B., Baebler, Š., Stitt, M. and Gruden, K. (2007) Adaptation of the MapMan ontology to biotic stress responses: application in solanaceous species. *Plant Methods* 3, 10.

Rüping, B., Ernst, A.M., Jekat, S.B., Nordzieke, S., Reineke, A.R., Müller, B., *et al.* (2010) Molecular and phylogenetic characterization of the sieve element occlusion gene family in *Fabaceae* and non-*Fabaceae* plants. *BMC Plant Biology* 10, 219.

Rutitzky, M., Ghiglione, H.O., Curá, J.A., Casal, J.J. and Yanovsky, M.J. (2009) Comparative genomic analysis of light-regulated transcripts in the Solanaceae. *BMC Genomics* 10, 60.

Salse, J., Abrouk, M., Bolot, S., Guilhot, N., Courcelle, E., Faraut, T., *et al.* (2009) Reconstruction of monocotelydoneous proto-chromosomes reveals faster evolution in plants than in animals. *Proceedings of the National Academy of Sciences* 106, 14908–14913.

Schultz, J., Milpetz, F., Bork, P. and Ponting, C.P. (1998) SMART, a simple modular architecture research tool: identification of signaling domains. *Proceedings of the National Academy of Sciences* 95, 5857–5864.

Schultz, L., Cogan, N.O.I., McLean, K., Dale, M.F.B., Bryan, G.J., Forster, J.W., *et al.* (2012) Evaluation and implementation of a potential diagnostic molecular marker for *H1*-conferred potato cyst nematode resistance in potato (*Solanum tuberosum* L.). *Plant Breeding* 131, 315–321.

Shah, S., Lee, Y.L., Hannapel, D.J. and Rao, A.G. (2011) Protein profiling of the potato petiole under short day and long day photoperiods. *Journal of Proteomics* 74, 212–230.

Shewry, P.R., Jones, H.D. and Halford, N.G. (2008) Plant biotechnology: transgenic crops. *Advances in Biochemical Engineering/Biotechnology* 111, 149–186.

Shin, D., Moon, S.J., Han, S., Kim, B.G., Park, S.R., Lee, S.K., Yoon, H.J., *et al.* (2011) Expression of *StMYB1R-1*, a novel potato single MYB-like domain transcription factor, increases drought tolerance. *Plant Physiology* 155, 421–432.

Sliwka, J., Jakuczun, H., Kaminski, P. and Zimnoch-Guzowska, E. (2010) Marker-assisted selection of diploid and tetraploid potatoes carrying *Rpi-phu1*, a major gene for resistance to *Phytophthora infestans*. *Journal of Applied Genetics* 51, 133–140.

Sohn, H.B., Lee, H.Y., Seo, J.S., Jung, C., Jeon, J.H., Kim, J.H., *et al.* (2011) Overexpression of jasmonic acid carboxyl methyltransferase increases tuber yield and size in transgenic potato. *Plant Biotechnology Reports* 5, 27–34.

Stensballe, A., Hald, S., Bauw, G., Blennow, A. and Welinder, K.G. (2008) The amyloplast proteome of potato tuber. *FEBS Journal* 275, 1723–1741.

Stupar, R.M., Bhaskar, P.B., Yandell, B.S., Rensink, W.A., Hart, A.L., Ouyang, S., *et al.* (2007) Phenotypic and transcriptomic changes associated with potato autopolyploidization. *Genetics* 176, 2055–2067.

Stushnoff, C., Ducreux, L.J.M., Hancock, R.D., Hedley, P.E., Holm, D.G., McDougall, G.J., *et al.* (2010) Flavonoid profiling and transcriptome analysis reveals new gene-metabolite correlations in tubers of *Solanum tuberosum* L. *Journal of Experimental Botany* 61, 1225–1238.

Tan, M.Y.A., Hutten, R.C.B., Visser, R.G.F. and van Eck, H.J. (2010) The effect of pyramiding *Phytophthora infestans* resistance genes *RPi-mcd1* and *RPi-ber* in potato. *Theoretical and Applied Genetics* 121, 117–125.

Tang, H.B., Wang, X.Y., Bowers, J.E., Ming, R., Alam, M. and Paterson, A.H. (2008) Unraveling ancient hexaploidy through multiply-aligned angiosperm gene maps. *Genome Research* 18, 1944–1954.

Upadhyaya, C.P., Venkatesh, J., Gururani, M.A., Asnin, L., Sharma, K., Ajappala, H., *et al.* (2011) Transgenic potato overproducing L-ascorbic acid resisted an increase in methylglyoxal under salinity stress via maintaining higher reduced glutathione level and glyoxalase enzyme activity. *Biotechnology Letters* 33, 2297–2307.

Urbany, C., Colby, T., Stich, B., Schmidt, L., Schmidt, J. and Gebhardt, C. (2012) Analysis of natural variation of the potato tuber proteome reveals novel candidate genes for tuber bruising. *Journal of Proteome Research* 11, 703–716.

Valkonen, J.P.T., Wiegmann, K., Hämäläinen, J.H., Marczewski, W. and Watanabe, K.N. (2008) Evidence for utility of the same PCR-based markers for selection of extreme resistance to *Potato virus Y* controlled by *Rysto* of *Solanum stoloniferum* derived from different sources. *Annals of Applied Biology* 152, 121–130.

van Dijk, J.P., Cankar, K., Scheffer, S.J., Beenen, H.G., Shepherd, L.V.T., Stewart, D., *et al.* (2009) Transcriptome analysis of potato tubers—effects of different agricultural practices. *Journal of Agricultural and Food Chemistry* 57, 1612–1623.

van Dijk, J.P., Leifert, C., Barros, E. and Kok, E.J. (2010) Gene expression profiling for food safety assessment: examples in potato and maize. *Regulatory Toxicology and Pharmacology* 58, S21–S25.

van Dijk, J.P., Cankar, K., Hendriksen, P.J.M., Beenen, H.G., Zhu, M., Scheffer, S., *et al.* (2012) The identification and interpretation of differences in the transcriptomes of organically and conventionally grown potato tubers. *Journal of Agricultural and Food Chemistry* 60, 2090–2101.

van Os, H., Andrzejewski, S., Bakker, E., Barrena, I., Bryan, G.J., Caromel, B., *et al.* (2006) Construction of a 10,000-marker ultradense genetic recombination map of potato: providing a framework for accelerated gene isolation and a genomewide physical map. *Genetics* 173, 1075–1087.

Veilleux, R.E. and De Jong, H. (2007) Potato. In: Singh, R.J. (ed.) *Genetic Resources, Chromosome Engineering, and Crop Improvement*, Vol 3. Boca Raton: CRC Press, Boca Raton, Florida, pp. 17–58.

Velásquez, A.C., Mihovilovich, E. and Bonierbale, M. (2007) Genetic characterization and mapping of major gene resistance to potato leafroll virus in *Solanum tuberosum* ssp *andigena*. *Theoretical and Applied Genetics* 114, 1051–1058.

Velculescu, V.E., Zhang, L., Vogelstein, B. and Kinzler, K.W. (1995) Serial analysis of gene expression. *Science* 270, 484–487.

Voith, M. (2010) BASF wins EC okay for transgenic potato. *Chemical and Engineering News* 88, 18.

Wang, Z., Gerstein, M. and Snyder, M. (2009) RNA-Seq: a revolutionary tool for transcriptomics. *Nature Reviews Genetics* 10, 57–63.

Watanabe, K.N., Kikuchi, A., Shimazaki, T. and Asahina, M. (2011) Salt and drought stress tolerances in transgenic potatoes and wild species. *Potato Research* 54, 319–324.

Waterer, D., Benning, N.T., Wu, G.H., Luo, X.M., Liu, X.J., Gusta, M., *et al.* (2010) Evaluation of abiotic stress tolerance of genetically modified potatoes (*Solanum tuberosum* cv. Desiree). *Molecular Breeding* 25, 527–540.

Watkinson, J.I., Hendricks, L., Sioson, A.A., Vasquez-Robinet, C., Stromberg, V., Heath, L.S., *et al.* (2006) Accessions of *Solanum tuberosum* ssp. *andigena* show differences in photosynthetic recovery after drought stress as reflected in gene expression profiles. *Plant Science* 171, 745–758.

Wei, Q., Wang, Q.Y., Feng, Z.H., Wang, B., Zhang, Y.F. and Yang, Q. (2012) Increased accumulation of anthocyanins in transgenic potato tubers by overexpressing the *3GT* gene. *Plant Biotechnology Reports* 6, 69–75.

Whitworth, J.L., Novy, R.G., Hall, D.G., Crosslin, J.M. and Brown, C.R. (2009) Characterization of broad spectrum Potato Virus Y resistance in a *Solanum tuberosum* ssp. *andigena*-derived population and select breeding clones using molecular markers, grafting, and field inoculations. *American Journal of Potato Research* 86, 286–296.

Wickramasinghe, W.M.D.K., Qu, X.S., Costanzo, S., Haynes, K.G. and Christ, B.J. (2009) Development of PCR-based markers linked to quantitative resistance to late blight in a diploid hybrid potato population of *Solanum phureja* × *S. stenotomum*. *American Journal of Potato Research* 86, 188–195.

Witek, K., Strzelczyk-Żyta, D., Hennig, J. and Marczewski, W. (2006) A multiplex PCR approach to simultaneously genotype potato towards the resistance alleles *Ry-fsto* and *Ns*. *Molecular Breeding* 18, 273–275.

Wu, F. and Tanksley, S.D. (2010) Chromosomal evolution in the plant family Solanaceae. *BMC Genomics* 11.

Yao, C.G., Song, B.T., Liu, J., Wu, C.J., Cheng, Q., Li, D.C., *et al.* (2011) Population improvement of resistance to late blight in tetraploid potato: a case study in combination with AFLP marker assisted background selection. *Agricultural Sciences in China* 10, 1177–1187.

Yang, X., Zhang, X.R., Zhang, M.J., Guo, W.C., Tian, Y.C., Zhong, Q., *et al.* (2011a) Transgenic potato overexpressing the *Amaranthus caudatus* agglutinin gene to confer aphid resistance. *Crop Science* 51, 2119–2124.

Yang, Y., Qiang, X., Owsiany, K., Zhang, S., Thannhauser, T.W. and Li, L. (2011b) Evaluation of different multidimensional LC-MS/MS pipelines for isobaric tags for relative and absolute quantitation (iTRAQ)-based proteomic analysis of potato tubers in response to cold storage. *Journal of Proteome Research* 10, 4647–4660.

Zeevi, V., Liang, Z.B., Arieli, U. and Tzfira, T. (2012) Zinc finger nuclease and homing endonuclease-mediated assembly of multigene plant transformation vectors. *Plant Physiology* 158, 132–144.

Zhang, L.H., Mojtahedi, H., Kuang, H., Baker, B. and Brown, C.R. (2007) Marker-assisted selection of Columbia root-knot nematode resistance introgressed from *Solanum bulbocastanum*. *Crop Science* 47, 2021–2026.

Zhang, N., Si, H.J., Wen, G., Du, H.H., Liu, B.L. and Wang, D. (2011) Enhanced drought and salinity tolerance in transgenic potato plants with a BADH gene from spinach. *Plant Biotechnology Reports* 5, 71–77.

18 Nutritional Characteristics of Potatoes

Duroy Navarre,[1,3]* Aymeric Goyer,[2] Raja Payyavula[3] and Hanjo Hellmann[4]

[1]USDA-ARS, Prosser, USA; [2]Oregon State University, Hermiston Agricultural Research and Extension Center, Hermiston, USA; [3]IAREC, Washington State University, Prosser, USA; [4]Plant Stress Physiology, Washington State University, Pullman, USA

Potatoes are the most consumed vegetable in much of the developed world, and their consumption is growing rapidly in the developing world. Over half of the global production now occurs in developing countries, where since the 1960s, growth in potato production has exceeded that of all other crops (International Potato Center, 2009). Staple foods like potatoes have a unique nutritional importance relative to foods consumed sparingly. Consequently, staple foods should be primary targets for nutritional enhancement because of their anticipated greater return on investment in terms of being able to positively impact a larger percentage of the global population. Because of their high consumption, potatoes are good vehicles to deliver meaningful amounts of dietary phytonutrients.

Providing food security for the growing global population will be challenging, with the United Nations projecting an increase in the global population from 6.8 billion currently to 9.2 billion in 2050. The Food and Agriculture Organization (FAO) projects global agricultural production will need to increase by 70% by 2050 (FAO, 2009). Contributing to this challenge is changing dietary patterns, in which an increasing global middle class consumes a more resource-intensive, Western-style diet; for example, greater meat consumption. Moving to this moderate diet will require 2.4 kg grain equivalent per person per day, about threefold higher than current levels (Poudel *et al.*, 2012). Challenges could also result from acreage taken out of food production and used for biofuels. In this context, it is an important point that potatoes yield more calories per land unit than any other major crop, including grains, rice, and soybean (Nunn and Qian, 2011). This claim for potato is often based on yielding 9.2 million calories per acre, but yield per acre depends on cultivar, environment, and management, and it can be much higher than the yields typically used to calculate calories per acre. For example, in Washington State, the typical family-owned farm growing russet potatoes produces 30 t $acre^{-1}$, or about 26 million calories $acre^{-1}$. Moreover, 40 t $acre^{-1}$ (35 million calories $acre^{-1}$) yields are not uncommon. While these yields will not be achievable everywhere, it does suggest a potential for increased yields elsewhere if crop management is optimized. Also contributing to their role in food security is that potatoes can grow in a wide range of climates and soils, and produce tubers in as quickly as 60 days that have generous amounts of phytonutrients, as described later in this chapter.

* E-mail: roy.navarre@ars.usda.gov

18.1 Basic Tuber Composition

Starch and fiber

Tubers are typically 20% solids and 80% water, although the dry matter can vary from 14 to 37% (Burton, 1966). Dry matter content can decrease along a stem- to bud-end gradient. Starch, a complex carbohydrate, accounts for 70–80% of the dry matter, with an amylopectin:amylose ratio of ~3:1 (Woolfe, 1987). The amylopectin:amylose ratio has a large effect on texture. Amylopectin is highly branched, digested more quickly, holds together better during cooking and forms less dense starch granules, whereas amylose is a straight chain molecule that is not as easily digested and separates easily when cooked. In general, food containing more amylose has a more floury texture when cooked is not digested as quickly and has lower glycemic index values.

Higher-starch potatoes (lower moisture, higher specific gravity) such as the Russets have greater proportions of amylose, reduced processing costs, lower oil absorption, and are used for French fries. The fluffier texture is also preferred by most for baking and mashed potatoes. Waxy potatoes have lower starch content, lower specific gravity, and higher amylopectin:amylose ratios that lessen the amount of sloughing during boiling, making these potatoes better for boiling. The FAO and World Health Organization (WHO) recommend that 55–75% of daily calories be consumed in the form of complex carbohydrates. Potatoes also contain dietarily desirable resistant starch that acts like fiber. Resistant starch increases if cooked potatoes are chilled. Potatoes contain fiber, most of which is insoluble. Eating a potato without the skin will reduce the amount of dietary fiber by over 30%. According to the USDA nutrition database, a large baked potato with skin provides 300 calories and 7 g of fiber, about 28% of the recommended dietary allowance (RDA). Potato starch and fiber content is affected by both genotype and environment (Bach *et al.*, 2012, 2013).

Protein

Potatoes contain ~2% protein on a fresh weight basis, and on a dry weight basis are comparable to cereals. In one study, potatoes were second only to soybean as a source of vegetable protein in the USA (Johnson and Lay, 1974). However, unlike cereals, potatoes have a high lysine content and provide a high-quality protein that is one of the few vegetable proteins that rival that of egg and other animal protein in quality (Woolfe, 1987; Lisinska and Leszczynski, 1989). A combination of potato and egg protein rates among the best nutritional quality proteins of all food proteins, including meat and milk. The major deficiency of potato protein is that it does not contain high amounts of methionine. The soluble proteins of potatoes are primarily of proteinase inhibitor proteins (Pouvreau *et al.*, 2001) that function in defense against insects and small animals, but when cooked are high-quality proteins. The protease inhibitor proteins include protease inhibitors I and II, cysteine protease, aspartate protease inhibitor, potato Kunitz-type protease inhibitor, and potato carboxypeptidase inhibitor. In addition to pest defense, these protease inhibitors may be involved in regulating tuber protein content. Potato proteins include the patatin glycoprotein family (Paiva *et al.*, 1983) that are the major storage proteins (Shewry, 2003). In tubers over 200 g, patatin accounts for ~40% of the protein (Paiva *et al.*, 1983).

The protein efficiency ratio (PER) based on weight gain per unit protein fed (Desborough *et al.*, 1981), averages about 2.3 compared to 1.76, 1.43, 0.77, and 1.3–1.9 respectively, for rice, maize, wheat, and soybean (Leiner, 1977; Jewell *et al.*, 1980). *Solanum* group Phureja and Andigena genotypes have been identified with higher amounts of protein, but yields were poor when crossed with *Solanum tuberosum* (Desborough, 1985). Transgenic potatoes overexpressing the protein amaranth albumin 1 had a 60% increase in total protein (Chakraborty *et al.*, 2010).

18.2 Potatoes and Health

While potatoes have been well regarded as a healthful, important food for hundreds of years in the West and thousands of years in the Andes, over the last decade the role of potatoes in a healthy diet has been disputed by some nutritionists in the USA. Whereas most nutritionists advocate a "consume in moderation" policy, a minority recommends that potatoes be eaten sparingly (Willett and Stampfer, 2003). This change in attitudes seems to be driven in large part by efforts to determine the cause of the alarming surge in obesity and diabetes, with some scientists asserting that carbohydrate consumption is responsible, although this is a highly contentious issue among nutritionists. A well-publicized epidemiological study reported that potatoes of all types, not just French fries and chips, were primary contributors to weight gain (Mozaffarian *et al.*, 2011). Notably, while obesity in the USA has been increasing rapidly (CDC, 2013), potato consumption has undergone a significant decrease, and is less than half of that reported in at least ten other countries (Fig. 18.1). Moreover, the USA and Mexico are the first and second most obese countries (OECD, 2013) yet rank 44th and 105th in potato consumption in the most recent FAO data (FAO, 2009). If, indeed, potatoes are a major contributor to obesity, it is interesting to note that obesity has not been problematic in European countries during periods where potato consumption was far higher than the present day. For example, German annual potato consumption in 1900 was 628 lbs per person and over 300 lbs in the 1950s (Heinzelmann, 2008). Such statistics by no means prove that potatoes do not contribute to obesity, but they are interesting to note and may suggest other factors are at play that are not accounted for in simplistic explanations that potatoes are a primary contributor.

The glycemic index is a measure of how quickly blood sugar increases following consumption of a food. Foods including potatoes, rice, and bread typically have high glycemic index values, which has contributed to a negative perception of potatoes. How potatoes are prepared and what they are eaten with markedly influences glycemic values; some cultivars give much lower values than others (Fernandes *et al.*, 2005). A "modest" association between potato consumption and the risk of type 2 diabetes has been reported (Halton *et al.*, 2006), while a Japanese study reported a diet rich in vegetables including potatoes was associated with a reduced risk (Morimoto *et al.*, 2012). The percentage of the US population with diabetes soared to 8.9% in 2011 (Fig. 18.1c; Center for Disease Control (CDC)). From the 1970s until the mid-1990s, the percentage varied by less than 1%, until beginning a steep rise starting around 1995, about the same time potato consumption started decreasing.

Potatoes contain bioactive compounds, and two human feeding studies with purple potatoes reported positive effects of potato supplementation. One study found pigmented potatoes reduced inflammation and DNA damage in adult males fed 150 g of potato a day for 6 weeks (Kaspar *et al.*, 2011). In another trial, subjects averaging 54 years in age had a significant drop in blood pressure without weight gain after consuming purple potatoes (Vinson *et al.*, 2012). Twenty Canadian breeding lines were evaluated for phenolic content, and extracts mildly increased low-density lipoprotein (LDL) cholesterol uptake in liver HepG2 cells and protected cortical neurons against cell death. Interestingly, the unknown bioactive components were not dependent on potato pigmentation (Ji *et al.*, 2012). Extracts from purple potatoes inhibited prostate cancer cell proliferation in a cell culture study (Reddivari *et al.*, 2007b). Other cell culture studies also reported that potatoes had inhibitory effects on cancer growth (Lee *et al.*, 2004; Madiwale *et al.*, 2012).

Unlike the complexities of interpreting epidemiological studies or determining the perfect diet, assessing the nutritional content of potatoes is straightforward, and is the focus of the rest of this chapter. Some foods are said to provide "empty calories", meaning they provide calories but little nutritive value. Potatoes are a nutrient-dense food; relative to the amount of calories consumed per serving, an equal or greater amount of the recommended daily allowance for various vitamins and minerals is provided. Figure 18.2 shows the percentage of recommended daily values of various nutrients in 100 g of baked potato with skin, which would provide 87 calories, or 4% of the daily value (data from USDA SR-21: http://www.ars.usda.gov/Services/docs.htm?docid=18880). As seen, this portion of potatoes provides a greater amount of various nutrients than it does calories. Furthermore, potatoes also provide carotenoids and phenolics, especially pigmented potatoes. While such compounds do not have recommended daily allowances like vitamins and minerals, they are none the less important for health, as discussed below. Many of the small molecules

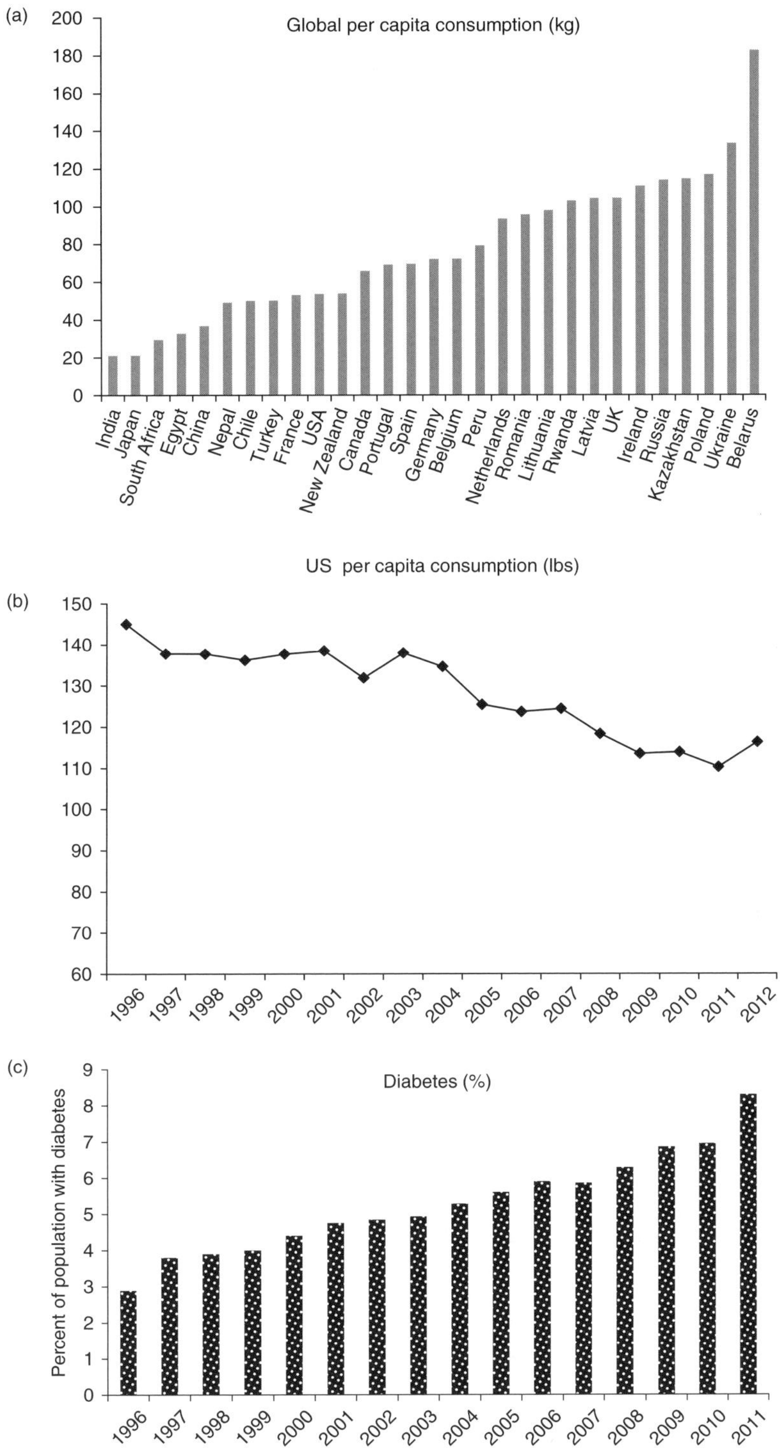

Fig. 18.1. (a) Global per capita potato consumption in 2009. (From FAO.) (b) Per capita potato consumption in the USA. (From USDA Economic Research Service.) (c) Percent of the US population with diabetes, 1996–2011. (From CDC.)

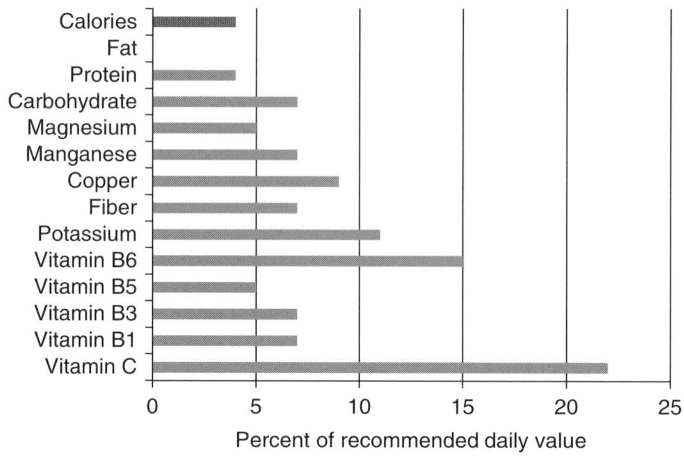

Fig. 18.2. Nutritional content of 100 g of baked potato. Note the amount of various nutrients relative to the amount of calories provided.

in tubers are known to affect health positively and are desirable in the diet (Katan and De Roos, 2004). A new study reports that potatoes and beans provide the most nutrients per dollar out of the 98 vegetables studied (Drewnowski and Rehm, 2013).

18.3 Minerals

Potatoes are a good source of minerals, with the amounts varying by genotype and environment. Typically, a serving of potatoes provides milligram amounts of phosphorus, magnesium, potassium, iron, and sodium, and microgram amounts of copper, manganese, and zinc. A survey of 74 Andean landraces found iron ranged from 29.87 to 157.96 μg g^{-1} DW, zinc from 12.6 to 28.83 μg g^{-1} DW, and calcium content 271.09 to 1092.93 μg g^{-1} DW (Andre *et al.*, 2007a). Location, stage of development, soil type, soil pH, soil organic matter, fertilization, and irrigation influence mineral content. Concentrations of 14 minerals varied markedly in two cultivars grown with different fertilizer regimes, and showed distinct gradients within the tuber, with some higher in the center and others showing stem- to bud-end gradients (LeRiche *et al.*, 2009). Data suggest that potato mineral content can be increased without compromising yield, independent of any yield dilution effect (White *et al.*, 2009).

Potassium (K)

Potatoes are a major source of potassium, an element commonly deficient in the diet. The RDA for K in adults is 4.7 g, but the median intake in the USA is less than 3.3 g. Potassium reduces risk of blood pressure and cardiovascular diseases, along with hypokalemia (low K blood levels), osteoporosis, strokes, inflammatory bowel disease (IBD), and kidney stones (Dyer *et al.*, 1994; Young *et al.*, 1995). K is also important for tuber starch synthesis. K levels in flesh among different genotypes varied from 3 to 8 mg g^{-1} FW (Sanchez-Castillo *et al.*, 1998; Casanas *et al.*, 2002), and decreased from bud end to stem end (Arteca *et al.*, 1980). The flesh contains about 90% of tuber potassium and the skin 10%.

Iron (Fe)

Iron deficiency is one of the most widespread health problems worldwide, especially for women. Like all the minerals, recommended amounts can vary by age, gender, and pregnancy, and is 8–18 mg day^{-1} for adults (IOM, 2011). Fe deficiency can lead to abnormally small red blood cells with reduced hemoglobin, resulting in decreased oxygen supply to tissues and anemia (Alton, 2005). Fe deficiency in pregnant women causes fetal growth retardation or low birth weight. In young children, it affects physical growth, cognitive development, and immunity (Alton, 2005). A wide range of iron content from 1 to 40 μg g^{-1} FW has been reported (Wills *et al.*, 1984; Andre *et al.*, 2007a; Burgos *et al.*, 2007). Potatoes lack phytic acid, a compound that decreases iron bioavailability in cereals, and has relatively high amounts of vitamin C and organic acids that are thought to increase bioavailability.

Calcium (Ca)

Potatoes are a significant source of calcium (RDA 1300 mg), a major constituent in bones and teeth (Dawson-Hughes *et al.*, 1997). Low Ca intake leads to osteoporosis, hypertension, and cardiac arrhythmias (Heaney, 2000; Clusin, 2003). Ca is concentrated in the cortex of the tuber (Bretzloff and McMenamin, 1971). It is more abundant in flesh than in peel, and ranges from 3 to 10 mg 100 g^{-1} FW (Luis *et al.*, 2011; Subramanian *et al.*, 2011). In the Andean cultivars, Ca content ranged from 22.7 to 109 mg 100 g^{-1} DW (Andre *et al.*, 2007a). Wild *Solanum* species vary in tuber calcium (Bamberg *et al.*, 1998).

Phosphorus (P)

Phosphorus, like calcium, is needed for healthy teeth and bones. It is also needed for the synthesis of phospholipids and acts as a buffer to maintain normal acid-base (pH) balance in the body. Phosphorus is abundant (RDA is 700 mg), and so deficiencies are rare. However, the effect of low phosphorus consumption includes anorexia, anemia, muscle weakness, bone pain, rickets, and osteomalacia (softening of the bones) (Lotz *et al.*, 1968; Takeda *et al.*, 2004). Potatoes are a good source of phosphorus, ranging from 30 to 60 mg 100 g^{-1} FW (Randhawa *et al.*, 1984; Sanchez-Castillo *et al.*, 1998).

Zinc (Zn)

Zinc deficiency leads to increased susceptibility to a variety of pathogens, impaired taste and smell, night blindness, diarrhea, and mental disturbance (Walsh *et al.*, 1994; Zalewski, 1996; Shankar and Prasad, 1998). For Zn, the RDA is 7–15 mg, and the tolerable upper limit is 40 mg (Anon., 2001). Potatoes accumulate 8–20 μg g^{-1} DW Zn (Burgos *et al.*, 2007; Subramanian *et al.*, 2011).

Copper (Cu)

Copper is incorporated in several proteins and metalloenzymes that perform essential metabolic functions and stimulate the immune system (Sherman, 1992). Copper in potatoes varies from 0.2 to 12 mg kg^{-1} FW (Randhawa *et al.*, 1984; Luis *et al.*, 2011). The RDA for Cu is 900 μg (IOM, 2011) .

Magnesium (Mg)

Potatoes are a moderate source of Mg, with ranges from 15 to 27 mg 100 g^{-1} FW, of which 90% is accumulated in flesh (Luis *et al.*, 2011; Subramanian *et al.*, 2011). It is involved in enzyme activities and energy metabolism, and is required for the synthesis of DNA and RNA (Wacker and Parisi, 1968; Rubin, 1975). Mg deficiency increases the risk of cardiovascular diseases (Arsenian, 1993). The RDA is 320–420 mg for adults (IOM, 2011)

Vitamins

Vitamins are essential nutrients needed for metabolism and stress tolerance in all living organisms. In contrast to humans and most animals, which often lack the biosynthetic machinery for vitamin biosynthesis, plants are capable of synthesizing nearly all vitamins *de novo*, and thus are a primary source for the vitamins on which we depend (Roje, 2007; Fitzpatrick *et al.*, 2012). While best known for vitamin C, potatoes are also an important source of various B vitamins.

18.4 B Vitamins

Originally, it was assumed that only one kind of B vitamin existed that had a critical role in preserving growth and health and that was required to prevent certain skin lesions, like pellagra in animals and humans (Birch *et al.*, 1935). However, it soon became clear with progressing research that vitamin B actually represents a group of compounds that are now collectively called the "vitamin B complex". B vitamins are classified nowadays as a group of eight water-soluble compounds that are chemically quite distinct, and to which belong vitamin B1 (thiamine), B2 (riboflavin), B3 (niacin), B5 (pantothenic acid), B6 (pyridoxine), B8 (biotin), B9 (folic acid), and B12 (various cobalamins) (Roje, 2007). While none of the B vitamins can be synthesized *de novo* by humans, plants have the necessary enzymatic machinery to synthesize these vitamins, excepting vitamin B_{12} (Fitzpatrick *et al.*, 2012). The following section will summarize the current knowledge about their biosynthesis and functions in plants and animals. It will also address to what extent potatoes represent a major resource for these vitamins, and what the potential is to increase B-vitamin contents in potato tubers as beneficial phytonutrients in the human diet.

Vitamin B_1 (thiamine)

Vitamin B_1, or thiamine, has critical roles in its phosphorylated form, thiamine-diphosphate (TDP; Fig. 18.3a), as a cofactor in carbohydrate and amino acid metabolism (Jordan, 2003; Settembre *et al.*, 2003; Nosaka, 2006), and in response to abiotic and biotic stresses (Gossert *et al.*, 2008). It is also called "aneurin" in reflection of neurological disorders that can occur when deficient (Wang and Yudkin, 1940). In general, severe vitamin B_1 deficiencies cause pleiotropic symptoms in humans, including in the gastrointestinal, muscular, and cardiovascular systems. The disease is commonly described as "beriberi", and affected people become lethargic and fatigued (Lonsdale, 2012). In plants, thiamine appears also to have critical roles in both biotic and abiotic stress responses, since biosynthesis is upregulated on such stresses and appears to be critical for stress adaptation (Wang *et al.*, 2006; Tunc-Ozdemir *et al.*, 2009; Rapala-Kozik *et al.*, 2012).

Thiamine biosynthesis in plants is not fully understood, and current knowledge indicates the pathway is related to that of bacterial organisms (Roje, 2007). Here, precursors include 2-methyl-4-amino-5-hydroxymethylpyrimidine diphosphate and 4-methyl-5-β-hydroxyethylthiazole phosphate, which are condensed by the activity of thiamine phosphate pyrophosphorylase to thiamine monophosphate (TMP). TMP is then further metabolized to the active cofactor by the concerted activities of a TMP phosphatase and a thiamine pyrophosphokinase (Rapala-Kozik *et al.*, 2008, 2012; Goyer, 2010).

The RDA values for thiamine consumption in the USA for adult women and men (ages 18 and older) are 1.1 and 1.2 mg day^{-1}, respectively (Table 18.1) (Otten *et al.*, 2006). The thiamine content in 100 g baked or boiled potato tubers provides ~10% RDA. Interestingly, potato chips have a significantly higher content, about 15% RDA in 100 g. This higher content in potato chips is consistent for all the other B vitamins discussed in this chapter, and is likely due to their dryness in comparison to potato products with high water content.

The first substantial screening of potato germplasm found a range of B_1 from 0.062 to 0.090 mg 100 g^{-1} FW (Augustin *et al.*, 1978; Holland *et al.*, 1996). More recent screening of cultivars or primitive germplasm identified genotypes containing up to 0.132 and 0.232 mg 100 g^{-1} FW of thiamine, respectively (Goyer and Sweek, 2011). Thiamine concentrations vary during tuber development, and are consistently higher in small tubers harvested early in the season than in large tubers harvested later in the season in most of the 54 varieties analyzed. Over a 2.4-fold increase between small and large tubers was observed for some varieties (Goyer and Haynes, 2011).

Stability and broad-sense heritability

Thiamine is one of the few metabolites for which stability and broad-sense heritability have been estimated in potato. One study showed that genetic variation accounted for about 50% of the thiamine variation observed in a group of 39 potato clones grown in four environments, as determined by calculating broad-sense heritability for this group of clones as 0.49 with a 95% confidence interval of 0.21–0.72 (Goyer and Haynes, 2011). There were significant genotype and genotype × environment effects. After accounting for environmental variation, 25 clones

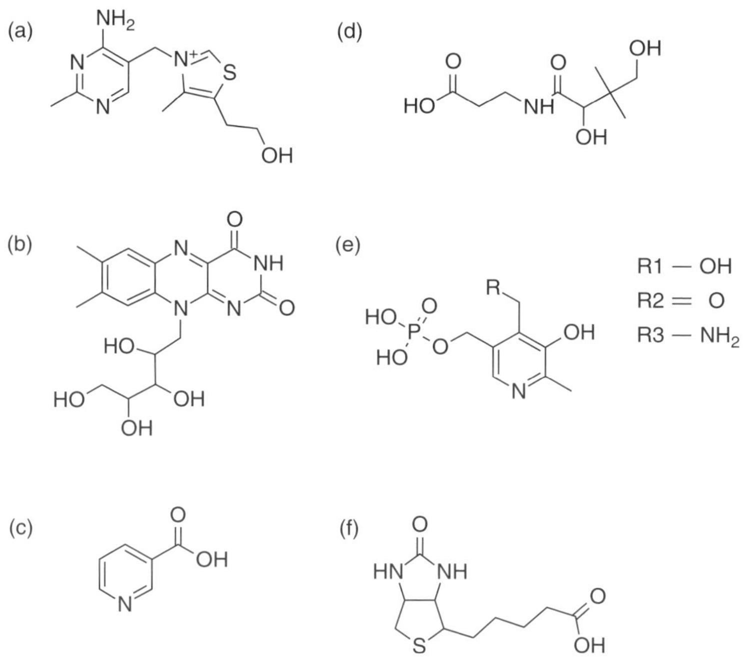

Fig. 18.3. Structures of B vitamins. (a) B_1, thiamine; (b) B_2, riboflavin; (c) B_3, niacin; (d) B_5, pantothenate; (e) B_6, pyridoxine; (f) B_8, biotin.

were unstable across environments. The moderate broad-sense heritability suggests that breeding for thiamine enhancement should be possible, but the identification of clones that perform well across environments is difficult because of the presence of genotype × environment interactions and the instability of many clones.

Postharvest and cooking

Storage at cold temperature does not lead to significant thiamine loss; instead, thiamine concentrations slightly increase in several genotypes when stored for a period of 3–8 months.

There is disparity in the literature regarding the effects of cooking and cooking methods on thiamine in potatoes. The USDA Nutrient Database reports that oven-baked white, russet, and red potatoes contain less thiamine (20% decrease) but more folate (75% increase) than raw potatoes, while boiled potatoes contain more thiamine (32% increase) but less folate (37% decrease). Thiamine losses (14%) in oven-baked potatoes have also been reported (Augustin *et al.*, 1978). Unlike the USDA Nutrient Database, Augustin *et al.* reported a 12% loss of thiamine in boiled (peeled or unpeeled) potatoes (Augustin *et al.*, 1978). Temperature and time of cooking used in these studies may account for these differences.

Vitamin B_2 (riboflavin)

Riboflavin is the precursor for flavin mononucleotide (FMN) and flavin adenine dinucleotide (FAD) (Fig. 18.3b). They are cofactors in various oxidative enzymes, also called flavoproteins, which participate in diverse biochemical reactions such as fatty acid oxidation, biosynthesis of other vitamins (B_6, B_9, B_{12}), or the mitochondrial electron transport chain (Roje, 2007). In plants, riboflavin is also specifically relevant for tetrapyrrole biosynthesis, with major end products being chlorophyll and heme (Hedtke *et al.*, 2012). RDA values for women and men in the USA are 1.1 and 1.3 mg day^{-1}, respectively (Table 18.1) (Otten *et al.*, 2006). Symptoms of riboflavin deficiency, also called "ariboflavinosis", include inflammation and cracks of the mouth and tongue, sore throat, and stunted growth.

Table 18.1. Daily intake recommendations for vitamins B_1, B_2, B_3, B_5, B_6, B_7 and B_9 in adults aged 18 and older. (From USDA National Nutrient Database for Standard Reference, Release 17, and National Institute of Health.)

	EAR	RDA	AI	UL
Vitamin B_1 (thiamine)				
Women	0.9	1.1	–	–
Men	1	1.2	–	–
Vitamin B_2 (riboflavin)				
Women	0.9	1.1	–	–
Men	1.1	1.3	–	–
Vitamin B_3 (niacin)				
Women	11	14	–	35
Men	12	16	–	35
Vitamin B_5 (pantothenate)				
Women	–	–	5	–
Men	–	–	5	–
Vitamin B_6 (pyridoxine)				
Women	1.1–1.3	1.3–1.5	–	100
Men	1.1–1.4	1.3–1.7	–	100
Vitamin B_7 (biotin)				
Women	–	–	0.03	–
Men	–	–	0.03	–
Vitamin B_9 (folate)				
Women	0.32	0.4	–	1
Men	0.32	0.4	–	1

Notes: EAR = estimated adequate intake; RDA = recommended dietary allowance; AI = adequate intake; UL = upper limit; – = no data available; all values given in [mg day^{-1}].

Shortages are common among developing countries, but are rare in Western societies, which have good access to riboflavin-fortified foods (Bates and Powers, 1985; Boisvert *et al.*, 1993; Hoey *et al.*, 2009).

Little is known about riboflavin content variation in different potato cultivars. A study on Australian retail potatoes and their nutrient content over a period of 12 months reported average riboflavin content in the two potato varieties, Pontiac and Sebago, was comparable and consistently measured at 0.3 mg 100 g^{-1} edible portion, which the authors defined as 85% of purchase weight (Wills *et al.*, 1984). Processed potato appeared to be a poor source for this vitamin, as 100 g of cooked or baked potatoes contained only around 2–4% (~0.02–0.04 mg 100 g^{-1}) of the recommended daily intake, suggesting a significant amount was lost through cooking. Because the overall amount of riboflavin is comparably low, breeding efforts may be desirable to enhance content in potatoes. An additional reason to enhance B_2 content in plants is given by recent findings from work on another nightshade, tobacco (Wu *et al.*, 2010). Transgenic tobacco plants overexpressing the riboflavin biosynthesis enzyme, lumazine synthase, had increased FMN and FAD levels. The corresponding tobacco plants grew better, had higher amounts of jasmonic acid, and showed upregulation of defense responses, leading to tobacco that was less susceptible to pathogen infection (Wu *et al.*, 2010). These findings point out that elevating riboflavin levels is technically feasible, and potentially may lead to superior plants with improved phytonutrient content, better growth, and enhanced performance when exposed to pathogens. Considering this, transgenic approaches or breeding efforts to increase B_2 content in potato are appealing. Unfortunately, no data about vitamin B_2 variation among a wider variety of potato germplasm are available, which clearly is a critical gap in knowledge.

Vitamin B_3 (niacin)

The term "niacin" includes nicotinamide or nicotinic acid, two pyridine derivatives that carry either a carboxamide or carboxyl group at the pyridine 3′-position (Fig. 18.3c). Both are precursors for the enzymatic cofactors NAD^+ and $NADP^+$ that are required for various redox reactions in context with carbohydrate and fatty acid metabolism (Wahlberg *et al.*, 2000). Niacin also plays roles in cellular signaling reactions, ADP-ribosylation, Ca signaling, or cell-cycle control (Lepiniec *et al.*, 1995; Doucet-Chabeaud *et al.*, 2001; Virag and Szabo, 2002; Hassa *et al.*, 2006; Monks *et al.*, 2006; Shiotani *et al.*, 2006; Pollak *et al.*, 2007; Adams-Phillips *et al.*, 2008; Lamb *et al.*, 2011). Consequently, deficiencies in humans have broad health implications, including mental disorders (poor concentration, apathy, depression) (Lanska, 2010), and pellagra, a disease where patients suffer from dementia, dermatitis, and diarrhea (Kohn, 1938). Niacin deficiencies can be severe in regions where people rely on a maize-based diet, which requires an elaborate processing known as "nixtamalization" before stored vitamins, amino acids, and proteins are accessible to the body.

The biosynthetic pathway of vitamin B_3 is not fully resolved in plants. An anticipated path runs from asparagine through α-aminosuccinate via the quinolinate and pyridine nucleotide cycles (Katoh *et al.*, 2006; Roje, 2007), and niacin biosynthetic activities have been reported recently for potato tuber (Katahira and Ashihara, 2009). Mammals can make vitamin B_3 from tryptophan in the liver (and some species can also produce it in the kidney) (Kurnasov *et al.*, 2003); however, humans cannot synthesize tryptophan *de novo* and need to have either the amino acid or the vitamin supplied with their food.

The RDA values for women and men are 14 and 16 mg, respectively (Otten *et al.*, 2006) (Table 18.1). A 100 g boiled or baked potato provides ~10% of the RDA (Table 18.2). However, French fries appear to be a relatively poor source (~ 7% RDA 100 g^{-1}), while chip snacks have a comparably high content (~15% RDA 100 g^{-1}; Table 18.2). Niacin is highly stable as a nutrient and comparably resistant to heat, light, air, acid, and alkali (Richardson, 1990).

Little information is available for niacin content variation in different potato germplasm; those that exist focus only on a few cultivars with no major variation (e.g. Pontiac and Sebago with 1.2 and 1.3 mg 100 g^{-1} edible portion; Wills *et al.*, 1984). No transgenic approaches have been published for generating potatoes with modified enzyme activities in the B_3 biosynthetic pathway, and only very little knowledge is established for plants in general. However, the few existing reports indicate that plants are decidedly sensitive to any deviations from normal B_3 biosynthetic activities. For instance, loss-of-function *Arabidopsis thaliana* mutants affected in the biosynthesis of quinolinate have reduced NAD^+ contents and senesce early (Schippers *et al.*, 2008). Upregulation of NAD^+ synthetase, a key enzyme in the pyridine nucleotide cycle, led to increased niacin and nicotinamide levels in Arabidopsis, but surprisingly not NAD^+. Again, however, the transgenic plants showed early senescence and also early flowering phenotypes, and were affected negatively in seed development (Shinnosuke *et al.*, 2010).

The importance of niacin in human nutrition makes increasing vitamin B_3 content in potato a beneficial trait for this crop, especially since nixtamalization is not required to have good access to the phytonutrients; however, the clear negative impacts on plant growth and development strongly suggest that it will be difficult to generate plants with elevated vitamin B_3 levels by transgenic approaches without significantly affecting other biological processes. It might be advantageous to investigate variation in existing germplasm to establish marker-assisted programs as an alternative approach.

Vitamin B_5 (pantothenate)

Vitamin B_5 (Fig. 18.3d) is the precursor for 4′-phosphopantetheine, a moiety of the coenzyme A (CoA) that serves as a prosthetic group to the acyl carrier protein and participates as an enzymatic cofactor in various biochemical reactions related mostly to lipid and fatty acid metabolism (Rebeille *et al.*, 2007). Pantothenate is synthesized *de novo* in bacteria and plants, and biochemical studies indicate similar pathways are used (Coxon *et al.*, 2005; Roje, 2007). Immediate precursors of pantothenate in both organisms appear to be β-alanine and l-pantoate, which are condensed to vitamin B_5 with ATP consumption by pantothenate synthetase (Chakauya *et al.*, 2008; Jonczyk *et al.*, 2008). Insufficient pantothenate appears to be uncommon among humans and is rarely reported, but symptoms may include burning foot syndrome, irritability, fatigue, and apathy (Tahiliani and Beinlich, 1991; Bender, 1999). Based on its critical function in lipid and fatty acid metabolism, it is not surprising that loss of the biosynthetic pathway is lethal, as shown by Arabidopsis deficient in pantothenate synthase that become arrested in embryo development (Jonczyk *et al.*, 2008).

Likely because malnutrition with vitamin B_5 is not a common problem, limited data about adequate intake are available, which is estimated to be 5 mg day^{-1} for adult women and men (Table 18.1). A 100 g serving of baked or boiled potatoes provides around 10% of the RDA (Table 18.2). Variation among potato germplasm is unknown. Recent findings showed that genetic engineering of the pantothenate content was possible in plants (Chakauya *et al.*, 2008). However, the vitamin B_5 levels in transgenic oilseed rape (*Brassica napus*) appear to correlate rather negatively with triacyl-glyceride

Table 18.2. B_1, B_2, B_3, B_5, B_6, B_7 and B_9 vitamin contents in selected processed potato foods. First value: [mg vitamin/100 g product], second values for B_1 B_2, B_3, B_6, and B_9: % RDA adult women/men, and for B_5 and B_7 % AI adult women/men. (From USDA National Nutrient Database for Standard Reference, Release 17. Note that B_7 is not included in this database. B_7 is from Staggs *et al.*, 2004.)

Vitamin	Fast food; French fried in vegetable oil	Snacks; potato chips	Baked, flesh and skin	Baked, flesh	Boiled, cooked in skin and flesh	Boiled, cooked in flesh w/o skin
B_1	0.08 (7.27/6.67)	0.17 (15.45/14.17)	0.067 (5.45/5)	0.105 (10/9.17)	0.106 (10/9.17)	0.098 (9.09/8.33)
B_2	0.04 (3.6/3.3)	0.197 (17.9/16.4)	0.048 (4.3/4)	0.021 (1.9/1.7)	0.02 (1.7/1.6)	0.019 (1.7/1.6)
B_3	0.28 (2/1.8)	3.83 (27.4/23.9)	1.348 10.1/8.8)	1.395 (9.9/8.7)	1.439 (10.3/9)	1.312 (9.4/8.2)
B_5	0.5 (10/10)	0.4 (8/8)	0.38 (7.6/7.6)	0.56 (11.2/11.2)	0.52 (10.4/10.4)	0.51 (10.2/10.2)
B_6	0.36 (24–27.69/ 21.18–27.69)	0.78 (52–60/ 45.88–60)	0.354 (20.7–23.85/ 18.24–23.85)	0.301 (20–23.08/ 17.65–23.08)	0.299 (20–23.08/ 17.65–23.08)	0.269 (18–20.77/ 15.88–20.77)
B_7	–	0.05 (167/167)	–	–	–	–
B_9	0.03 (7.5/7.5)	0.075 (18.75/18.75)	0.026 (6.5/6.5)	0.009 (2.25/2.25)	0.01 (2.5/2.5)	0.009 (2.25/2.25)

contents (Chakauya *et al.*, 2008); therefore, it may not be desired to increase pantothenate levels in plants.

Vitamin B_6 (pyridoxine)

Vitamin B_6 is the generic term for a group of three chemically related compounds, pyridoxine (PN), pyridoxal (PL), and pyridoxamine (PM), which only differ in a variable group at the 4′ position. PN carries a hydroxyl group, while PL and PM have an aldehyde and an amino group, respectively (Fig. 18.3e). All three B_6 derivatives can be phosphorylated at their 5′ position, which is a required step to convert them to their biologically active forms as cofactors in enzymatic reactions. Pyridoxal-5-phosphate (PLP) is thought to be the primary isoform (Amadasi *et al.*, 2007).

Vitamin B_6 is required for more than 140 biochemical reactions in the cell that are related mainly to amino acid and carbohydrate metabolism (Percudani and Peracchi, 2009), but it also plays a role in the biosynthesis of other vitamins like B_8, B_9, and B_{12} (Battersby and Leeper, 1998; Basset *et al.*, 2004; Pinon *et al.*, 2005). Because of its versatile functions as a cofactor, it is not surprising that the vitamin is critical for human health. Insufficient supply of the vitamin can lead to neurological disorders, diabetes, and skin diseases like pellagra (for a comprehensive overview on vitamin B_6 and human health issues see Hellmann and Mooney, 2010). Noteworthy is that the vitamin is also an antioxidant that rivals other vitamins like C and E in its antioxidative capacity (Ehrenshaft *et al.*, 1998; Bilski *et al.*, 2000).

Two different biosynthetic pathways are known in plants. One is the "salvage pathway" that allows "recycling" or conversion of B_6 vitamers to their active phosphorylated forms, and which depends on the concerted activities of vitamin B_6 dehydrogenases and kinases (Denslow *et al.*, 2005; di Salvo *et al.*, 2011). The second *de novo* biosynthesis pathway has only recently been characterized in plants (Tambasco-Studart *et al.*, 2005), but since then has been studied intensively by several groups (Chen and Xiong, 2005, 2009; Wagner *et al.*, 2006; Leuendorf *et al.*, 2010). It employs a pentose-phosphate (either ribulose-5-phosphate or ribose-5-phosphate), a triose-phosphate (either glycerinaldehyde-3-phosphate or dihydroxyacetone-phosphate), and glutamate to synthesize PLP. The pathway depends on two enzymes, PDX1 and PDX2 (for pyridoxine biosynthesis 1 and 2), which assemble to a multimeric PLP synthase complex of 12 PDX1 and 12 PDX2 proteins (Strohmeier *et al.*, 2006).

A recent study indicated that B_6 deficiency was relatively common in the USA (Morris *et al.*, 2008). Potato is a very good source of vitamin B_6, and analysis of B_6 content in processed food demonstrates that the vitamin can also withstand high temperatures like baking or cooking. USDA nutritional values indicate that 100 g of baked or boiled potatoes provide ~20% of the RDA values for adult women and men (1.3–1.5 and 1.3–1.7 mg day^{-1}, respectively; Table 18.1) (Otten *et al.*, 2006). A recent analysis of vitamin B_6 variation among different genotypes of immature (baby) potatoes showed a 30% variation from 16.9 μg vitamin B_6 g^{-1} DW (PORO07PG63-1) to 22 μg (Ruby Crescent). In addition, analysis of mature tuber genotypes revealed a similar degree of variation from 18.6 μg (PORO07PG63-1) to 27 μg (Clearwater Russet) (Mooney *et al.*, 2013). This study underscores not only the existing variation of vitamin B_6 in different cultivars, and that potato is a very good nutritional resource for vitamin B_6, but also the potential to use these genotypes for breeding efforts to improve phytonutrient content in potato. Increasing vitamin B_6 content in plants by transgenic approaches has been demonstrated recently by several groups (Chen and Xiong, 2009; Leuendorf *et al.*, 2010; Raschke *et al.*, 2011). Interestingly, increased B_6 content also correlated positively with seed size (Raschke *et al.*, 2011), and it would be interesting to see whether similar results could be accomplished with potatoes. Based on the comparably easy transformation of potato, it would be practical to generate transgenic plants with *dominant* traits that lead to elevated or reduced B_6 contents; this would also bypass the comparably difficult genetic situation of this crop. An interesting aspect about B_6 is its involvement in α-glucan phosphorylase activity, a PLP-dependent key enzyme in starch degradation (Mori *et al.*, 1991). It would be informative to investigate whether reduced B_6 content in the tuber can affect low-temperature sweetening of the tubers positively under long-term storage conditions.

Vitamin B_8 (biotin)

Biotin (Fig. 18.3f) is only reluctantly considered to be a vitamin (Food and Nutrition Board, 1998), and the nomenclature for this compound is inconsistent as some sources refer to it as B_7, B_8, or H (Roje, 2007; Bailey *et al.*, 2010; Fitzpatrick *et al.*, 2012). It participates as a cofactor in carboxylation reactions of some enzymes involved in carbohydrate and fatty acid metabolism (Lin and Cronan, 2011) and in cell signaling and biotinylation of histones via biotinidase (Kobza *et al.*, 2005; Zempleni, 2005). Like pantothenate, vitamin B_8 deficiency is relatively rare in humans, and occurs mostly in individuals that carry genetic defects in the biosynthetic pathway or biotin metabolism, as well as sometimes in women during pregnancy (Said, 2002, 2012). The recommended AI value for adults over 18 years is 0.03 mg day^{-1}, according to the National Institute of Health. Data about the extent of variation in potato germplasm are not available; however, limited data from chips show they contain high concentrations (167% of AI in 100 g; Table 18.2) (Staggs *et al.*, 2004), which would indicate that potato is an excellent dietary source. Interestingly, overexpression of biotin-binding proteins is currently discussed as a method to increase resistance against potato tuber moth (*Phthorimaea operculella*) (Meiyalaghan *et al.*, 2005). In some areas, the insect represents a severe pest that damages foliage and tubers in the field and in non-refrigerated stores, and causes significant loss in yield (Nault *et al.*, 2001; Basavaraju *et al.*, 2009). Experiments in tobacco showed that overexpressed biotin-binding proteins in the vacuole resulted in a 100% lethality of feeding larvae (Murray *et al.*, 2010), likely by interfering with the insect's metabolism on consuming high amounts of these biotin-binding proteins. Since the insect cannot synthesize biotin, biotin-binding proteins will, over time, deplete biotin availability, causing death of the larvae. The plant is not affected, as both proteins are retained in the vacuole, and thus do not interfere with the plant's biotin metabolism. It will be interesting to follow this approach in the future and see whether it will be used in potatoes.

Vitamin B_9 (folate)

Folate is involved in DNA biosynthesis, the methylation cycle (synthesis of methionine and the universal methyl group donor, *S*-adenosylmethionine), and the metabolism of iron–sulfur clusters (Cossins and Chen, 1997; Scott *et al.*, 2000; Waller *et al.*, 2010). Folate has a crucial role in photorespiration, a process unique to plants (Cossins and Chen, 1997; Scott *et al.*, 2000).

Because of their high consumption, potatoes are important sources of folate. Brussaard *et al.* (1997) reported that potatoes were the most important vegetable source of folate in the diet, supplying 10% of the total folate intake in the Netherlands (Brussaard *et al.*, 1997). Another study estimated that potatoes were the third most important overall source of folate in the Dutch diet, providing 7% of the total folate intake (Konings *et al.*, 2001). In Norway, potatoes provided 9–12% of the total folate intake (Brevik *et al.*, 2005), while in Finland, potatoes supplied ~10% of the total folate intake (Alfthan *et al.*, 2003). Analysis of a Greek population found increased consumption of potatoes was associated with decreased risk of low serum folate (Hatzis *et al.*, 2006).

Folate screening found a range of 12–37 μg 100 g^{-1} FW in cultivars (Holland *et al.*, 1996; Vahteristo *et al.*, 1997). More recently, systematic screening of potato cultivars, advanced breeding lines, and primitive and wild species extended the range of folate concentrations (Goyer and Navarre, 2007; Goyer and Sweek, 2011). Primitive cultivated species contained up to 41.6 μg 100 g^{-1} FW of folate, and the wild species, *Solanum boliviense*, contained 115 μg 100 g^{-1} FW of folate, which was more than double the amount found in any modern variety. Thus, this may be useful material to integrate into potato breeding programs.

Effect of tuber enlargement and maturity

Folate content was lower in small tubers harvested early in the season compared to large tubers harvested later in the season (Goyer and Navarre, 2009), with decreases as high as 3.4-fold.

However, when small and large tubers were harvested from the same plant under the same environmental conditions, folate concentrations were not significantly different, suggesting that tuber maturity, but not size, was the main factor responsible for changes in folate content.

Postharvest and cooking

Losses of folate from 17 to 40% on a dry weight basis during an 8-month storage period were reported by Augustin *et al.* (Augustin *et al.*, 1978), while up to 87% increase in folate concentrations over a 7-month period were reported in some genotypes (Goyer and Navarre, 2007). This discrepancy may be due to differences in storage conditions (e.g. humidity), genotypes used, and the physiological stage of the tubers (e.g. dormant versus sprouting).

Loss of folate after boiling has been reported in several studies (Augustin *et al.*, 1978; Konings *et al.*, 2001; McKillop *et al.*, 2002), with overall loss percentage between 19 and 25 percent. An overall 29% loss of folate in oven-baked potatoes, with loss as high as 52%, was reported, but the overall retention of folate after cooking exceeded 70% (Augustin *et al.*, 1978). Thermal processing can increase the digestibility of proteins and carbohydrates, and therefore the release of folates from the food matrix.

Forms and bioavailability

Raw potato tubers contained 21 μg 100 g^{-1} FW of 5-methyl-tetrahydrofolate (THF), 3 μg 100 g^{-1} FW of THF, and traces of the oxidation product, 10-formyl-folic acid (Vahteristo *et al.*, 1997). More than 90% of potato folate was polyglutamylated and 95% of folates were present as the 5-methyl-THF derivative, with the rest composed of 10-formyl-folic acid and folic acid, (Konings *et al.*, 2001). Polyglutamates must be hydrolyzed to monoglutamylated forms by folate deconjugase before intestinal absorption, but it is controversial whether polyglutamates are less bioavailable than monoglutamates (McNulty and Pentieva, 2004).

In addition to cooking, other factors affecting the bioavailability of folate include the instability of labile folate derivatives during digestion, the food matrix, the presence of food constituents that enhance folate stability, and the efficiency of deconjugation of polyglutamylated folate to monoglutamates for normal absorption in the small intestine. Folate covalently bound to the food matrix must be released before absorption by the intestine. It is not clear how much matrix-bound folate is in potato. The addition of protease and amylase did not increase folate content in potato significantly (Konings *et al.*, 2001), but we found that protease treatment followed by amylase and conjugase resulted in folate values ~20% higher, indicating that a significant amount of protein-bound folate was present in tubers (Goyer and Navarre, unpublished data).

Bioavailability is influenced by the presence of certain plant molecules, some of which protect folate from degradation during processing, cooking, and digestion. Folate bound to folate-dependent proteins greatly improves folate stability (Rebeille *et al.*, 1994), as do antioxidants such as ascorbic acid (McNulty and Pentieva, 2004). Potatoes can contain high amounts of vitamin C and other antioxidants, as detailed later in this chapter.

Vitamin B summary

As the information listed above shows, potatoes are a good source for the B vitamins. Data on processed potatoes indicate tubers contain either modest (B_1, B_3, B_5) or even high (B_6, B_7) amounts of the vitamins, with the exception of vitamins B_2 and B_9, which appear to be only poorly available in potato. Vitamin B_7 (biotin) clearly had the highest values, but surprisingly little data are currently available for this vitamin in processed potato.

Potato chips consistently have the highest content of the vitamins shown in Table 18.2, probably due to the low moisture content of chips, which effectively concentrates phytonutrient content relative to fresh potatoes. Consequently, baked chips with reduced fat and salt could be promoted as a comparably healthy snack food.

Given the RDA values for the different vitamins in combination with their content in processed potatoes, a major effort could be in focusing on increasing B_2 and B_9 contents in potato tubers, because deficiencies of these two vitamins are a current problem in human populations. It has been proven that their content can be increased through bioengineering without broadly affecting plant biology. Startlingly little work has been done to investigate vitamin content variation among potato germplasm for these and most other vitamins. This leaves one to speculate about what potential there is to enhance vitamin B uptake by breeding or choosing different extant cultivars. A reasonable approach would be to generate detailed vitamin profiles in a wide range of germplasm to understand the potential of breeding efforts for enhancing their vitamin B content. The efficiency of this approach has already been demonstrated for vitamins B_1 and B_6, but is missing for other B vitamins. It may also be of interest for the industry to use this information for consumers, to let them know that a specific consumed potato variety is very rich in these beneficial vitamins and to help consumers understand more fully the nutritional benefits of potatoes.

Another poorly investigated aspect is to engineer genetically the biosynthetic pathway of B vitamins in potato or to establish proof-of-concept that changing B-vitamin content in this crop is possible. Such attempts would provide critical knowledge about the potential advantages and disadvantages of altered vitamin biosynthesis in potato, verify the potential of biofortification, and show how this affects plant development and performance under normal and stressful conditions. Given the current focus in Western societies about "healthy foods", providing more detailed information about the B vitamins in potato would be beneficial for the processing industry as well as for consumers.

18.5 Vitamin C

Potatoes may be best known nutritionally for providing high amounts of two nutrients, vitamin C and potassium. The RDA for vitamin C is 75–90 mg, and plants are the primary dietary source. A red-skinned potato (173 g) provides about 36% of the RDA according to the USDA. Vitamin C deficiency causes scurvy, which in severe cases is typified by loss of teeth, liver spots, and bleeding. Vitamin C is an important cellular antioxidant and is a cofactor for numerous enzymes.

Like many antioxidants, vitamin C concentrations in plants are subject to environmental control, and perhaps consequently are quite variable. Some genotypes have more consistent vitamin C concentrations than others across multiple years or when grown in different locations (Love *et al.*, 2004). This study of 75 genotypes found concentrations ranged from 11.5 to 29.8 mg 100 g^{-1} FW. Analysis of 33 cultivars grown in three locations around Europe found 13–30.8 mg 100 g^{-1} FW (Dale *et al.*, 2003).

Vitamin C levels decrease rapidly during cold storage of potatoes, and can decrease 60% (Keijbets and Ebbenhorst-Seller, 1990). Vitamin C in the cultivar Bintje decreased when stored at 7°C, but increased at 16 and 28°C (Linnemann *et al.*, 1985). Conversely, in a more recent study, levels decreased at the same rate in Ranger Russet stored at 4 or 32°C, but decreased less rapidly when stored under reduced oxygen (Blauer *et al.*, 2013). The decrease in cold storage varies by genotype, from 20 to 60% among 33 genotypes, leading authors to suggest that breeding efforts to increase vitamin C should focus on post-storage content more than on concentrations at harvest (Dale *et al.*, 2003). Another study examined the effect of cold storage on vitamin C content in 12 potato genotypes after 2, 4, and 7 months (Kulen *et al.*, 2013). All showed a substantial loss after 4 months. Interestingly, several of the genotypes did not show a significant loss after 2 months of cold storage. It would be of interest to know whether those potatoes showing no loss after 2 months would consistently show no loss in repeated experiments from different years and locations, because a genotype capable of showing no loss after months of storage would be highly desirable.

Potatoes peeled, blanched, and then fried lost only 10% of their vitamin C after 6 months of storage at −18°C (Tosun and Yücecan, 2008). Vitamin C losses during cooking can be substantial, but need not be. Skin-on potatoes prepared using microwaving, steaming, baking, and boiling had a negligible loss of vitamin C (Navarre *et al.*, 2010).

Wounding increases vitamin C levels. Potatoes stored for 2 days following slicing or bruising underwent a 400% increase in vitamin C in sliced tubers, but a 350% decrease in bruised tubers (Mondy and Leja, 1986). Vitamin C increased in fresh-cut potatoes stored in air, but decreased when stored frozen or under modified atmosphere (Tudela *et al.*, 2002b).

18.6 Glycoalkaloids

Potatoes and many other *Solanaceae* contain glycoalkaloids (GAs), compounds that can be toxic when present in too high a concentration, causing vomiting, nausea, and other ill effects (McMillan and Thompson, 1979; Hopkins, 1995). Most of the effects of GAs are due to their disruption of cell membranes and cholinesterase inhibition. They occur in higher concentrations in fruits, leaves, and sprouts than in tubers, with GA concentrations of 18 g kg^{-1} FW reported in sprouts (Valkonen *et al.*, 1996). In the USA, potatoes must contain less than 20 mg 100 g^{-1} FW of GAs, but the guidelines established in other countries vary (Wilson, 1959).

GAs are steroidal alkaloids with a heterocyclic nitrogen, and a C27 steroid conjugated to a sugar, typically a tri- or tetrasaccharide. GAs are derived from the mevalonate pathway via cholesterol (Johnson *et al.*, 1963; Heftmann, 1983), but the full pathway has not yet been worked out (Ginzberg *et al.*, 2012). The heterocyclic nitrogen is derived from arginine (Kaneko *et al.*, 1976). Glycosylation steps are involved in GA biosynthesis, and some of the glycosyltransferases have been identified (Stapleton *et al.*, 1991; Zimowski, 1991; Moehs *et al.*, 1997; McCue *et al.*, 2007, 2011). Recently, several genes in the pathway were identified, and it was shown that silencing of GAME4 greatly reduced GA accumulation in tubers (Itkin *et al.*, 2013).

Interestingly, recent studies show that some GAs can have health-promoting effects, such as inhibition of mice sarcoma tumors by a solamarine (Kupchan *et al.*, 1965), whereas solasodine may protect against skin cancer (Cham, 1994). Tomatine, solanine, and chaconine inhibited growth of human colon and liver cancer cells in cell culture assays (Lee *et al.*, 2004; Friedman *et al.*, 2005). Tomatine was effective against cancer when provided in a feeding study using rainbow trout (Friedman *et al.*, 2007). Tomatidine is a chemosensitizing agent, increasing the effectiveness of cancer chemotherapy (Lavie *et al.*, 2001). Likewise, solamargine increased the susceptibility of human lung cancer cell lines to anticancer drugs (Liang *et al.*, 2008). In addition to chemopreventive effects, GAs can boost the immune response. Mice fed GAs were more resistant to Salmonella infection (Gubarev *et al.*, 1998), while tomatine potentiated the mice immune response to vaccines (Rajananthanan *et al.*, 1999). GAs are reported to inactivate herpes viruses (Chataing *et al.*, 1997).

Future directions for glycoalkaloid research

These findings demonstrate that assessing the value of potato GAs is complex. GAs contribute to pest and pathogen resistance, so reducing the amount in potatoes can lead to plants that are more susceptible to pests and require more pesticide use. Although not easily accomplished with the current state of the art, it may be desirable to have high concentrations of GAs in foliage but low amounts in tubers to provide at least foliar pest resistance without impacting food safety. However, GAs are also involved in tuber resistance to various pests, so even plants with normal levels in leaves but lower amounts in tubers may still face increased pest pressure. Solanine and chaconine typically make up over 90% of the total GA complement of commercial cultivars, with chaconine often more abundant than solanine (Griffiths *et al.*, 1997; Sotelo and Serrano, 2000). However, potatoes can contain a much greater diversity of GAs because wild species and primitive germplasm likely contain over 100 different GAs (Friedman and McDonald, 1997; Shakya and Navarre, 2008; Mweetwa *et al.*, 2012). Such diversity offers opportunities to develop new cultivars with desired types and quantities of GAs. Some GAs have markedly less toxic effects than others, and if they have health-promoting effects or contribute to pest resistance, then they might be more desirable to have in tubers than chaconine. In general, solanidanes are more toxic than spirosolanes. Chaconine is more toxic than solanine, and both become less toxic with the loss of sugars, with the aglycone the least toxic (Friedman and McDonald, 1997). Friedman suggests replacing solanidine and chaconine with tomatine (Friedman, 2002), which also has health-promoting properties and is less toxic. The predominance of solanine and chaconine in commercial cultivars is likely due to the relatively small percentage of the available potato germplasm utilized in breeding programs. In order to identify what the ideal GA profile would be for potatoes, more medical knowledge is needed. Likewise, more knowledge is needed about the consequence on crop yield of eliminating or reducing GAs.

18.7 Tuber Phenylpropanoids

Phenylpropanoids are a diverse group of plant secondary metabolites that number in the tens of thousands and include phenolic acids, flavonols, and anthocyanins. Many phenylpropanoids have health-promoting effects, including on longevity, mental acuity, cardiovascular disease, eye health, and many others (Parr and Bolwell, 2000; Manach *et al.*, 2004; Scalbert *et al.*, 2005). Phenolics are metabolized by digestive and hepatic enzymes, by intestinal microflora, and have a widely varying bioavailability (Manach *et al.*, 2004). Phenylpropanoids are the major source of dietary antioxidants. Consumers have become increasingly aware of phenylpropanoids due to high-profile news coverage, such as about the health benefits of antioxidants, green tea, coffee, red wine and resveratrol, and the general perception that colored fruits and vegetables are desirable in the diet. Perhaps less widely understood by consumers is that potatoes are an important source of phenylpropanoids. Indeed, potatoes were the third largest contributor of antioxidants in the American diet, trailing only oranges and apples (Chun *et al.*, 2005). This study used readily available potatoes from grocery stores, so they were almost certainly white potatoes. This is an important point, as it shows that even standard white potatoes are an important source of antioxidants, but it also highlights the potential of potatoes to contribute even greater amounts of antioxidants, because pigmented potato genotypes are available that have much higher amounts than white potatoes.

A slew of recent human feeding trials and cell culture studies have shown health-promoting effects of pigmented potatoes that are likely due at least in part to phenylpropanoids. However, the major phenylpropanoid in most potatoes is chlorogenic acid, a colorless compound, as are other tuber hydroxycinnamic acids. Thus, it is important to emphasize that phenolics and strategies to increase phenolics are also relevant for white-fleshed cultivars, which are the consumer preferred type of potato in many countries, whereas yellow potatoes are more consumed in others. Red- or purple-flesh potatoes might not be as readily accepted by consumers as white or yellow potatoes, and are currently a niche market.

Historically, assessment of the nutritional benefits of potatoes has focused only on the most commonly grown potato cultivars, and it has not been fully appreciated by either nutritionists, consumers, and perhaps not even by many plant biologists, exactly how much natural variation in vitamin and phytonutrient content can vary among different cultivars of the same species. This greatly complicates slapping a simple, generic food label on fruits and vegetables that lists a constant amount of vitamins or minerals for a crop, but the facts are that amounts can vary greatly by cultivar, and are also influenced strongly by environment. Thus, phytonutrient amounts in potatoes and other vegetables are moving targets, not fixed amounts. A survey of Andean potato landraces found an 11-fold variation among 74 genotypes in total phenolics that correlated strongly with total antioxidant capacity (Andre *et al.*, 2007a). Twelve pigmented Chilean landraces evaluated for total phenolics were found to have 8- and 11-fold more, respectively, than Desiree and Shepody, common light yellow or white potato cultivars (Ah-Hen *et al.*, 2012).

We have analyzed over 500 genotypes and found over a 15-fold difference in the amount of phenolics. Among the white cultivars we have screened, Russet Norkotah has high amounts of total phenolics, about 3.5–4 mg g^{-1} DW, whereas the typical white potato has about 2 mg g^{-1} DW. Several purple or red potato lines had over 15 mg g^{-1} DW (Fig. 18.4). The potato with the highest amounts of total phenolics that we have analyzed, and to our knowledge by far the highest ever reported in potatoes, is RN27.01 from the Phureja group, which had over 49 mg g^{-1} DW of total phenolics (Pillai *et al.*, 2013). This finding is partly mitigated by the fact that this potato has unusually low solids.

We compared the amount of antioxidants (ORAC) in two common white potato cultivars, Russet Burbank and Norkotah Russet, and one

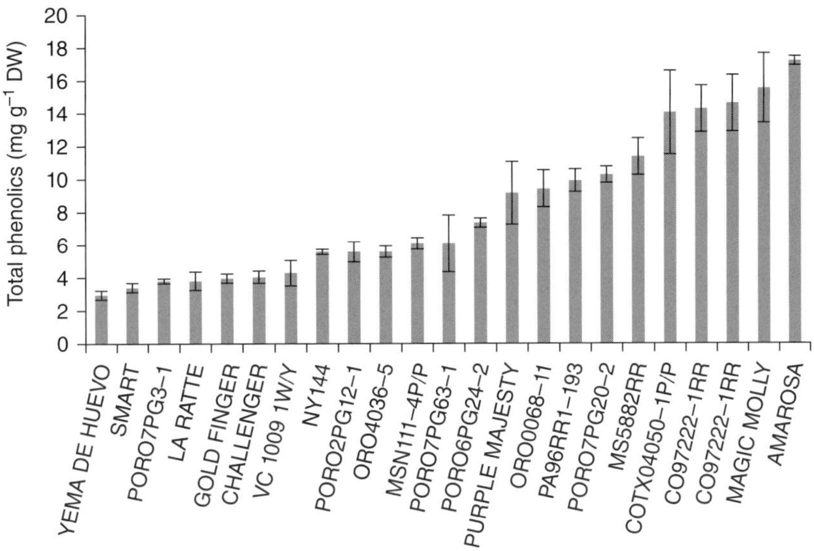

Fig. 18.4. Total phenolics in baby potatoes grown in Washington State, USA, in 2010–2011. Phenolics were measured by the FC method using freeze-dried samples. Potatoes were harvested 60–80 days after planting.

purple potato, Magic Molly, to 15 other vegetables (Navarre *et al.*, 2011). Russet Burbank had an ORAC value of ~5 μmol Trolox equivalents (TE) g^{-1} FW, which was comparable or superior to 8 other vegetables, whereas Norkotah (~9 μmol TE g^{-1} FW) had comparable or greater amounts than 11 other vegetables, rivaled spinach, and was double that of tomato. Magic Molly (~34 μmol TE g^{-1} FW) had the highest amounts of all the vegetables measured, surpassing both broccoli and Brussels sprouts. More recently, we have identified potatoes with even higher amounts of antioxidants than Magic Molly (Navarre, data not shown). These antioxidant values are not mentioned to show that potatoes are superior to other vegetables, but rather to emphasize the fact that potatoes can hold their own with other vegetables, and the perception by some that they are just "empty calories" or "provide only starch" ignores a lot of data that contradict such misperceptions.

Chlorogenic acid

Typically, tuber skin has a greater concentration of phenylpropanoids than the flesh, but because the flesh comprises a much greater percentage of tuber mass than the periderm, the flesh contains a majority of soluble phenylpropanoids. The most abundant phenolics in tubers are hydroxycinnamic acids, and chlorogenic acid can account for up to 90% of a tuber's total soluble phenylpropanoids in white potatoes (Malmberg and Theander, 1985). Several pathways exist for chlorogenic acid (CGA) biosynthesis, but the major pathway in *Solanaceae* is via the enzyme, hydroxycinnamoyl-CoA:quinate hydroxycinnamoyl transferase (HQT) (Niggeweg *et al.*, 2004). Red and purple potatoes usually have considerably greater amounts of CGA than white potatoes, but because of the presence of anthocyanins, CGA accounts for a smaller percentage of the total phenylpropanoid content than in white potatoes (Navarre *et al.*, 2011). The R2R3 transcription factor, *StAN1*, appears to mediate CGA expression, in addition to regulating anthocyanins (Payyavula *et al.*, 2013).

CGA is highly bioavailable in humans (Monteiro *et al.*, 2007; Farah *et al.*, 2008), can protect animals against degenerative, age-related diseases, has antiviral and bacterial properties, and can reduce the risk of some cancers and heart disease (Nogueira and do Lago, 2007). CGA is also thought to be antihypertensive (Yamaguchi *et al.*, 2007), but whether it is the cause of the hypotensive effect in people eating purple potatoes is unknown (Vinson *et al.*, 2012).

Given the studies mentioned earlier in this chapter that suggest potato consumption contributes to obesity and can increase the risk of diabetes, it is especially notable that chlorogenic acid may decrease the risk of type 2 diabetes (Legrand and Scheen, 2007) and can slow the release of glucose into the bloodstream (Bassoli *et al.*, 2008). If so, then high-CGA potatoes might have lower glycemic index values. CGA also exhibited anti-obesity effects in mice, where it reduced body weight, improved lipid profiles and decreased obesity-related hormones, possibly via activation of an adenosine monophosphate (AMP)-dependent kinase (Cho *et al.*, 2010; Ong *et al.*, 2013), but another study failed to replicate these results (Mubarak *et al.*, 2013). Other polyphenols, including some found in potatoes, may also have anti-obesity properties (Williams *et al.*, 2013).

Anthocyanins and flavonols

Red- and purple-flesh cultivars can contain substantial amounts of anthocyanins, compounds that are antioxidants and have other health-promoting effects. A potato anthocyanin extract had anticancer properties (Reddivari *et al.*, 2007b). Anthocyanins from purple potatoes were found to be more stable at low pH and temperatures, and protected against chromium toxicity, as measured by protecting bovine serum albumin conformation (Zhao *et al.*, 2011). Lewis *et al.* screened 26 colored-fleshed cultivars for anthocyanin content and found up to 7 mg g^{-1} FW in the skin and 2 mg g^{-1} FW in the flesh (Lewis *et al.*, 1998). Another study evaluated 31 colored genotypes and found a range of 0.5–3 mg g^{-1} FW in the skin and up to 1 mg g^{-1} FW in the flesh (Jansen and Flamme, 2006). Brown evaluated several genotypes for anthocyanins and found whole tubers that contained up to 4 mg g^{-1} FW and that anthocyanin concentration correlated with antioxidant value (Brown *et al.*, 2005). The Andean cultivar, Guincho Negra, from the Andigenum group, contained 16 mg g^{-1} DW of anthocyanins (Andre *et al.*, 2007b), whereas RN27.01 of the Phureja group contained 41 mg g^{-1} DW (Pillai *et al.*, 2013). We have found baby potatoes from advanced breeding lines that have 15–19 mg g^{-1} DW of anthocyanins (Navarre, unpublished data).

Tuber anthocyanin synthesis has been studied more in the periderm, where it is controlled by at least three loci, *D*, *P*, and *R*. *P* and *R* are structural genes, whereas *D* encodes an R2R3 MYB (Jung *et al.*, 2005, 2009; Zhang *et al.*, 2009). The *D* locus mapped to a chromosome 10 region that harbored *StAN1* (Jung *et al.*, 2009). *StAN1* expression correlated with anthocyanins in drought-stressed potatoes (André *et al.*, 2009). A novel approach to dissect tuber anthocyanin regulation used purple potatoes that had incomplete pigmentation, with the flesh containing both white and purple sectors. Microarray analysis of the white versus purple tissue from the same tuber, along with tubers from white or purple genotypes, was able to reduce 1817 differentially expressed genes down to 27 genes differentially expressed in purple tissue in all sets (Stushnoff *et al.*, 2010).

Potatoes contain small amounts of flavonols including rutin and kaempferol, but are not regarded as important sources of dietary flavonols. It is unclear to what extent levels can be increased in potatoes by directed efforts. Various studies suggest quercetin and related flavonols have health-promoting effects, including reduced risk of heart disease, lowered risk of asthma, bronchitis, and emphysema, and reduced risk of prostate and lung cancer. A survey of Andean potatoes found a range from zero to 191 and 222 μg g^{-1} DW for rutin and kaemp-3-rut, respectively (Andre *et al.*, 2007b), while Phureja group tubers contained up to 3000 μg g^{-1} DW (Pillai *et al.*, 2013). Advanced breeding lines containing over 430 μg g^{-1} DW are reported (Navarre *et al.*, 2011). Flavonols increased in fresh-cut tubers up to 140 μg g^{-1} FW, suggesting that the use of specific processing methods may be a way to increase tuber flavonols (Tudela *et al.*, 2002a).

Tuber discoloration

A potential complication of high-phenolic white potatoes could be the potential for unacceptable levels of browning or after-cooking darkening (ACD), as documented by older literature. However, recent studies showed that amounts of total phenolics or chlorogenic acid did not correlate with the amount of browning observed in fresh-cut potatoes (Cantos *et al.*, 2002). Another study using a quantitative trait locus (QTL) approach found no correlation (correlation coefficient 0.13) between browning and chlorogenic acid (Werij *et al.*, 2007). An association genetics analysis found correlation between enzymatic browning and unexpected traits including specific gravity and a class III lipase (Urbany *et al.*, 2011). Chlorogenic acid and other phenolics interact non-enzymatically with iron to cause ACD (Hughes and Evans, 1969; Wang-Pruski, 2007). However, other tuber compounds mediate this reaction and organic acids like citrate ameliorate the reaction, as does vitamin C and phosphate (Silva *et al.*, 1991). Dip treatments such as EDTA or sodium acid sulfate also reduce ACD (Calder *et al.*, 2012). Other minerals including P, S, and Ca may influence the severity of ACD (LeRiche *et al.*, 2009).

Given the health benefits of polyphenols, it is strongly desired to have cultivars that provide generous amounts of these compounds. This may be particularly important for potatoes, given the negative view of some nutritionists. Moreover, the major tuber phenylpropanoid, chlorogenic acid, has been implicated in reducing the risk of diabetes, obesity, and hypertension, some of the things for which potatoes have been criticized. Red and purple potatoes not only contain higher amounts of polyphenols but also are less affected by discoloration than white potatoes, because their natural color would mask browning. Moreover, key roles of polyphenols in potatoes include improving the plant's resistance to biotic and abiotic stresses. Decreasing these compounds could require increased use of pesticides. A strategy to reduce discoloration by decreasing polyphenols may be short-sighted. A preferable approach is to minimize discoloration by other methods, focusing on other genetic and environmental factors that influence discoloration.

Baby potatoes

Baby potatoes, also referred to as new potatoes or gourmet potatoes, are immature potatoes harvested from young plants, typically 60–80 days after planting. Baby potatoes could be important for the potato industry because they have unique advantages. Evidence is accumulating that they have higher amounts of phytonutrients than mature potatoes, including B1 as mentioned above. Tuber development has a strong effect on phenylpropanoid metabolism and various other phytonutrients (Navarre *et al.*, 2013). Swelling stolons and developing tubers had higher concentrations of carotenoids than in mature tubers (Morris *et al.*, 2004). Folate in tuber flesh and anthocyanins in tuber periderm decreased during tuber development, as did vitamin C (Hung *et al.*, 1997; Goyer and Navarre, 2009; Blauer *et al.*, 2013). Products of both early and late phenylpropanoid pathways decreased in field-grown tubers harvested at intervals throughout the season, starting about 60 days after planting. The magnitude of the phenylpropanoid decrease between immature and mature potatoes varied by genotype, but in some cases was up to threefold. CGA was higher in immature tubers, as was expression of various genes involved in early phenylpropanoid metabolism. Late phenylpropanoid pathways were also affected, with total anthocyanins decreasing from 10 mg g^{-1} DW to 6.3 mg g^{-1} DW, and *dihydroflavonol reductase* (*DFR*) expression decreased almost 70% (Payyavula *et al.*, 2013). Sucrose, fructose, and glucose concentrations were highest earlier in tuber maturation, as was expression of key genes and enzymes in the glycolytic and shikimate pathways, suggesting sugars may modulate tuber phenylpropanoids. The strong correlation between sucrose and total phenolics during development ($r = 0.91$ in Russet Burbank) and the higher expression of *sucrose synthase* (*SUSY*) was consistent with a role for sugars in modulating phenylpropanoid synthesis.

Historically, baby potatoes command a price premium because consumers perceive them to have superior taste. Potato products that appeal to people with more disposable income are an important need for the potato industry, which is seeing declining consumption in much of the developed world. Reasons for the decline are complex and multifaceted, but economics, nutritional perception, and a decrease in the number of sit-down, home-cooked meals likely contribute. In the USA, potato consumption has particularly decreased among the demographic with more disposable income, including much of the middle class. Surveys have found that some consumers regard potatoes as "boring" or "old-fashioned". The combination of more disposable income and a much greater variety of readily available fresh produce than in the past creates more choices for shoppers and little incentive to purchase anything regarded as "boring". Feeding into this attitude is the perception by some that potatoes are "empty calories". Consumers with more income tend to be better educated, and more likely to be aware of the negative comments made by some nutritionists about potatoes. Moreover, consumers with higher disposable income are also more likely to actively pursue a healthy diet. Baby potatoes are perceived as a more gourmet, exciting product and are thus more likely to appeal to affluent consumers. Likewise, as the public becomes more aware of the phytonutrient content of baby potatoes, this should influence perception positively.

Another advantage of baby potatoes is that they are cooked and eaten with their skin, thereby retaining all of the phytonutrients in the skin and reducing leaching of phytonutrients during cooking. Another benefit is that, because of their small size, baby potatoes cook faster, which, along with not having to be peeled, contributes to faster and less laborious preparation. With changing lifestyles such as both couples working full-time jobs and fewer family sit-down dinners, ease of preparation is more important than it once was. Thus, "baby" or "new" potatoes are a good example of a product with consumer-oriented traits, namely, superior taste, visually appealing, faster preparation, and more phytonutrients.

Effect of cooking on potato phytonutrients

Research directed towards increasing potato phytonutrients should focus on those that survive cooking. The effect of multiple cooking methods on new potatoes was measured to see how well selected phytonutrients survive cooking. Baby potatoes from the cultivars, Piccolo, Bintje, and Purple Majesty, were baked, boiled, microwaved, steamed, or stir-fried. Figure 18.5 shows the effect of cooking on chlorogenic acid in Bintje, Piccolo, and Purple Majesty. None of the four cooking methods significantly decreased the amount of chlorogenic acid, rutin, or vitamin C (Navarre *et al.*, 2010). In fact, compared to uncooked potatoes, there was a trend for amounts of these compounds to increase after cooking, especially after baking. This was likely due to the phytonutrients becoming more extractable from the food matrix after cooking. A slight decrease occurred in chlorogenic acid and vitamin C in potatoes overcooked by baking for 15 additional minutes. Similar results were seen with mature cooked potatoes, where multiple cooking methods resulted in an increase in various polyphenols (Blessington *et al.*, 2010). Conversely, a different study reported a marked decrease in polyphenols and antioxidant activity in potatoes cooked by various methods (Perla *et al.*, 2012). The reasons for such different findings between studies are not clear, but such conflicting results are common among the great many published studies of the effect of cooking on phytonutrients (Navarre *et al.*, 2010). Whatever the reason for the discrepancies, it is clear that cooking can decrease the phytonutrient content of potatoes, but it is equally clear that it need not if leaching and overcooking are avoided. Also clear is that heating alone does not necessarily decrease amounts, so additional degradative mechanisms must be at play.

Efforts have been made to assess the effect on phytonutrients of different commercial processing methods. The effect of freeze drying (FD), drum drying (DD), and refractive window (RW) drying on a purée of purple potatoes showed no significant loss of total phenolics or total antioxidant capacity by any method, but a 23–45% loss of anthocyanins (Nayak *et al.*, 2011). The authors suggested purple potato flour could be used as an ingredient in nutritionally enhanced food products and snack foods high in antioxidants. Similar results were reported in another study using these three drying methods (Kaspar *et al.*, 2012). Unlike the previous study where FD caused the most loss of anthocyanins, FD resulted in less loss of anthocyanins and carotenoids in processed potatoes than DD or RW drying. Drum drying caused up to a 48% loss of anthocyanins and 30% loss of carotenoids. Despite the better results, FD may not be cost-effective on a commercial scale. Curiously, both studies showed little or no loss in total antioxidant capacity by the different methods, despite the significant decreases in anthocyanins.

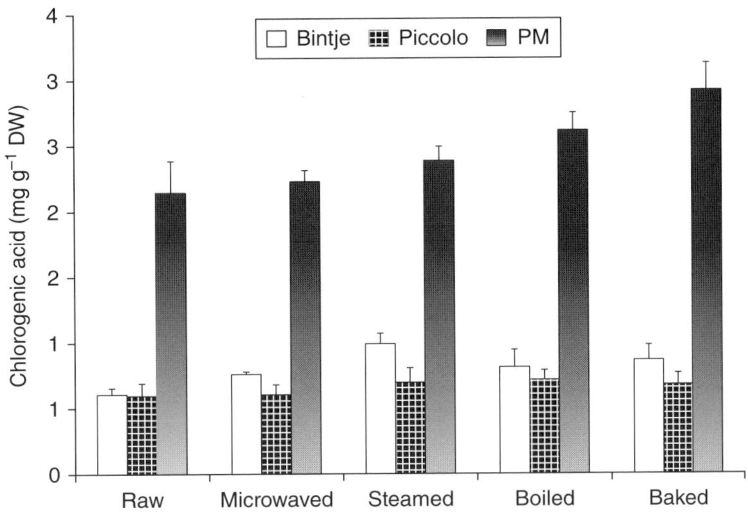

Fig. 18.5. Baby potatoes from three cultivars were cooked as shown and chlorogenic acid measured by HPLC. Raw potatoes were the control.

Another study showed baked potatoes had similar antiproliferative and pro-apoptotic properties to fresh potatoes, whereas chipping caused a reduction in bioactives (Madiwale *et al.*, 2012).

Environmental effects on phenylpropanoids

Phenylpropanoid expression in plants is well known to be influenced by environment (Dixon and Paiva, 1995), which is not surprising given that one of their primary roles is to allow the plant to cope better with stresses, including drought, disease, and photooxidative stress. Thus, it is expected that tuber phenylpropanoid profiles will vary by location and year, in addition to genotype. Anthocyanins were found to be higher in potatoes grown at higher altitudes in Colorado than in the same cultivars grown in Texas (Reyes *et al.*, 2004; Reddivari *et al.*, 2007a). Analysis of the total antioxidants and total phenolics in the same cultivars grown in nine different states showed that environment had a significant effect, but the majority of the variation was due to genotype (Nzaramba *et al.*, 2013). An up to ~35% difference in total phenolics was observed among the same 11 potato cultivars grown in Canada in 2008 and 2009, with a mean difference of 16% (Hu *et al.*, 2012). We grew the purple cultivar, Magic Molly, in Texas, Florida, and four locations in Alaska, including in the Arctic Circle, and found total phenolics varied by 28%. Chlorogenic acid was twofold higher in the potatoes grown in the Arctic Circle versus Florida. Anthocyanins were higher in potatoes grown in Alaska (6.3–7.3 mg g^{-1}) than in Texas or Florida (4.4 mg g^{-1}). Petunidin-3-coum-rutinoside-5 glucoside was the most abundant anthocyanin, ranging from 4.9 mg g^{-1} in Alaska to 2.3 mg g^{-1} in Magic Molly in Texas. Over a tenfold difference between locations occurred in the expression of some phenylpropanoid genes, and sizeable differences were present in both early and late phenylpropanoid genes.

18.8 Carotenoids

Carotenoids are synthesized in plastids from isoprenoids and function to help plants resist photo- and oxidative stress and influence scent and flavor (Dellapenna and Pogson, 2006; Cazzonelli and Pogson, 2010). Phytoene biosynthesis is a rate-limiting step in carotenogenesis, and *PSY* (phytoene synthase) expression can be induced by light, photoperiod, drought, and temperature (Cazzonelli and Pogson, 2010). *PSY1* expression in tubers was correlated negatively with carotenoids, which might indicate negative feedback control (Diretto *et al.*, 2010). Mixed results have been reported about the extent that tuber carotenoid expression is controlled at the transcriptional level (Goo *et al.*, 2009; Zhou *et al.*, 2011; Payyavula *et al.*, 2012). Substantial differences were observed in gene–metabolite relationships in leaves versus tubers (Diretto *et al.*, 2010). Recent evidence suggests the expression of some carotenoid genes may be influenced by tuber sugar concentrations (Zhou *et al.*, 2011).

Carotenoids have provitamin A activity and decrease the risk of several diseases (Fraser and Bramley, 2004). The most abundant potato carotenoids, lutein and zeaxanthin, are important for eye health, and reduce the risk of age-related macular degeneration (Chucair *et al.*, 2007; Tan *et al.*, 2008). The type and quantity of potato carotenoids varies by cultivar, and the yellow to almost orange flesh color found in some potatoes is due to carotenoids. Lutein amounts influence the intensity of yellow coloration, and zeaxanthin influences the orange coloration (Brown *et al.*, 1993). Analysis of eleven *Chy2* (β-carotene hydroxylase 2) alleles in potato showed that just one dominant allele was responsible for changing white flesh to yellow flesh (Wolters *et al.*, 2010). Five *Zep* (zeaxanthin epoxidase) alleles were detected, one of which was recessive and responsible for the orange flesh color, but only in the presence of the dominant *Chy2*. The authors suggested the reduced Zep activity in orange-flesh genotypes was not due to reduced enzymatic activity but to reduced amounts due to inefficient splicing (Wolters *et al.*, 2010). An additional level of control over tuber carotenoid accumulation may be exerted by the suborganellar compartmentation, as Chy2 and PSY2 were localized to chloroplasts in leaves, but vesicular structures inside amyloplasts in tubers (Pasare *et al.*, 2013). Tuber carotenoid concentrations are also controlled by catabolism; reducing the expression of a carotenoid cleavage dioxygenase resulted in a two- to fivefold increase in carotenoids (Campbell *et al.*, 2010).

Some cultivated diploid potatoes derived from *Solanum stenotomum* and *Solanum phureja* contained 2000 μg 100 g^{-1} FW of zeaxanthin (Brown *et al.*, 1993). Total carotenoid concentrations in 74 Andean landraces ranged from 3 to 36 μg g^{-1} DW (Andre *et al.*, 2007a). A screen of 24 Andean cultivars identified genotypes with almost 18 μg g^{-1} DW each of lutein and zeaxanthin and over 2 μg g^{-1} DW of β-carotene (Andre *et al.*, 2007b). Sixty mostly Spanish cultivars were found to contain 0.50–15.5 μg g^{-1} DW, but curiously no zeaxanthin was found, possibly due to prolonged storage in the dark (Fernandez-Orozco *et al.*, 2013). White-flesh potatoes contain less carotenoids than the yellow or orange cultivars. One study found white cultivars had 27–74 μg 100 g^{-1} FW of carotenoids (Iwanzik *et al.*, 1983).

Burgos *et al.*, reported that violaxanthin and antheraxanthin in tubers decreased after boiling, but lutein and zeaxanthin did not (Burgos *et al.*, 2012). In addition to carotenoid degradation, substantial isomerization was observed in boiled tubers (Burmeister *et al.*, 2011). Declines in carotenoids during cold storage can occur (Morris *et al.*, 2004; Griffiths *et al.*, 2007).

The broad-sense heritability in potato was 0.96 for total carotenoids, suggesting that breeding should be able to increase the amounts significantly (Haynes *et al.*, 2010). Besides breeding, transgenic approaches can increase potato carotenoids substantially. A bacterial *PSY* overexpressed in tubers increased carotenoids from 5.6 to 35 μg g^{-1} DW and β-carotene concentrations from trace amounts to 11 μg g^{-1} DW (Ducreux *et al.*, 2005). Overexpression of the cauliflower *Or* gene in Desiree resulted in a sixfold increase in tuber carotenoids to about 20–25 μg g^{-1} DW (Lu *et al.*, 2006). Overexpression of three bacterial genes in the cultivar Desiree resulted in a 20-fold increase in total carotenoids to 114 μg g^{-1} DW and an impressive 3600-fold increase in β-carotene to 47 μg g^{-1} DW (Diretto *et al.*, 2007).

18.9 Conclusion

Since their introduction from the Andes, the cultivation of potatoes in the West has helped curtail hunger for over 200 years, and they have been the staple food of generations of people. Potatoes have even been linked to marked population growth in Europe in the 17–1800s. Today, the potato finds itself in unfamiliar territory, as some nutritionists question the healthfulness of carbohydrate-rich staples like potatoes and speculate based only on epidemiological data that potatoes are linked to increasing diabetes and obesity. However, as described in this chapter, potatoes contain numerous health-promoting compounds in addition to complex carbohydrates. Furthermore, potatoes are a very efficient crop, which in addition to its nutritional merits, suggests that potatoes will be a key part of the solution needed to feed the projected increase in global population. In the US Pacific Northwest, potatoes provide about 26 million calories acre^{-1}. The white cultivar, Norkotah Russet, would provide about 150,000 g of potassium acre^{-1}, 6700 g of vitamin C, and 19,000 g of phenolics. Some have estimated that agricultural production will have to double by 2050 to maintain food security (Ray *et al.*, 2013). In view of their nutritional content and yields per acre, potatoes will almost certainly have a key role in "feeding the future". This role for potatoes is ironic in light of a minority of nutritionists advocating that potatoes be eaten sparingly and replaced in the diet with other foods. Leaving aside the abundant published scientific evidence documenting the nutritional value of potatoes, those advocating such a policy have not provided data documenting exactly what the replacement cost would be. For example, what would the replacement for potatoes be and would it be as affordable as potatoes? How much more land would be needed to provide an equivalent amount of energy and nutrition?

Efforts to maximize the nutritional potential of potatoes are in their early stages. The rapid pace of scientific progress over the last decade is facilitating the development of new cultivars with even higher nutritional value. This ability, along with increasing information about exactly what plant components are health promoting and desirable in the diet creates new opportunities for plant biologists to develop the next generation of cultivars.

References

Adams-Phillips, L., Wan,J., Tan, X., Dunning, F.M., Meyers, B.C., Michelmore, R.W., *et al.* (2008) Discovery of ADP-ribosylation and other plant defense pathway elements through expression profiling of four different Arabidopsis-Pseudomonas R-avr interactions. *Molecular Plant–Microbe Interactions* 21, 646–657.

Ah-Hen, K., Fuenzalida, C., Hess, S., Contreras, A., Vega-Gálvez, A. and Lemus-Mondaca, R. (2012) Antioxidant capacity and total phenolic compounds of twelve selected potato landrace clones grown in Southern Chile. *Chilean Journal of Agricultural Research* 72, 3–9.

Alfthan, G., Laurinen, M.S., Valsta, L.M., Pastinen, T. and Aro, A. (2003) Folate intake, plasma folate and homocysteine status in a random Finnish population. *European Journal of Clinical Nutrition* 57, 81–88.

Alton, I. (2005) Iron deficiency anemia. In: Stang, J. and Story, M. (eds) *Guidelines for Adolescent Nutrition Services*. Center for Leadership, Education and Training in Maternal and Child Nutrition, University of Minnesota, Minneapolis, Minnesota, pp. 101–108.

Amadasi, A., Bertoldi, M., Contestabile, R., Bettati, S., Cellini, B., di Salvo, M.L., *et al.* (2007) Pyridoxal 5'-phosphate enzymes as targets for therapeutic agents. *Current Medicinal Chemistry* 14, 1291–1324.

Andre, C.M., Ghislain, M., Bertin, P., Oufir, M., del Rosario Herrera, M., Hoffmann, L., *et al.* (2007a) Andean potato cultivars (*Solanum tuberosum* L.) as a source of antioxidant and mineral micronutrients. *Journal of Agricultural and Food Chemistry* 55, 366–378.

Andre, C.M., Oufir, M., Guignard, C., Hoffmann, L., Hausman, J.F., Evers, D., *et al.* (2007b) Antioxidant profiling of native Andean potato tubers (*Solanum tuberosum* L.) reveals cultivars with high levels of beta-carotene, alpha-tocopherol, chlorogenic acid, and petanin. *Journal of Agricultural and Food Chemistry* 55, 10839–10849.

André, C.M., Schafleitner, R., Legay, S., Lefèvre, I., Aliaga, C.A.A., Nomberto, G., *et al.* (2009) Gene expression changes related to the production of phenolic compounds in potato tubers grown under drought stress. *Phytochemistry* 70, 1107–1116.

Anon. (2001) *Dietary Reference Intakes for Vitamin A, Vitamin K, Arsenic, Boron, Chromium, Copper, Iodine, Iron, Manganese, Molybdenum, Nickel, Silicon, Vanadium, and Zinc*. National Acadamy Press, Washington, DC.

Arsenian, M.A. (1993) Magnesium and cardiovascular disease. *Progress in Cardiovascular Diseases* 35, 271–310.

Arteca, R., Poovaiah, B. and Hiller, L. (1980) Electron microprobe and neutron activation analysis for the determination of elemental distribution in hollow heart potato tubers. *American Journal of Potato Research* 57, 241–247.

Augustin, J., Johnson, S.R., Teitzel, C., Toma, R.B., Shaw, R.L., True, R.H., *et al.* (1978) Vitamin composition of freshly harvested and stored potatoes. *Journal of Food Science* 43, 1566–1570, 1574.

Bach, S., Yada, R.Y., Bizimungu, B. and Sullivan, J.A. (2012) Genotype by environment interaction effects on fibre components in potato (*Solanum tuberosum* L.). *Euphytica* 187, 77–86.

Bach, S., Yada, R.Y., Bizimungu, B., Fan, M. and Sullivan, J.A. (2013) Genotype by environment interaction effects on starch content and digestibility in potato (*Solanum tuberosum* L.). *Journal of Agricultural and Food Chemistry* 61, 3941–3948.

Bailey, L.M., Wallace, J.C. and Polyak, S.W. (2010) Holocarboxylase synthetase: correlation of protein localisation with biological function. *Archives of Biochemistry and Biophysics* 496, 45–52.

Bamberg, J.B., Palta, J.P., Peterson, L.A., Martin, M. and Krueger, A.R. (1998) Fine screening potato (Solanum) species germplasm for tuber calcium. *American Journal of Potato Research* 75, 181–186.

Basavaraju, B.S., Chakravarthy, A.K., Doddabasappa, B. and Krishnamurthy, N. (2009) Yield loss estimation due to major insect and mite pests on potato in Karnataka. *Karnataka Journal of Agricultural Science* 22, 597–600.

Basset, G.J., Quinlivan, E.P., Ravanel, S., Rebeille, F., Nichols, B.P., Shinozaki, K., *et al.* (2004) Folate synthesis in plants: the p-aminobenzoate branch is initiated by a bifunctional PabA-PabB protein that is targeted to plastids. *Proceedings of the National Academy of Sciences* 101, 1496–1501.

Bassoli, B.K., Cassolla, P., Borba-Murad, G.R., Constantin, J., Salgueiro-Pagadigorria, C.L., Bazotte, R.B., *et al.* (2008) Chlorogenic acid reduces the plasma glucose peak in the oral glucose tolerance test: effects on hepatic glucose release and glycaemia. *Cell Biochemistry and Function* 26, 320–328.

Bates, C.J. and Powers, H.J. (1985) A simple fluorimetric assay for pyridoxamine phosphate oxidase in erythrocyte haemolysates: effects of riboflavin supplementation and of glucose 6-phosphate dehydrogenase deficiency. *Human Nutrition. Clinical Nutrition* 39, 107–115.

Battersby, A.R. and Leeper, F.J. (1998) Biochemistry of vitamin B12. *Topics in Current Chemistry* 195, 143–193.

Bender, D.A. (1999) Optimum nutrition: thiamin, biotin and pantothenate. *Proceedings of the Nutrition Society* 58, 427–433.

Bilski, P., Li, M.Y., Ehrenshaft, M., Daub, M.E. and Chignell, C.F. (2000) Vitamin B6 (pyridoxine) and its derivatives are efficient singlet oxygen quenchers and potential fungal antioxidants. *Photochemistry and Photobiology* 71, 129–134.

Birch, T.W., Gyorgy, P. and Harris, L.J. (1935) The vitamin B(2) complex. Differentiation of the antiblacktongue and the "P.-P." factors from lactoflavin and vitamin B(6) (so-called "rat pellagra" factor). Parts I–VI. *Biochemical Journal* 29, 2830–2850.

Blauer, J.N., Kumar, M.G.N., Knowles, L.O., Dingra, A. and Knowles, N.R. (2013) Changes in ascorbate and associated gene expression during development and storage of potato tubers. *Postharvest Biology and Technology* 78, 76–91.

Blessington, T., Nzaramba, M.N., Scheuring, D.C., Hale, A.L., Redivari, L. and Miller, J.C., Jr (2010) Cooking methods and storage treatments of potato: effects on carotenoids, antioxidant activity, and phenolics. *American Journal of Potato Research* 87, 479–491.

Boisvert, W.A., Castaneda, C., Mendoza, I., Langeloh, G., Solomons, N.W., Gershoff, S.N., *et al.* (1993) Prevalence of riboflavin deficiency among Guatemalan elderly people and its relationship to milk intake. *American Journal of Clinical Nutrition* 58, 85–90.

Bretzloff, C. and McMenamin, J. (1971) Some aspects of potato appearance and texture. III. Sampling tubers for cation analysis. *American Journal of Potato Research* 48, 246–254.

Brevik, A., Vollset, S.E., Tell, G.S., Refsum, H., Ueland, P.M., Loeken, E.B., *et al.* (2005) Plasma concentration of folate as a biomarker for the intake of fruit and vegetables: the Hordaland Homocysteine Study. *American Journal of Clinical Nutrition* 81, 434–439.

Brown, C.R., Edwards, C.G., Yang, C.P. and Dean, B.B. (1993) Orange flesh trait in potato: inheritance and carotenoid content. *Journal of the American Society for Horticultural Science* 118, 145–150.

Brown, C.R., Culley, D., Yang, C.P., Durst, R. and Wrolstad, R. (2005) Variation of anthocyanin and carotenoid contents and associated antioxidant values in potato breeding lines. *Journal of the American Society for Horticultural Science* 130, 174–180.

Brussaard, J.H., Lowik, M.R., van den Berg, H., Brants, H.A. and Goldbohm, R.A. (1997) Folate intake and status among adults in the Netherlands. *European Journal of Clinical Nutrition* 51 Suppl 3, S46–50.

Burgos, G., Amoros, W., Morote, M., Stangoulis, J. and Bonierbale, M. (2007) Iron and zinc concentration of native Andean potato cultivars from a human nutrition perspective. *Journal of the Science of Food and Agriculture* 87, 668–675.

Burgos, G., Amoros, W., Salas, E., Munoa, L., Sosa, P., Diaz, C., *et al.* (2012) Carotenoid concentrations of native Andean potatoes as affected by cooking. *Food Chemistry* 133, 1131–1137.

Burmeister, A., Bondiek, S., Apel, L., Kuhne, C., Hillebrand, S. and Fleischmann, P. (2011) Comparison of carotenoid and anthocyanin profiles of raw and boiled *Solanum tuberosum* and *Solanum phureja* tubers. *Journal of Food Composition Analysis* 24, 865–872.

Burton, W.G. (1966) *The Potato* (2nd edn). Drukkerij Veeninan BV, Wageningen, the Netherlands.

Calder, B.L., Cowles, E.A., Davis-Dentici, K. and Bushway, A.A. (2012) The effectiveness of antibrowning dip treatments to reduce after-cooking darkening in potatoes. *Journal of Food Science* 77, S342–S347.

Campbell, R., Ducreux, L.J.M., Morris, W.L., Morris, J.A., Suttle, J.C., Ramsay, G., *et al.* (2010) The metabolic and developmental roles of carotenoid cleavage dioxygenase4 from potato. *Plant Physiology* 154, 656–664.

Cantos, E., Tudela, J.A., Gil, M.I. and Espin, J.C. (2002) Phenolic compounds and related enzymes are not rate-limiting in browning development of fresh-cut potatoes. *Journal of Agricultural and Food Chemistry* 50, 3015–3023.

Casanas, R., Gonzalez, M., Rodriguez, E., Marrero, A. and Diaz, C. (2002) Chemometric studies of chemical compounds in five cultivars of potatoes from Tenerife. *Journal of Agricultural and Food Chemistry* 50, 2076–2082.

Cazzonelli, C.I. and Pogson, B.J. (2010) Source to sink: regulation of carotenoid biosynthesis in plants. *Trends in Plant Science* 15, 266–274.

CDC (Centers for Disease Control and Prevention) (2013) http://www.cdc.gov/obesity/data/adult.html (accessed 23 May 2014).

Chakauya, E., Coxon, K.M., Wei, M., Macdonald, M.V., Barsby, T., Abell, C., *et al.* (2008) Towards engineering increased pantothenate (vitamin B(5)) levels in plants. *Plant Molecular Biology* 68, 493–503.

Chakraborty, S., Chakraborty, N., Agrawal, L., Ghosh, S., Narula, K., Shekhar, S., *et al.* (2010) Next-generation protein-rich potato expressing the seed protein gene AmA1 is a result of proteome rebalancing in transgenic tuber. *Proceedings of the National Academy of Sciences* 107, 17533–17538.

Cham, B.E. (1994) Solasodine glycosides as anti-cancer agents: pre-clinical and clinical studies. *S Asia Pacific Journal of Pharmacology* 9, 113–118.

Chataing, B., Concepcion, J.L., de Cristancho, N.B. and Usubillaga, A. (1997) Estudio clinico de la efectividad de extractos alcaloides obtenidos de los frutos del Solanum americanum Miller sobre el Herpes simplex, Herpes Zoster y Herpes genitalis. *Revista de la Facultad de Farmacia* 32, 18–25.

Chen, H. and Xiong, L. (2005) Pyridoxine is required for post-embryonic root development and tolerance to osmotic and oxidative stresses. *Plant Journal* 44, 396–408.

Chen, H. and Xiong, L. (2009) Enhancement of vitamin B6 levels in seeds through metabolic engineering. *Plant Biotechnology Journal* 7, 673–681.

Cho, A.S., Jeon, S.M., Kim, M.J., Yeo, J., Seo, K.I., Choi, M.S., *et al.* (2010) Chlorogenic acid exhibits anti-obesity property and improves lipid metabolism in high-fat diet-induced-obese mice. *Food and Chemical Toxicology* 48, 937–943.

Chucair, A.J., Rotstein, N.P., Sangiovanni, J.P., During, A., Chew, E.Y. and Politi, L.E. (2007) Lutein and zeaxanthin protect photoreceptors from apoptosis induced by oxidative stress: relation with docosahexaenoic acid. *Investigative Ophthalmology and Visual Science* 48, 5168–5177.

Chun, O.K., Kim, D.O., Smith, N., Schroeder, D., Han, J.T. and Lee, C.Y. (2005) Daily consumption of phenolics and total antioxidant capacity from fruit and vegetables in the American diet. *Journal of the Science of Food and Agriculture* 85, 1715–1724.

Clusin, W.T. (2003) Calcium and cardiac arrhythmias: DADs, EADs, and Alternans. *Critical Reviews in Clinical Laboratory Sciences* 40, 337–375.

Cossins, E.A. and Chen, L. (1997) Folates and one-carbon metabolism in plants and fungi. *Phytochemistry* 45, 437–452.

Coxon, K.M., Chakauya, E., Ottenhof, H.H., Whitney, H.M., Blundell, T.L., Abell, C., *et al.* (2005) Pantothenate biosynthesis in higher plants. *Biochemical Society Transactions* 33, 743–746.

Dale, M.F.B., Griffiths, D.W. and Todd, D.T. (2003) Effects of genotype, environment, and postharvest storage on the total ascorbate content of potato (*Solanum tuberosum*) tubers. *Journal of Agricultural and Food Chemistry* 51, 244–248.

Dawson-Hughes, B., Harris, S.S., Krall, E.A. and Dallal, G.E. (1997) Effect of calcium and vitamin D supplementation on bone density in men and women 65 years of age or older. *New England Journal of Medicine* 337, 670–676.

Dellapenna, D. and Pogson, B. (2006) Vitamin synthesis in plants: tocopherols and carotenoids. *Annual Review of Plant Biology* 57, 711–738.

Denslow, S.A., Walls, A.A. and Daub, M.E. (2005) Regulation of biosynthetic genes and antioxidant properties of vitamin B_6 vitamers during plant defense responses. *Physiological and Molecular Plant Pathology* 66, 244–255.

Desborough, S.L. (1985) *Potato Proteins*. Academic Press, Michigan.

Desborough, S.L., Liener, I.E. and Lulai, E.C. (1981) The nutritional quality of potato protein from intraspecific hybrids. *Plant Foods for Human Nutrition* 31, 11–20.

Diretto, G., Al-Babili, S., Tavazza, R., Papacchioli, V., Beyer, P. and Giuliano, G. (2007) Metabolic engineering of potato carotenoid content through tuber-specific overexpression of a bacterial mini-pathway. *PLoS ONE* 2, e350.

Diretto, G., Al-Babili, S., Tavazza, R., Scossa, F., Papacchioli, V., Migliore, M., *et al.* (2010) Transcriptional-metabolic networks in beta-carotene-enriched potato tubers: the long and winding road to the Golden phenotype. *Plant Physiology* 154, 899–912.

di Salvo, M.L., Contestabile, R. and Safo, M.K. (2011) Vitamin B(6) salvage enzymes: mechanism, structure and regulation. *Biochimica et Biophysica Acta* 1814, 1597–1608.

Dixon, R.A. and Paiva, N.L. (1995) Stress-induced phenylpropanoid metabolism. *Plant Cell* 7, 1085–1097.

Doucet-Chabeaud, G., Godon, C., Brutesco, C., de Murcia, G. and Kazmaier, M. (2001) Ionising radiation induces the expression of PARP-1 and PARP-2 genes in Arabidopsis. *Molecular Genetics and Genomics* 265, 954–963.

Drewnowski, A. and Rehm, C.D. (2013) Vegetable cost metrics show that potatoes and beans provide most nutrients per penny. *PLoS One* 8, e63277.

Ducreux, L.J.M., Morris, W.L., Hedley, P.E., Shepherd, T., Davies, H.V., *et al.* (2005) Metabolic engineering of high carotenoid potato tubers containing enhanced levels of beta-carotene and lutein. *Journal of Experimental Botany* 56, 81–89.

Dyer, A.R., Elliot, P. and Shipley, M. (1994) Urinary electrolyte excretion in 24 hours and blood pressure in the INTERSALT Study. II. Estimates of electrolyte–blood pressure associations corrected for regression dilution bias. *American Journal of Epidemiology* 139, 940–951.

Ehrenshaft, M., Jenns, A.E., Chung, K.R. and Daub, M.E. (1998) SOR1, a gene required for photosensitizer and singlet oxygen resistance in Cercospora fungi, is highly conserved in divergent organisms. *Molecular Cell* 1, 603–609.

FAO (Food and Agriculture Organization) (2009) Food Security and Agricultural Mitigation in Developing Countries: Options for Capturing Synergies (http://www.fao.org/docrep/012/i1318e/i1318e00.pdf, accessed 23 May 2014).

Farah, A., Monteiro, M., Donangelo, C.M. and Lafay, S. (2008) Chlorogenic acids from green coffee extract are highly bioavailable in humans. *Journal of Nutrition* 138, 2309–2315.

Fernandes, G., Velangi, A. and Wolever, T.M. (2005) Glycemic index of potatoes commonly consumed in North America. *Journal of the American Dietetic Association* 105, 557–562.

Fernandez-Orozco, R.B., Gallardo-Guerrero, L. and Hornero-Mendez, D. (2013) Carotenoid profiling in tubers of different potato (Solanum sp) cultivars: accumulation of carotenoids mediated by xanthophyll esterification. *Food Chemistry* 141, 2864–2872.

Fitzpatrick, T.B., Basset, G.J., Borel, P., Carrari, F., Dellapenna, D., Fraser, P.D., *et al.* (2012) Vitamin deficiencies in humans: can plant science help? *Plant Cell* 24, 395–414.

Food and Nutrition Board, Institute of Medicine (1998) Biotin. In*: Dietary Reference Intakes: Thiamin, Riboflavin, Niacin, Vitamin B6, Vitamin B12, Pantothenic Acid, Biotin, and Choline*. National Academy Press, Washington, DC, pp. 374–389.

Fraser, P.D. and Bramley, P.M. (2004) The biosynthesis and nutritional uses of carotenoids. *Progress in Lipid Research* 43, 228–265.

Friedman, M. (2002) Tomato glycoalkaloids: role in the plant and in the diet. *Journal of Agricultural and Food Chemistry* 50, 5751–5780.

Friedman, M. and McDonald, G.M. (1997) Potato glycoalkaloids: chemistry, analysis, safety, and plant physiology. *Critical Reviews in Plant Sciences* 16, 55–132.

Friedman, M., Lee, K.R., Kim, H.J., Lee, I.S. and Kozukue, N. (2005) Anticarcinogenic effects of glycoalkaloids from potatoes against human cervical, liver, lymphoma, and stomach cancer cells. *Journal of Agricultural and Food Chemistry* 53, 6162–6169.

Friedman, M., McQuistan, T., Hendricks, J.D., Pereira, C. and Bailey, G.S. (2007) Protective effect of dietary tomatine against dibenzo[a,l]pyrene (DBP)-induced liver and stomach tumors in rainbow trout. *Molecular Nutrition and Food Research* 51, 1485–1491.

Ginzberg, I., Thippeswamy, M., Fogelman, E., Demirel, U., Mweetwa, A.M., Tokuhisa, J., *et al.* (2012) Induction of potato steroidal glycoalkaloid biosynthetic pathway by overexpression of cDNA encoding primary metabolism HMG-CoA reductase and squalene synthase. *Planta* 235, 1341–1353.

Goo, Y.-M., Kim, T.-W., Ha, S.-H., Back, K.-W., Bae, J.-M., Shin, Y.-W., *et al.* (2009) Expression profiles of genes involved in the carotenoid biosynthetic pathway in yellow-fleshed potato cultivars (*Solanum tuberosum* L.) from South Korea. *Journal of Plant Biology* 52, 49–55.

Gossert, A.D., Bettendorff, P., Puorger, C., Vetsch, M., Herrmann, T., Glockshuber, R., *et al.* (2008) NMR structure of the *Escherichia coli* type 1 pilus subunit FimF and its interactions with other pilus subunits. *Journal of Molecular Biology* 375, 752–763.

Goyer, A. (2010) Thiamine in plants: aspects of its metabolism and functions. *Phytochemistry* 71, 1615–1624.

Goyer, A. and Haynes, K.G. (2011) Vitamin B1 content in potato: effect of genotype, tuber enlargement, and storage, and estimation of stability and broad-sense heritability. *American Journal of Potato Research* 88, 374–385.

Goyer, A. and Navarre, D.A. (2007) Determination of folate concentrations in diverse potato germplasm using a trienzyme extraction and a microbiological assay. *Journal of Agricultural and Food Chemistry* 55, 3523–3528.

Goyer, A. and Navarre, D.A. (2009) Folate is higher in developmentally younger potato tubers. *Journal of the Science of Food and Agriculture* 89, 579–583.

Goyer, A. and Sweek, K. (2011) Genetic diversity of thiamin and folate in primitive cultivated and wild potato (Solanum) species. *Journal of Agricultural and Food Chemistry* 59, 13072–13080.

Griffiths, D.W., Bain, H. and Dale, M.F.B. (1997) The effect of low-temperature storage on the glycoalkaloid content of potato (*Solanum tuberosum*) tubers. *Journal of the Science of Food and Agriculture* 74, 301–307.

Griffiths, D.W., Dale, M.F., Morris, W.L. and Ramsay, G. (2007) Effects of season and postharvest storage on the carotenoid content of *Solanum phureja* potato tubers. *Journal of Agricultural and Food Chemistry* 55, 379–385.

Gubarev, M.I., Enioutina, E.Y., Taylor, J.L., Visic, D.M. and Daynes, R.A. (1998) Plant derived glycoalkaloids protect mice against lethal infection with *Salmonella typhimurium*. *Phytotherapy Research* 12, 79–88.

Halton, T.L., Willett, W.C., Liu, S., Manson, J.E., Stampfer, M.J. and Hu, F.B. (2006) Potato and french fry consumption and risk of type 2 diabetes in women. *American Journal of Clinical Nutrition* 83, 284–290.

Hassa, P.O., Haenni, S.S., Elser, M. and Hottiger, M.O. (2006) Nuclear ADP-ribosylation reactions in mammalian cells: where are we today and where are we going? *Microbiology and Molecular Biology Reviews* 70, 789–829.

Hatzis, C.M., Bertsias, G.K., Linardakis, M., Scott, J.M. and Kafatos, A.G. (2006) Dietary and other lifestyle correlates of serum folate concentrations in a healthy adult population in Crete, Greece: a cross-sectional study. *Nutrition Journal* 5, 5.

Haynes, K.G., Clevidence, B.A., Rao, D., Vinyard, B.T. and White, J.M. (2010) Genotype × environment interactions for potato tuber carotenoid content. *Journal of the American Society for Horticultural Science* 135, 250–258.

Heaney, R.P. (2000) Calcium, dairy products and osteoporosis. *Journal of the American College of Nutrition* 19, 83S–99S.

Hedtke, B., Alawady, A., Albacete, A., Kobayashi, K., Melzer, M., Roitsch, T., *et al.* (2012) Deficiency in riboflavin biosynthesis affects tetrapyrrole biosynthesis in etiolated Arabidopsis tissue. *Plant Molecular Biology* 78, 77–93.

Heftmann, E. (1983) Biogenesis of steroids in Solanaceae. *Phytochemistry* 22, 1843–1860.

Heinzelmann, U. (2008) *Food Culture in Germany*. Greenwood, Westport, Connecticut.

Hellmann, H. and Mooney, S. (2010) Vitamin B6: a molecule for human health? *Molecules* 15, 442–459.

Hoey, L., McNulty, H. and Strain, J.J. (2009) Studies of biomarker responses to intervention with riboflavin: a systematic review. *American Journal of Clinical Nutrition* 89, 1960S–1980S.

Holland, B., Unwin, I.D. and Buss, D.H. (1996) *Vegetables, Herbs and Spices. Fruits and Nuts.* Royal Society of Chemistry, Cambridge, UK.

Hopkins, J. (1995) The glycoalkaloids: naturally of interest (but a hot potato?). *Food and Chemical Toxicology* 33, 323–328.

Hu, C., Tsao, R., Liu, R., Sullivan, J.A. and McDonald, M.R. (2012) Influence of cultivar and year on phytochemical and antioxidant activity of potato (*Solanum tuberosum* L.) in Ontario. *Canadian Journal of Plant Science* 92, 485–493.

Hughes, J.C. and Evans, J.L. (1969) Studies on after-cooking blackening. V. Changes in aftercooking blackening and the chemistry of Majestic and Ulster Beacon tubers during the growing season. *European Potato Journal* 12, 26–40.

Hung, C.-Y., Murray, J.R., Ohmann, S.M. and Tong, C.B.S. (1997) Anthocyanin accumulation during potato tuber development. *Journal of the American Society for Horticultural Science* 122, 20–23.

International Potato Center (2009) Potato Facts and Figures (http://cipotato.org/potato/facts, accessed 23 May 2014).

IOM (2011) Dietary Reference Intakes for Vitamin A, Vitamin K, Arsenic, Boron, Chromium, Copper, Iodine, Iron, Manganese, Molybdenum, Nickel, Silicon, Vanadium, and Zinc (http://www.iom.edu/Reports/2001/Dietary-Reference-Intakes-for-Vitamin-A-Vitamin-K-Arsenic-Boron-Chromium-Copper-Iodine-Iron-Manganese-Molybdenum-Nickel-Silicon-Vanadium-and-Zinc.aspx, accessed 27 May 2014).

Itkin, M., Heinig, U., Tzfadia, O., Bhide, A.J., Shinde, B., Cardenas, P.D., *et al.* (2013) Biosynthesis of antinutritional alkaloids in solanaceous crops is mediated by clustered genes. *Science* 341, 175–179.

Iwanzik, W., Tevini, M., Stute, R. and Hilbert, R. (1983) Carotinoidgehalt und -zusammensetzung verschiedener deutscher Kartoffelsorten und deren Bedeutung fur die Fleischfarbe der Knolle. *Potato Research* 26, 149–162.

Jansen, G. and Flamme, W. (2006) Coloured potatoes (*Solanum Tuberosum* L.) – anthocyanin content and tuber quality. *Genetic Resources and Crop Evolution* 53, 1321–1331.

Jewell, D.K., Kendrick, J.G. and Satterlee, L.D. (1980) The DC-PER assay: a method for predicting protein quality solely from amino acid compositional data. *Nutrition Reports International* 21, 25–38.

Ji, X., Rivers, L., Zielinski, Z., Xu, M., MacDougall, E., Stephen, J., *et al.* (2012) Quantitative analysis of phenolic components and glycoalkaloids from 20 potato clones and *in vitro* evaluation of antioxidant, cholesterol uptake, and neuroprotective activities. *Food Chemistry* 133, 1177–1187.

Johnson, D.F., Bennett, R.D. and Heftmann, E. (1963) Cholesterol in higher plants. *Science* 140, 198–199.

Johnson, V.A. and Lay, C.L. (1974) Genetic improvement of plant protein. *Journal of Agricultural and Food Chemistry* 22, 558–566.

Jonczyk, R., Ronconi, S., Rychlik, M. and Genschel, U. (2008) Pantothenate synthetase is essential but not limiting for pantothenate biosynthesis in Arabidopsis. *Plant Molecular Biology* 66, 1–14.

Jordan, F. (2003) Current mechanistic understanding of thiamin diphosphate-dependent enzymatic reactions. *Natural Product Reports* 20, 184–201.

Jung, C.S., Griffiths, H.M., De Jong, D.M., Cheng, S., Bodis, M. and De Jong, W.S. (2005) The potato *P* locus codes for flavonoid 3',5'-hydroxylase. *Theoretical and Applied Genetics* 110, 269–275.

Jung, C.S., Griffiths, H.M., De Jong, D.M., Cheng, S., Bodis, M., Kim, T.S., *et al.* (2009) The potato developer (*D*) locus encodes an R2R3 MYB transcription factor that regulates expression of multiple anthocyanin structural genes in tuber skin. *Theoretical and Applied Genetics* 120, 45–57.

Kaneko, K., Tanaka, M.W. and Mitsuhashi, H. (1976) Origin of nitrogen in the biosynthesis of solanidine by *Veratrum grandiflorum*. *Phytochemistry* 15, 1391–1393.

Kaspar, K.L., Park, J.S., Brown, C.R., Mathison, B.D., Navarre, D.A. and Chew, B.P. (2011) Pigmented potato consumption alters oxidative stress and inflammatory damage in men. *Journal of Nutrition* 141, 108–111.

Kaspar, K.L., Park, J.S., Mathison, B.D., Brown, C.R., Massimino, S. and Chew, B.P. (2012) Processing of pigmented-flesh potatoes (*Solanum tuberosum* L.) on the retention of bioactive compounds. *International Journal of Food Science and Technology* 47, 376–382.

Katahira, R. and Ashihara, H. (2009) Profiles of the biosynthesis and metabolism of pyridine nucleotides in potatoes (*Solanum tuberosum* L.). *Planta* 231, 35–45.

Katan, M.B. and De Roos, N.M. (2004) Promises and problems of functional foods. *Critical Reviews in Food Science and Nutrition* 44, 369–377.

Katoh, A., Uenohara, K., Akita, M. and Hashimoto, T. (2006) Early steps in the biosynthesis of NAD in Arabidopsis start with aspartate and occur in the plastid. *Plant Physiology* 141, 851–857.

Keijbets, M.J.H. and Ebbenhorst-Seller, G. (1990) Loss of vitamin C (L-ascorbic acid) during long-term cold storage of Dutch table potatoes. *Potato Research* 33, 125–130.

Kobza, K., Camporeale, G., Rueckert, B., Kueh, A., Griffin, J.B., Sarath, G., *et al.* (2005) K4, K9 and K18 in human histone H3 are targets for biotinylation by biotinidase. *FEBS Journal* 272, 4249–4259.

Kohn, H.I. (1938) The concentration of coenzyme-like substance in blood following the administration of nicotinic acid to normal individuals and pellagrins. *Biochemical Journal* 32, 2075–2083.

Konings, E.J., Roomans, H.H., Dorant, E., Goldbohm, R.A., Saris, W.H. and van den Brandt, P.A. (2001) Folate intake of the Dutch population according to newly established liquid chromatography data for foods. *American Journal of Clinical Nutrition* 73, 765–776.

Kulen, O., Stushnoff, C. and Holm, D.G. (2013) Effect of cold storage on total phenolics content, antioxidant activity and vitamin C level of selected potato clones. *Journal of the Science of Food and Agriculture* 93, 2437–2444.

Kupchan, S.M., Barboutis, S.J., Knox, J.R. and Cam, C.A. (1965) Beta-solamarine: tumor inhibitor isolated from *Solanum dulcamara*. *Science* 150, 1827–1828.

Kurnasov, O., Goral, V., Colabroy, K., Gerdes, S., Anantha, S., Osterman, A., *et al*. (2003) NAD biosynthesis: identification of the tryptophan to quinolinate pathway in bacteria. *Chemistry and Biology* 10, 1195–1204.

Lamb, R.S., Citarelli, M. and Teotia, S. (2011) Functions of the poly(ADP-ribose) polymerase superfamily in plants. *Cellular and Molecular Life Sciences* 69, 175–189.

Lanska, D.J. (2010) Chapter 30: historical aspects of the major neurological vitamin deficiency disorders: the water-soluble B vitamins. *Handbook of Clinical Neurology* 95, 445–476.

Lavie, Y., Harel-Orbital, T., Gaffield, W. and Liscovitch, M. (2001) Inhibitory effect of steroidal alkaloids on drug transport and multidrug resistance in human cancer cells. *Anticancer Research* 21, 1189–1194.

Lee, K.R., Kozukue, N., Han, J.S., Park, J.H., Chang, E.Y., Baek, E.J., *et al*. (2004) Glycoalkaloids and metabolites inhibit the growth of human colon (HT29) and liver (HepG2) cancer cells. *Journal of Agricultural and Food Chemistry* 52, 2832–2839.

Legrand, D. and Scheen, A.J. (2007) Does coffee protect against type 2 diabetes? *Revue Medicale de Liege* 62, 554–559.

Leiner, I.E. (1977) Nutritional aspects of soy protein products. *Journal of the American Oil Chemists Society* 54, 454A–472A.

Lepiniec, L., Babiychuk, E., Kushnir, S., Van Montagu, M. and Inze, D. (1995) Characterization of an *Arabidopsis thaliana* cDNA homologue to animal poly(ADP-ribose) polymerase. *FEBS Letters* 364, 103–108.

LeRiche, E.L., Wang-Pruski, G. and Zheljazkov, V. (2009) Distribution of elements in potato (*Solanum tuberosum* L.) tubers and their relationship to after-cooking darkening. *HortScience* 44, 1866–1873.

Leuendorf, J.E., Osorio, S., Szewczyk, A., Fernie, A.R. and Hellmann, H. (2010) Complex assembly and metabolic profiling of *Arabidopsis thaliana* plants overexpressing vitamin B biosynthesis proteins. *Molecular Plant* 3, 890–903.

Lewis, C.E., Walker, J.R.L., Lancaster, J.E. and Sutton, K.H. (1998) Determination of anthocyanins, flavonoids and phenolic acids in potatoes. I. Coloured cultivars of *Solanum tuberosum L. Journal of the Science of Food and Agriculture* 77, 45–57.

Liang, C.H., Shiu, L.Y., Chang, L.C., Sheu, H.M., Tsai, E.M. and Kuo, K.W. (2008) Solamargine enhances HER2 expression and increases the susceptibility of human lung cancer H661 and H69 cells to trastuzumab and epirubicin. *Chemical Research in Toxicology* 21, 393–399.

Lin, S. and Cronan, J.E. (2011) Closing in on complete pathways of biotin biosynthesis. *Molecular Biosystems* 7, 1811–1821.

Linnemann, A.R., Es, A.v. and Hartmans, K.J. (1985) Changes in the content of L-ascorbic acid, glucose, fructose, sucrose and total glycoalkaloids in potatoes (cv. Bintje) stored at 7, 16 and 28 degrees C. *Potato Research* 28, 271–278.

Lisinska, G. and Leszczynski, W. (1989) *Potato Science and Technology*. Springer, New York.

Lonsdale, D. (2012) Thiamin(e): the spark of life. *Sub-Cellular Biochemistry* 56, 199–227.

Lotz, M., Zisman, E. and Bartter, F.C. (1968) Evidence for a phosphorus-depletion syndrome in man. *New England Journal of Medicine* 278, 409–415.

Love, S.L., Salaiz, T., Shafii, B., Price, W.J., Mosley, A.R. and Thornton, R.E. (2004) Stability of expression and concentration of ascorbic acid in North American potato germplasm. *HortScience* 39, 156–160.

Lu, S., Van Eck, J., Zhou, X., Lopez, A.B., O'Halloran, D.M., Cosman, K.M., *et al*. (2006) The cauliflower *Or* gene encodes a DnaJ cysteine-rich domain-containing protein that mediates high levels of beta-carotene accumulation. *Plant Cell* 18, 3594–3605.

Luis, G., Rubio, C., González-Weller, D., Gutiérrez, A.J., Revert, C. and Hardisson, A. (2011) Comparative study of the mineral composition of several varieties of potatoes (*Solanum tuberosum* L.) from different countries cultivated in Canary Islands (Spain). *International Journal of Food Science and Technology* 46, 774–780.

McCue, K.F., Allen, P.V., Shepherd, L.V.T., Blake, A., Maccree, M.M., Rockhold, D.R., *et al*. (2007) Potato glycosterol rhamnosyltransferase, the terminal step in triose side-chain biosynthesis. *Phytochemistry* 68, 327–334.

McCue, K.F., Rockhold, D.R., Chhan, A. and Belknap, W.R. (2011) Structure of two *Solanum tuberosum* steroidal glycoalkaloid glycosyltransferase genes and expression of their promoters in transgenic potatoes. *American Journal of Potato Research* 88, 485–492.

McKillop, D.J., Pentieva, K., Daly, D., McPartlin, J.M., Hughes, J., Strain, J.J., *et al.* (2002) The effect of different cooking methods on folate retention in various foods that are amongst the major contributors to folate intake in the UK diet. *British Journal of Nutrition* 88, 681–688.

McMillan, M. and Thompson, J.C. (1979) An outbreak of suspected solanine poisoning in schoolboys: examinations of criteria of solanine poisoning. *Quarterly Journal of Medicine* 48, 227–243.

McNulty, H. and Pentieva, K. (2004) Folate bioavailability. *Proceedings of the Nutrition Society* 63, 529–536.

Madiwale, G.P., Reddivari, L., Stone, M., Holm, D.G. and Vanamala, J. (2012) Combined effects of storage and processing on the bioactive compounds and pro-apoptotic properties of color-fleshed potatoes in human colon cancer cells. *Journal of Agricultural and Food Chemistry* 60, 11088–11096.

Malmberg, A. and Theander, O. (1985) Determination of chlorogenic acid in potato tubers. *Journal of Agricultural and Food Chemistry* 33, 549–551.

Manach, C., Scalbert, A., Morand, C., Remesy, C. and Jimenez, L. (2004) Polyphenols: food sources and bioavailability. *American Journal of Clinical Nutrition* 79, 727–747.

Meiyalaghan, S., Takla, M.F., Jaimess, O., Yongjin, S., Davidson, M.M., Cooper, P.A., *et al.* (2005) Evaluation of transgenic approaches for controlling tuber moth in potatoes. *Communications in Agricultural and Applied Biological Sciences* 70, 641–650.

Moehs, C.P., Allen, P.V., Friedman, M. and Belknap, W.R. (1997) Cloning and expression of solanidine UDP-glucose glucosyltransferase from potato. *Plant Journal* 11, 227–236.

Mondy, N.I. and Leja, M. (1986) Effect of mechanical injury on the ascorbic acid content of potatoes. *Journal of Food Science* 51, 355–357.

Monks, T.J., Xie, R., Tikoo, K. and Lau, S.S. (2006) Ros-induced histone modifications and their role in cell survival and cell death. *Drug Metabolism Reviews* 38, 755–767.

Monteiro, M., Farah, A., Perrone, D., Trugo, L.C. and Donangelo, C. (2007) Chlorogenic acid compounds from coffee are differentially absorbed and metabolized in humans. *Journal of Nutrition* 137, 2196–2201.

Mooney, S., Chen, L., Kuhn, C., Navarre, R., Knowles, N.R. and Hellmann, H. (2013) Genotype-specific changes in vitamin B 6 content and the PDX family in potato. *Biomedical Research International*, 389723.

Mori, H., Tanizawa, K. and Fukui, T. (1991) Potato tuber type H phosphorylase isozyme. Molecular cloning, nucleotide sequence, and expression of a full-length cDNA in *Escherichia coli*. *Journal of Biological Chemistry* 266, 18446–18453.

Morimoto, A., Ohno, Y., Tatsumi, Y., Mizuno, S. and Watanabe, S. (2012) Effects of healthy dietary pattern and other lifestyle factors on incidence of diabetes in a rural Japanese population. *Asia Pacific Journal of Clinical Nutrition* 21, 601–608.

Morris, M.S., Picciano, M.F., Jacques, P.F. and Selhub, J. (2008) Plasma pyridoxal 5'-phosphate in the US population: the National Health and Nutrition Examination Survey, 2003–2004. *American Journal of Clinical Nutrition* 87, 1446–1454.

Morris, W.L., Ducreux, L., Griffiths, D.W., Stewart, D., Davies, H.V. and Taylor, M.A. (2004) Carotenogenesis during tuber development and storage in potato. *Journal of Experimental Botany* 55, 975–982.

Mozaffarian, D., Hao, T., Rimm, E.B., Willett, W.C. and Hu, F.B. (2011) Changes in diet and lifestyle and long-term weight gain in women and men. *New England Journal of Medicine* 364, 2392–2404.

Mubarak, A., Hodgson, J.M., Considine, M.J., Croft, K.D. and Matthews, V.B. (2013) Supplementation of a high-fat diet with chlorogenic acid is associated with insulin resistance and hepatic lipid accumulation in mice. *Journal of Agricultural and Food Chemistry* 61, 4371–4378.

Murray, C., Markwick, N.P., Kaji, R., Poulton, J., Martin, H. and Christeller, J.T. (2010) Expression of various biotin-binding proteins in transgenic tobacco confers resistance to potato tuber moth, *Phthorimaea operculella* (Zeller) (fam. Gelechiidae). *Transgenic Research* 19, 1041–1051.

Mweetwa, A.M., Hunter, D., Poe, R., Harich, K.C., Ginzberg, I., Veilleux, R.E., *et al.* (2012) Steroidal glycoalkaloids in *Solanum chacoense*. *Phytochemistry* 75, 32–40.

Nault, B.A., Speese, J., 3rd, Kennedy, G.G., Linduska, J. and Dively, G.P. (2001) Response of potato tuber yield to stem injury by European corn borer (*Lepidoptera: Crambidae*) in the Mid-Atlantic United States. *Journal of Economic Entomology* 94, 1162–1169.

Navarre, D.A., Shakya, R., Holden, J. and Kumar, S. (2010) The effect of different cooking methods on phenolics and vitamin C in developmentally young potato tubers. *American Journal of Potato Research* 87, 350–359.

Navarre, D.A., Pillai, S., Shakya, R. and Holden, M.J. (2011) HPLC profiling of phenolics in diverse potato genotypes. *Food Chemistry* 127, 34–41.

Navarre, D.A., Payyavula, R.S., Shakya, R., Knowles, N.R. and Pillai, S. (2013) Changes in potato phenylpropanoid metabolism during tuber development. *Plant Physiology and Biochemistry* 65, 89–101.

Nayak, B., Berrios, J.D.J., Powers, J.R., Tang, J. and Ji, Y. (2011) Colored potatoes (*Solanum tuberosum* l.) dried for antioxidant-rich value-added foods. *Journal of Food Processing and Preservation* 35, 571–580.

Niggeweg, R., Michael, A.J. and Martin, C. (2004) Engineering plants with increased levels of the antioxidant chlorogenic acid. *Nature Biotechnology* 22, 746–754.

Nogueira, T. and do Lago, C.L. (2007) Determination of caffeine in coffee products by dynamic complexation with 3,4-dimethoxycinnamate and separation by CZE. *Electrophoresis* 28, 3570–3574.

Nosaka, K. (2006) Recent progress in understanding thiamin biosynthesis and its genetic regulation in *Saccharomyces cerevisiae*. *Applied Microbiology and Biotechnology* 72, 30–40.

Nunn, N. and Qian, N. (2011) The potato's contribution to population and urbanization: evidence from a historical experiment. *Quarterly Journal of Economics* 126, 593–650.

Nzaramba, M.N., Scheuring, D.C., Koym, J.W. and Miller, J.C. Jr (2013) Relationships among antioxidant activity, total phenolic content and specific gravity in several potato cultivars grown in different environments. *American Journal of Potato Research* doi:10.1007/s12230-013-9326-z.

OECD (Organisation for Economic Co-operation and Development) (2013) http://www.oecd.org/ (accessed 23 May 2014).

Ong, K.W., Hsu, A. and Tan, B.K. (2013) Anti-diabetic and anti-lipidemic effects of chlorogenic acid are mediated by ampk activation. *Biochemical Pharmacology* 85, 1341–1351.

Otten, J.J., Pitzi Hellwig, J. and Meyers, L.D. (2006) *Dietary Reference Intakes: The Essential Guide to Nutrient Requirements*. The National Academies, Washington, DC.

Paiva, E., Lister, R.M. and Park, W.D. (1983) Induction and accumulation of major tuber proteins of potato in stems and petioles. *Plant Physiology* 71, 161–168.

Parr, A.J. and Bolwell, G.P. (2000) Phenols in the plant and in man. The potential for possible nutritional enhancement of the diet by modifying the phenols content or profile. *Journal of the Science of Food and Agriculture* 80, 985–1012.

Pasare, S., Wright, K., Campbell, R., Morris, W., Ducreux, L., Chapman, S., *et al.* (2013) The sub-cellular localisation of the potato (*Solanum tuberosum* L.) carotenoid biosynthetic enzymes, CrtRb2 and PSY2. *Protoplasma* doi:10.1007/s00709-013-0521-z.

Payyavula, R.S., Navarre, D.A., Kuhl, J.C., Pantoja, A. and Pillai, S.S. (2012) Differential effects of environment on potato phenylpropanoid and carotenoid expression. *BMC Plant Biology* 12, 39.

Payyavula, R.S., Singh, R. and Navarre, D.A. (2013) Transcription factors, sucrose and sucrose metabolic genes interact to regulate potato phenylpropanoid metabolism. *Journal of Experimental Botany* 64, 5115–5131.

Percudani, R. and Peracchi, A. (2009) The B6 database: a tool for the description and classification of vitamin B6-dependent enzymatic activities and of the corresponding protein families. *BMC Bioinformatics* 10, 273.

Perla, V., Holm, D.G. and Jayanty, S.S. (2012) Effects of cooking methods on polyphenols, pigments and antioxidant activity in potato tubers. *Food Science and Technology* 45, 161–171.

Pillai, S., Navarre, D.A. and Bamberg, J.B. (2013) Analysis of polyphenols, anthocyanins and carotenoids in tubers from *Solanum tuberosum* group Phureja, Stenotomum and Andigena. *American Journal of Potato Research* 90, 440–450.

Pinon, V., Ravanel, S., Douce, R. and Alban, C. (2005) Biotin synthesis in plants. The first committed step of the pathway is catalyzed by a cytosolic 7-keto-8-aminopelargonic acid synthase. *Plant Physiology* 139, 1666–1676.

Pollak, N., Dolle, C. and Ziegler, M. (2007) The power to reduce: pyridine nucleotides – small molecules with a multitude of functions. *Biochemical Journal* 402, 205–218.

Poudel, B.N., Paudel, K.P., Timilsina, G. and Zilberman, D. (2012) Providing numbers for a food versus fuel debate: an analysis of a future biofuel production scenario. *Applied Economics Perspective and Policy* 34, 637–668.

Pouvreau, L., Gruppen, H., Piersma, S.R., Broek, L.A.M. v.d., Koningsveld, G.A. v. and Voragen, A.G.J. (2001) Relative abundance and inhibitory distribution of protease inhibitors in potato juice from cv. Elkana. *Journal of Agricultural and Food Chemistry* 49, 2864–2874.

Rajananthanan, P., Attard, G.S., Sheikh, N.A. and Morrow, W.J. (1999) Novel aggregate structure adjuvants modulate lymphocyte proliferation and Th1 and Th2 cytokine profiles in ovalbumin immunized mice. *Vaccine* 18, 140–152.

Randhawa, K.S., Sandhu, K.S., Kaur, G. and Singh, D. (1984) Studies of the evaluation of different genotypes of potato *Solanum tuberosum* for yield and mineral contents. *Plant Foods for Human Nutrition* 34, 239–242.

Rapala-Kozik, M., Kowalska, E. and Ostrowska, K. (2008) Modulation of thiamine metabolism in *Zea mays* seedlings under conditions of abiotic stress. *Journal of Experimental Botany* 59, 4133–4143.

Rapala-Kozik, M., Wolak, N., Kujda, M. and Banas, A.K. (2012) The upregulation of thiamine (vitamin B1) biosynthesis in *Arabidopsis thaliana* seedlings under salt and osmotic stress conditions is mediated by abscisic acid at the early stages of this stress response. *BMC Plant Biology* 12, 2.

Raschke, M., Boycheva, S., Crevecoeur, M., Nunes-Nesi, A., Witt, S., Fernie, A.R., *et al.* (2011) Enhanced levels of vitamin B(6) increase aerial organ size and positively affect stress tolerance in Arabidopsis. *Plant Journal* 66, 414–432.

Ray, D.K., Mueller, N.D., West, P.C. and Foley, J.A. (2013) Yield trends are insufficient to double global crop production by 2050. *PLoS One* 8, e66428.

Rebeille, F., Neuburger, M. and Douce, R. (1994) Interaction between glycine decarboxylase, serine hydroxymethyltransferase and tetrahydrofolate polyglutamates in pea leaf mitochondria. *Biochemical Journal* 302, 223–228.

Rebeille, F., Ravanel, S., Marquet, A., Mendel, R.R., Webb, M.E., Smith, A.G., *et al.* (2007) Roles of vitamins B5, B8, B9, B12 and molybdenum cofactor at cellular and organismal levels. *Natural Product Reports* 24, 949–962.

Reddivari, L., Hale, A.L. and Miller, J.C., Jr (2007a) Genotype, location, and year influence antioxidant activity, carotenoid content, phenolic content, and composition in specialty potatoes. *Journal of Agricultural and Food Chemistry* 55, 8073–8079.

Reddivari, L., Vanamala, J., Chintharlapalli, S., Safe, S.H. and Miller, J.C., Jr (2007b) Anthocyanin fraction from potato extracts is cytotoxic to prostate cancer cells through activation of caspase-dependent and caspase-independent pathways. *Carcinogenesis* 28, 2227–2235.

Reyes, L., Miller, J. and Cisneros-Zevallos, L. (2004) Environmental conditions influence the content and yield of anthocyanins and total phenolics in purple- and red-flesh potatoes during tuber development. *American Journal of Potato Research* 81, 187–193.

Richardson, D.P. (1990) Food fortification. *Proceedings of the Nutrition Society* 49, 39–50.

Roje, S. (2007) Vitamin B biosynthesis in plants. *Phytochemistry* 68, 1904–1921.

Rubin, H. (1975) Central role for magnesium in coordinate control of metabolism and growth in animal cells. *Proceedings of the National Academy of Sciences* 72, 3551–3555.

Said, H.M. (2002) Biotin: the forgotten vitamin. *American Journal of Clinical Nutrition* 75, 179–180.

Said, H.M. (2012) Biotin: biochemical, physiological and clinical aspects. *Sub-Cellular Biochemistry* 56, 1–19.

Sanchez-Castillo, C.P., Dewey, P.J.S., Aguirre, A., Lara, J.J., Vaca, R., Barra, P.L. d.l., *et al.* (1998) The mineral content of Mexican fruits and vegetables. *Journal of Food Composition and Analysis* 11, 340–356.

Scalbert, A., Manach, C., Morand, C., Remesy, C. and Jimenez, L. (2005) Dietary polyphenols and the prevention of diseases. *Critical Reviews in Food Science and Nutrition* 45, 287–306.

Schippers, J.H., Nunes-Nesi, A., Apetrei, R., Hille, J., Fernie, A.R. and Dijkwel, P.P. (2008) The Arabidopsis onset of leaf death5 mutation of quinolinate synthase affects nicotinamide adenine dinucleotide biosynthesis and causes early ageing. *Plant Cell* 20, 2909–2925.

Scott, J., Rebeille, F. and Fletcher, J. (2000) Folic acid and folates: the feasibility for nutritional enhancement in plant foods. *Journal of the Science of Food and Agriculture* 80, 795–804.

Settembre, E., Begley, T.P. and Ealick, S.E. (2003) Structural biology of enzymes of the thiamin biosynthesis pathway. *Current Opinion in Structural Biology* 13, 739–747.

Shakya, R. and Navarre, D.A. (2008) LC-MS analysis of solanidane glycoalkaloid diversity among tubers of four wild potato species and three cultivars (*Solanum tuberosum*). *Journal of Agricultural and Food Chemistry* 56, 6949–6958.

Shankar, A. and Prasad, A. (1998) Zinc and immune function: the biological basis of altered resistance to infection. *The American Journal of Clinical Nutrition* 68, 447S–463S.

Sherman, A.R. (1992) Zinc, copper, and iron nutriture and immunity. *Journal of Nutrition* 122, 604–609.

Shewry, P.R. (2003) Tuber storage proteins. *Annals of Botany* 91, 755–769.

Shinnosuke, H., Kentaro, T., Hirabayashi, T., Kawai-Yamada, M., Shoji, K., Goto, F., *et al.* (2010) Metabolome analysis of Arabidopsis plants over-expressing NAD synthetase gene. Poster 8, Gene Regulation and Molecular Biology. In: *Plant Biology 2010*, 31 July–4 August 2010, Montreal Canada. American Society of Plant Biologists, Rockville, Maryland, No 08021.

Shiotani, B., Kobayashi, M., Watanabe, M., Yamamoto, K., Sugimura, T. and Wakabayashi, K. (2006) Involvement of the ATR- and ATM-dependent checkpoint responses in cell cycle arrest evoked by pierisin-1. *Molecular Cancer Research* 4, 125–133.
Silva, G.H., Chase, R.W., Hammerschmidt, R. and Cash, J.N. (1991) After-cooking darkening of Spartan Pearl potatoes as influenced by location, phenolic acids, and citric acid. *Journal of Agricultural and Food Chemistry* 39, 871–873.
Sotelo, A. and Serrano, B. (2000) High-performance liquid chromatographic determination of the glycoalkaloids alpha-solanine and alpha-chaconine in 12 commercial varieties of Mexican potato. *Journal of Agricultural and Food Chemistry* 48, 2472–2475.
Staggs, C.G., Sealey, W.M., McCabe, B.J., Teague, A.M. and Mock, D.M. (2004) Determination of the biotin content of select foods using accurate and sensitive HPLC/avidin binding. *Journal of Food Composition Analysis* 17, 767–776.
Stapleton, A., Allen, P.V., Friedman, M. and Belknap, W.R. (1991) Purification and characterization of solanidine glycosyltransferase from the potato (*Solanum tuberosum*). *Journal of Agricultural and Food Chemistry* 39, 1187–1193.
Strohmeier, M., Raschle, T., Mazurkiewicz, J., Rippe, K., Sinning, I., Fitzpatrick, T.B., *et al.* (2006) Structure of a bacterial pyridoxal 5'-phosphate synthase complex. *Proceedings of the National Academy of Sciences* 103, 19284–19289.
Stushnoff, C., Ducreux, L.J., Hancock, R.D., Hedley, P.E., Holm, D.G., McDougall, G.J., *et al.* (2010) Flavonoid profiling and transcriptome analysis reveals new gene-metabolite correlations in tubers of *Solanum tuberosum* L. *Journal of Experimental Botany* 61, 1225–1238.
Subramanian, N.K., White, P.J., Broadley, M.R. and Ramsay, G. (2011) The three-dimensional distribution of minerals in potato tubers. *Annals of Botany* 107, 681–691.
Tahiliani, A.G. and Beinlich, C.J. (1991) Pantothenic acid in health and disease. *Vitamins and Hormones* 46, 165–228.
Takeda, E., Taketani, Y., Sawada, N., Sato, T. and Yamamoto, H. (2004) The regulation and function of phosphate in the human body. *BioFactors* 21, 345–355.
Tambasco-Studart, M., Titiz, O., Raschle, T., Forster, G., Amrhein, N. and Fitzpatrick, T.B. (2005) Vitamin B6 biosynthesis in higher plants. *Proceedings of the National Academy of Sciences* 102, 13687–13692.
Tan, J.S., Wang, J.J., Flood, V., Rochtchina, E., Smith, W. and Mitchell, P. (2008) Dietary antioxidants and the long-term incidence of age-related macular degeneration: the Blue Mountains Eye Study. *Ophthalmology* 115, 334–341.
Tosun, B.N. and Yücecan, S. (2008) Influence of commercial freezing and storage on vitamin C content of some vegetables. *International Journal of Food Science and Technology* 43, 316–321.
Tudela, J.A., Cantos, E., Espin, J.C., Tomas-Barberan, F.A. and Gil, M.I. (2002a) Induction of antioxidant flavonol biosynthesis in fresh-cut potatoes. Effect of domestic cooking. *Journal of Agricultural and Food Chemistry* 50, 5925–5931.
Tudela, J.A., Espin, J.C. and Gil, M.I. (2002b) Vitamin C retention in fresh-cut potatoes. *Postharvest Biology and Technology* 26, 75–84.
Tunc-Ozdemir, M., Miller, G., Song, L., Kim, J., Sodek, A., Koussevitzky, S., *et al.* (2009) Thiamin confers enhanced tolerance to oxidative stress in Arabidopsis. *Plant Physiology* 151, 421–432.
Urbany, C., Stich, B., Schmidt, L., Simon, L., Berding, H., Junghans, H., *et al.* (2011) Association genetics in *Solanum tuberosum* provides new insights into potato tuber bruising and enzymatic tissue discoloration. *BMC Genomics* 12, 7.
Vahteristo, L., Lehikoinen, K., Ollilainen, V. and Varo, P. (1997) Application of an HPLC assay for the determination of folate derivatives in some vegetables, fruits and berries consumed in Finland. *Food Chemistry* 59, 589–597.
Valkonen, J.P.T., Keskitalo, M., Vasara, T. and Pietila, L. (1996) Potato glycoalkaloids: a burden or a blessing? *Critical Reviews in Plant Sciences* 15(1), 1–20.
Vinson, J.A., Demkosky, C.A., Navarre, D.A. and Smyda, M.A. (2012) High-antioxidant potatoes: acute *in vivo* antioxidant source and hypotensive agent in humans after supplementation to hypertensive subjects. *Journal of Agricultural and Food Chemistry* 60, 6749–6754.
Virag, L. and Szabo, C. (2002) The therapeutic potential of poly(ADP-ribose) polymerase inhibitors. *Pharmacological Reviews* 54, 375–429.
Wacker, W. and Parisi, A. (1968) Magnesium metabolism. *New England Journal of Medicine* 45, 658–663.
Wagner, S., Bernhardt, A., Leuendorf, J.E., Drewke, C., Lytovchenko, A., Mujahed, N., *et al.* (2006) Analysis of the Arabidopsis rsr4-1/pdx1-3 mutant reveals the critical function of the PDX1 protein family in metabolism, development, and vitamin B6 biosynthesis. *Plant Cell* 18, 1722–1735.

Wahlberg, G., Adamson, U. and Svensson, J. (2000) Pyridine nucleotides in glucose metabolism and diabetes: a review. *Diabetes/Metabolism Research and Reviews* 16, 33–42.

Waller, J.C., Alvarez, S., Naponelli, V., Lara-Nunez, A., Blaby, I.K., Da Silva, V., *et al.* (2010) A role for tetrahydrofolates in the metabolism of iron-sulfur clusters in all domains of life. *Proceedings of the National Academy of Sciences* 107, 10412–10417.

Walsh, C.T., Sandstead, H.H., Prasad, A.S., Newberne, P.M. and Fraker, P.J. (1994) Zinc: health effects and research priorities for the 1990s. *Environmental Health Perspectives* 102, 5–46.

Wang, G., Ding, X., Yuan, M., Qiu, D., Li, X., Xu, C., *et al.* (2006) Dual function of rice OsDR8 gene in disease resistance and thiamine accumulation. *Plant Molecular Biology* 60, 437–449.

Wang, Y.L. and Yudkin, J. (1940) Assessment of the level of nutrition. Urinary excretion of aneurin at varying levels of intake. *Biochemical Journal* 34, 343–352.

Wang-Pruski, G. (2007) The canon of potato sciene: 47. After-cooking darkening. *Potato Research* 50, 403–406.

Werij, J.S., Kloosterman, B., Celis-Gamboa, C., de Vos, C.H., America, T., Visser, R.G., *et al.* (2007) Unravelling enzymatic discoloration in potato through a combined approach of candidate genes, QTL, and expression analysis. *Theoretical and Applied Genetics* 115, 245–252.

White, P.J., Bradshaw, J.E., Finlay, M., Dale, B., Ramsay, G., Hammond, J.P., *et al.* (2009) Relationships between yield and mineral concentrations in potato tubers. *HortScience* 44, 6–11.

Willett, W.C. and Stampfer, M.J. (2003) Rebuilding the food pyramid. *Scientific American* 288, 64–71.

Williams, D.J., Hamernig, I., Jian, L., James, A.P., Johnson, S.K. and Tapsell, L.C. (2013) Vegetables containing phytochemicals with potential anti-obesity properties: a review. *Food Research International* 52, 323–333.

Wills, R.B.H., Lim, J.S.K. and Greenfield, H. (1984) Variation in nutrient composition of Australian retail potatoes over a 12-month period. *Journal of the Science of Food and Agriculture* 35, 1012–1017.

Wilson, G.S. (1959) A small outbreak of solanine poisoning. *Monthly Bulletin of the Ministry of Health and the Public Health Laboratory Service* 18, 207–210.

Wolters, A.M., Uitdewilligen, J.G., Kloosterman, B.A., Hutten, R.C., Visser, R.G. and van Eck, H.J. (2010) Identification of alleles of carotenoid pathway genes important for zeaxanthin accumulation in potato tubers. *Plant Molecular Biology* 73, 659–671.

Woolfe, J. (1987) *The Potato in the Human Diet*. Cambridge University Press, Cambridge, UK.

Wu, T., Guo, A., Zhao, Y., Wang, X., Wang, Y., Zhao, D., *et al.* (2010) Ectopic expression of the rice lumazine synthase gene contributes to defense responses in transgenic tobacco. *Phytopathology* 100, 573–581.

Yamaguchi, T., Chikama, A., Mori, K., Watanabe, T., Shioya, Y., Katsuragi, Y., *et al.* (2007) Hydroxyhydroquinone-free coffee: a double-blind, randomized controlled dose-response study of blood pressure. *Nutrition, Metabolism and Cardiovascular Disease* 18, 408–414.

Young, D.B., Lin, H. and McCabe, R.D. (1995) Potassium's cardiovascular protective mechanisms. *American Journal of Physiology – Regulatory, Integrative and Comparative Physiology* 268, R825–R837.

Zalewski, P. (1996) Zinc and immunity: implications for growth, survival and function of lymphoid cells. *Journal of Nutritional Immunology* 4, 39–80.

Zempleni,J. (2005) Uptake, localization, and noncarboxylase roles of biotin. *Annual Review of Nutrition* 25, 175–196.

Zhang, Y., Cheng, S., De Jong, D., Griffiths, H., Halitschke, R. and De Jong, W. (2009) The potato *R* locus codes for dihydroflavonol 4-reductase. *Theoretical and Applied Genetics* 119, 931–937.

Zhao, X., Sheng, F., Zheng, J. and Liu, R. (2011) Composition and stability of anthocyanins from purple *Solanum tuberosum* and their protective influence on Cr(VI) targeted to bovine serum albumin. *Journal of Agricultural and Food Chemistry* 59, 7902–7909.

Zhou, X., McQuinn, R., Fei, Z., Wolters, A.M., van Eck, J., Brown, C., Giovannoni, J.J. and Li, L. (2011) Regulatory control of high levels of carotenoid accumulation in potato tubers. *Plant, Cell and Environment* 34, 1020–1030.

Zimowski, J. (1991) Occurrence of a glucosyltransferase specific for solanidine in potato plants. *Phytochemistry* 30, 1827–1831.

19 Potato Flavor

Mark A. Taylor*
Cell and Molecular Sciences, The James Hutton Institute, Invergowrie, UK

Within the European Union (EU), potato is the major vegetable crop – in 2007 tuber yield was 61 million tonnes (Mt) on 2.2 million hectares (Mha), with yield increasing year on year (http://ec.europa.eu/agriculture/potatoes/index_en.htm). Despite this, potato consumption in many parts of the EU is falling—for example, in Denmark consumption has halved during the last generation to 57 kg year^{-1} per capita (Seefeldt *et al.*, 2011). There are many reasons for this decline, including the time for preparation, competition from rice and pasta as major carbohydrate sources, and a perception by increasingly sophisticated customers that potato does not offer the variability or variety of other foods and is somewhat passé. However, one of the most important factors in consumer purchase is product quality. "Quality" per se is a nebulous term that encompasses a wide range of traits, but the most important aspects are the organoleptic drivers: appearance, texture, and flavor (McGregor, 2007). Within a price range, appearance drives the initial purchase, while cooking performance, including the flavor components, taste, texture, and volatile aroma compounds, underpin continued purchase. In addition to being a staple source of carbohydrate, the nutritional value of potato is being increasingly appreciated (Navarre *et al.*, 2009). Potato is a good source of vitamin C (Love *et al.*, 2004), and contains nutritionally significant levels of potassium (Burrowes and Ramer, 2008). Some potato germplasm produce tubers containing significant levels of vitamins and micronutrients including folate, vitamin B6, carotenoids, and anthocyanins (Navarre *et al.*, 2009). Potatoes with improved flavor would encourage consumption, potentially having a significant impact on nutrition.

Despite the increasing importance of potato flavor to consumers, there is still much to be learned about the complex flavor trait and the key components that contribute to it. A particular problem is how to breed for good tuber flavor. With the emphasis on breeders to produce cultivars that yield well and are disease resistant, there has been less attention paid to potato flavor. In fact, breeders lack the tools to assess flavor, as this trait is very difficult to phenotype (Klee, 2010). Flavor assessments depend on quantitative descriptive analysis carried out by trained taste panels. Taste panels are expensive and have a relatively low throughput, and so flavor is generally assessed only in the later stages of a breeding program after selection for more easily quantifiable traits. In fact, most of the potential flavor and texture improvements are likely to be discarded, and to a large extent, the marketplace determines whether a new cultivar is acceptable to consumers (Wang and Kays, 2003). Means of determining flavor analytically

* E-mail: mark.taylor@hutton.ac.uk

that can be applied efficiently to germplasm are obvious requirements for improvement of the potato crop. Additionally, breeders require clear target phenotypes for flavor, as preferences are likely to vary depending on consumer groups, market classes, and cooking methods. In this chapter, we shall review the progress that has been made in understanding these complex traits, and the opportunities that are arising from advances in analytic chemistry, sensory science, and genomic and metabolomic technologies that are now being applied to potato.

19.1 Flavor Perception

The perception of flavor involves the integration of signals emanating from taste and olfactory receptors. However, while taste receptors clearly respond to relatively few cues, olfactory receptors respond to thousands of chemicals, and as such are thought to be responsible for the vast diversity of unique food flavors (Goff and Klee, 2006; Tieman *et al.*, 2006). In cooked potato, as with other food, it is thus necessary to consider both non-volatile taste compounds, as well as volatile components, as contributing to flavor. Although the flavor of processed potatoes is important to consumers, the blend of volatile and non-volatile metabolites that impact on flavor attributes is not well defined. With the advent of high-throughput analytical techniques and advances in sensory science, we are starting to arrive at a picture of the key flavor components, including both volatile and non-volatile components.

19.2 Non-volatile Compounds

The umami taste

In combination with the volatile compounds produced on cooking potato tubers, soluble cellular constituents are also likely to be important in flavor (Halpern, 2000). The four taste qualities, sweet, bitter, salty, and sour, were recognized traditionally as those on which human taste was based. However, a fifth taste quality was proposed by Ikeda in 1909, based on the dominant taste of a seaweed-based soup that was clearly distinct from the four previously considered basic tastes (Lindemann and Ogiwara, 2002). This fifth taste was called "umami", and was usually described as meaty, broth-like, or savory. The acceptance of umami as a basic taste has received impetus from the discovery of a taste receptor, expressed on the surface of taste cells, that can interact specifically with umami compounds (Li *et al.*, 2002; Nelson *et al.*, 2002). Umami compounds generally enhance flavor and mouth feel, giving the impression of creaminess and viscosity to savory dishes (Halpern, 2000).

Initially, the umami taste was attributed solely to the amino acid, glutamate. However, more recently, adenosine-5′-monophosphate (5′-AMP), inosine-5′-monophosphate (5′-IMP) and guanosine-5′-monophosphate (5′-GMP) were also demonstrated to show umami-like sensory characteristics (Ninomiya, 1998; Yamaguchi and Ninomiya, 1998). Glutamate is naturally present in most foods, including meat, poultry, seafood, and vegetables. The two kinds of nucleotides that contribute most to the umami taste, 5′-IMP and 5′-GMP, are also present in many foods, with the former found primarily in meat, whereas the latter is more abundant in plants. Another nucleotide, 5′-AMP, is abundant in fish and shellfish. Generally recognized umami-rich food includes seaweed, Parmesan cheese, sardines, cod, prawns, and oysters. Vegetables recognized as having significant umami levels include tomatoes, mushrooms, truffles, soybeans, and potatoes (Ninomiya, 1998).

An important and unusual characteristic of umami compounds is their mutual taste synergism. Thus, umami taste intensity increases exponentially when glutamate or its analogs interact with 5′-ribonucleotides (Yamaguchi *et al.*, 1971). The synergistic effect between certain free amino acids and 5′-nucleotides can be measured using the equivalent umami calculation developed by Yamaguchi *et al.* (Yamaguchi *et al.*, 1971):

$$Y = \Sigma a_i b_i + 1218(\Sigma a_i b_i)(\Sigma a_j b_j)$$

where Y is the equivalent umami content of the mixture; a_i is the concentration of each umami amino acid (Glu or Asp); a_j is the concentration of each umami 5′-nucleotide (5′-GMP or 5′-AMP); b_i is the relative umami concentration (RUC) for each umami amino acid to Glu (Glu = 1, Asp = 0.077); b_j is the RUC for each umami 5′-nucleotide to 5′-IMP (5′-GMP = 2.3,

5′-AMP = 0.18), and 1218 is a synergistic constant. All concentrations must be in g 100 g^{-1}.

Glutamate is the most potent umami amino acid, with aspartate showing only 7% of the taste activity of glutamate. 5′-GMP is the most potent common 5′-nucleotide, having a 2.3-fold greater taste activity than 5′-IMP (Yamaguchi *et al.*, 1971). Common chemical structural motifs appear to be necessary for a taste-enhancing effect; in particular, two negative charges at a distance of 3–9, or preferably 4–6, carbon atoms are essential (Ney, 1971). In addition to glutamate and aspartate, the C4-dicarboxylic acids, succinic acid and tartaric acid, fulfill this criterion and also exhibit some umami taste, probably as a result of the negative charge distance (Ney, 1971; Velisek *et al.*, 1978). More controversial is the question of whether small peptides can also give rise to an umami taste. However, several glutamate glycoconjugates formed during processing by the Maillard reaction have umami-like properties, and these compounds may contribute to overall flavor (Beksan *et al.*, 2003).

Umami in potato

Umami was first considered as a component of potato flavor in a paper published in 1970 (Buri *et al.*, 1970). This report detailed levels of glutamate and 5′-nucleotides in tuber samples from two potato cultivars, Bintje and Ostara. Using a taste panel, it was reported that the higher levels of glutamate and 5′-nucleotides in the Bintje sample corresponded to an "enhanced" taste. Following on from this, it was suggested that the flavor of boiled potato was due largely to the natural mixture of glutamic acid and other amino acids in combination with guanosine-5′-monophosphate (GMP) and other 5′-ribonucleotides produced on cooking (Solms, 1971). Thus, chemicals that impart an umami flavor are likely to be an important component of potato flavor. Indeed, some authors claim that there is only a small contribution from volatile (olfactory) components, and that chemicals representing the so-called sweet, sour, salty, and bitter tastes do not make a major contribution to potato flavor (Solms, 1971; Solms and Wyler, 1979). Thus, these authors suggest that the presence of salt, sugars, or glycoalkaloids do not enhance potato flavor, although their presence at high levels may decrease palatability. With the discovery of umami-tasting glutamate glycoconjugates, produced by a Maillard reaction (Beksan *et al.*, 2003), the potential indirect effects of sugars on potato flavor should not be overlooked

5′-nucleotides in potato tubers

Early work on umami in potato addressed the question of the origin of the 5′-nucleotides (Buri and Solms, 1971). Barely detectable levels of 5′-nucleotides could be measured in raw tuber samples, whereas appreciable levels (for example 64.4 μmoles kg^{-1} fresh weight 5′-guanosine monophosphate) accumulate in steamed samples. It was concluded that the origin of the 5′-nucleotides was RNA that had been degraded enzymatically by the action of RNases or phophodiesterases. More detailed analysis of 5′-nucleotide formation during cooking of potato tubers followed (Dumelin and Solms, 1976). There were no clear-cut conclusions from this work, apart from the recognition that potato tubers contained a complex enzyme system that was able to attack ribonucleic acid and to form different intermediates and end products of degradation. The interactions of phosphodiesterase I, phosphodiesterase II, and ribonucleases are likely to be important, but details of the interactions could not be elucidated from this study. However, temperature and pH conditions are of great importance to these reactions.

Glutamate in potato tubers

As well as 5′- nucleotides, the levels of amino acids, most importantly glutamate but also to a lesser degree aspartate, are contributors to the umami taste. The amino acid biosynthetic networks are complex and heavily regulated, although most of the biosynthetic genes have been identified (Halford *et al.*, 2004). The levels of amino acids in potato tubers have been the subject of several studies—for example, Brierley *et al.* surveyed tuber amino acid content in several cultivars at harvest and during storage (Brierley *et al.*, 1997). Asparagine and glutamine were the major amino acids, although the levels of glutamic and aspartic acids were significant. During storage, the level of glutamate increased (at 10°C the increase was *c.*fourfold).

Amino acid metabolism is a central part of primary metabolism, the various pathways of which are intimately interrelated. Recently, these complex pathways were analysed by state-of-the-art metabolomic approaches. Thus, Roessner-Tunali *et al.* demonstrated an inverse correlation between sucrose content and free amino acid level in the tuber (Roessner-Tunali *et al.*, 2004). Later, the same group published methods, based on [U-^{13}C] glucose labeling of potato tuber discs and gas chromatography–mass spectrometry (GC–MS), to investigate metabolic fluxes to different primary metabolites including amino acids (Roessner-Tunali *et al.*, 2004). Despite these advances, there are still important questions to be addressed concerning potato tuber amino acid content. For example, it is still unclear what proportions of amino acids are synthesised *de novo* in the tuber compared with those imported from the phloem (Fischer *et al.*, 1998), but clearly amino acid transporters could have a key role (Koch *et al.*, 2003). It is well known that glutamate and aspartate and their respective amides are found in the phloem sap of many species; however, the relative contribution to final tuber contents remains to be elucidated (Koch *et al.*, 2003). Recent work has also identified protein kinases involved in the regulation of carbon and amino acid metabolism (Halford *et al.*, 2004); however, the role of these regulators in potato tubers remains unknown.

Levels of umami compounds in different potato genotypes

In order to extend studies of potato flavor, a useful approach has been to compare potential flavor metabolite levels in potato germplasm with very different sensory properties. A valuable resource in this regard is *Solanum tuberosum* group Phureja, differentiated from *S. tuberosum* group Tuberosum on the basis of a number of important tuber quality traits such as flavor, texture, color, and reduced tuber dormancy (De Maine *et al.*, 1993, 1998; Dobson *et al.*, 2004; Morris *et al.*, 2004; Ghislain *et al.*, 2006). Despite being able to differentiate the Phureja group of landraces based on geographical origin, this group is very similar genetically to the Tuberosum group (Spooner *et al.*, 2007). Whatever the final taxonomic outcome, because many of the Phureja group can be clearly differentiated based on tuber quality trait parameters, they form a useful resource for the identification of key metabolites, and ultimately genes that underpin these traits. The study of Morris *et al.* has addressed one aspect of this issue, by measuring the levels of umami compounds in tuber samples that had been subjected to detailed sensory analysis by a trained panel (Morris *et al.*, 2007). The free levels of the major umami amino acids, glutamate and aspartate, and the 5′-nucleotides, 5′-GMP, 5′-AMP, were measured in potato samples during the cooking process. Tubers were sampled several times during the growing season. The levels of both glutamate and 5′-nucleotides were significantly higher in mature tubers of two *S. tuberosum* group Phureja clones compared with two *S. tuberosum* group Tuberosum cultivars. The equivalent umami concentration was calculated for five clones/cultivars, and there was a strong positive correlation with the acceptability scores derived from a trained evaluation panel, suggesting that umami was an important component of potato flavor (Fig. 19.1).

There was a large effect of the developmental stage on the level of cooked 5′-nucleotides. Interestingly, as tuber development progressed, the effect on 5′-nucleotide levels was different for the Phureja and Tuberosum cultivars. In small developing tubers of both cultivars, 5′-nucleotide levels were generally lower, and levels in the Tuberosum cultivars were not significantly different to those in the Phureja clones (Stage H1 in Fig. 19.2). Conversely, at tuber maturity, whereas low levels of 5′-GMP could be detected in the Tuberosum tubers, significantly higher levels were present in cooked Phureja tubers (Stage H2 in Fig. 19.2). In addition, levels of 5′-GMP remained higher in the Phureja tubers, compared with Tuberosum, after 6 weeks storage at 4°C (Stage H3 in Fig. 19.2). The implications of this finding, in view of the demonstrated association of umami level and taste, is that these compounds contribute to taste differences of tubers at different developmental stages (Fig. 19.2).

Previously, it has been suggested that 5′-nucleotides accumulate due to the action of nucleases during cooking processes, particularly due to RNA degradation (Buri and Solms, 1971). Ribonucleases are active under the pH and temperature conditions that occur during heating, being particularly active at around 50°C (Singh *et al.*, 1991). As the temperature

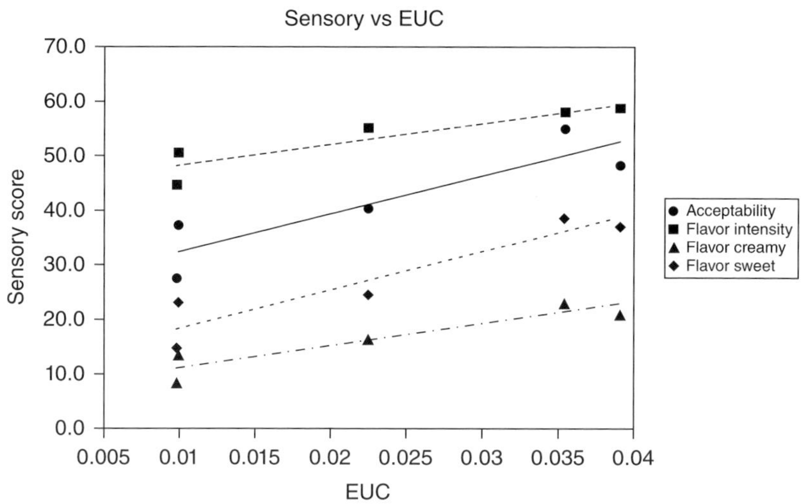

Fig. 19.1. Correlation of sensory evaluation scores with equivalent umami concentration (EUC) of potato. *Solanum tuberosum* cultivars, Maris Piper (MP) and Record, were compared with *Solanum phureja* clones, DB333-16 and DB257-28, and cultivar Mayan Gold (MG). Sensory scores are on a scale of 0–100. (From Morris *et al.*, 2007.)

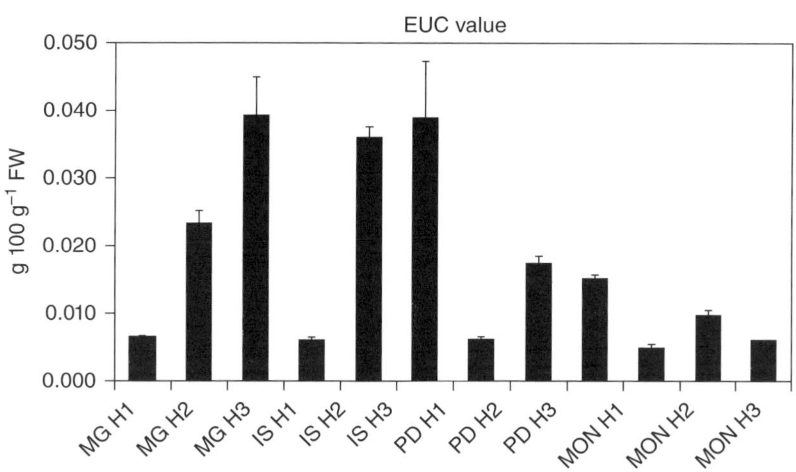

Fig. 19.2. Effect of tuber developmental stage on equivalent umami concentrations (EUC) in cooked potato cultivars, Mayan Gold (MG), Inca Sun (IS), Pentland Dell (PD), and Montrose (MON). H1 = harvest 1; H2 = harvest 2; H3 = harvest 3 (see text for details). Error bars represent the SEM (*n* = 3).

of potato tissues increases slowly from 40 to 60°C during some cooking processes (for example boiling), nuclease activity may be significant (Solms and Wyler, 1979). Despite the proposed involvement of RNases, no differences in RNase, or phosphohydrolytic enzyme, activities could be detected in extracts from Tuberosum and Phureja (Morris *et al.*, 2007). It may be that more subtle differences in these activities, which escape detection in the extraction and assay methods used in this study, account for the difference. Alternatively, the higher rate of 5′-nucleotide accumulation in Phureja may reflect textural differences between Phureja and

Tuberosum, so that the substrate (RNA) and degradative enzymes come into contact more readily during cooking of the Phureja tubers.

Other recent work has investigated the effects of adding umami compounds to different model food matrices and measuring the hedonic effects (Barylko-Pikielna and Kostyra, 2007). Glutamate, either on its own or in combination with 5′-IMP and 5′-GMP, was added to a range of soups and broths, or to mashed potato. There were significant increases in the degree of acceptability of most of the foods when the umami compounds were added, as assessed by a taste panel. For mashed potato, the increase was smaller than for the soups and broths, and the authors of this report concluded that potato as a food matrix was insensitive to supplementation with umami substances. Although this approach was interesting, little further analysis of the potato samples was carried out, and there was no comparison of different potato cultivars.

Glycoalkaloids and flavor

Potato tubers are well known to contain glycoalkaloids, the major forms being α-chaconine and α-solanine, which represent approximately 95% of the total (Slanina, 1990). The level of tuber glycoalkaloids varies according to cultivar (Ramsay *et al.*, 2004), and is also influenced by growth and storage conditions (Sengul *et al.*, 2004). Tuber glycoalkaloids are toxic to humans if present at levels in excess of 20 mg 100 g^{-1} (Osman, 1983). Additionally, high levels of glycoalkaloids are responsible for off-flavors in potato tubers (Zitnak, 1961; Sinden *et al.*, 1976; Ross *et al.*, 1978). The emerging consensus is that potato glycoalkaloids at elevated levels are responsible for flavors described as bitter, burning, scratchy, or acrid, and are thus generally undesirable components of flavor. In one study, taste panels correlated glycoalkaloid content in excess of 10 mg 100 g^{-1} fresh weight with a burning taste. However, at levels between 0.76 and 4.98 mg 100 g^{-1}, no correlation was found (Ross *et al.*, 1978). Other taste panel studies described how tubers with glycoalkaloid contents greater than 14 mg 100 g^{-1} had a bitter taste (Sinden *et al.*, 1976).

19.3 Volatile Compounds

The volatiles produced by raw and cooked potatoes have been studied extensively (reviewed by Maga, 1994), and over 250 compounds have been identified in potato volatile fractions. Since then, attempts have been made to discriminate which of these components are important for potato flavor and which are specific to the method of cooking, cultivar differences, the effects of agronomic conditions, and the effects of storage (Duckham *et al.*, 2001, 2002; Oruna-Concha *et al.*, 2001, 2002; Dobson *et al.*, 2004). Although many different volatiles can be detected in cooked potato, it is necessary to consider the odor activity value (OAV) for each when assessing whether a compound is likely to impact on sensory assessment. OAV is the ratio of concentration of the odorant in the material to the odor threshold (the lowest concentration of a vapor in air that can be detected by smell). In some foods, only a small fraction of volatile compounds contribute to aroma (Grosch, 2001). For example, although more than 400 aroma volatiles have been identified in tomato, only 15–20 are present in sufficient quantities to have an impact on flavor (Buttery, 1993; Baldwin *et al.*, 2000). These compounds are derived from a diverse set of amino acid, carotenoid, and fatty acid precursors (Klee, 2010).

Recent advances in analytical technology have facilitated the study of flavor volatiles greatly. In particular, relatively high throughput methods for the profiling of flavor volatiles using GC–MS have been reported (Thybo *et al.*, 2006; Shepherd *et al.*, 2007). A particularly useful and sensitive method is the analysis of head-space volatiles by solid phase micro extraction. In this process, samples of tuber are cooked in a sealed vial prior to analysis (Morris *et al.*, 2010).

The origins of potato flavor volatiles

The volatile profile obtained from cooked potatoes contains many process-derived compounds originating from a range of metabolite precursors (Maarse, 1991). Many flavor volatiles, including aldehydes, alcohols, and alkyl furans, are derived from lipids in reactions that have both enzymic and non-enzymic steps. Enzymic reactions in lipid degradation are likely to be important if results from tomato fruit can be extrapolated to potato. Lipoxygenases catalyze the dioxygenation of polysaturated fatty acids containing *cis*, *cis*-1,4-pentadiene moieties (such as linoleic and linolenic acids), converting them into corresponding fatty-acid hydroperoxides (HPOs). In tomato, five lipoxygenase genes have been identified, and one (*TomloxC*) has a critical role in the generation of important six-carbon aldehyde and alcohol fatty-acid-derived flavor volatiles (Chen *et al.*, 2004). Autooxidation (Frankel, 1998) and enzymic (hydroperoxide lyase) action on hydroperoxides derived from linoleic and α-linolenic acids (Gardner, 1995) are also involved in the production of lipid-derived, flavor-active volatiles such as aldehydes, ketones, alcohols, and alkyl furans.

Thermally driven reactions are important in the development of many flavor volatiles. Major examples include the Maillard reaction between reducing sugars and amino acids, and the Strecker degradation of methionine to produce methional. The Maillard reaction takes place when compounds possessing a carbonyl group, typically reducing sugars, react with components with a free amino group such as amino acids to form heterocyclic compounds. The Maillard reaction occurs under more forcing conditions such as baking or roasting (Shepherd *et al.*, 2007). The Strecker degradation of methionine involves the interaction of α-dicarbonyl compounds, intermediates in the Maillard reaction, with methionine, resulting in the formation of methional (Lindsay, 1996). The other major classes of cooked potato volatiles include methoxypyrazines and terpenes. In addition, several specific compounds have been related directly to desirable flavor and aroma characteristics with methional (Lindsay, 1996), methoxypyrazines (Murray and Whitfield, 1975), and the lipid degradation product, *cis*-4-heptenal (Josephson and Lindsay, 1987), all reported to exhibit "a cooked potato odor". This provides some evidence that these compounds may be significant contributors to potato flavor. To generalize the findings of several studies, in boiled potatoes, the most important volatile constituents are likely to be those derived from lipid degradation, the Maillard reaction, and the Strecker reaction, and include methional, aliphatic alcohols and aldehydes, thiols and sulfides and methoxypyrazines.

The types and amounts of volatiles derived from potato tubers vary significantly according to the processing method. Distinctive quantitative and qualitative differences in volatile profile were observed in a comparison of tubers (cultivars Estima and Maris Piper) cooked by boiling, baking, or microwaving (Oruna-Concha *et al.*, 2002). For the three methods of cooking, the main sources of flavor compounds were from lipid degradation and the Maillard reaction and/or sugar degradation; however, there was a characteristically different ratio of the contribution of each class to the total volatile fraction. For boiled potatoes, volatiles formed by lipid degradation were judged to be the most important, whereas for baked potatoes, those arising from amino acids contributed more significantly (Oruna-Concha *et al.*, 2002). A much lower amount of volatile aroma compounds was produced from microwave-cooked potatoes than for the other cooking methods, possibly indicating that microwave-cooked potato products are blander in flavor than potato products produced by boiling or conventional baking. The cooking-dependent differences in volatile profiles have been attributed to variations in heat and mass transfer processes associated with the different cooking methods (Wilson *et al.*, 2002a,b).

In order to determine which, of the many volatiles present in cooked potato, are the main contributors to flavor, comparisons have been made between potato germplasm with different sensory attributes. As with studies on umami compounds, accessions of Phureja have been compared with Tuberosum. The study of Winfield *et al.* showed that, in general, boiled tubers from a range of Phureja accessions were more acceptable than those from a wide range of Tuberosum (Winfield *et al.*, 2005). Head-space volatile analysis from these samples identified 70 compounds, and principal component analysis revealed that hexenal, pentanal, pentyl-furan, and the sesquiterpene, α-copaene, accounted for most of the variability in the data set.

This type of approach was extended to investigate in detail the correlation between volatile and non-volatile tuber components and sensory attributes. The resulting correlation matrix revealed new associations between flavor attributes, most notably between flavor intensity, flavor creaminess, and flavor savoriness (Fig. 19.3). Interestingly, there was an almost inverse relationship

Metabolite	Aroma	Intensity	Sweetness	Savory	Creaminess	Off
Methylsalicylate	−0.30	0.95	−0.76	0.86	0.74	−0.39
Hexanal	−0.57	0.93	−0.54	0.80	0.89	−0.19
Pentanal	−0.26	0.87	−0.86	0.82	0.58	−0.40
Z-(or *E*)-2- (2-pentenyl)furan	−0.35	0.82	−0.71	0.67	0.61	−0.05
3-Ethyl-2-methyl-1,3-hexadiene	−0.64	0.80	−0.87	0.68	0.57	0.04
3-or-2-Methylbutanoic acid methyl ester	−0.55	0.78	−0.51	0.76	0.67	0.07
AMP g 100 g^{-1} FW	−0.63	0.76	−0.61	0.71	0.63	0.11
GMP g 100 g^{-1} FW	−0.69	0.74	−0.57	0.70	0.65	0.12
1-Pentanol	−0.29	0.73	−0.97	0.72	0.39	−0.34
Unknown (RRI 1499)	−0.14	0.67	−0.85	0.58	0.34	−0.33
EUC	−0.51	0.65	−0.14	0.60	0.72	−0.14
2-Heptenal	−0.40	0.60	−0.59	0.47	0.38	0.21
2-Methylfuran	−0.43	0.60	−0.55	0.60	0.38	0.23
2-Pentanal	−0.69	0.59	−0.30	0.45	0.61	0.31
1-Pentanal-3-ol	−0.59	0.57	−0.79	0.66	0.29	0.04
2-Propylfuran	−0.49	0.54	−0.43	0.50	0.39	0.33
Octanal	−0.32	0.51	−0.45	0.38	0.30	0.30
2-Ethylfuran	−0.27	0.48	−0.35	0.53	0.25	0.21
Decanal	−0.09	0.38	−0.41	0.30	0.10	0.22
Sucrose mg g^{-1} FW	−0.03	0.34	0.27	0.42	0.52	−0.53
5- Methylhexanal	−0.14	0.18	0.01	0.10	0.10	0.41
a-Copaene	−0.16	0.15	0.00	0.13	0.05	0.45
Propanal	−0.45	0.12	−0.01	0.10	0.09	0.63
Nonanal	0.13	0.04	−0.30	−0.02	−0.28	0.44
b-Cubebene or other sesquiterpene	0.30	0.00	−0.21	0.03	−0.33	0.14
Glutamate g 100g^{-1} FW	0.09	−0.02	0.53	−0.05	0.19	−0.22
Dimethylsulfide	−0.15	−0.02	−0.39	−0.12	−0.27	0.70
Fructose mg g^{-1} FW	0.03	−0.03	0.54	−0.03	0.30	−0.39
2-Heptanone	−0.03	−0.04	−0.11	0.14	0.00	−0.44
Pentane	−0.18	−0.10	−0.43	0.01	−0.27	0.21
Aspartate g 100g^{-1} FW	0.31	−0.22	0.61	−0.23	0.02	−0.43
2-Octenal	−0.49	−0.30	0.41	−0.35	0.06	0.40
2-Butylfuran	−0.21	−0.36	0.51	−0.28	−0.02	0.11
Glucose mg g^{-1} FW	−0.36	−0.36	0.40	−0.31	−0.08	0.44
Ethylacetate	0.01	−0.36	−0.16	−0.46	−0.38	0.16
2-Propanone	0.76	−0.39	0.21	−0.37	−0.44	−0.43
Dimethyltrisulfide	0.58	−0.43	0.26	−0.43	−0.40	−0.43
2-Hexenal	−0.40	−0.44	0.48	−0.37	−0.22	0.64
Furfural	0.84	−0.45	0.37	−0.47	−0.49	−0.27
Benzaldehyde	0.64	−0.51	0.19	−0.36	−0.78	0.11
Heptanal	−0.09	−0.52	0.50	−0.45	−0.22	0.12
2-Pentylfuran	0.02	−0.55	0.56	−0.47	−0.23	0.05
Dimethyldisulfide	0.59	−0.57	0.49	−0.44	−0.47	−0.44
E-(or *Z*)-2-(2-pentenyl)furan	0.24	−0.66	0.62	−0.51	−0.42	0.02
Methional	0.46	−0.67	0.43	−0.60	−0.55	−0.21
Benzeneacetaldehyde	0.30	−0.70	0.18	−0.55	−0.76	0.12
E-(or *Z*)-2-(2-pentenyl)furan	0.10	−0.70	0.64	−0.59	−0.42	0.20
1-Penten-3-one	0.26	−0.72	0.91	−0.69	−0.38	0.15
4-Heptanal	0.19	−0.74	0.60	−0.64	−0.51	0.10
Unknown (RRI 845)	0.59	−0.75	0.31	−0.67	−0.92	0.37
Furan	0.81	−0.78	0.57	−0.71	−0.76	−0.09
Unknown (RRI 1219)	0.13	−0.79	0.81	−0.71	−0.47	0.33
2,4-heptadienal	0.07	−0.80	0.78	−0.73	−0.51	0.48
2-Nonenal	0.03	−0.81	0.43	−0.79	−0.64	0.42
3-Methylbutanal	0.55	−0.81	0.44	−0.71	−0.79	0.00
2-Methylbutanal	0.53	−0.84	0.39	−0.72	−0.85	0.07
2-Methylpropanal	0.56	−0.87	0.49	−0.75	−0.83	0.10
Chaconine mg 100 g^{-1} FW	0.47	−0.92	0.61	−0.90	−0.89	0.64
Solanine mg 100 g^{-1} FW	0.57	−0.95	0.62	−0.92	−0.91	0.48

Fig. 19.3. Heat map showing the correlation analysis between quantitative descriptive analysis scores of potato flavor attributes and levels of metabolites in cooked tubers. Regions in dark gray and light gray indicate positive or negative correlations between traits, respectively. (From Morris *et al.*, 2010.) The flavor descriptors were aroma intensity, flavor intensity, flavor savoriness, flavor creaminess, flavor sweetness and off (unpleasant) flavor.

between aroma intensity and the three flavor attributes, so that metabolites such as the umami compounds, AMP and GMP, hexanal, and 2-methylbutanoic acid methyl ester were strongly correlated negatively with aroma intensity but correlated positively with flavor intensity, flavor creaminess, and flavor savoriness. Conversely, metabolites associated positively with aroma intensity, such as 2-methylbutanal, 3-methylbutanal, and furan, are strongly associated negatively with flavor intensity, flavor savoriness, and flavor creaminess.

This study also emphasized the differences in flavor between cooked tubers from Tuberosum and Phureja. Principal component analysis (PCA) identified that the largest differences in flavor were due to stronger flavor intensity, flavor creaminess, and flavor savoriness in the Phureja genotypes, and that the metabolites associated most closely with these traits were the non-volatile compounds, AMP and GMP, and the volatiles, 2-methylbutanoic acid methylester, hexanal, and methylsalicylate. Previous work (Winfield *et al.*, 2005) also identified the methylester of methylbutanoic acid and methylsalicylate as being significantly different between Phureja and Tuberosum genotypes, but also identified a further ten volatiles that were different between the two potato types. Esterified branched-chain carboxylic acids such as 2-methylbutanoic acid methyl ester are derived from branched-chain amino acids such as valine, leucine, and isoleucine, in a reaction pathway that is thought to involve branched-chain amino acid aminotransferase (BCAAT; Beck, 2005), which catalyzes the rate-limiting step. Methyl salicylate (MeSA) is a volatile compound found in the floral scents of more than 100 species of 30 different families (Effmert *et al.*, 2005). In some plants, MeSA acts as a pollinator attractant, is involved in tritrophic interactions, or is emitted in response to herbivore damage (Effmert *et al.*, 2005; Zhu and Park, 2005). Additionally, MeSA is an important flavor constituent of many fruits, including strawberries and apples, as well as processed foods such as breads, cheeses, wines, and beer (reviewed in Goff and Klee, 2006). MeSA is formed by the action of salicylic acid carboxyl methyltransferase on salicylic acid.

19.4 Tuber Flavor: Environment and Agronomy

Interactions between genotype, environment, and agronomy in the development of potato flavor can be important (Faulks and Giffiths, 1983; Arvanitoyannis *et al.*, 2008; Jansky, 2008). Some potential flavor compounds are affected significantly by growing environment. For example, the level of the volatile methional, derived from the sulfur-containing amino acid methionine, is sensitive to sulfur application rate (Duckham *et al.*, 2002). Tuber potassium level may enhance umami taste intensity, and so the effects of potassium fertilization may also be important (Morris *et al.*, 2007). Tuber nitrogen content correlates with the levels of organic acids (Cieslik, 1997), as well as amides and amines, compounds that can potentially give rise to off flavors (Thybo *et al.*, 2006; Jansky, 2008). Several studies have compared organic versus conventional production systems on sensory attributes. In one study, sensory panellists could distinguish between boiled samples of the cultivar, Dark Red Norland, if the samples were not peeled. However, if the tubers were peeled prior to boiling, the panel could not detect any significant flavor differences (Wszelaki *et al.*, 2005). The studies of Woese *et al.* and Hajslova *et al.* also made comparisons of cooked tuber flavor in samples from organic and conventional production (Woese *et al.*, 1997; Hajslova *et al.*, 2005). As with the Wszelaki study, there was no evidence that the production system gave rise to flavor differences. Studies have been published investigating the effects of environment on the quantitative descriptive analysis of cooked tuber attributes (Jansky, 2008, 2010; Seefeldt *et al.*, 2011). Seefeldt *et al.* compared 11 potato varieties grown in two locations (one with a sandy soil type, the other having clay soil) used for three culinary preparations (Seefeldt *et al.*, 2011). Attributes related to tuber texture and yellowness were affected most strongly by location, and there was a relatively low variation in taste and flavor. Environmental effects on baked tuber flavor were reported by Jansky, with the textural attribute, mealiness, showing the most variation (Jansky, 2008).

19.5 Flavor and Tuber Storage

A key feature of the potato crop is its storability, enabling the product to be supplied to both the fresh and processing industries for several months postharvest. Surprisingly, few studies have addressed changes in flavor quality during storage, presumably because it has been difficult to determine the key metabolites that contribute to this trait. It is known that there are significant changes in amino acid content including the major umami amino acids, glutamate and aspartate (Brierley *et al.*, 1997). The significant effects of storage time on the volatile flavor components of baked potatoes have also been investigated (Duckham *et al.*, 2002). As for amino acids, other flavor precursors, including fatty acids and sugars, change in levels during storage. Although potato lipids account for only 0.8–1.3 mg g^{-1} dry weight (Galliard, 1973), 70–75% of lipids are the relatively reactive polyunsaturated linoleic and linolenic acids, precursors of a wide range of volatile compounds (Galliard, 1973). Some studies report an increase in total fatty acid levels on storage (Cotrufo and Lunsetter, 1964; Cherif and Ben Abdelkader, 1970), whereas others report genotypic variations in fatty acid levels, depending on the duration of storage (Dobson *et al.*, 2004). Changes in tuber reducing sugar during storage are well characterized (for example, Finglas and Faulks, 1984; Brown *et al.*, 1990) and generally increase during storage, although the extent of the increase is cultivar dependent (Blenkinsop *et al.*, 2002).

Quantitative descriptive analysis of tuber samples from Tuberosum and Phureja germplasm stored for 3 months revealed significant changes in flavor. Principal component analysis enabled the identification of the main metabolites driving these changes. Among the volatiles that were found at enhanced levels following storage were several aldehydes; propanol, 5-methylhexanal, and 2-hexenal, all known to be derived from fatty acids via enzymatic and non-enzymatic processes (Dobson *et al.*, 2004).

19.6 Genes Involved in Potato Flavor Compound Biosynthesis

In order to breed potatoes with enhanced flavor, knowledge of the genes involved in the metabolic pathways that lead to flavor compound biosynthesis is required. As reviewed in this chapter, knowledge of the key potato tuber flavor constituents is still in its infancy; however, some progress has been made, and candidate flavor compounds are emerging. In other fruit and vegetables, there is more detailed knowledge of flavor metabolites, and it is of relevance to consider approaches to isolating the biosynthetic genes in these species. In tomato fruit, for example, several approaches have been taken to identify flavor genes. Using a genetics approach, it has been shown that the number of genes that impact on tomato flavor is likely to be quite large. In tomato, *c.*100 quantitative trait loci (QTLs) affecting volatiles and their precursors have been identified (Causse *et al.*, 2001; Saliba-Colombani *et al.*, 2001; Schauer *et al.*, 2006; Tieman *et al.*, 2006; Mathieu *et al.*, 2008). Similar approaches in potato have been undertaken, and the results should be published in the near future (G.J. Bryan, UK, 2011, personal communication). The relatively large number of QTLs that affect tomato flavor emphasizes the complex nature of flavor development, and indicates that multiple pathways and control mechanisms conspire to develop flavor attributes.

With the knowledge of the key metabolites and the genes involved in their biosynthesis and turnover, the opportunity arises to use transgenic approaches to manipulate the expression level of key genes and determine the effects on the level of the target metabolite, and ultimately organoleptic quality. For example, in tomato fruit, transgenic expression of a geraniol synthase gene from basil altered the profile of terpenoid volatiles significantly, and the transgenic fruit could be distinguished by taste panellists, 60% of whom preferred them to the non-transgenic control.

In potato, several studies have led to the production of transgenic lines that contain altered levels of potential flavor compounds. For example, overexpression of an *Arabidopsis* cystathionine γ-synthase gene in potato tubers led to a sixfold increase in tuber methionine content (Di *et al.*, 2003). The increased methionine content resulted in an increase in methional content (up to 4.4-fold) in baked tubers; however, sensory evaluation of these transgenic lines has not been reported. Other transgenic experiments have demonstrated that tuber fatty acid levels can be increased by overexpressing an *Arabidopsis* acetyl-CoA carboxylase gene (Klaus *et al.*, 2004). In this study, overexpression led to a *c.*40% increase in total fatty acid content and changes in the ratios of linoleic and α-linolenic acids. Again, the effects on the organoleptic quality of cooked tubers from these lines have not been reported.

One approach to identifying candidate genes that underpin flavor differences between potato germplasm is to use transcriptomics. Transcriptomic studies in potato have benefitted from the development of a near whole-transcriptome microarray produced by the Potato Oligo Chip Initiative (POCI) consortium that contains gene probes based on 42,034 potato unigene sequences, using the custom Agilent platform. A detailed description of the array has recently been published (Kloosterman *et al.*, 2008). Following the elucidation of the potato genome, a more comprehensive version of the POCI array has been developed (P.E. Hedley, UK, 2011, personal communication), and in the near future RNAseq methodologies will ensure transcriptomic studies have comprehensive coverage of the potato transcriptome. Using the POCI microarray, we were able to identify consistent differences in gene expression profiles between Phureja and Tuberosum cultivars, including genes likely to impact on flavor and other quality traits. As an example of the utility of the approach, functional analysis of a sesquiterpene synthase gene that was expressed consistently at higher levels in Phureja tubers demonstrated that this gene encoded an α-copaene synthase.

As α-copaene was one of the major volatile differences between Phureja- and Tuberosum-type potatoes, and it was established that α-copaene had high odor impact in lettuce and carrots (Nielsen and Poll, 2007), the effects of overexpressing α-copaene synthase on potato flavor were investigated. In fact, the α-copaene-producing transgenic lines could not be distinguished from parental control by taste panelists (Morris *et al.*, 2011), indicating that α-copaene was probably not a major factor in potato flavor, a conclusion that concurred with the correlation matrix developed by Morris *et al.* (Morris *et al.*, 2010; Fig. 19.3).

The results of the transcriptomic comparison of Phureja and Tuberosum germplasm identified other potential flavor genes. Esterified branched-chain carboxylic acids such as 2-methylpropionic acid methyl ester and 2-methylbutanoic acid methyl ester are other volatiles produced in much higher levels from cooked Phureja tubers than from Tuberosum (Shepherd *et al.*, 2007). These esters are derived from branched-chain amino acids such as valine, leucine, and isoleucine, in a reaction pathway that is thought to involve branched-chain amino acid aminotransferase (BCAAT; Beck, 2005). A BCAAT gene was expressed as much higher levels in Phureja than in Tuberosum tubers, and again suggests that a transgenic route in which the gene is overexpressed or silenced may resolve the issue of its contribution to overall volatile level and flavor.

Glutamate is likely to be a determinant of flavor in potato tubers, due to its contribution to the umami taste (Morris *et al.*, 2007). Genes encoding glutamate ammonia ligase and glutamine synthetase I (enzymes involved in glutamate biosynthesis) were expressed at higher levels in Phureja cultivars. Conversely, another gene involved in glutamate biosynthesis, γ-aminobutyric acid transaminase, was expressed at a higher level in Tuberosum cultivars. Further investigation of gene expression profiles of these and other genes involved in tuber glutamate biosynthesis may reveal those that account for higher levels of glutamate in Phureja tubers.

19.7 Future Directions

Compared with other fruit and vegetables such as tomatoes, there are still significant gaps in our knowledge linking metabolites to flavor metabolites. The analytical tools are now available to address this issue; however, there are still relatively few studies linking metabolites and quantitative descriptive analysis. The use of germplasm that shows wide variation in flavor characteristics such as Phureja compared with Tuberosum has proved useful, but only begins to investigate the full diversity available within potato germplasm collections. Producers, processors, and consumers alike wish to have potato tubers that show consistent and high-quality flavor properties, and to achieve this goal we need to link the metabolites with quality traits. The interaction between genotype, environment, and agronomy in developing potato flavor is an area that requires urgent attention. This will result in a metabolite "blueprint" for sensory quality that can be used to predict product consistency and acceptability of different genotypes.

Once the key genes and metabolites that impact on flavor have been identified, the question arises as to the best way of improving potato germplasm. As illustrated in this chapter, transgenic approaches provide a rapid means of determining the impact of a candidate flavor gene. However, despite the continued growth in the production of transgenic crops, public opinion in many markets is still vehemently opposed to genetically modified (GM) solutions. It may be argued that the tide of public opinion is changing as more and more GM produce is consumed; however, the regulatory processes for getting a GM product to market still impose serious economic constraints on commercialization. Traits that have a major economic advantage are likely to be the ones developed for GM crops in the near future, and quality traits such as flavor are likely to lag behind.

More realistically, marker-assisted breeding approaches are likely to be the way forward in improving potato tuber flavor. It has been seen already that germplasm derived from Phureja potatoes has enhanced flavor, and it would be of interest to investigate the flavor metabolites in germplasm collections.

The transfer of complex traits, controlled by many genes, into a modern, high-yielding cultivar represents a serious challenge to potato breeders and scientists alike. However, recent advances in genome and transcriptome sequencing, better understanding of the metabolic pathways, together with high-throughput molecular marker screening now makes breeding for flavor much more realistic. Stacking of the multiple, independent genes likely to be needed for flavor quality improvement is now technically possible. The identification of the appropriate genes and alleles needed is within our grasp, and represents an exciting possibility to restore the dwindling popularity of potato consumption.

References

Arvanitoyannis, I.S., Vaitsi, O. and Mavromatis, A. (2008) Physicochemical and sensory attributes in conjunction with multivariate analysis of two potato (*Solanum tuberosum* L.) cultivars after 90 days of storage: an exploratory authentication study. *International Journal of Food Science and Technology* 43, 1960–1970.

Baldwin, E.A., Scott, J.W., Shewmaker, C.K. and Schuch, W. (2000) Flavor trivia and tomato aroma: biochemistry and possible mechanisms for control of important aroma components. *Hortscience* 35, 1013–1022.

Barylko-Pikielna, N. and Kostyra, E. (2007) Sensory interaction of umami substances with model food matrices and its hedonic effect. *Food Quality and Preference* 18, 751–758.

Beck, H.C. (2005) Branched-chain fatty acid biosynthesis in a branched-chain amino acid aminotransferase mutant of *Staphylococcus carnosus*. *FEMS Microbiology Letters* 243, 37–44.

Beksan, E., Schieberle, P., Robert, F., Blank, I., Fay, L.B., Schlichtherle-Cerny, H., *et al.* (2003) Synthesis and sensory characterization of novel umami-tasting glutamate glycoconjugates. *Journal of Agricultural and Food Chemistry* 51, 5428–5436.

Blenkinsop, R.W., Copp, L.J., Yada, R.Y. and Marangoni, A.G. (2002) Changes in compositional parameters of tubers of potato (*Solanum tuberosum*) during low-temperature storage and their relationship to chip processing quality. *Journal of Agricultural and Food Chemistry* 50, 4545–4553.

Brierley, E.R., Bonner, P.L.R. and Cobb, A.H. (1997) Aspects of amino acid metabolism in stored potato tubers (cv. Pentland Dell). *Plant Science* 127, 17–24.

Brown, J., Mackay, G.R., Bain, H., Griffith, D.W. and Allison, M.J. (1990) The processing potential of tubers of the cultivated potato, *Solanum tuberosum* L. after storage at low temperatures. 2. Sugar concentration. *Potato Research* 33, 219–227.

Buri, R. and Solms, J. (1971) Ribonucleic acid – a flavour precursor in potatoes. *Naturwissenschaften* 58, 56.

Buri, R., Signer, V. and Solms, J. (1970) Die Bedeutung von freien aminosauren und nukleotiden fur den flavour von gekochten karkoffeln. *Lebensmittel-Wissenschaft und-Technologie* 3, 63–66.

Burrowes, J.D. and Ramer, N.J. (2008) Changes in potassium content of different potato varieties after cooking. *Journal of Renal Nutrition* 18, 530–534.

Buttery, R.G. (1993) Quantitative and sensory aspects of flavour of tomato and other vegetables and fruits. In: Acree, T.E. and Teranishi, R. (eds) *Flavor Science: Sensible Principles and Techniques*. American Chemical Society, Washington, DC, pp. 259–286.

Causse, M., Saliba-Colombani, V., Lesschaeve, I. and Buret, M. (2001) Genetic analysis of organoleptic quality in fresh market tomato. 2. Mapping QTLs for sensory attributes. *Theoretical and Applied Genetics* 102, 273–283.

Chen, G., Hackett, R., Walker, D., Taylor, A., Lin, Z. and Grierson, D. (2004) Identification of a specific isoform of tomato lipoxygenase (TomloxC) involved in the generation of fatty acid-derived flavour compounds. *Plant Physiology* 136, 2641–2651.

Cherif, A. and Ben Abdelkader, A. (1970) Quantitative analysis of the fatty acid present in different parts of the potato tuber: variation during storage at 10°C. *Potato Research* 13, 284–295.

Cieslik, E. (1997) Effect of the levels of nitrate and nitrites on the nutritional and sensory quality of potato tubers. *Hygiene and Nutrition in Foodservice and Catering* 1, 225–230.

Cotrufo, C. and Lunsetter, P. (1964) The fatty acids of potato tubers (*Solanum tuberosum*). *American Potato Journal* 41, 18–22.

De Maine, M., Carrol, C.P. and Torrance, C.J.W. (1993) Culinary quality of tubers derived from *Solanum phureja* and *Solanum tuberosum* × *Solanum phureja* hybrids. *Journal of Agricultural Science* 120, 213–217.

De Maine, M., Lees, A. and Bradshaw, J. (1998) Soft-rot resistance combined with other tuber characters in long day-adapted *Solanum phureja*. *Potato Research* 41, 69–82.

Di, R., Kim, J., Martin, M.N., Leustek, T., Jhoo, J., Ho, C.T., *et al.* (2003) Enhancement of the primary flavour compound methional in potato by increasing the level of soluble methionine. *Journal of Agricultural and Food Chemistry* 51, 5695–5702.

Dobson, G., Griffiths, D.W., Davies, H.V. and McNicol, J.W. (2004) Comparison of fatty acid and polar lipid contents of tubers from two potato species, *Solanum tuberosum* and *Solanum phureja*. *Journal of Agricultural and Food Chemistry* 52, 6306–6314.

Duckham, S.C., Dodson, A.T., Bakker, J. and Ames, J.M. (2001) Volatile flavour components of baked potato flesh: a comparison of eleven potato cultivars. *Nahrung* 45, 317–323.

Duckham, S.C., Dodson, A.T., Bakker, J. and Ames, J.M. (2002) Effect of cultivar and storage time on the volatile flavour components of baked potato. *Journal of Agricultural and Food Chemistry* 50, 5640–5648.

Dumelin, E. and Solms, J. (1976) The formation of 5′-nucleotides during cooking of potatoes. *Potato Research* 19, 215–228.

Effmert, U., Saschenbrecker, S., Ross, J., Negre, F., Fraser, C.M., Noel, J.P., *et al.* (2005) Floral benzenoid carboxyl methyltransferases: from *in vitro* to *in planta* function. *Phytochemistry* 66, 1211–1230.

Faulks, R.M. and Griffiths, N.M. (1983) Influence of variety, site and storage on physical, sensory and compositional aspects of mashed potato. *Journal of the Science of Food and Agriculture* 34, 979–986.

Finglas, P.M. and Faulks, R.M. (1984) Nutritional composition of UK retail potatoes, both raw and cooked. *Journal of the Science of Food and Agriculture* 35, 1347–1356.

Fischer, W.N., Andre, B., Rentsch, D., Krolkiewicz, S., Tegeder, M., Breitkreuz, K., *et al.* (1998) Amino acid transport in plants. *Trends in Plant Science* 3, 188–195.

Frankel, E.N. (1998) Hydroperoxide decomposition. In: *Lipid Oxidation*. The Oily Press, Dundee, UK, pp. 55–77.

Galliard, T. (1973) Lipids of potato tubers. Lipid and fatty acid composition of tubers from different varieties of potato. *Journal of the Science of Food and Agriculture* 24, 617–622.

Gardner, H.W. (1995) Biological roles and biochemistry of the lipoxygenase pathway. *HortScience* 30, 197–205.

Ghislain, M., Andrade, D., Rodriguez, F., Hijmans, R. and Spooner, D. (2006) Genetic analysis of the cultivated potato *Solanum tuberosum* L. Phureja group using RAPDs and nuclear SSRs. *Theoretical and Applied Genetics* 113, 1515–1527.

Goff, S.A. and Klee, H.J. (2006) Volatile compounds: sensory cues for health and nutritional value? *Science* 311, 815–819.

Grosch, W. (2001) Evaluation of the key odorants of foods by dilution experiments, aroma models and omission. *Chemical Senses* 26, 533–545.

Hajslova, J., Schulzova, V., Slanina, P., Janne, K., Hellenas, K.E. and Andersson, C. (2005) Quality of organically and conventionally grown potatoes: four-year study of micronutrients, metals, secondary metabolites, enzymic browning, and organoleptic qualities. *Food Additives and Contaminants* 22, 514–534.

Halford, N.G., Hey, S., Jhurreea, D., Laurie, S., McKibbin, R.S., Zhang, Y., *et al.* (2004) Highly conserved protein kinases involved in the regulation of carbon and amino acid metabolism. *Journal of Experimental Botany* 55, 35–42.

Halpern, B.P. (2000) Glutamate and the flavour of foods. *Journal of Nutrition* 130, 910–914.

Jansky, S. (2008) Genotypic and environmental contributions to baked potato flavour. *American Journal of Potato Research* 85, 455–465.

Jansky, S.H. (2010) Potato flavor. *American Journal of Potato Research* 87, 209–217.

Josephson, D.B. and Lindsay, R.C. (1987) c4-Heptenal: an influential volatile compound in boiled potato flavour. *Journal of Food Science* 52, 328–331.

Klaus, D., Ohlrogge, J.B., Neuhaus, H.E. and Dormann, P. (2004) Increased fatty acid production in potato by engineering of acetyl-CoA carboxylase. *Planta* 219, 389–396.

Klee, H.J. (2010) Improving the flavour of fresh fruits: genomics, biochemistry, and biotechnology. *New Phytologist* 187, 44–56.

Kloosterman, B., De Koeyer, D., Griffiths, R., Flinn, B., Steuernagel, B., Scholz, U., *et al.* (2008) The potato transcriptome: a new look at transcriptional changes during tuber development using the POCI array. *Comparative and Functional Genomics* 8, 329–340.

Koch, W., Kwart, M., Laubner, M., Heineke, D., Stransky, H., Frommer, W.B., *et al.* (2003) Reduced amino acid content in transgenic potato tubers due to antisense inhibition of the leaf H^+/amino acid symporter StAAP1. *Plant Journal* 33, 211–220.

Li, X., Staszewski, L., Xu, H., Durick, K., Zoller, M. and Adler, E. (2002) Human receptors for sweet and umami taste. *Proceedings of the National Academy of Science* 99, 4692–4696.

Lindemann, B. and Ogiwara, Y. (2002) The discovery of umami. *Chemical Senses* 27, 843–844.

Lindsay, R.C. (1996) Flavors. In: Fennema O.R. (ed.) *Food Chemistry*, 3rd edn. Dekker, New York, pp. 723–785.

Love, S.L.S.T., Shafii, B., Price, W.J., Moslet, A.R. and Thornton, R.E. (2004) Stability of expression and concentration of ascorbic acid in North American potato germplasm. *HortScience* 39, 156–160.

Maarse, H. (1991) *Volatile Compounds in Foods and Beverages*. Marcel Dekker, New York.

McGregor, I. (2007) The fresh potato market. In: Vreugdenhil, D. (ed.) *Potato Biology and Biotechnology: Advances and Perspectives*. Elsevier, Amsterdam, pp. 3–26.

Maga, J.A. (1994) Potato flavor. *Food Reviews International* 10, 1–48.

Mathieu, S., Dal Cin, V., Fei, Z., Li, H., Bliss, P., Taylor, M.G., *et al.* (2008) Flavor compounds in tomato fruits: identification of loci and potential pathways affecting volatile composition. *Journal of Experimental Botany* 60, 325–337.

Morris, W.L., Ducreux, L., Griffiths, D.W., Stewart, D., Davies, H.V. and Taylor, M.A. (2004) Carotenogenesis during tuber development and storage in potato. *Journal of Experimental Botany* 55, 975–982.

Morris, W.L., Ross, H.A., Ducreux, L.J.M., Bradshaw, J.E., Bryan, G.J. and Taylor, M.A. (2007) Umami compounds are a determinant of the flavour of potato (*Solanum tuberosum* L.). *Journal of Agricultural and Food Chemistry* 55, 9627–9633.

Morris, W.L., Shepherd, T., Verrall, S.R., McNicol, J.W. and Taylor, M.A. (2010) Relationships between volatile and non-volatile metabolites and attributes of processed potato flavour. *Phytochemistry* 71, 1765–1773.

Morris, W.L., Ducreux, L.J.M., Shepherd, T., Lewinshohn, E., Davidovich-Rikanati, R., Sitrit, Y., *et al.* (2011) Utilisation of the MVA pathway to produce elevated levels of the sesquiterpene α-copaene in potato tubers. *Phytochemistry* 72, 2288–2293.

Murray, K.E. and Whitfield, F.B. (1975) Occurrence of 3-alkyl-2-methoxypyrazines in raw vegetables. *Journal of the Science of Food and Agriculture* 26, 973–986.

Navarre, D.A., Goyer, A. and Shakya, R. (2009) Nutritional value of potatoes: vitamin, phytonutrient and mineral content. In: Singh, J. and Kaur, L. (eds) *Advances in Potato Chemistry and Technology*. Academic Press, Burlington, Massachusetts, pp. 395–425.

Nelson, G., Chandrashekar, J., Hoon, M.A., Feng, L., Zhao, G., Ryba, N.J.P., *et al.* (2002) An amino acid taste receptor. *Nature* 416, 199–202.

Ney, K.H. (1971) Flavor enhancing effect of L-glutamate and similar compounds. *Zeitschrift fur Lebensmittel-Untersuchung und -Forschung* 146, 141–143.

Nielsen, G.S. and Poll, L. (2007) Determination of odour active aroma compounds in a mixed product of fresh cut iceberg lettuce, carrot and green bell pepper. *Developments in Food Science* 43, 517–520.

Ninomiya, K. (1998) Natural occurance. *Food Reviews International* 14, 177–211.

Oruna-Concha, M.J., Duckham, S.C. and Ames, J.M. (2001) Comparison of volatile compounds isolated from the skin and flesh of four potato cultivars after baking. *Journal of Agricultural and Food Chemistry* 49, 2414–2421.

Oruna-Concha, M.J., Bakker, J. and Ames, J.M. (2002) Comparison of the volatile components of two cultivars of potato cooked by boiling, conventional baking and microwave baking. *Journal of the Science of Food and Agriculture* 82, 1080–1087.

Osman, S.F. (1983) Glycoalkaloids in potatoes. *Food Chemistry* 11, 235–247.

Ramsay, G., Griffiths, D.W. and Deighton, N. (2004) Patterns of solanidine gycoalkoloid variation in four gene pools of the cultivated potato. *Genetic Resources and Crop Evolution* 51, 805–813.

Roessner-Tunali, U., Lui, J., Leisse, A., Balbo, I., Perez-Melis, A., Willmitzer, L., *et al.* (2004) Kinetics of labelling of organic and amino acids in potato tubers by gas chromatography-mass spectrometry following incubation in ^{13}C labelled isotopes. *Plant Journal* 39, 668–679.

Ross, H., Pasemann, P. and Nitzsche, W. (1978) Glycoalkaloid content of potatoes and its relationship to location, year and taste. *Zeitschrift Fur Pflanzenzuchtung* 80, 64–79.

Saliba-Colombani, V., Causse, M., Langlois, D., Philouze, J. and Buret, M. (2001) Genetic analysis of organoleptic quality in fresh market tomato. 1. Mapping QTLs for physical and chemical traits. *Theoretical and Applied Genetics* 102, 259–272.

Schauer, N., Semel, Y., Roessner, U., Gur, A., Balbo, I., Carrari, F., *et al.* (2006) Comprehensive metabolic profiling and phenotyping of interspecific introgression lines for tomato improvement. *Nature Biotechnology* 24, 447–454.

Seefeldt, H.F., Tonning, E. and Thybo, A.K. (2011) Exploratory sensory profiling of three culinary preparations of potatoes (*Solanum tuberosum* L.). *Journal of the Science of Food and Agriculture* 91, 104–112.

Sengul, M., Keles, F. and Keles, M.S. (2004) The effect of storage conditions (temperature, light, time) and variety on the glycoalkaloid content of potato tubers and sprouts. *Food Control* 15, 281–286.

Shepherd, T., Dobson, G., Verrall, S.R., Conner, S., Griffiths, D.W., McNicol, J.W., *et al.* (2007) Potato metabolomics by GC-MS: what are the limiting factors? *Metabolomics* 3, 475–488.

Sinden, S.L., Deahl, K. and Aulenbach, B.B. (1976) Effect of glycoalkaloids and phenolics on potato flavour. *Journal of Food Science* 41, 520–523.

Singh, A., Ai, Y. and Kao, T.-h. (1991) Characterization of ribonuclease activity of three S-allele-associated proteins of *Petunia inflata*. *Plant Physiology* 96, 61–68.

Slanina, P. (1990) Solanine (glycoalkaloids) in potatoes – toxicological evaluation. *Food Chemistry and Toxicology* 28, 759–761.

Solms, J. (1971) Nonvolatile compounds and flavour. In: Olhloff, G. and Thomas, A.F. (eds) *Gustation and Olfaction*. Academic Press, London, pp. 92–110.

Solms, J. and Wyler, R. (1979) Taste components of potatoes. In Boudreau, J.C. (ed.) *Food Taste Chemistry. ACS Symposium Series 115*. American Chemical Society, Washington, DC, pp. 175–184.

Spooner, D.M., Nunez, J., Trujillo, G., del Rosario Herrera, M., Guzman, F. and Ghislain, M. (2007) Extensive simple sequence repeat genotyping of potato landraces supports a major re-evaluation of their gene pool structure and classification. *Proceedings of the National Academy of Sciences* 104, 19398–19403.

Thybo, A.K., Christiansen, J., Kaack, K. and Petersen, M.A. (2006) Effect of cultivars, wound healing and storage on sensory quality and chemical components in pre-peeled potatoes. *Lebensmittel-Wissenschaft und -Technologie* 39, 166–176.

Tieman, D.M., Zeigler, M., Schmelz, E.A., Taylor, M.G., Bliss, P., Kirst, M., *et al.* (2006) Identification of loci affecting flavour volatile emissions in tomato fruits. *Journal of Experimental Botany* 57, 887–896.

Velisek, J., Davidek, J., Kubelka, V., Thu, T.T.B. and Hajslova, J. (1978) Succinic acid in yeast autolysates and its sensory properties. *Nahrung* 22, 735–743.

Wang, Y. and Kays, S.J. (2003) Analytically directed flavour selection in breeding food crops. *Journal of the American Society of Horticultural Science* 128, 711–720.

Wilson, W.D., Mackinnon, I.M.R. and Jarvis, M.C. (2002a) Transfer of heat and moisture during microwave baking of potatoes. *Journal of the Science of Food and Agriculture* 82, 1070–1073.

Wilson, W.D., Mackinnon, I.M.R. and Jarvis, M.C. (2002b) Transfer of heat and moisture during oven baking of potatoes. *Journal of the Science of Food and Agriculture* 82, 1074–1079.

Winfield, M., Lloyd, D., Griffiths, D.W., Bradshaw, J.E., Muir, D., Nevison, I.M., *et al.* (2005) Assessing organoleptic attributes of *Solanum tuberosum* and *S. phureja* potatoes. *Aspects of Applied Biology* 76, 127–135.

Woese, K., Lange, D., Boess, C. and Bögl, K.W. (1997) A comparison of organically and conventionally grown foods—results of a review of the relevant literature. *Journal of the Science of Food and Agriculture* 74, 281–293.

Wszelaki, A.L., Delwiche, J.F., Walker, S.D., Liggett, R.E., Scheerens, J.C. and Kleinhenz, M.D. (2005) Sensory quality and mineral and glycoalkaloid concentrations in organically and conventionally grown redskin potatoes (*Solanum tuberosum*). *Journal of the Science of Food and Agriculture* 85, 720–726.

Yamaguchi, S. and Ninomiya, K. (1998) What is umami? *Food Reviews International* 14,123–138.

Yamaguchi, S., Yoshikawa, T., Ikeda, S. and Ninomiya, T. (1971) Measurement of the relative taste intensity of some L-α-amino acids and 5′-nucleotides. *Journal of Food Science* 36, 846–849.

Zhu, J. and Park, K.-C. (2005) Methyl salicylate, a soybean aphid-induced plant volatile attractive to the predator *Coccinella septempunctata*. *Journal of Chemical Ecology* 31, 1733–1746.

Zitnak, A. (1961) Occurrence and distribution of free alkaloid solanidine in netted gem potatoes. *Canadian Journal of Biochemistry and Physiology* 39, 1257.

Index

abscisic acid (ABA) 56, 65, 70–71
acrylamide 258, 277–278
adenosine-5′-monophosphate (5′-AMP) 346, 347, 348–350
Africa
 diseases/pests 154, 167, 203
 potato market 31, 32, 256
 seed potatoes 116
age of tubers 66–67, 88, 263
Alternaria spp. (early blight/brown leaf spot) 185–188, 190
Amflora potato (GM) 37
amino acids 311, 347–348
 chemical reactions on cooking 38, 246, 257–258, 277, 351
 and flavor 346–347, 350, 354, 355
amylases 260
amylopectin 311
amyloplasts 59, 259–260
amylose 311
Andean region
 origin of the potato 1, 5, 20, 22
 viruses indigenous to 205–206, 207–208
anthocyanins 181, 250, 328, 330, 331
anticancer activity 312, 325, 328
antimicrobial agents 182
antioxidants
 carotenoids 331–332
 phenylpropanoids 312, 326–331
 vitamins 321, 324–325
aphids 138–139, 298
 as vectors of disease 129, 139, 209, 215
apical dominance 66, 67–68
aroma 348, 350–353
ascorbic acid (vitamin C) 324–325, 330
Asia
 potato market 8, 30–31, 32, 33, 36, 39, 41
 seed potatoes 119
 storage 256
 TPS use 116
asparagine 277
aspartate 347
aster leafhopper (*Macrosteles fascifrons*) 140
Australia 207
auxin (IAA) 54–55
available soil water (ASW) 93

B vitamins 315–324
baby potatoes 327, 329–330
 see also petite potatoes
Bacillus subtilis fungicide 189
backcrossing 274
bacterial biocontrol agents 182, 189
bacterial diseases 86, 167, 169, 183
 common scab 180–182
 diagnosis 190–191
 phytoplasmas 140
 in seed potatoes 117, 122, 172
 soft rot 124, 178–180
 zebra chip 141–142, 183, 278
baked potatoes 38, 314, 317, 320, 323, 330, 351
balance sheet scheduling of irrigation 94
bar-coding (genomic) 25
bed planting 89
beet leafhopper (*Circulifer tenellus*) 140
begomoviruses 207
Belgium 33
biocontrol agents 159, 182, 189
bioinformatics 301
biotin (vitamin B_8) 322

bitter taste 1, 78, 240, 258, 350
black dot (*Colletotrichum coccodes*) 176–177
black scurf (*Rhizoctonia solani*) 173, 177, 183, 297
blackheart 246
blackspot bruise 98, 245
blue potatoes 41
boiled potatoes 39, 40, 311, 317, 320, 323, 330, 332, 351
boscalid 190
bottleneck tubers 238
breeding programs
  molecular methods 274–275, 290–301
  targets 7–8, 41–42, 276–279
    drought resistance 108–109, 297
    flavor 345–346, 354–355, 356
    pathogen resistance 182, 215–216, 276, 291–292
    pest resistance 158–159, 159–160, 278, 297–298
    phytonutrient levels 277, 316–317, 318, 319, 321, 322, 324, 332
  traditional 285–286
    history 5–8, 272–273
    methods 273, 279–285
    and polyploidy 8, 273–275
    sources of germplasm 8, 275–276, 322, 324
brown center 243
brown leaf spot (*Alternaria alternata*) 185–188
browning, enzymatic 298, 328–329
bruising 98–99, 245–246, 257, 295
bulk storage 180, 256–257, 264, 266–267

calcium
  and physiological disorders 244, 245, 248
  tuber content 315
Canada
  fungicide resistance 175, 190
  potato market 33, 35, 36
  seed potato certification 118, 119
Candidatus Liberibacter solanacearum (zebra chip disease) 141–142, 183, 278
canopy size/growth 72–73, 75–76, 105
canopy temperature 108
carbohydrate
  and cooking characteristics 38, 39, 40, 41, 311
  metabolism 59, 247, 259–262
  nutritional values 311
  see also sugars
carbon dioxide (CO_2)
  and climate change 105–106, 209–210
  during storage 264
carotenoids 331–332
CBSG project (Centre of Biosystem Genomics) 16–17
CDF proteins (cycling DOF factor) 49
cellar storage 257
certification of seed potatoes 87–88, 115–116, 117–124, 172, 284
CGA (chlorogenic acid) 327–328, 329, 330, 331
chaconine 325–326
chain tubers 241–242
Chile 4, 5, 22
China 8, 31, 33, 36, 39, 116
chips (crisps) 36, 42, 261
  vitamin content 316, 320, 323
chips (French fries) *see* French fries
chitting (pre-sprouting) 68–69, 88
chlorogenic acid (CGA) 327–328, 329, 330, 331
chloroplast genome 24
chloropropam 266
Cicadellidae (leafhoppers) 140
CIP (International Potato Center, Peru) 19, 39, 65
circadian clock 49, 50–51, 52
clamps 256
classification *see* taxonomy
clethodim 232–233
click beetles (wireworm, Elateridae) 135–137, 144
climate, ideal 64, 83
climate change 103, 105–106, 168, 184, 189–190, 208–210, 216
clover 229
cold injuries 249–250, 258, 295
Colletotrichum coccodes (black dot) 176–177
Colorado potato beetle (*Leptinotarsa decemlineata*) 133, 134–135, 144
common potato scab 180–182
comparative genomics 298–301
consumer attitudes 32, 42, 329–330
consumption, global 31, 32, 35–36, 41, 42, 313
cooking
  carotenoids 332
  flavor 347, 348–350, 351–353
  Maillard reaction 38, 246, 257–258, 277, 351
  phenylpropanoids 328–329, 330–331
  starch content and variety 38, 39, 40, 41, 311
  vitamin content 317, 318, 320, 323, 324
α-copaene 355
copper (Cu) 315
corky patch syndrome 243
corky ringspot (spraing) 153, 207
cortex, tuber 67, 152
Cousens rectangular hyperbola 227–228
cover crops 229
cracked tubers 238–239
crib storage 256–257
crisps (potato chips) 36, 42, 261
  vitamin content 316, 320, 323
crop rotation 85
  for pest control 144, 155–156, 187
  for weed control 229

crop water stress index (CWSI) 108
cultural management 79, 83–99
control of
bacteria 179–180, 181–182
fungi 187
insects 144
nematodes 155–156, 229
physiological disorders 238, 239, 240, 242, 244, 246, 248, 249, 250
weeds 229–231
growth phase 72–73, 92–98
harvest 98–99, 179, 239, 242, 246
planting 87–92, 129, 229
pre-planting 85–87, 128–129, 230
seed potatoes 128–130
cup planters 91
cuticle 106
cytokinin (CK) 55–56, 65–66

Dammer Diker® tillage 230
day length 47–48
Defender cultivar 278–279
deficit irrigation (DI) 106–107
defoliation of vines 78–79, 98, 230–231, 250
dehydrated products 36–37, 330
depth of planting 69, 90–91
desiccation of vines 78–79, 98, 230–231, 250
determinant (early) cultivars 70, 104, 156
developing countries
diseases 167
potato market 8, 29–31, 32, 33, 39, 41–42
seed potatoes 37, 116, 119
TPS use 65, 116
diabetes (type 2) 312, 328
Dickeya spp. (soft rot) 124, 178–180
diet *see* nutritional qualities of potatoes
difenoconazole 175
dimethenamid-P 231–232
disease management *see* pest management
disinfection
seed potato production 124, 127, 129
of stored tubers 189
Ditylenchus destructor (potato rot nematode) 154, 156
domestication 1–2, 21–22
dormancy (tuber) 47, 65–66, 127
drought 92–93, 103–106, 108–109, 247, 297
dry matter content 77, 311
dry rot (*Fusarium*) 174–175, 189, 283–284, 293
drying methods (processing) 330
dumb-bell tubers 238

early blight (*Alternaria solani*) 185–188, 190
Early Rose variety 5–6, 272
early-generation stock production
asexual 124–128
sexual 281–284
early-maturing varieties 70, 104, 156
economic costs 143–144
of fungal/bacterial diseases 167, 178, 184
of nematodes 148, 160
of viral diseases 216
of weeds 225
electrical resistance blocks 93–94
elephant hide 240
environmental conditions
causing physiological disorders 238, 239, 240, 241, 243, 244, 245, 246, 247
and flavor 353–354
genotype × environment interactions 284
good for growing potatoes 64, 83
phenylpropanoid levels 331
see also climate change; temperature
EPTC (herbicide) 232
ESTs (expressed sequence tags) 293
ethaluralin 231
ethylene 65, 263–264, 266
Europe
climate change 208
introduction of the potato 2–4, 22
potato market 30, 31–32, 33
seed potato certification 118, 119, 122
export and import 32–33, 34, 36, 38–39, 41
eyes 7, 47, 67, 68, 70

false root-knot nematode (*Nacobbus aberrans*) 151, 159
fast food *see* French fries
female flowers 280
fertilizer use 76, 85, 94–97, 279
and flavor 353
fiber 311
field characteristics 85–87, 128–129
see also soil
field production of seed potatoes 127–130
fingerlings 40–41
flavonols 328
flavor 345–356
environmental effects 353–354
genetics 354–355
off-tastes 1, 78, 240, 350, 353
and storage 354
umami 346–350
and variety 40, 41
volatile compounds 350–353, 354, 355
flooding 108
flour 37
flowers
flowering signal 48
morphology 14, 121
sexual reproduction 279–280

fludioxonil 175
flumioxazin 232
folate (vitamin B_9) 322–323
fomesafen 232
forward contracts 84
France 3, 33, 119
freeze drying 330
freezing injury/necrosis 249–250, 258
French fries
  desirable tuber characteristics 8, 39, 40, 84
  market 32, 35–36, 41
  sugar content affects end product 38, 76, 92–93, 246–247, 257–258, 261
  vitamin content 320
fresh potato market 29, 35, 42, 329–330
frozen products 32–33, 35–36
  see also French fries
fructose 76, 260, 261
fruit 280
fumigation of soil 157, 177
fungal biocontrol agents 159
fungal diseases 86, 93, 170–171
  black dot 176–177
  black scurf 173, 177, 183, 297
  in breeding programs 283–284
  diagnosis 190–191
  dry rot 174–175, 189, 283–284, 293
  early blight/brown leaf spot 185–188, 190
  late blight *see* late blight (*Phytophthora infestans*)
  management 129, 175, 177, 178, 183, 185, 187–188
  postharvest 188–189
  resistance in transgenic potatoes 297
  seed potatoes 117, 129, 172, 174, 175, 177, 183
  transmission 173–178, 184
fungal vectors of viral disease 209
fungicides 86, 139, 175, 178, 183, 187–188, 189
  resistance 175, 189, 190
Fusarium dry rot 174–175, 189, 283–284, 293

GA *see* gibberellic acid/gibberellin (GA)
galls, nematode 150, 151
Gelechiidae (potato tuber moths) 137–138, 322
gene banks 19–20, 275
genetically modified (GM) varieties 295, 296–298
  disease/pest resistance 145, 159–160, 294, 297–298
  flavor 354–355
  phytonutrient content 324, 332
  starch content 37
genetics and genomics
  anthocyanin synthesis 328
  candidate gene approach 295–296
  carotenoid synthesis 332
  comparative genomics 298–301
  disease diagnosis 121, 191
  flavor 354–355
  gene expression profiles 56–57, 293–294, 301, 355
  gene mapping 290–291, 292
  genome sequencing 24–25, 290, 300–301
  marker-assisted selection 291–293
  polyploidy 8, 13, 21–23, 273–275, 300–301
  source of cultivated varieties 4, 5, 7
  and taxonomy 13, 17, 24–25
  tuberization 48–54, 56–57, 301
  virus strains 210–214, 215
  see also breeding programs
Germany 3, 33
germination of TPS 281
gibberellic acid/gibberellin (GA)
  secondary growth 241
  sprouting 66
  tuber bulking 73
  tuber initiation 51, 52–54, 70–71
global markets *see* market for potatoes
Globodera nematodes 149–150, 155, 156, 157
  resistance to 158, 298
glucose 76, 260, 261
glutamate 346–347, 347–348, 350, 355
glycemic index 212, 328
glycoalkaloids 1, 240, 325–326, 350
grass weeds 225, 228, 231
gravimetric soil sampling 94
Great Britain 3–4, 118, 256
green foxtail 228
green manure 85, 156, 229
green peach aphid (*Myzus persicae*) 139
greenhouse gases 105–106, 209–210
greening of tubers 239–240
ground storage 256
growth 64–79
  dormancy 47, 65–66, 127
  physiological disorders 237–244
  propagation 64–65, 87–88, 115
  source–sink balance 72–73, 75–76, 97
  stem and leaves 67–70, 72, 88, 105
  tuber age 66–67, 88
  tuber bulking/maturation 73–75, 76–79, 98, 105
  tuberization 45–60, 70–72, 92, 105, 301
growth cracks 238–239
Guamán Poma de Ayala, Felipe 1–2
guanosine-5′-monophosphate (5′-GMP) 346, 347, 348–350

hairy vetch 229
harvest
  avoiding tuber damage 98–99, 179–180, 239, 242, 246
  and respiration rate 263

seed potatoes 129–130
timing 77–78, 78, 156
harvest index (HI) 72, 75–76
Hawkes, J.G. 13, 15, 16, 20
health issues 42, 277, 311–314, 328, 329
heat sprouts 241
herbicides 86, 228, 231–233
hilling (ridging) 92, 228, 230, 240
history 1–9
breeding 5–8, 272–273
taxonomy 12–14, 20–21
hollow heart 243–244
hopper burn 140
hormones
secondary growth 241
tuber bulking 73
tuber dormancy 65–66
tuber initiation 51, 52–56, 70–71
humidity management during storage 265
hybridization 16, 22
in breeding programs 275, 276, 279–280
hydrogen peroxide 189

IAA (auxin) 54–55
IBCN/ICNCP nomenclature 21, 22–24
Illumina Infinium 8303 Potato Array 291
import and export 32–33, 34, 36, 38–39, 41
in vitro propagation 124–127, 284
indeterminant (late) cultivars 70, 187
India 8, 22, 31, 119, 207, 256
industrial uses 37, 38, 298
inosine-5′-monophosphate (5′-IMP) 346, 347, 348–350
in-row spacing 88–89
insecticides 86, 129, 136, 141, 144–145, 215
insects
beneficial 86, 142, 145
and climate change 209
pests 86, 133–142
management 129, 142–145
transgenic potatoes resistant to 145, 297–298, 322
as vectors of disease 129, 139, 140, 141, 209–210, 215, 278
integrated pest management (IPM) 86, 142–144, 160–161, 187, 229
integrated weed management 225–233
internal brown spot/heat necrosis/rust spot 248–249
International Potato Center (CIP) (Peru) 19, 39, 65
introgression 16, 158
invertase 260–261
Irish Potato Famine 4–5
iron (Fe) 314–315
irrigation systems 85, 92–94, 106–108, 109
and common scab 181–182
fertigation/chemigation 85, 97, 232

Japan 36
jasmonic acid 56
jelly end 247

Kenya 116, 256
knobby tubers 237

late blight (*Phytophthora infestans*) 167, 173–174, 183–185, 190
appearance 186
epidemics 5, 6, 21
and potato gene expression 293, 294
resistance to 5, 6–7, 17, 278–279, 297
late-maturing varieties (indeterminant) 70, 187
leafhoppers 140
leaves
canopy size/growth 70, 72, 75–76, 105
cuttings used to produce minitubers 128
defoliation 78–79, 98, 230–231, 250
drought 105–106
early blight/brown leaf spot 185–186
emergence 69–70, 90
insect damage 135, 137, 141
morphology 14, 19, 69, 105
senescence 76, 77, 78, 106
lenticels 67, 265
enlarged 239
less developed countries *see* developing countries
light, tuber exposure to 239–240
lipids, and flavor 350–351, 354, 355
lipoxygenases 56, 351
little tuber 242
long-day adaptation 5, 6, 22
low-temperature injuries 249–250, 258, 295
lutein 332

magnesium (Mg) 315
mahogany browning 249
Maillard reaction 38, 246, 257–258, 277, 351
male sterility 6, 280
malformed tubers 237–238
mancozeb 183
market for potatoes 29–42
consumer attitudes 32, 42, 329–330
global production and consumption 8, 29–33, 35–36, 41–42, 167–168, 313
new potatoes 329–330
product types 33–38, 84
storage 256
varieties 38–41, 84–85, 285
mashed potatoes 38, 311
Megistacroloba spp. 19
Meloidogyne nematodes (root-knot) 150–151, 155, 156
resistance to 158–159

metam sodium 177
methionine 354
methyl salicylate 353
S-metolachlor 231–232
metribuzin 231, 232
Mexican species 6–7, 13
micro RNA (*miR172*) 51
microarray analysis 56–57, 293, 355
micropropagation 125–126
microtubers 127
microwaved potatoes 41, 330, 351
minerals
  and physiological disorders 244, 245, 248–249
  required for growth 94, 95–96
  in tubers 314–315
minitubers 126–127
models
  irrigation 94, 107–108
  late blight 184–185
  nematode management 160
  PVY 210, 215
mosaic viruses *see individual viruses*

Nacobbus aberrans (false root-knot nematode) 151, 159
NemaDecidePlus™ 160
nematicides 86, 157, 177
nematodes 148–154
  management 154–161, 177, 229, 298
  seed potatoes 128, 153, 155, 157
Netherlands 33, 118, 119, 123
neutron probes 93
new potatoes 327, 329–330
  see also petite potatoes
niacin (vitamin B_3) 318–319
nightshade weeds 156, 225, 228, 232
nitrogen 76–77, 97
nitrogen fertilizer 71, 76, 85, 94–97, 279
  and flavor 353
nomenclature
  potato (*Solanum* spp.) 12–25
  potato viruses 210–214, 215
North America *see* Canada; USA
Northwest Potato Variety Development Program (NPVDP) (USA) 282
nutrients, for crops *see* fertilizer use
nutrient status (plants and soil) 96, 99
nutritional qualities of potatoes 310–333
  carotenoids 331–332
  consumer knowledge of 42, 330
  fiber 311
  glycoalkaloids 240, 325–326, 350
  and a healthy diet 42, 311–314, 328, 329
  minerals 314–315
  new potatoes 329–330
  phenylpropanoids 326–331
  productivity (in calories) 30, 310, 333
  protein 311
  starch 311
  vitamins 315–325, 329, 330

obesity 312, 328
olfaction 346
  volatile compounds 350–353
oomycete diseases 169–170
  see also late blight (*Phytophthora infestans*)
orange potatoes 332
organically grown potatoes 84, 145, 189, 266, 353
osmotic adjustment (OA) 106
oxygen deficiency 108, 239, 246
Ozette variety 4

packaging 35
pallet box storage 257
pantothenate (vitamin B_5) 319–321
Paratrichodorus spp. (stubby-root nematode) 153–154, 157
Parmentier, A.A. 3
partial root zone drying (PRD) 107
patatin 60, 311
PCR (polymerase chain reaction) 121, 191
Pectobacterium spp. (soft rot) 124, 178–180
peeling, ease of 7–8
pendimethalin 231
Pepino mosaic virus (PepMV) 208
periderm, tuber 47, 67, 242
  dietary fiber 311
  physiological disorders 238–239, 240, 242, 243
  scab 180, 181
  skin set 78–79, 98, 265
  skinning 78, 98, 242, 264
  and water loss 264–265
Peru
  origin of the potato 1, 15, 20
  seed potatoes 116
pest management 178
  bacteria 179–180, 181–182
  diagnosis 121–122, 151, 154, 190–191, 216
  fungi 129, 175, 177, 183, 187–188, 189
  insects 129, 142–145
  IPM 86, 142–144, 160–161, 187, 229
  late blight 184–185, 278–279
  nematodes 154–161, 177, 229, 298
  pre-planting 85, 86
  PVY 214–215
  see also resistance in potato plants; weed management
pesticides *see* fungicides; herbicides; insecticides; nematicides
petiole analysis 96
petite potatoes 41
  see also baby potatoes

Petota Dumort (*Solanum* section) taxonomy 12–25
phenylpropanoids 181, 326–331
  anthocyanins 181, 250, 328, 330, 331
phosphorous acid 189
phosphorus (P) 95–96, 315
photoperiod 47–48, 51
photosynthesis 106
phylogeny/phylogenetics
  potato 13, 16, 17, 24
  potato viruses 210–214, 215
physiological disorders of the tuber 237–250, 257–258, 295
phytochromes (PHYA, PHYB) 48, 49, 51
Phytophthora infestans see late blight (*Phytophora infestans*)
phytoplasmas 140
pick planters 91
pink eye 242–243
pink rot (*Phytophthora erythroseptica*) 173, 189
pith, tuber 67
  physiological disorders 243–244, 246
Planetary Biodiversity Inventory (PBI) program 17
plant nutrition 94–96
plant population 64, 72, 73, 75, 79, 88–89, 91
Plant Variety Protection (PVP) 285
planting 69, 87–92, 129, 229
planting depth 69, 72, 90–91, 99
PLRV (*Potato leaf roll virus*) 203, 209, 210, 215, 276
PMTV (*Potato mop-top virus*) 178, 207, 209
pointed-end tubers 238
Poland 31–32
pollen 280
polymerase chain reaction (PCR) 121, 191
polyploidy 13, 21–23, 300–301
  and breeding programs 8, 273–275
postharvest period
  certification testing of seed potatoes 122–123
  diseases 174–175, 180, 188–189, 283–284
  see also storage
potassium (K) 95–96, 314, 353
potato aphid (*Macrosiphum euphorbiae*) 139
potato cyst nematode (PCN, *Globodera* spp.) 149–150, 155, 156, 157
  resistance to 158, 298
Potato deforming mosaic virus 207
potato early die complex 153, 177
Potato Genome Sequencing Consortium (PGSC) 24, 57
potato late blight *see* late blight (*Phytophthora infestans*)
Potato leaf roll virus (PLRV) 203, 209, 210, 215, 276
potato leafhopper (*Empoasca fabae*) 140
Potato mop-top virus (PMTV) 178, 207, 209
Potato Oligo Chip Initiative (POCI) 293, 355
potato psyllid (*Bactericera cockerelli*) 141–142, 183, 278
potato rot (or tuber) nematode (*Ditylenchus destructor*) 154, 156
potato tuber moth (Gelechiidae) 137–138, 322
potato tuber necrotic ringspot disease (PTNRD) 207, 212
potato tuberworm (*Phthorimaea operculella*) 133, 144, 298
Potato virus S (PVS) 203, 210–211
Potato virus X (PVX) 203, 208, 209, 211
Potato virus Y (PVY) 203, 207
  control measures 214–215
  resistance 297
  in seed potatoes 121–122, 282
  strains 208, 211–214, 215
POTH1 (potato homeo box 1) 51, 54
Pratylenchus nematode (root-lesion) 151–153, 177
pre-planting preparation 85–87, 128–129, 230
pre-sprouting 68–69, 88
precision irrigation (PI) 107–108
preconditioning 261, 267
pressure bruise 246, 257
pricing 84
processed products 32–33, 35–37, 77
  see also chips (crisps); French fries
production
  global markets 8, 29–32, 35–36, 41, 167–168
  of seed potatoes 124–130
  sustainability 278–279
propagation 64–65, 87–88, 115
  in vitro 124–127, 284
protein, tuber 59–60, 311
proteomics 294–295
psyllid yellows disease 141
psyllids 141–142, 183, 278
purple potatoes 41, 312, 326–327, 328, 330, 331
purple top disease 140
PVS (*Potato virus S*) 203, 210–211
PVX (*Potato virus X*) 203, 208, 209, 211
PVY *see Potato virus Y*
pyrethroids 145
pyridoxine (vitamin B_6) 321–322

quackgrass 228
quarantine 122, 149, 150, 151, 154, 172
quinone outside inhibitor (QoI) fungicides 187, 190

red potatoes 40, 326, 328
remote sensing systems 94
resistance in potato plants
  genomic studies 291–292, 301
  glycoalkaloids 325
  to bacterial diseases 181, 182, 278
  to fungi 187, 297
  to insects 145, 297–298, 322
  to late blight 5, 6–7, 17, 278–279, 297
  to nematodes 157–160, 298
  to viruses 215–216, 276, 297
  transgenic varieties 145, 159–160, 294, 297–298, 322

resistance to pesticides 144–145, 175, 189, 190
respiration in the tuber 66–67, 108, 246, 262–264
Rhizoctonia solani (black scurf) 173, 177, 183, 297
riboflavin (vitamin B_2) 317–318
ribonucleotides, umami flavor 346, 347, 348–350
ridging (hilling) 92, 228, 230, 240
rimsulfuron 232
RNA
 regulatory 51, 159
 transcriptomics 56–57, 293–294, 301, 355
roguing 122, 129
root-knot nematode (*Meloidogyne* spp.) 150–151, 155, 156
 resistance to 158–159
root-lesion nematode (*Pratylenchus* spp.) 151–153, 177
roots 69, 104
Rough Purple Chili variety 5, 272
row planting 89–92
row width 88
Russet varieties 35, 38, 39, 285
 Russet Burbank 5–6, 38, 285, 326–327
 Russet Norkotah 104, 326–327
Russia 32, 33
rye 229

salt tolerance/sensitivity 295, 297
sanitation in seed potato production 124, 127, 129
scald 242
Scotland 118, 119
second growth 241–242
seed potatoes 115–130
 age 66, 88
 breaking dormancy 47, 65–66
 certification 87–88, 115–116, 117–124, 172, 284
 climate change 210
 cut pieces 67–68, 70, 88
 diseases/pests 117, 120, 124, 129, 172
  fungal 174, 175, 177, 183
  insects 144
  nematodes 128, 153, 155, 157
  viruses 121–122, 123, 210, 214–215, 282
 market 37, 38–39
 planting 87–92, 129
 pre-sprouting 68–69, 88
 production 124–130
 storage 66, 123, 127, 129–130, 242
 treatments 66, 129–130, 157, 175, 183
seedbanks (weed) 226–227
seedbeds 86–87
seedlings/seedling tubers 281–282
seeds *see* true potato seed (TPS)
senescence sweetening 258, 262
senescence of vines/leaves 76, 77, 78, 106
sethoxydim 232–233
sexual reproduction 6, 273, 279–280
  see also true potato seed (TPS)
shatter bruise 245–246
shipping inspections 123–124
short-day tuberization 47–48
silver scurf 175, 189
single-hill selections 281–282
single-nucleotide polymorphisms (SNPs) 290–291, 292, 296
skin, tuber *see* periderm, tuber
skinning 78, 98, 242, 264
smell 346
 volatile compounds 350–353
snacks *see* chips (crisps); French fries
soft rot (*Pectobacterium*/*Dickeya* spp.) 124, 178–180
soil
 fumigation 157, 177
 nutrients 94
 pathogens in 149–154, 175–178, 180, 182
 plant growth 69, 94
 pre-planting 85, 86–87, 128
 tuber maturation 78
 water monitoring and measurement 93–94
 see also water utilization and management
SOL Genomics Network 301
Solanaceae, comparative genomics 298–299
Solanaceae Source website 13, 17, 19, 21
solanine 325–326
Solanum (*Solanum* L. section *Petota* Dumort) 12–25
Solanum astleyi 19
Solanum berthaultii 16, 275–276
Solanum boliviense 19, 322
Solanum brevicaule complex 1, 15, 17–19, 21
Solanum bulbocastanum 13–14, 24
Solanum demissum 6, 13
Solanum etuberosum 275–276
Solanum phureja 23, 348, 353, 355
Solanum sisymbriifolium 156
Solanum stenotomum 21–22
Solanum stoloniferum 13–14, 15
Solanum tarijense 16
Solanum tuberosum (taxonomy) 20–25
Sol*R*gene database 17, 291
source–sink balance 72–73, 75–76, 97
South America
 origin of the potato 1–2, 5, 20, 22
 seed potatoes 116
 viruses indigenous to 205–206, 207–208
spacing of plants 89, 129, 229
 seedling tubers 281
Spain, colonial 2–3, 4
spraing (corky ringspot) 153, 207
sprouting 65–68, 88
 heat sprouts 241
 prevention during storage 265–266

sprouts used for seed potato production 128
starch
  bulk production 37
  and cooking characteristics 38, 39, 40, 41, 311
  tuber content 311
  tuber metabolism 59, 247, 259–260
StBEL5/StBEL5 51, 54
StCDF1/StCDF1 49–50
StCO/StCO (CONSTANS) 49
stem and stolon canker (*Rhizoctonia solani*) 173, 177, 183, 297
stem-end chip defect 258
stems 69–70, 74, 77, 88
StGA2ox1/StGA2ox1 52
StGA3ox2/StGA3ox2 52
StGA20ox1/StGA20ox1 51, 53, 54
stolons 45, 52, 55, 69, 70, 301
stomata 105–106
storage
  carbohydrate metabolism 259–262
  diseases 174–175, 180, 188–189, 283–284
  and flavor 354
  methods 180, 256–257, 264, 266–267
  physiological disorders 239, 240, 246, 249, 257–258, 295
  purpose 167–168, 255–256
  respiration 67, 246, 263–264, 265
  of seed potatoes 66, 123, 127, 129–130, 242
  sprout inhibition 265–266
  sugar content 261–262
  vitamin content 317, 323, 324
  water loss 264–265
StPOTH1/StPOTH1 51, 54
Strecker degradation 351
Streptomyces spp. (common scab) 180–182
strigolactones 55
StSP6A/StSP6A 48–49
stubby-root nematode (*Paratrichodorus* spp. and *Trichodorus* spp.) 153–154, 157
suberization 78, 88, 264
subspecies 15–16, 20–21, 22
sucrose 59, 247, 260–262
sucrose synthase (SuSy) 59, 261, 298
sugar end 76, 92–93, 246–247, 257–258
sugars
  at harvest 76, 78
  breeding for low levels 277
  and flavor 78, 351
  metabolism 59, 247, 259–262
  in stored tubers 261–262
  and suitability for frying 38, 40, 41
sulfur 94
sulfuric acid 78, 98
sunlight, tuber damage 240
sustainability 278–279
sweetening of potatoes 258, 262, 295
systematics *see* taxonomy
table-stock market 29, 35, 42, 329–330
taste *see* flavor
taxonomy
  cultivated potato 20–25
  wild species 12–20
temperature
  at harvest 98, 180, 239, 246
  at planting 87
  of the canopy 108
  and physiological disorders 239, 241, 242, 245
    low temperature 249–250, 258, 295
  storage 256, 257, 267
    disease management 180, 265
    dormancy of seed potatoes 66
    respiration 263, 265
    sprout inhibition 266
    sugar management 261–262
    vitamin C content 324
  tuber maturation 78
  tuber respiration 67, 263, 265
  and viruses 208–209
tensiometers 93
tetraploidy 273–275
texture 38, 40, 311
thiabendazole (TBZ) 175, 189
thiamine (vitamin B_1) 316–317
thrips 209, 210
thumbnail cracks 239
tillage 85, 87, 92, 230
tillage radish 229
timing
  harvest 77–78, 78, 156
  planting 87
tissue culture 124–126, 284
Tobacco rattle virus (TRV) 153, 155, 207
tomatine 325
Tomato spotted wilt virus (TSWV) 203, 207
tospoviruses 203, 207, 209
toxic compounds (glycoalkaloids) 1, 240, 325, 350
TPS *see* true potato seed
trace elements 315
trade 32–33, 172
transcriptomics 56–57, 293–294, 301, 355
transgenic varieties 295, 296–298
  disease/pest resistance 145, 159–160, 294, 297–298, 322
  flavor 354–355
  phytonutrient content 324, 332
  starch content 37
translucent end 246–248
transpiration 105
trap cropping 156–157
Tri-State Potato Variety Trials (TSPVT) (USA) 284
Trichodorus spp. (stubby-root nematode) 153–154, 157
trifluralin 231

true potato seed (TPS) 64–65, 87, 116–117, 273, 280–281
  viruses 208, 216
TRV (*Tobacco rattle virus*) 153, 155, 207
tuber age 66–67, 88, 263
tuber anatomy 67
tuber bulking 64, 72–75, 78, 92, 95, 105
tuber color 40, 41, 328, 332
  enzymatic browning 298, 328–329
  see also purple potatoes
tuber damage
  bacterial 142, 180
  fungal 173–175, 176, 177, 184, 186, 187
  insects 136, 137–138
  mechanical 98–99, 179, 245
  nematodes 151, 152, 154
  physiological disorders 237–250, 257–258, 295
  viral 153, 203, 207, 212, 239
tuber dormancy 47, 65–66, 127
tuber initiation *see* tuberization
tuber maturation 75, 76–79, 98, 322–323, 348
  early-maturing varieties 70, 104, 156
  late-maturing varieties 70, 187
tuber number per plant 71–72
tuber set 66, 68, 70–73, 89
tuber size 7–8, 73–75, 83, 283
  special varieties 40, 41
  see also baby potatoes
tuberization 45–60, 70–72, 92, 105, 301
12-hill (12H) selection 282–283
two-spotted spider mite (*Tetranychus urticae*) 145

umami taste 346–350
USA
  breeding programs 282, 284–285
  climate change 184, 190, 208
  crop rotation 229
  insect pests 134, 139, 140, 141
  introduction of the potato 4, 9
  late blight 167, 184, 190
  potato market 30–31, 32, 33, 35, 36, 41, 256, 313
    varieties 36, 38
  seed potato certification 118, 119, 122, 123, 172, 284

vaccine production 298
vaculoar acid invertase (VINV) 260–261
varieties, range 38–41, 42, 84–85
vascular discoloration 250
ventilation during storage 257, 264, 267
verticillium wilt 153, 177, 297
viral diseases 86, 202–216
  and breeding programs 282, 283
  diagnosis 121–122, 216
  growth cracks 239
  nomenclature 210–214, 215
  PVY *see Potato virus Y*
  resistance to 215–216, 276, 297
  in seed potatoes 117, 121–122, 123, 129, 210, 214–215, 282
  TRV 153, 155, 207
  vectors 129, 139, 153, 209–210, 215
vitamins 315–325, 329, 330

Waldensians 3
water utilization and management
  and common scab 181–182
  drought 92–93, 103–106, 108–109, 247, 297
  excessive water 93, 108
  irrigation 85, 93–94, 106–108, 109, 181–182
  nutrition productivity index 30
  water loss during storage 264–265
weed management 156, 225–233
Western Regional Potato Variety Trial (WRPVT) (USA) 284
white knot (or spot) bruise 245
white potatoes 40, 326–327
whitefly (*Bemisia tabaci, Trialurodes vaporariorum*) 209, 210
whole-genome duplication (WGD) 300–301
wild potatoes 1, 6–7
  in breeding programs 8, 275–276, 322, 324
  taxonomy 12–20
wireworm (Elateridae) 135–137, 144
wounding
  entry points for pathogens 175, 179, 188
  healing 88, 98, 264–265
  vitamin C content 324–325

yam nematode (*Scutellonema bradys*) 154
yellow nutsedge 232
yellow potatoes 40, 332
yield
  in breeding programs 7, 282, 283
  and crop management 72–73, 75–77
  fertilizer use 95–96
  loss due to weeds 227–228

zeaxanthin 332
zebra chip disease 141–142, 183, 278
zinc (Zn) 315